Zoological Monographs

Volume 2

Series editors
Heike Feldhaar
Tierökologie I, Universität Bayreuth, Bayreuth, Germany

Andreas Schmidt-Rhaesa
Wirbellose Tiere 1, Zoologisches Museum, Centrum für Naturkunde, Hamburg, Germany

The book series *Zoological Monographs* is devoted to the study of the animal kingdom and covers all aspects of animals, both living and extinct, including taxonomy, systematics and evolution, biodiversity and biogeography, ecology, morphology and anatomy, physiology and behavior. The volumes in the series either discuss the biology of a single animal group, approach a topic from an interspecies level, or present recent methodologies. The series is intended for graduate students, university teachers, scientists and professionals.

More information about this series at http://www.springer.com/series/15188

Ralph O. Schill
Editor

Water Bears: The Biology of Tardigrades

Springer

Editor
Ralph O. Schill
Institute of Biological Materials
and Biomolecular Systems
University of Stuttgart
Stuttgart, Germany

ISSN 2523-3904 ISSN 2523-3912 (electronic)
Zoological Monographs
ISBN 978-3-319-95701-2 ISBN 978-3-319-95702-9 (eBook)
https://doi.org/10.1007/978-3-319-95702-9

Library of Congress Control Number: 2018962674

This Springer imprint is published by the registered company Springer Nature Switzerland AG
The registered company address is: Gewerbestrasse 11, 6330 Cham, Switzerland

Preface

Some of the most delightful hours of my scientific career have been spent studying tardigrades. And even after many years, I observe these little animals enthusiastically under my microscope. They are true survival artists that occur in the most diverse habitats on earth. I discovered them in dry desert regions, on glaciers, in rivers and lakes, in cold and tropical seas, and at my front door in the German forest.

Everything started with Johann August Ephraim Goeze, a protestant pastor in Quedlinburg, Germany. He noted that in his hours of rest he preferably refreshed himself by the microscopic examination of the boundless wealth of nature in the water. And that led to the first discovery and description of a small animal which looked similar to a little bear. Therefore, he gave him the name "kleiner Wasserbär" (small water bear) in 1773.

A continuously growing number of scientists became interested over centuries in these fascinating animals, but there are no comprehensive summaries of their history, morphology, phylogeny and taxonomy, biogeography, paleontology, cytology, ecology, and adaptive strategies. The idea for this comprehensive tardigrade book was born during the 11th International Symposium on Tardigrada in 2009 at the University of Tübingen, Germany. This symposium series goes back to the inaugural meeting in Pallanza, Italy, in 1974. Subsequent meetings were held every three years in various locations around the world.

I'm extremely grateful to my friends and colleagues who immediately agreed to work on this important task, to share ideas, provide references, and review chapters. Therefore, I would like to take this opportunity to thank all the authors named below in the order of the chapters: Hartmut Greven, Nadja Møbjerg, Aslak Jørgensen, Reinhardt Møbjerg Kristensen, Ricardo C. Neves, Sandra J. McInnes, P. J. A. Pugh, Roberto Guidetti, Roberto Bertolani, Lorena Rebecchi, Diane R. Nelson, Paul J. Bartels, Noemi Guil, Tiziana Altiero, Atsushi C. Suzuki, Steffen Hengherr, K. Ingemar Jönsson, Eliana B. Levine, Andrzej Wojcik, Thomas C. Boothby, and Peter Degma. I also wish to thank our project coordinator at Springer, Andrea Schlitzberger, for her support, encouragement, and patience. It has been a pleasure to work with her.

Lastly, my deepest thanks go to the families of the authors, for their support, and especially for their help and understanding during the time this comprehensive and at the moment unique book about tardigrades evolved. I hope this work will serve as a useful guide for everyone interested in these fascinating animals.

Tübingen, Germany
2018

Ralph O. Schill

Contents

Chapter 1
From Johann August Ephraim Goeze to Ernst Marcus: A Ramble Through the History of Early Tardigrade Research (1773 Until 1929)

Hartmut Greven

There is a time for everything...
Ecclesiastes 3:1

Abstract A survey is presented about the early history of tardigrade research spanning the time from 1773, when the first description of a tardigrade was published by Goeze, until 1929, when the most comprehensive monographic approach by E. Marcus, unsurpassed today, was published. Almost from the beginning, two topics dominated "tardigradology", i.e. phylogeny and systematics as well as cryptobiosis, especially anhydrobiosis, but also other issues (e.g. morphology, development and life history) have followed successfully with ongoing technical and preparatory improvements.

1.1 Introduction

Occasionally it seems useful to remember the origins and pioneers of scientific disciplines, not to look amused from our present point of view on their curious views or their ignorance (this does happen not only to ancient authors like Aristotle[1]

[1]**Aristotle** (Gr. Aristotélēs; 384–322 BC), Greek philosopher, who joined Plato's Academy in Athens. His writings cover physics, biology, zoology, metaphysics, logic, ethics, aesthetics, poetry, etc. He was the teacher of Alexander the Great (356–323 BC). Aristotle is the earliest natural historian, whose work has survived, among others his writings on natural science such as (translated

H. Greven (✉)
Department of Biology, Heinrich-Heine-University, Düsseldorf, Germany
e-mail: grevenh@uni-duesseldorf.de

R. O. Schill (ed.), *Water Bears: The Biology of Tardigrades*, Zoological Monographs 2,
https://doi.org/10.1007/978-3-319-95702-9_1

or Pliny the Elder[2]), but to acknowledge their work, their forward-looking ideas, often completely forgotten today, sometimes consciously ignored (however, nobody will and can go back to Aristotle with his citations) or distributed by succeeding researchers as their own ideas, and last but not least to understand their errors caused by a general poor knowledge of the matter at the time, by insufficient techniques or by ideologically narrow-minded views.

Tardigrada, which belong to the so-called minor phyla, were known for decades only to a small group of insiders commonly called tardigradologists.[3] Nevertheless, the early studies in tardigradology[4] have quite a few surprises in store. From the few surveys on the history of tardigrade research, the summarizing chapter in Ernst Marcus's's[5] remarkable monograph on Tardigrada published in 1929 is still the most concise, critical and informative overview (see Marcus 1929b). Thereafter limited historical information can be found in specific monographs (e.g. Cuénot[6] 1924; Greven 1980; Maucci 1982; Ramazzotti and Maucci 1983; Kinchin 1994) and more recently on various websites. Due to general interest, early studies on tardigrades have also been reviewed repeatedly in the literature on anabiosis,[7] in particular detail by Keilin (1959) and less detailed by Wright et al. (1992) and Rebecchi et al. (2007).

---

into Latin) the "Historia animalium" [History of animals], "De generatione animalium" [On the generation of animals] and "De partibus animalium" [On the parts of animals].

[2]Gaius **Plinius Secundus** (Pliny the Elder) (ca. 23–79 AD), Roman naval and army commander, as well as author and naturalist, wrote an encyclopaedia entitled "Naturalis Historiae" [On Natural History] that comprises 37 books, into which he collected much of the knowledge of his time.

[3]Commonly assigned to people predominantly concerned with research on tardigrades. However, at all times knowledge of tardigrades has been promoted by people with very different backgrounds and various provenances including "heroes" of natural sciences such as L. Spallanzani, laymen and even people, who never have seen a living tardigrade (especially in modern times, in which sophisticated methods need the experience of specialists). Most researchers mentioned in the present essay were engaged in studying tardigrades only for a certain period of their life, often only during their doctoral dissertation and sometimes only along the way.

[4]Composed of the Latin words "tardus" (= slow) and "gradi" (= to walk) and the ancient Greek lógo*s* (= word, study, research). The term Tardigrada comes from Spallanzani's "il tardigrado" (= sluggard). See footnote 15.

[5]Ernst Gustav Gotthelf **Marcus** (1893–1968), German zoologist, worked closely with his wife Eveline Du Bois-Reymond (1901–1990), granddaughter of the famous physiologist Emil Heinrich **Du Bois-Reymond** (1818–1896). Due to his Jewish origin, Ernst M. moved in 1936 from Berlin to Brazil and became professor at the University of São Paulo, where he taught and researched for 37 years (until 1963). Eveline M. did all the drawings for the publications of her husband and her own work. Ernst Marcus's research covered a very wide spectrum of topics and organisms. In his doctoral thesis, he dealt with lamellicorn beetles.

[6]Lucien Claude Jules Marie **Cuénot** (1866–1951), French biologist, zoologist and geneticist. In 1898 he became chair of zoology in Nancy (France).

[7]The term anabiosis (aná, Gr. = upwards; bíos, Gr. = life) was introduced by Preyer (1880, 1891) (see note 89). Keilin (1959) replaced this term with cryptobiosis (kryptós, Gr. = hidden) to describe a specific type of hypobiosis (hypó, Gr. = under), viz. "the state of an organism when it shows no visible signs of life and when its metabolic activity becomes hardly measurable, or comes reversibly to a standstill" (p. 166). He distinguished cryobiosis (krýos, Gr. = ice) induced by cooling, anhydrobiosis (a(n), Gr. = a denial prefix; hýdor, Gr. = water) induced by loss of water, osmobiosis

Initially, I planned to give an overview of the history of tardigradology at least until the First International Symposium on the Tardigrada. This Symposium, held June 17–19, 1974, at the Istituto Italiano di Idrobiologia in Pallanza (Italy) in honour of Giuseppe Ramazzotti[8] on the occasion of his 75th birthday, marked a turning point in tardigradology. After that tardigradologists began to organize themselves and to arrange further symposia (see Bertolani and Nelson 2011). However, a presentation even of this period would result in a simple listing of what has been done. I therefore dismissed this idea in favour of the present essay, which attempts to delineate the history of tardigrade research from the first published description of a water bear by Goeze (see Figs. 1.1 and 1.2 top left) to the above-mentioned monograph by Marcus (see Figs. 1.1, bottom right, and 1.23) by means of examples (in part subjectively selected) from the literature. It was also my intention to familiarize the reader with a period in which tardigradology was primarily a European matter and in which results were published in various languages, with which many contemporaries may be unfamiliar. Early studies outside of Europe were restricted to finding, describing and naming of species (α-taxonomy) and were performed largely by European tardigradologists. However, the first mention of tardigrades written by a US scientist was previously published in 1850 (Bailey 1850).[9]

Further, I included several "historical" images of tardigrades and their organization to give the reader an impression of how differently and subjectively these animals were visualized, whether or not authors had an appropriate microscopy equipment. I excluded those articles with pure α-taxonomy and local faunas, which only have increased the number of species of already known taxa, as well as the treatment of tardigrades in text books of zoology and popular books. These subjects may be worth investigating separately.

Undoubtedly, Marcus's monograph is a milestone in tardigradology that, apart from the valuable and critical overview of previous literature, includes observations not published elsewhere and provides important ideas for future research in nearly all fields of tardigradology. Strictly speaking, the present essay follows closely the chapter "Geschichte der Tardigradenforschung" [History of Tardigrade Research] in this monograph. However, some parts I have shortened and others I have

---

(osmós, Gr. = impulse) induced by high levels of concentration and anoxybiosis (oxýs, Gr. = acidic) induced by absence of oxygen or any combination of these factors. Wright et al. (1992) suggested not to restrict the term cryptobiosis to organisms surviving in an ametabolic state (as Keilin did), but "to define cryptobiosis as a collective term for those quiescent states in which metabolism may be reversibly arrested" (p. 22) and to exclude anoxybiosis and encystation, since both "qualify for quiescence but not cryptobiosis" (p. 23).

[8]Giuseppe **Ramazzotti** (1898–1986), Italian engineer, enthusiastic collector of pipes, began research of tardigrades in 1938 and presented in 1962 a comprehensive review of the literature focussing on the description and determination keys of the known species. The second edition was published in 1972 and the third (with W. Maucci as coauthor) in 1983.

[9]Jacob Whitman **Bailey** (1811–1857), US American naturalist, professor of chemistry, mineralogy and geology at West Point, pioneer in microscopic research in America. He corresponded with Ehrenberg (see footnote 46). Bailey noted in 1850 "The waters in which I detected the species above recorded, also abounded in many other forms of microscopic life; as, Entomostraca, Tardigradi, Anguilluli, &c., &c. Of these I have made no record, as I did not possess sufficient knowledge concerning them." (p. 44).

Fig. 1.1 Portraits of influential scientists studying tardigrades. Top left: Johann August Ephraim Goeze (1731–1793), painting attributed to Fiedrich Schlüter from around 1780. With kind permission of Gleimhaus Halberstadt—Museum of German Enlightenment. Top right: Lazzaro Spallanzani (1729–1799). http://ihm.nlm.nih.gov/images/B24049. Bottom left: Karl August Sigismund Schultze (1795–1877), painting by Wilhelm Titel from 1837. From Schmitt and Schultze (1931). Bottom right: Ernst Marcus (1893–1968), drawing by JB Santos in 1981 according to an earlier photo. With kind permission of the Department of Zoology, Institute of Biosciences, University of São Paulo

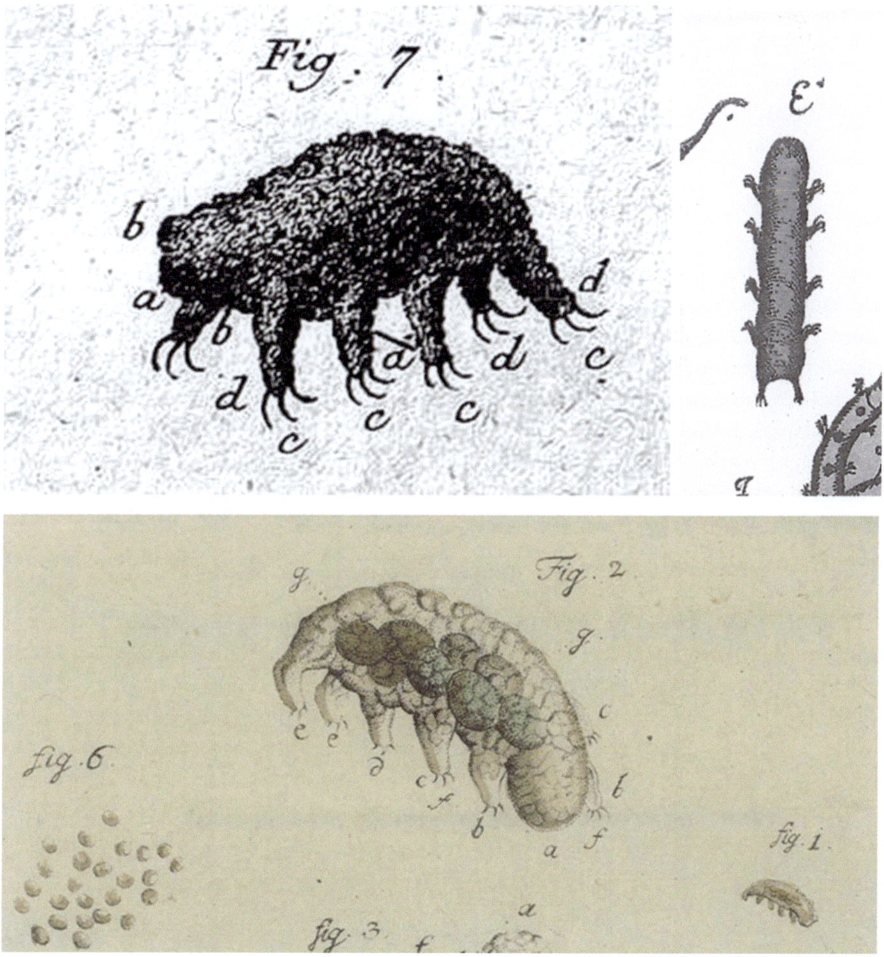

Fig. 1.2 Early pictorial presentations of tardigrades. Top left: The little water bear. From Goeze (1773), section from Plate 4 in the appendix of the second part. Top right: The water bear. From Eichhorn (1775), Plate 7, fig. E. Bottom: The bear animalcule. From Müller (1785), section from Plate 36 (with kind permission of the Zentralbibliothek Zürich Scan NNN827)

extended. For that I consulted again the original sources, as available, and, contrary to Marcus, provided herein some literal citations from the texts of earlier authors including the English translation (mostly in endnotes), to give the reader an idea why tardigradologists of earlier times were fascinated about these microscopically small life forms and how they studied them.

To keep this essay reasonably short, I decided not to describe the current state of knowledge. Only in a few cases do I refer to articles published after 1929. Recent findings that may modify or disprove the interpretations and conclusions of the data reported herein are, at least in part, summarized in the following chapters of the present book.

The endnotes also contain some explanations (including etymological ones) and biographical data of the researchers quoted (primarily year of birth and year of death, research activity and main profession). The latter were collected by means of online encyclopaedias (not cited) and checked in some cases with other sources (cited, but some only in the endnotes). In some cases, detailed bibliographic data could not be ascertained.

Besides α-taxonomy, which started, albeit with a certain delay, soon after the first discovery of tardigrades and which continues unabatedly, tardigrade research was dominated almost from the beginning by two topics, i.e. (1) anabiosis or abiosis or, the more modern term, cryptobiosis and here especially anhydrobiosis (the ability many tardigrades share with other organisms such as the metazoan taxa rotifers and nematodes, in which this phenomenon was detected earlier than in tardigrades[10]) and (2) possible relationships to other organisms; almost every author has discussed relationships broadly recapping previous opinions either to endorse or to modify them or to work out new proposals. These two subjects occasionally appeared to overshadow other findings such as functional morphology, development and life history data, which have suffered from the lack of adequate equipment and techniques for a long time.

In the following text, individual sections deal with arbitrarily specified periods. It would certainly be useful to further subdivide these sections into specific subject matters, but I preferred to present information in chronological order, so far as possible, to avoid too much repetition, as earlier authors often covered more than one subject in their articles. I have retained the species names originally used by the authors, although most have changed in the course of time. Moreover, many could not be assigned to a specific species by later researchers.

1.2 The Early Findings (Eighteenth Century)

The first mention and descriptions of tardigrades (only eutardigrades) were published in quick succession within a period of 9 years (Goeze 1773; Corti 1774; Eichhorn 1775; Spallanzani 1776; Müller 1785; the latter note was published posthumously). Considering (1) that publications needed a certain lead time, (2) that the authors had different backgrounds—Eichhorn[11] was an amateur scientist

---

[10]This phenomenon was known in rotifers observed by the Dutch tradesman, amateur scientist and "father of microbiology" Antonie Philips **van Leeuwenhoek** (1632–1723) (see Leeuwenhoek 1702) and in nematodes observed by the English biologist and Roman Catholic priest John Turberville **Needham** (1713–1781) (see Needham 1743).

[11]Johann Conrad **Eichhorn** (1728–1790), German protestant pastor and amateur researcher, who described microorganisms including a tardigrade in the moat and other water bodies near his home city Danzig, now Gdansk (Poland). In 1775 he published the booklet "Beyträge zur Natur-Geschichte der kleinsten Wasser-Thiere" [Contributions to the natural history of the smallest aquatic animals], primarily for lovers of natural history, which was reprinted in 1781.

throughout his life and Goeze[12] was just at the start of his career as a respected zoologist[13] (see Greven 2015), whereas Corti,[14] Spallanzani[15] and Müller[16] were trained scientists—(3) that authors probably did not know each other (only Müller cites Spallanzani and Eichhorn) and (4) that both Eichhorn and Müller claimed that they had seen a water bear long before they published their observations (see Greven 2015; unpublished manuscript), it is most likely that tardigrades have been independently discovered several times. However, formally Goeze is the first describer of a tardigrade, probably a *Hypsibius* species (Goeze 1773; see Fig. 1.2 top left). This description was published under the heading "Von einigen merkwürdigen Wasserinsekten" [On some strange water insects] in an appendix of the German translation of Bonnet's[17] "Traite d'Insectologie". The appendix contains personal observations made by Goeze in the environment of his hometown Quedlinburg. Goeze underlined the rarity of this animalcule (later he qualified this statement; see Müller 1785) and compared its appearance with that of a bear, giving it the name "water bear"[18]; his illustration reflects this bearish appearance, somewhat exaggerated (see Fig. 1.2 top left). Goeze found water bears in duckweeds (especially in winter), mistook the pharynx for an "egg sac", saw the floating body cavity cells, considered the lateral folds as tracheae and was surprised at the eight short feet,

---

[12]Johann August Ephraim **Goeze** (1731–1793), protestant pastor in Quedlinburg (Germany). He began his career as a translator of various scientific and philosophical works and became a highly respected and productive zoologist, especially helminthologist, only in the last third of his life. The description of the little water bear is considered as one of the first steps of an interested amateur. http://www.deutsche-biographie.de/ppn116745509.html. For further details see Greven (2015).

[13]"In meinen Erholungsstunden erquicke ich mich am liebsten durch mikroskopische Betrachtungen der unermesslichen Reichtümer der Natur im Wasser" [in my hours of rest I preferably refresh myself by the microscopic examination of the boundless wealth of nature in the water] (Goeze 1773, p. 361).

[14]Bonaventura **Corti** (1729–1813), Italian botanist, studied under L. Spallanzani (see Footnote 15), detected the plasma flow and investigated stoneworts (Characeae), jelly fungi (*Tremella* sp.), rotifers, ciliates and their reviviscence after dehydration (see footnote 20).

[15]Lazzaro **Spallanzani** (1729–1799), Italian polymath, catholic priest and naturalist, studied bodily functions, animal reproduction, echolocation, biogenesis, preservation of food, generation of microbes, etc. Stages in his life were lived in Reggio (professor of logic, metaphysic, Greek), Modena and Pavia (chair of natural history and director of the museum). "No biological phenomenon of any interest ever escaped the attention of Spallanzani (...)" (Keilin 1959, p. 153).

[16]Otto Frederick **Müller** (1730–1784), Danish naturalist (botanist and zoologist), studied later primarily Infusoria (= collective term for minute aquatic organisms including Protozoa and small "invertebrates"); he used for the first time Linné's binomial classification.

[17]Charles **Bonnet** (1720–1793), Swiss naturalist and philosopher, detected as a young man parthenogenesis in aphids and wanted to name the science of insects "insectology" (s. also Goeze 1773, p. 39).

[18]"Seltsam ist dieses Thierchen, weil der ganze Bau seines Körpers ausserordentlich und seltsam ist, und weil es in seiner äusserlichen Gestalt, dem ersten Anblicke nach, die gröste Aehnlichkeit mit einem Bäre im Kleinen hat. Das hat mich auch bewogen, ihm den Namen des kleinen Wasserbärs zu geben". [Strange is this little creature, because the whole organisation of its body is extraordinary and strange and because of its external appearance. At the first glance, has the closest similarity to a little bear. This also led me to give it the name little water bear.] (Goeze 1773, p. 368).

where he noticed three sharp, curved claws. Further he saw the exuviae with developing eggs and compared the reproduction of the water bear with that of the green freshwater alga *Volvox*.[19] Eichhorn's description and figure of the water bear with ten legs (Fig. 1.2 top right) do not compare with Goeze's presentation; the drawing and description are poor and virtually worthless and, therefore, are omitted here (for the translation of Eichhorn's text, see Greven 2015).

The next researchers to be acknowledged are the Italian scientists Corti and Spallanzani. Corti observed anhydrobiosis in jelly fungi, rotifers and some other organisms and Spallanzani in rotifers and nematodes. Both researchers occasionally observed tardigrades during their studies. Although not depicted and insufficiently described, the small caterpillars ["brucolini"] Corti (1774) found in rain gutters are most probably tardigrades.[20] Thus, Corti was the first who discovered anhydrobiosis in tardigrades; he wondered whether anhydrobiotic animals were dead or not, but he fully agreed with his academic teacher Spallanzani, who believed them to be dead.[21]

Spallanzani (1776) also studied tardigrades from the sandy substrate of the gutters; he compared their movement with that of a tortoise and called this animal "il Tardigrado". He described the animals with six (!) legs (considering the posterior pair of legs to be hooked filaments), noticed the position of the midgut and the pharyngeal bulb that he called the oesophagus but failed to breed them. Curiously, he compared its overall appearance with the testicle of a cock. The figures—Spallanzani employed an illustrator—are far below the quality of his observations (see Fig. 1.5). Further, he accurately described how tardigrades enter anhydrobiosis, pointing out

---

[19]"Sie hatten die größte Aehnlichkeit mit der Lage und Gestalt der sogenannten jungen Kugelthiere, die auf eben die Art in den Alten eingeschlossen sind". [Concerning overall appearance, they had the greatest similarity with the so-called globular animalcules, which are encased in the old ones in just the same way] (Goeze 1773, pp. 374/375).

[20]"Intorno al risorgere, questa proprietà è singolare, ma non è delle sole Tremelle: è gia stata osservata in altri animali, e in altri piante . . . Anch'io in bagnando della polvere delle grondaje ho veduto tornare a vita non solamente i *rotiferi*, ma ancora certi animaluzzi, cui ho chiamati i *brucolini* della polvere delle grondaje, a cagione di qualche somiglianza, che hanno coi bruchi;. . ." [With regard to reviviscence, this property is not only unique to the jelly fungi (this was considered in detail on the previous pages; translator's note), but was also observed in other animals and plants already. . . .I also saw in the dust of the gutter I had moistened, not only reviving rotifers, but also certain creatures that I have called the caterpillars of the gutter dust, because of some similarity they had with caterpillars.] (Corti 1774, p. 97).

[21]"Il punto più rilevante si è quello di decidere se le Tremelle, e gli accennati animalucci risorgenti sieno veracemente morti, oppure soltanto in apparenza: poi supposta la morte verace, come accada il loro risorgimento. Il Sig. Abate Spallanzani è di opinione, che il *rotifero*, e gli altri animaluzzi, i quali seccati in pria ritornano agli usati movimenti per mezzo dell'acqua, sieno rigorosamente morti, e ne soggiugne prove degne di lui, che è quanto dire da bravissimo filosofo naturalista". [The most relevant point is to decide whether the jelly fungi and the resurging little creatures in question are in fact truly dead, or only seem to be dead: assuming that they are truly dead, how does it happen that they come to life again? Signor Abate Spallanzani thinks that the rotifer as well as other little animals, which have been previously desiccated and are thus definitely dead, obtain their previous ability to move through contact with water. And he adds proofs that are worthy of his reputation as an exceptional natural philosopher.] (Corti 1774, p. 97/98).

that dehydration should be gradual to survive[22] (this observation has been ignored for a long time or has been forgotten), and emphasized that these animals cannot desiccate any number of times. The secondary (summarizing and rather popular) literature of this time appeared to accept this interpretation by writing that the sloth belongs to those animals that suffer death but are able to rise from the dead several times (Senebier 1795).[23]

The well-illustrated note (Fig. 1.2 bottom) on the "Bärthierchen" [bear animalcule] by Müller (1785), obviously published posthumously, contains corrections of Goeze's folds (tracheal tubes) and Eichhorn's descriptions (number of legs) and points out the similarity with Spallanzani's "sloth". Among other things Müller identified eggs in the animal, observed egg deposition during moulting, noted that the shed cuticle may protect the developing offspring, observed that bear animalcules are herbivores feeding on disintegrated duckweeds and vividly described their clumsy movements.[24] Müller was the first who classified tardigrades and assumed

[22]"I fenomeni del morire, mancando l'acqua, e del risorgere, sosituendone della nuova, succedono nel Tardigrado al modo sesso, che nel Rotifero. Il moto in lui via via si va perdendo, le gambe ſi ritirano, e s' internano totalmente dentro del corpo, questo rimpicciolisce assaissimo, si secca affatto, ed acquista forma globosa (...). Ed il contrario del fin quì narrato accade vivificando il Tardigrado con acqua novella. E siccome il Rotifero è limitato nelle volte, che può risorgere, così interviene al Tardigrado. Sembra però questo portarsi meglio in ciò, che quantunque l'arena conferisca per gran maniera al suo risorgere, non è però sì strettamente richiesa, come veduto abbiam nel Rotifero.

Que' gradi di calore, che son fatali ai Rotiferí risorti o da risorgere, lo sono ai Tardigradi, e lo stesso vuol dirsi degli odori, e dei liquori. Il freddo all'oppoſito, per quantunque aspro che sia, nulla può contra di essi, onde anche in ciò si accordano coi Rotiferi". (Spallanzani 1776, p. 225/226).

[The phenomena of its death, from the want of water, and of resurrection when water is supplied, are precisely the same with those of the wheel animal. Motion gradually ceases: the limbs are contracted and drawn entirely within the body, which diminishes very much, is completely dried, and assumes a globular figure, ... The reverse succeeds when the sloth is revived by supplying water. As the wheel animal can only revive a certain number of times, so it is with the sloth. And, although sand is necessary for its resurrection, it does not appear as essential as for the wheel animal.

The degrees of heat, fatal to revived or dead wheel animals, are also fatal to sloths; and the same must be said of odours and liquors. Cold, however intense, does them not harm, and in this they likewise coincide with wheel animals] (Spallanzani 1803, pp. 162–163; translated by J. G. Dalyell).

[23]Jean **Senebier** (1742–1809), Swiss reformed pastor, naturalist and bibliographer, studied vegetable physiology, mainly the influence of light on vegetation. The above quoted statement is to be read in the German translation (from French) "...die, wie das Kugelthier, das Räderthier, das Faulthierchen (le Tardigrade) den Tod leiden und mehrmals wieder auferstehen können". [...which, as the globular animalcule, the wheel animals, the little sloth, meet death and are able to rise again several times] (Senebier 1795, p. 41/42).

[24]"Der kleine Bär ist ein schwerfälliges, kaltblütiges und sanftes Thierchen, er lässet die Mitbewohner seines Tropfens mit gleicher Gleichgültigkeit als der Löwe das Hündgen um und an sich fahren". [The little bear is a clumsy, cold-blooded and gentle animalcule; he let himself be touched by the occupants of his drop with the same indifference as the lion the dog] (Müller 1785, p. 28).

their relationship with mites[25]; mites were considered as insects at that time, i.e. tardigrades were included into the much later created taxon Arthropoda.[26] Müller (l.c.) gave the bear animalcule the binomen *Acarus ursellus*[27] and added a short Latin diagnosis.[28] The binomen and some further remarks were included under the heading insects (mites) in the 13th edition of Linné's "Systema Naturae" (Linné 1790; Fig. 1.3).[29]

As a side note, Dalyell,[30] the translator of Spallanzani's "Opuscoli" (see Spallanzani 1803), noticed in his comments on the translation that he also observed a few sloths, similar to caterpillars. He considered them as new species and named them *Tardigradus Octopdalis* (sic!), i.e. the tardigrade with eight legs, and *Tardigradus Italicus* (see Spallanzani 1803) but qualified this statement in an added note after he had read Müller's article (see Greven 2015).

In brief, at the threshold of the nineteenth century, a few tardigrades (exclusively eutardigrades) were known. One species received a binomen that was included in the 13th edition of Linné's "Systema Naturae" (Fig. 1.3). Further, anhydrobiosis of tardigrades and some important parameters that enabled the animals to withstand dehydration were described, but generally anhydrobiotic animals were considered dead.

---

[25]". . .den Milben, die in die Klasse der Insecten gehören, kömmt es in der Gestalt, der Zahl und dem Gebrauch seiner Füße am nächsten" [Concerning the general appearance, and the number and use of his feet, he is most similar to mites that belong to the class of insects] (Müller 1785, p. 26).

[26]The term "Arthropoda" was used for the first time in 1848 by the German physiologist and zoologist Karl (Carl) Theodor Ernst **von Siebold** (1804–1885) in his "Lehrbuch der Vergleichenden Anatomie der wirbellosen Thiere" [Textbook of Comparative Anatomy of Invertebrates]. Arthropods included the classes Crustacea, Arachnida and Insecta. Tardigrades were considered as arachnids: "(. . .) die Tiere sind wohl nirgends passender unterzubringen als gerade hier, nur müssen sie obenan gestellt werden, da sie den Uebergang von den Arachniden zu den Annulaten bilden" [. . .the animals may be placed nowhere better than just here, but they must be placed at the top, since it forms the transition from the arachnids to Annulaten more appropriate than just here, but they must be placed first as they form the transition from arachnids to annelids] (von Siebold 1848, p. 506). According to Haeckel (1896, p. 597), the full separation of the annelids and arthropods was "einer der größten Rückschritte der neueren Systematik und hat 30 Jahre hindurch das Verständniss des Articulaten-Stammes in hohem Maasse erschwert; (. . .)" [was one of the biggest steps backwards of the recent systematics and has extremely complicated the conception of the phylum Articulata for 30 years]. See footnote 130.

[27]ácari, Gr. = mite; urséllus is the diminutive of úrsus, Lat. = bear.

[28]"Dieses in den allgemeinen Thiergeschichten noch nicht angeführte Thierchen kann mit folgenden Namen bezeichnet werden. Acarus Ursellus corpore rugoso, pedibus conicis" [This animalcule, not yet recorded in general histories of animals, can be described with the following words. Acarus Ursellus with wrinkled body and conical feet] (Müller 1785, p. 30).

[29]Carl **Linné** (Carolus Linnaeus), after his ennoblement in 1761. Carl von Linné (1707–1778), Swedish botanist, professor of medicine and botany at the University of Uppsala (Sweden), was the inventor of the binomial system. The first edition of his "Systema Naturae" was published in 1738. Apparently there were 13 editions, of which Linnaeus wrote only 5 (Usinger 1964).

[30]Sir John Graham **Daylyell** (1775–1851), Scottish antiquary and naturalist, translated Spallanzani's "Opuscoli" from Italian to English and provided the second translated edition with comments and his own observations (see Spallanzani 1776, 1803).

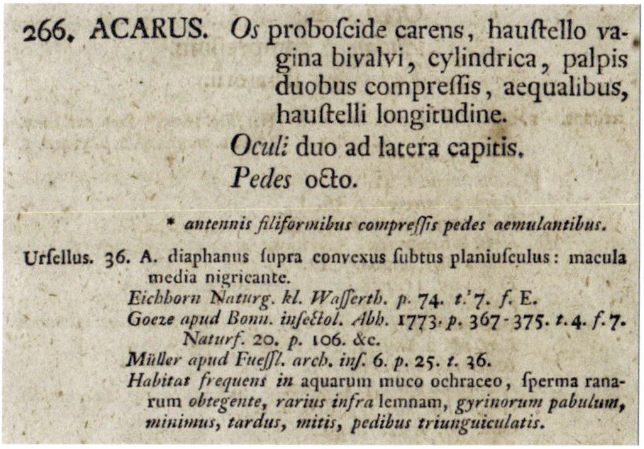

266. ACARUS. *Os* proboſcide carens, hauſtello vagina bivalvi, cylindrica, palpis duobus compreſſis, aequalibus, hauſtelli longitudine.
Oculi duo ad latera capitis.
Pedes octo.

* *antennis filiformibus compreſſis pedes aemulantibus.*

Urſellus. 36. A. diaphanus ſupra convexus ſubtus planiuſculus: macula media nigricante.
Eichhorn Naturg. kl. Waſſerth. p. 74. t. 7. f. E.
Goeze apud Bonn. inſectol. Abh. 1773. p. 367-375. t. 4. f. 7.
Naturf. 20. p. 106. &c.
Müller apud Fueſſl. arch. inſ. 6. p. 25. t. 36.
Habitat frequens in aquarum muco ochraceo, ſperma ranarum *obtegente, rarius infra* lemnam, *gyrinorum pabulum, minimus, tardus, mitis, pedibus triunguiculatis.*

Fig. 1.3 Page 2924 from the 13th edition of Linné's "Systema Naturae" (1790) showing the entry of the water bear *Acarus ursellus* ([Ursellus 36. translucent *Acarus* on the dorsal side convex, on the ventral side slightly flattened: with a blackish spot in the middle. Habitat: Common in the yellowish slime covering the sperm of frogs in water bodies, rarer under *Lemna* (duckweed), food for the tadpoles, very small, slow, peaceful, with three-clawed feet]. This is exactly what one can read in Müller (1785) and the notes added to this article by Goeze)

1.3 Moves in Various Directions: The First Half of the Nineteenth Century

In 1803 Paula von Schrank[31] briefly introduced a new species, *Arctiscon*[32] *tardigradum*, which he described as caterpillar-like and equipped with two short antennae (cephalic papillae of *Milnesium*?). He did not provide a drawing, but criticized the figures in Spallanzani's (bad) and Eichorn's work (not better), which were the only authors he cited. Schrank belonged to those researchers who doubted reviviscence of anhydrobiotic animals including tardigrades. His fault was that he did not allow them to dry together with sand,[33] a prerequisite for the success of such an experiment as was already emphasized by Spallanzani (1776).

---

[31]Franz **Paula von Schrank** (1747–1835), German Jesuit priest, botanist and entomologist and professor at the University Ingolstadt and Landshut (Germany); first director of the botanical garden in Munich (Germany), editor of a multivolume Fauna of Bavaria (1803), one of the most important botanists of Bavaria (http://www.deutsche-biographie.de/pnd11861066X.html).

[32]árktos, Gr. = bear.

[33]"Es ist völlig falsch, was Senebier nach flüchtigen Beobachtungen behauptet..., dass dieses Thierchen, das Kugelthier, und das Räderthier den Tod öfter leiden, und mehrmals wieder aufleben können. Alle diese Thierchen platzen, wann der Wassertropfen abdunstet, und sind dann unwiederbringlich verloren. Die Naturgeschichte bedarf der angeblichen Wunder nicht, sie hat der wahren genug". [It is quite incorrect to assert, as Senebier does based on superficial observations... that this animalcule (tardigrade in water bodies, translator's note), the spherical sphere, and the wheel suffer death more often, and are able to revive many times. All these little creatures burst

Thereafter remarkable studies were issued in quick succession. The systematic position of tardigrades changed their place mostly within "arthropods" several times, and publications attempting classification based on superficial similarities often appeared almost simultaneously. Researchers, most of them trained scientists (arranged according to the year of their publications with a note on their classification proposal), were Paula von Schrank (1803; wingless insect between fleas (*Pulex*) and mites (*Acarus*)), Dutrochet[34] (1812, 1837; insects or larvae of mites; see Fig. 1.4 top) and Blainville[35] (1826; larvae of beetles). The works of Schultze[36] (1834a, b), Perty[37] (1834), Dujardin[38] (1838, 1851) and Doyère[39] (1840, 1842a, b, c) are considered in more detail below.

In 1834 Schultze published two relatively short studies of almost identical content, focussing on the new species and its capacity to survive dehydration with just a single relevant quotation[40] paying tribute to Spallanzani.[41] Schultze (1834a, b)

---

when the water droplet evaporates, and are then lost forever. The natural history does not require the alleged miracles, it has true ones enough] (Paula von Schrank 1803, p. 195/196).

[34]René Joachim Henri **Dutrochet** (1776–1847), French physician, botanist and physiologist with a broad spectrum of scientific interests, particularly known for his studies on osmosis, respiration, embryology, and the effect of light on plants. The two volumes of the "Memoires…" (Essays…) from 1837 are a collection of his most important biological articles.

[35]Henri Marie Ducrotay de **Blainville** (1777–1850), French zoologist and anatomist at the Faculté des Sciences at the Sorbonne (Paris) and later successor of Georges Cuvier (1769–1832) to the chair for Comparative Anatomy at the Muséum National d'histoire naturelle in Paris.

[36]Carl August Sigismund **Schultze** (1795–1877), German physiologist and anatomist. He was from 1812 onwards professor at the University of Freiburg im Breisgau (Germany) and from 1931 professor at the University of Greifswald (Germany). He wrote four articles about tardigrades (Schultze 1834a, b, 1840, 1861), numerous medical reports and a textbook of comparative anatomy. The two publications from 1834 are of similar content, one of them was issued as Festschrift dedicated to his distinguished contemporary, the physician Christoph Wilhelm Hufeland (1762–1836) on the occasion of the 50th anniversary of his doctorate (Schultze 1834a).

[37]Josef Anton Maximilian **Perty** (1804–1884), German naturalist and natural philosopher, professor at the University of Bern (Switzerland) as of 1834, published among other things about rotifers, Infusoria and insects, mainly Coleoptera. http://www.deutsche-biographie.de/pnd116092386.html?anchor=adb

[38]Félix **Dujardin** (1801–1860), French biologist, professor at the University of Rennes, researched on protozoan and other "invertebrates", e.g. echinoderms, helminths and cnidarians, and wrote a "Histoire naturelle des zoophytes" (Dujardin 1841).

[39]Louis Michel François **Doyère** (1811–1863), student of H. Milne-Edwards (see footnote 56), French agriculturist, (1850–1852) professor of natural history (1842), 1850–1852 professor of Applied Zoology at the "Institute agronomique de Versailles" (see Maire 1892) and then at the "École centrale des arts et manufactures" (a well-known school of engineering founded in1829). He developed a process for conservation of grain silage. His thesis was published in three parts (Doyère 1840, 1842b, c) and is also available as a separate volume (thesis) (Doyère 1842a).

[40]This is in contrast to his later article (Schultze 1861), where he considered the literature very detailed.

[41]"Quae quidem descriptio, sicut alia diligentissimi naturae scrutatoris inventa, oblivioni omnino tradita esse videtur; certo a nemine confirmata est". [Of course, his description as well as other findings of this very diligent student of nature seems to have been entirely forgotten; certainly, nobody has confirmed them.] (Schultze 1834b, p. 5).

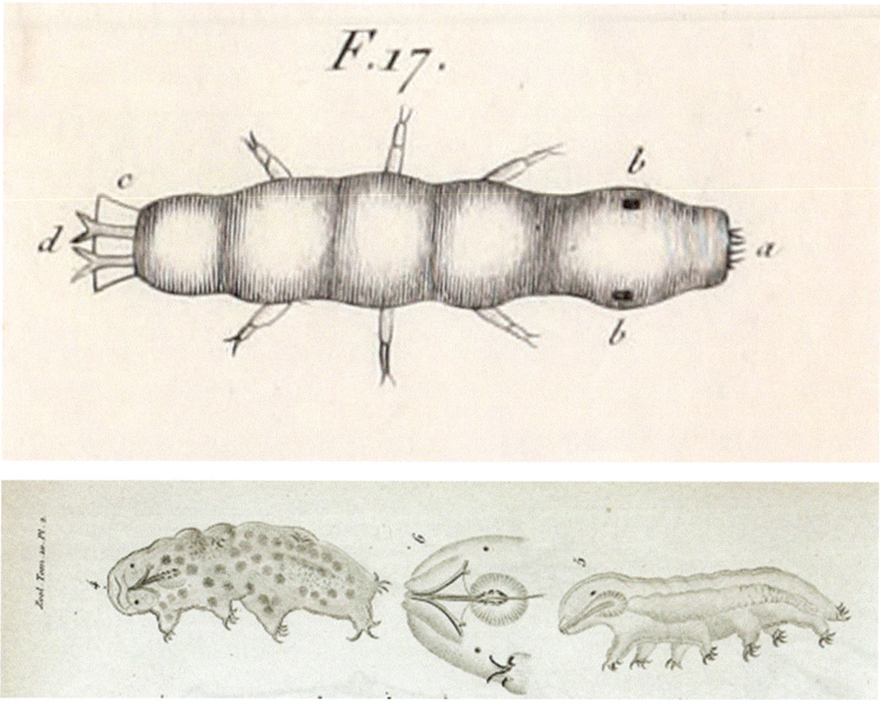

Fig. 1.4 Early pictorial presentations of tardigrades (without species name). Top: From Du Trochel (1812); detail from Plate 18. Bottom: From Dujardin (1838); detail from Plate 2

named his new species *Macrobiotus hufelandii*[42] (to my knowledge the first scientific name of a tardigrade dedicated to a person, which is still valid today). He noted the correct number of claws, the different parts of the digestive tract (but was wrong as far as the function is concerned), the position of the ovary and the sculpturing of the egg chorion (see Fig. 1.5). He classified this animal as a crustacean taxon based on its relatively strong integument divided into segments like a suit of armour, jointed (!) legs with claws and blood vessels; the absence of a heart and respiratory

[42]The genus name refers to the famous book of the German physician Christoph Wilhem **Hufeland** (1762–1836) entitled "Makrobiotik oder die Kunst das menschliche Leben zu verlängern" [Makrobiotik or the art of prolonging human life] that was reprinted several times and translated in several languages. The title is derived from the Greek word makróbios = long living. In this book Hufeland laid down the principle: " Je weniger intensiv das Leben eines Geschöpfs und je geringer seine innere und äußere Consumtion, desto dauerhafter ist es" [The less intense life of a creature and its internal and external consumption, the longer its life will last.] (Hufeland 1826, p. 82). Hufeland believed, however, that this rule does not apply to human beings due to their exceedingly great portion of spiritual power.

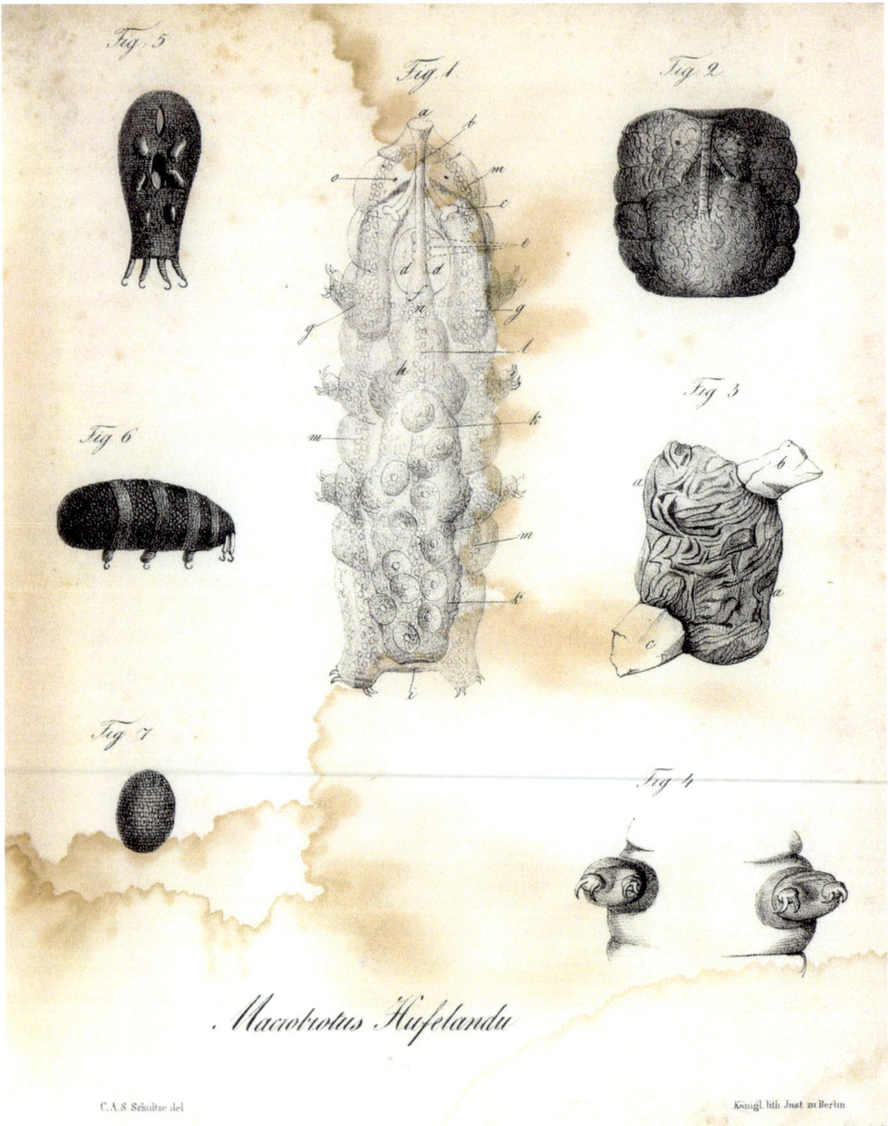

Fig. 1.5 Plate from Schultze (1834a) showing (1) the organization of *Macrobiotus hufelandii*. (2) *M. hufelandii* contracted by evaporation of water. (3) The same with adhering sand grains after complete dehydration. (4) Forelegs, from below. (5–7) Figures from Spallanzani's treatise showing the animalcule from below (5), the side (6) and desiccated (7)

organs he considered as a sign of primitiveness.[43] Concerning anhydrobiosis, Schultze (re)described the "tun" (figured by Spallanzani; see Fig. 1.5) of *M. hufelandii* and considered anhydrobiotic tardigrades as being in a state of long-lasting suspended animation ("mortis simulatione diuturna insigne", p. 2, the article is not paginated) or—even more provocative at that time—assumed that animals during evaporation of water are able to contract but then unable to pick up stimuli, being brought into a death-like state retaining the ability to restore all phenomena of life, i.e. a state that might considerably prolong their life.[44] This wording contrasted with previous views assuming a resurrection (from death). This way of looking at anhydrobiosis provoked distinguished contemporary scholars to disagree vehemently. An influential opponent, for example, was Ehrenberg.[45] Disregarding previous studies, he believed that rotifers and nematodes were not able to revive due to their complex structure, and he considered their reanimation (and that of tardigrades) an illusion, assuming that the individuals seemingly animated after years are the great-great-grandchildren of the formerly dried individuals (Ehrenberg 1834). This was the prevailing opinion of all those who denied "latent life" or "vita minima" (not only in tardigrades) in the sense of an extremely retarded or even arrested regular metabolism. Even the already disproved abiogenesis ("generatio spontanea")[46] was

---

[43]"Macrobiotus simplici structurà ac parvitate Protozoiset Annulatis affinis est. Attamen duriusculo corporis tegumento, in speciem loricae e segmentis composito, pedibus articulatis ungues ferentibus, truncis vasorum sanguiferorum Crustaceorum signa offert. Qua in classe autem nonnisi infimum occupare potest, cum organa respiratoria propria et cor eum deficere videantur." [Due to its simplicity and smallness Macrobiotus is near the Protozoa and Annulata. However, with the rather tough integument, composed of segments like armour, with jointed legs that bear claws and with trunks supplied with blood they show traits of crustaceans. But in this class, it can hold only the lowest place, as it seems to lack own respiratory organs and a heart.] (Schultze 1834b, p. 5).

[44]"Maxime memorabilis est indoles animalis nostri, aqua omni evaporante sese contrahendi (...) omissa ipsa stimulos percipiendi facultate in statum morti simillimum transeundi, solamque vim restitutionis omnium vitae phaenomenorum retinendi. In hoc statu devinctarum virium vitalium (...) Macrobiotus non per breve solum tempus, sed per plures annos durare potest, multo igitur diutius, quam si vita ejus non turbata fuisset." [Most curious is the ability of our animal to contract (...), and to transform in a state very similar to the death after having lost the ability to be receptive to stimuli, and to retain only the power to restore again all phenomena of life. In this state of bound vitality (...) Macrobiotus is able to persist not only a short time, but many years, i.e. much longer than if its life would not be interrupted.] (Schultze 1834b, p. 4).

[45]Christian Gottfried **Ehrenberg** (1795–1876), a very productive and influential German zoologist, microgeologist and micropalaeontologist, professor of medicine at the University Berlin, friend of Alexander von Humboldt (1769–1859). He published on Hydrozoa, molluscs, coral polyps, rotifers, etc. but later concentrated on microscopic organisms (Protozoa, microalgae and bacteria) and founded the science of micropaleontology. He employed the term "Infusoria" for a wide range both of animal and vegetable life, i.e. he did not yet separate the multicellular organisms from the unicellular ones. One of his most famous monographs is "Die Infusionsthierchen als vollkommene Organismen" [The Infusoria as complete organisms] in 1838. He was one of the strongest opponents of the concept of resuscitation of desiccated "Infusoria".

[46]The first scientists to challenge the theory of spontaneous generation, i.e. formation of living organisms from non-living matter, were Francesco **Redi** (1626–1697) and Lazzaro Spallanzani (1729–1799; see footnote 15).

brought into play, e.g. by Purkinje.[47] Later Schultze erected the genus *Echiniscus*[48] and described two species of this genus (Schultze 1840, 1861). Interestingly, he also observed *Arctiscon* (=*Milnesium*) species, but did not describe them in detail.[49]

Perty (1834), who had at that time no other involvement with tardigrades but later published faunistic articles on Swiss tardigrades, summarized most of the existing taxonomic literature noting that Schultze (1834a) and Ehrenberg (1834; this is a letter commenting on Schultze's experiments; see below) had ignored the observations of Paula von Schrank (see above). He asserted the "right of the earlier observer" and, therefore, preferred the genus name *Arctiscon*, distinguishing five species and giving each the name of the first observer as the epithet, e.g. *Arctiscon mülleri*, *A. schrankii*, *A. hufelandii*, *A. spallanzanii* and *A. dutrochetii*.[50] Further, he included tardigrades in the newly created Xenomorphidae,[51] a family of crustaceans, which he positioned between Lernaeidae (Copepoda, Crustacea) and Annelida with similarities to Infusoria Rotatoria.[52] However, he did not provide reasons for that.

---

[47]See footnote 49.

[48]échinos, Gr. = hedgehog.

[49]A letter by Schultze read aloud at the 15th Meeting of the Society of German Researchers and Physicians in Prague says under point 31 1. "Uiber den *Macrobiotus Hufelandi* und noch vier andere Species dieser merwürdigen Krusterfamilie. Zwei davon sind mit Fühlhörnern, eine mit zwei, die andere mit vier, und mit Fressspitzen versehen, welche, sowohl die Fühlhörner als Fressspitzen, der Gattung *Macrobiotus* fehlen. Das mit zwei Fühlhörnern versehene Thierchen dürfte das von Schrank beschriebene *Arctiscon tardigradum* seyn; auch legt dieses Thier seine Eier (7-11) immer in die abgelegte Haut, was der *Macrobiotus* niemals thut. Hr. Hofr. Schulze (sic!) hat ein kleines Päckchen Sand, welcher diese Thierchen enthält, eingesendet, den er seit sechs Monaten trocken aufbewahrt hat, worin gewiß einige Exemplare dieses *Arctiscon* durch Benetzung mit frischem Regenwasser oder destillirtem Wasser werden sich beleben lassen. (...). "[About *Macrobiotus Hufelandi* and four other species of this strange family of crustaceans. Two of them are equipped with antennae, one with two, the other with four, and with feeding tips (= peribuccal papillae, translator's note); both, the antennae and feeding tips, are lacking in the genus *Macrobiotus*. The animalcule with the two antennae is probably *Arctiscon tardigradum* described by Schrank; in addition, this animalcule lays its eggs (7–12) in the shed skin, which *Macrobiotus* never does. Hofrat Schulze (sic!) has sent a small parcel with sand containing these animalcules, which he has stored in a dry place for six months; certainly it will be possible to reanimate some specimens of this *Arctiscon* by wetting <the sand> with rain water or distilled water].

On the same page under point 32, Mr Purkinje, professor of physiology at the University of Breslau (at that time Germany), informs "(...) dass man die Entstehung des *Macrobiotus* und ähnlicher Thiere in seiner Gewalt habe. Man braucht nämlich Sand mit Wasser zu benetzen und stehen zu lassen, wo dann nach einigen Tagen solche Thiere sich zeigen" [that the origin of Macrobiotus and similar animals can be mastered. You have only to leave sand moistened with water, where then such animals emerge after a few days] (Sternberg and von Kromholz 1838, p. 187).

[50]Except for *Macrobiotus hufelandii*, none of these names, neither genus nor epithet, survived due to the impossibility to identify the species in question.

[51]xénos, Gr. = strange; morphé, Gr. = form, shape

[52]Friedrich Siegmund **Voigt** (1781–1850), professor of medicine and botany and director of the Botanic Garden at the University Jena (Germany), translated the second edition of "Le règne animal...." by the famous French naturalist Georges **Cuvier** (1769–1832) into German and extended the text considerably (Voigt 1843). Voigt classified tardigrades as the first order of the Infusoria, in which also Rotatoria, Polygastrica (Infusoria sensu Ehrenberg) and Spermatozoa were included.

Concerning anhydrobiosis (resurrection versus reawakening), he agreed with Schultze, using the term "latent life".[53]

Dujardin (1838) corrected some data given by Schultze (1834a), noting that tardigrades have folds and wrinkles instead of segments, non-jointed legs, eyes like those of planarians and "blood corpuscles" that do not float in vessels, but freely in the body cavity. Further, he described the organization and the function of the buccal apparatus and added drawings from living tardigrades (see Fig. 1.4). His notes on reproduction (egg deposition, moulting) and anhydrobiosis were also correct. He considered rotifers and tardigrades as Systolides,[54] a class of organisms with a strong ability to contract, with a relatively tough outer covering of the body, a simple intestine and a "jaw apparatus" equipped with articulated elements and specific muscles. Tardigrades were called "Systolides marcheurs", i.e. the walking Systolides, lacking cilia. They were suggested to form the transition between helminths on the one side and annelids and arachnids on the other side (Dujardin 1841).

The unanimous opinion of all later researchers is that Doyère's dissertation "Memoire sur les Tardigrades" is an indisputable milestone in tardigradology, providing a profound basis for subsequent studies in anatomy and physiology. In 144 pages, he contributed to a wide spectrum of topics, e.g. taxonomy, morphology, reproduction and anhydrobiosis (Fig. 1.6), based on deep knowledge of the relevant literature. The thesis was published in three parts (Doyère 1840, 1842b, c), but is available, although rarely, as single-volume thesis (1842a; see Fig. 1.6). In the first part (Doyère 1840), he started with a broad historical introduction, following Dujardin's view concerning "Systolides broyeurs" (crushing systolids, i.e. rotifers) and "Systolides suceurs" (sucking systolids, i.e. tardigrades) and their relations to crustaceans and annelids and distinguishing three genera, the genus *Emydium*[55] (three species; see above), the still valid *Milnesium*[56] (one species), both created by him, and *Macrobiotus* (four species). He detected asphyxia[57] and described in

[53]The term was adopted from physics (e.g. latent heat) and introduced in 1834 for physiology by the German physician and natural philosopher Carl Gustav **Carus** (1789–1869) (see Carus 1834).

[54]systolé, Gr. = contraction.

[55]emýs, Gr. = turtle. In an annex of his thesis, Doyére noted the similarity of Schultze's *Ech. bellermanni* with his *Em. testudo* and accepted the priority of the term *Echiniscus* (see Doyére 1842c).

[56]Dedicated to Henri Milne-**Edwards** (1800–1885), eminent French zoologist, at that time professor of entomology at the Muséum National d'Histoire Naturelle and at the faculty of Sciences at the Sorbonne in Paris and later chair of zoology

[57]asphyxía, Gr. = stopping of the pulse. "L'asphyxie est le moyen qui réussit le mieux, celui qui donne les plus beaux résultats. Je prends des Tardigrades vivans, je les place dans un tube en verre plein d'eau préalbablement privée d'air par l'ebulliton, et au-dessus de laquelle j'ai le soin de mettre une couche d'huile pour la séparer de l'atmosphère. Après vingt-quatre heures l'engourdissement est complet, il est plus complet et plus durable après deux, trois, quatre jours; ce n'est qu'après cinq à six jours que les Tardigrades perdent la faculté de revenir à la vie." [Asphyxia is the most successful way, one that gives the best results. I take living Tardigrades, I place them in a glass tube filled with water, deoxygenated beforehand by boiling, and cover it carefully with a layer of oil to separate it from the atmosphere. After twenty-four hours, numbness is complete, it is even more complete and lasting after two, three or four days; it was only after five or six days that Tardigrades lose their ability to come back to life.] (Doyère 1840, p. 333). This technique was simplified later and was the method of choice for decades to study tardigrades (e.g. Greeff 1865; Plate 1889; Basse 1905).

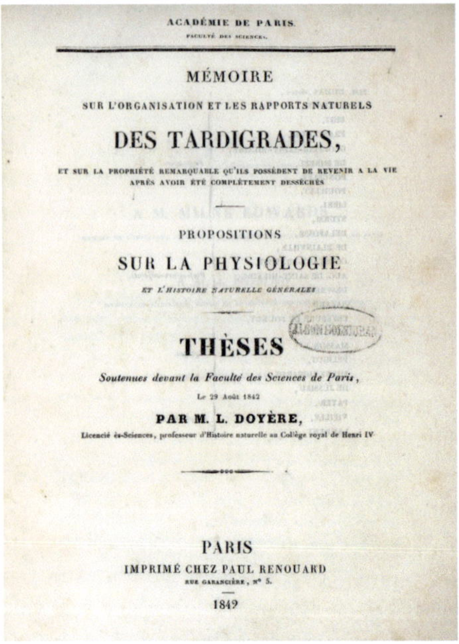

ACADÉMIE DE PARIS.
FACULTÉ DES SCIENCES.

MÉMOIRE
SUR L'ORGANISATION ET LES RAPPORTS NATURELS
DES TARDIGRADES,
ET SUR LA PROPRIÉTÉ REMARQUABLE QU'ILS POSSÈDENT DE REVENIR A LA VIE APRÈS AVOIR ÉTÉ COMPLÈTEMENT DESSÉCHÉS

PROPOSITIONS
SUR LA PHYSIOLOGIE
ET L'HISTOIRE NATURELLE GÉNÉRALES

THÈSES
Soutenues devant la Faculté des Sciences de Paris,
Le 29 Août 1842
PAR M. L. DOYÈRE,
Licencié ès-Sciences, professeur d'Histoire naturelle au Collège royal de Henri IV

PARIS
IMPRIMÉ CHEZ PAUL RENOUARD
RUE GARANCIÈRE, N° 5.
1849

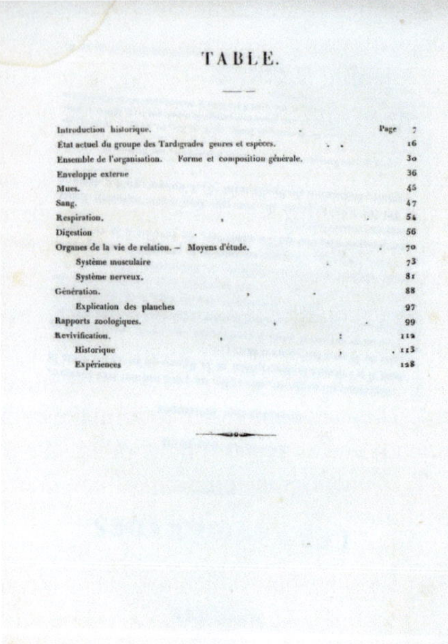

TABLE.

Fig. 1.6 Title page and table of contents of Doyère's thesis (Doyère 1842a)

detail how to put tardigrades in this state that he found very effective to study their anatomy. Thus, he dealt with myology, neurology and reproduction (he recognized spermatids) and described some stages of oogenesis, egg deposition and much more. Despite the poor optic equipment (by today's standards), the quality of his observations is excellent (irrespective of whether his interpretations are correct), and the reader can only be amazed by the many details he saw, for example, when he described that newly hatched *Echiniscus testudo* possess only two claws per leg.

Obviously he was also the first who saw cysts, but he did not realize their true nature.[58] He also discovered "anomalies" in the buccopharyngeal apparatus (see below). His accuracy is perhaps best illustrated (without going into detail) by the plates added to the thesis, from which four are shown in Fig. 1.7.

Starting with Dujardin's Systolides, the second part (Doyère 1842b) contains a further, broad discussion about the relationships of tardigrades considering the previous existing literature and his own morphological findings. However, concerning the class Systolidae as well as its relations to other "articulated" animal classes, he did not come to a clear decision.

[58]"J'eus d'abord quelque peine à reconnaître l'animal, dans la petite masse, inert, en apparence granuleuse et amorphe que je rencontrais parfois à l'interieure de certains peaux qui me semblait abandonnés. C'était le Tardigrade lui-méme. Déjà dépouillé, mais non encore sorti de sa dépouille." [Initially I had some difficulty to recognize the animal, in the small mass, inert, granular and amorphous in appearance, which I sometimes found within a skin. It was the tardigrade itself. I thought it had already been shed. Already shed, but not yet outside of its skin.] (Doyère 1840, p. 308).

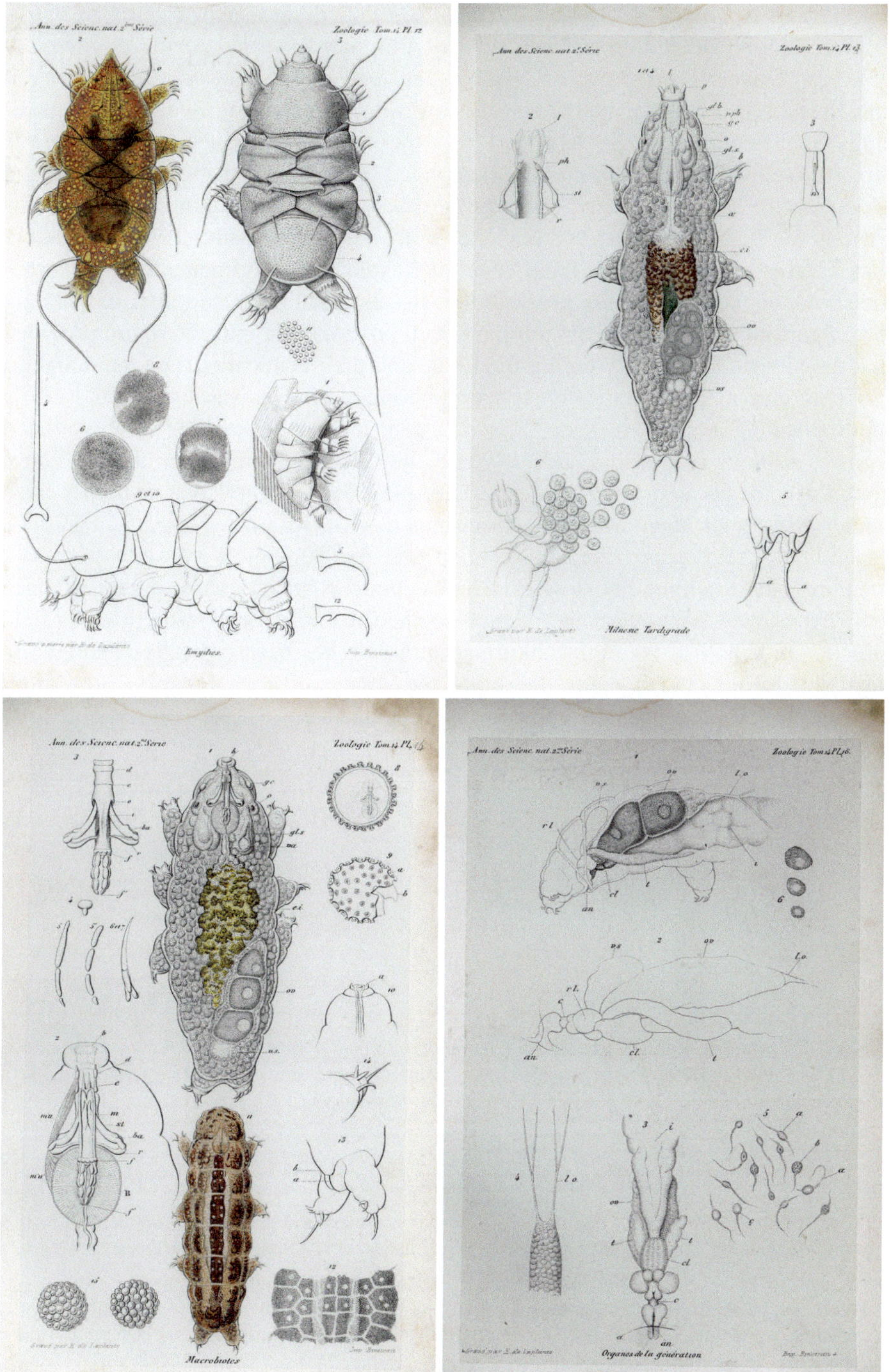

Fig. 1.7 Four plates from Doyère's thesis. Top left: Plate 12, counting of the plates starts with 12, *Emydium testudo*. Top right: Plate 13, *Milnesium tardigradum* (six squeezed buccal apparatus

The last part (Doyère 1842c) contains the dehydration experiments on tardigrades. Among other things the results of this part have led to the controversy between Doyère and Pouchet[59] (see below). In the historical review on anhydrobiosis, he highly appreciated Leeuwenhoek, Needham, Schultze and especially Spallanzani. Then he described his experiments (mainly desiccation and the effect of elevated temperature on desiccated tardigrades) using *Emydium* and *Macrobiotus* and emphasized the worthlessness of a "desiccation à nu" (means literally translated "nude desiccation", i.e. direct desiccation without an environment that releases moisture slowly) and concluded from his experiments that the "organized matter" contains life *in potentia* that passes into life *in actu* after moistening the anhydrobiotic animals, i.e. their pass from potential life into active life provided that the molecular integrity within the tissue and their connections are maintained.

The above-mentioned dispute concerning anhydrobiosis was submitted 1859 to the Société de Biologie in Paris[60] for decision; it can be presented in abbreviated form as follows (see also Lance 1896; Keilin 1959). Both parties agreed that a completely desiccated body is dead. The question was, however, whether such bodies can regain their lost life. According to Doyère a dried organism has the capability to revive, provided desiccation and moisturization are slow (e.g. in a medium releasing humidity slowly), whereas Pouchet stated that an organism, once fully dehydrated, cannot escape the final death. The 156-page report written by Paul Broca[61] on behalf of the high-ranking commission has been critically examined in detail by Keilin (1959), who also corrected some of Broca's errors. Both sides carried out experiments; the control experiments made by the commission fully

Fig. 1.7 (continued) of *Macrobiotus hufelandii*). Bottom left: Plate 14, *Macrobiotus hufelandii* (1–11) and *Macrobiotus oberhaeuser* (11–15). Bottom right: Plate 16, reproductive organs of *Macrobiotus hufelandii* (1, 4, 5, 6) and *Milnesium tardigradum* (2–4). From Doyère (1842a)

[59]Félix-Archimède **Pouchet** (1800–1872, French physician and naturalist, founder of Rouen Museum of Natural History (1828), from 1838 professor at the School of Medicine at Rouen, a leading proponent of spontaneous generation of life from non-living materials.

[60]Société de biologie, a learned society founded in Paris (France) in 1848. The members of the society held regular meetings; proceedings were published in the scientific journal "Comptes rendus de la Société de Biologie" first issued in 1849.

[61]Pierre Paul **Broca** (1824–1880), French physician and anthropologist, known for his research on a region of the frontal lobe involved with language that was named after him (= Broca's area).

confirmed Doyère's view.[62] Negative results[63] were attributed to fluctuating humidity during anhydrobiosis as alternating swelling and drying might cause mechanical damages. Animals in anhydrobiosis tolerated the highest temperature. This was not accepted by some contemporaries and "intellectual latecomers" (Marcus 1929a). However, the controversial issue was no longer the ability to undergo anhydrobiosis, but how to explain it, i.e. whether the metabolism was extremely reduced (vita minima, see above) or suspended without dying.

In several publications issued between 1848 and 1862, Ehrenberg promoted knowledge of α-taxonomy and distribution of terrestrial tardigrades especially in high altitudes (Alps, Himalaya), in which he suspected specific forms. He distinguished tardigrades from the Acaridae mainly by the absence of metamorphosis in the presence of moulting. Among others he described several *Milnesium* species, i.e. *M. alpigenum* (Ehrenberg 1853a, b; see Fig. 1.8), and even a new genus *Acrophanes* related to *Milnesium* with *A.* (*Milnesium*) *schlagintweitii* as species from the Himalaya (Ehrenberg 1859), using, among other distinguishing features (e.g. claw configuration), the number of the "tentacles" (=the laterally positioned cephalic papilla) and peribuccal papilla.[64]

---

[62]The key conclusion of the commission was: "La résistance des tardigrades et des rotifères aux températures élevées paraît s'accroitre d'autant plus qu'ils ont été plus complétement desséchés d'avance. Les rotifères peuvent se ranimer après avoir séjourné quatre-vingt-deux jours dans le vide sec et subi imdiatement aprés une température de 100° pendant trente minutes. Par conséquent, des animaux desséchés successivement à froid dans le vide sec, puis à 100° sous a pression atmosphérique, c'est-à-dire amenés au degré de desiccation le plus complet qu'on puisse réaliser dans ces conditions et dans l'état actuel el a science, peuvent conserver encore la propriété de se ranimer au contact de l'eau." [The resistance of tardigrades and rotifers to high temperature appears to increase, the more they had been completely dehydrated before. Rotifers can be become alive again after having stayed for eighty-two days in a dry vacuum and immediately thereafter being exposed to a temperature of 100° for thirty minutes. Therefore, animals which had been dried first in a cold dry vacuum and then at 100° under atmospheric pressure, i.e. brought to the most complete degree of desiccation we can achieve under these conditions and in the present state of science, still retain the ability to revive in contact with water] (Broca 1860, p. 139).

[63]In this context, a short note from 1858 by the French physician Casimir **Davaine** (1812–1882), co-discoverer of *Bacillus anthracis*, should be mentioned. He showed that, in contrast to moss-dwelling tardigrades and rotifers, fully aquatic species of both are not able to withstand dehydration (Davaine 1858). Later Marcus (1928a, 1929b) thought that such differences depend to a large extent on the habitat rather than on the general inability of the species in question to tolerate desiccation. He further suggested to leave up the distinction between freshwater and terrestrial tardigrades in favour of strongly and less strongly hygrophilous (hygrós, Gr. = wet; phílos, Gr. = friend, loving) species and rejected the term xerophilous (xéros, Gr. = trocken) as all tardigrades can only be active in a medium containing humidity.

[64]*Milnesium* was considered over decades as a monospecific genus with a single cosmopolitan species, which currently, however, contains more than 30 species. No single tardigrade species described by Ehrenberg has been revalidated so far. However, more recently *Milnesium alpigenum* appears to be revitalized using animals from a parthenogenetic lab strain collected near Tübingen (Germany) (Morek et al. 2016), but Ehrenberg collected his material in the Mone Rosa massif in the Pennine Alps at approx. 3600 m above sea level (Ehrenberg 1853b). According to Ehrenberg, Doyère's *M. tardigradum* and his *M. alpigenum* differ in the claw configuration and also in the number of peribuccal papilla, i.e. three in the former and six in the latter. However, also

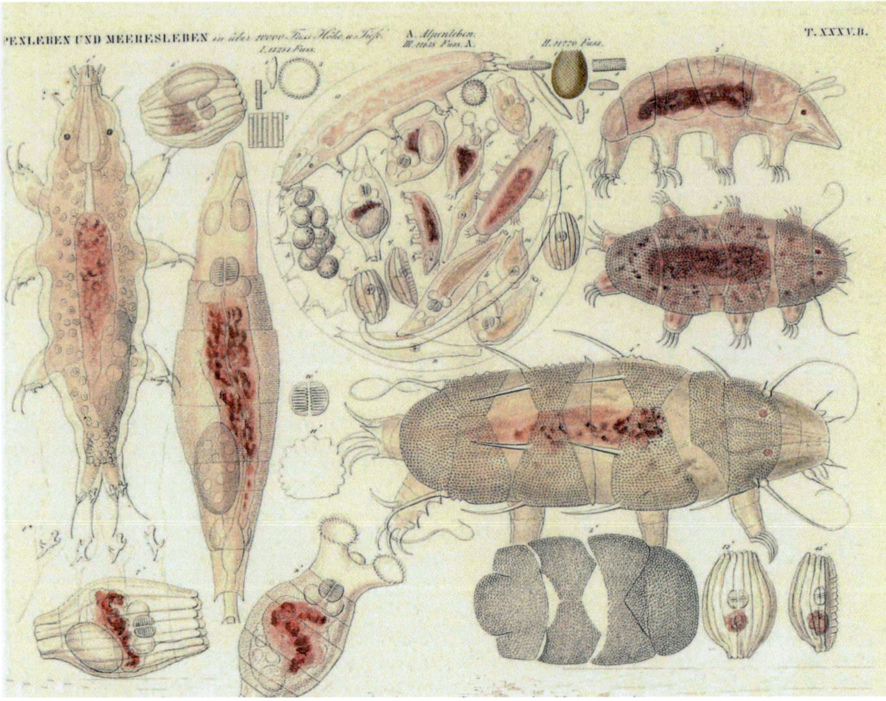

Fig. 1.8 Various rotifers and tardigrade. Detail of Plate 35B from Ehrenberg (1854). Tardigrades are *Milnesium alpigenum* (left side); various "Infusoria" including tardigrades (middle above). *Echiniscus suillus*, *Echiniscus arctomys*, *Echiniscus victor*, *Echiniscus altissimus* (right side, top down)

In brief, in the first half of the nineteenth century, significant progress was achieved with regard to anatomy and physiology, although some structures were misinterpreted because of the small size of the items and lack of suitable preparation methods and optical equipment. Regarding systematics and phylogeny, the opinions based on plausible morphological similarities differed widely, but generally "arthropods" sensu lato appeared to be favoured, i.e. mites, fleas and crustaceans with relationships to worms.

M. tardigradum has six papilla as clearly noted by the author: "bouche entourèe de six petits palpes" (Doyère 1840, p. 283).

1.4 More and More Studies: The Second Half of the Nineteenth Century

In this period marine tardigrades were discovered, the first one not by the describer himself, but by one of his students.[65] This minute probably subadult animal, named *Lydella*[66] without a species epithet, seemed to be equipped with jointed legs and bifurcated head appendages (see Dujardin 1851). This animal was never found again (see Fig. 1.14a). Dujardin was so much perplexed by this animal[67] that he revoked the previously suggested connection of tardigrades with rotifers (see above). In the same article, he described two obligatory freshwater tardigrades (*Macrobiotus lacustris*, *Macrobiotus macronyx*).[68] In the subsequent years, a further marine species was discovered independently by M. Schultze[69] and Greeff (see Greeff 1866) and named *Echiniscus sigismundi* by Schultze, who explicitly pointed out the relationship between habitat and organization of this tardigrade (Schultze 1865). This species was later renamed *Echiniscoides sigismundi* by Plate (1889). Further, Cuénot (1893) detected the strange epizoic, sessile and dorsoventrally strongly

---

[65]"Un des auditeurs les plus zélés de la Faculté des sciences, M. Boulengey, qui m'a souvent aidé dans la recherche des animaux microscopiques, m'apporta, au mois d'août 1849, un petit animal qu'il avait vu ramper à la paroi de ses vases d'eau de mer, et qu'il avait bien reconnu pour un Tardigrade, malgré sa petitesse extrême. En effet, cette Lydella, qui, à la vérité, n'était peut-être pas adulte, n'a souvent qu'un vingtième de millimètre, et rarement son corps dépasse un dixième de millimètre (...)" [One of the most zealous auditors of the Faculty of Science, Mr. Boulengey, who often helped me in the search for microscopic animals, brought me in August 1849 a small animal that he saw crawling on the wall of its sea water vessels, and that he had recognized as tardigrade despite its extreme smallness. Indeed, this *Lydella*, which, to be true, probably was not an adult, often measures only a twentieth of a millimeter, and its body rarely exceeds one tenth of a millimeter (...)] (Dujardin 1851, p. 164).

[66]*Lydella* was later renamed in *Microlyda* by Hay (1907), because *Lydella* is a genus of flies in the family Tachinidae since 1830 up to this day.

[67]"En résume, il me semble que les charactères de cette nouvelle espèce de Tardigrade, malgré son apparente ressemblance avec certains animaux articulés, concourent avec ce que nous savons des autres espèces, pour montrer l'indépendance du type auquel elles appartiennent. Est-ce au groupe des Articulés, est-ce au groupe des Vers qu'on doit les rattacher désormais? La question, je crois, est au moins indécise, quoique je penche encore pour cette dernière opinion." [In summary, it seems that the characteristics of this new species of Tardigrade, despite its apparent resemblance to some articulated animals, concur with what we know from other species, to show the independence of the type to which they belong. Is it the articulated group, is it the group of worms to which we should link them from now on? The question, I think, is at least undecided, though I still incline towards the latter view.] (Dujardin 1851, p. 166).

[68]lacus, Lat. = water (body); makrós, Gr. = long, big; ónyx, Gr. = claw.

[69]Max Johann Sigismund **Schultze** (1825–1874), German anatomist, professor at Halle (Germany) and Bonn (Germany), dedicated the new species to his father C.A.S. Schultze (see footnote 36). M. Schultze was known for his work on cell theory. He published also about Turbellaria and Foraminifera.

flattened *Tetrakentron*[70] *synaptae*, an ectoparasite[71] on the holothurian *Synapta galliennei* (1893).

At that time the development of a tardigrade (*Macrobiotus Dujardin* (sic!)) was examined from deposition of eggs until hatching by means of whole preparations of intact living eggs (see Kaufmann[72] 1851). Although the descriptions and images contained errors and misinterpretations due to the inadequate techniques, later researchers acknowledged the ideas and drawings of the author (see Fig. 1.9), confirming the total-equal cleavage of the egg.[73] Moreover, Kaufmann presumed fertilization in the ovary, considered tardigrades as hermaphrodites and noted the resistance of the tardigrade cuticle to caustic potash (potassium hydroxide). From the latter, he concluded that the cuticle consisted of chitin, a conclusion that many later authors took for granted.[74] With previous and new arguments (e.g. he thought he had demonstrated a "germinal disc" and the chitinous nature of the cuticle), Kaufmann considered tardigrades as arthropods, which, as the most primitive group of arachnids, might represent the transition from annelids to pycnogonids[75] and acarines.

A diverging opinion was expressed (perhaps somewhat carelessly; see Richters and Krumbach 1926) by Graff.[76] In his broad study on myzostomids,[77] he created

[70]tetra, Gr. = four; kéntron, Gr. = thorn, sting.

[71]Parasitism has been only assumed by Cuénot, but was (indirectly) shown not earlier than 1980 (Kristensen 1980).

[72]Joseph **Kaufmann**, studied at the University of Zurich (Switzerland), further details not identified.

[73]Interestingly, a textbook of zoology previously issued contains the note: "Bei Macrobiotus Hufelandii beobachtete ich deutlich, dass die von der abgeworfenen Haut umhüllten Eier einen totalen Furchungsprocess durchmachten". [I clearly saw in Macrobiotus Hufelandii that the eggs covered by the shed skin undergo a total cleavage.] (von Siebold 1848, p. 552 note 2) (see Footnote 26).

[74]Later Marcus (1928a) tried to demonstrate chitin in the tardigrade cuticle with zinc iodine chloride that at that time was considered to be specific for the detection of chitin (today it is used for the rapid detection of cellulose), but he failed and suggested the cuticle might consist of a swellable albumin-like substance with high alkali resistance. Cuénot (1924) only says that the cuticle seems to be chitinous; he relies on its resistance against caustic potash but also mentions Marcus's view. Clear evidence that the tardigrade cuticle contains chitin was shown not until 1972, when Baccetti and Rosati (1971) extracted ultrathin sections with chitinase.

[75]Pycnogonida (pyknós Gr, = thick, densely packed; goné, Gr. =generation, progeny) or Pantopoda (pántos, Gr. = complete; pus, gen. podós, Gr. = foot), i.e. sea spiders are arthropods assigned to the Arachnoidea at that time (see Claus 1887); today a place outside the Arachnomorpha is also discussed.

[76]Ludwig Bartholomäus **Graff** de Pancsova (1851–1924) Austro-Hungarian zoologist, professor at the "Königlich Bayerische Forstlehranstalt Aschaffenburg" [Royal Bavarian Academy of Forestry at Aschaffenburg] and later professor of Zoology at the University Graz (Austria), especially known for his studies on turbellarian and as an expert on myzosotmids.

[77]Myzostomida (mýzein, Gr. = suckle; stoma, Gr. = mouth), parasitic marine "worms" considered as basal lineage of Bilateria. Affinities with Annelida and Platyhelminthes.

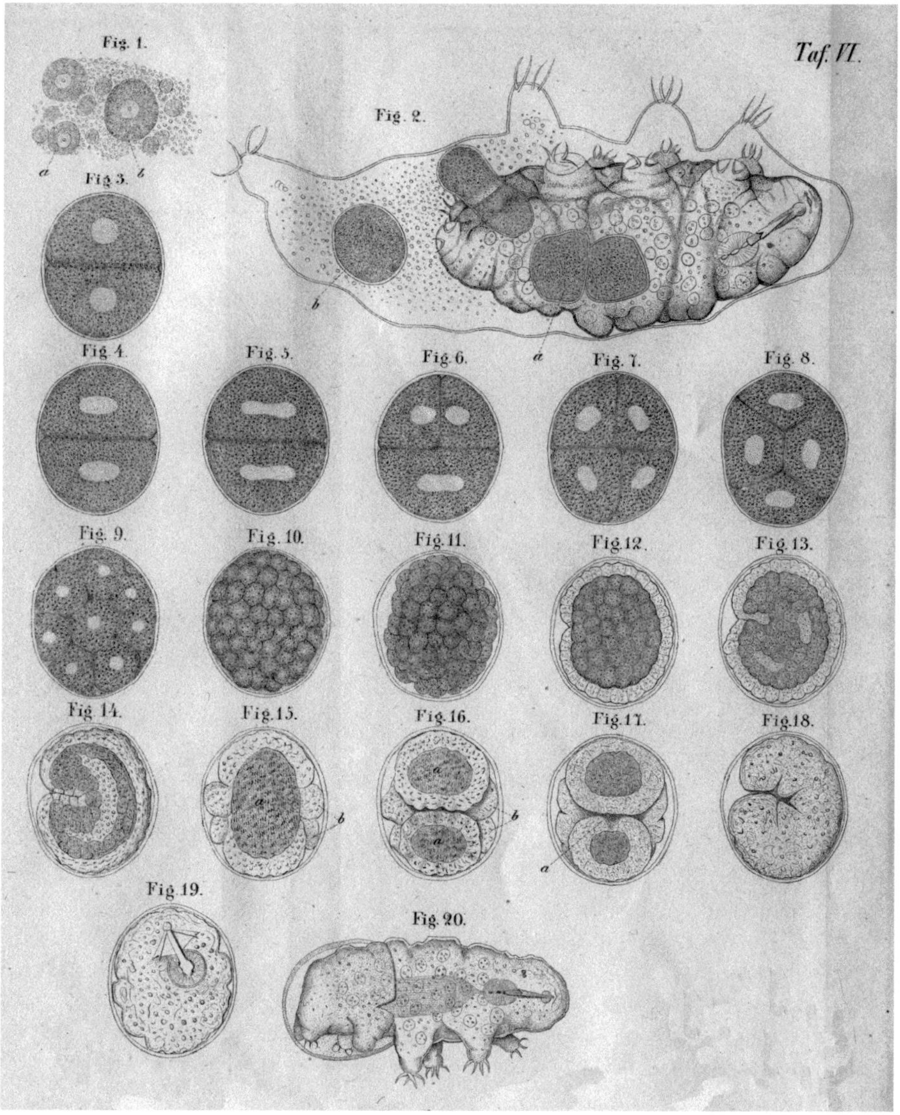

Fig. 1.9 Development of *Macrobiotus dujardin*. Female laying her eggs in the shed cuticle (2). Several stages of cleavage (5–8) and the supposed (10–13) morula (10), the developing (11) and the fully developed germinal disc (12) and the formation of the archenteron (13). From Kaufmann (1851), details of Plate 6

the Stelechopoda[78] that included the Tardigrada, Linguatulida[79] and Myzostomida, suggesting a position between Vermes and Arthropoda (Graff 1877). Later Haeckel[80] (1896) considered Stelechopoda to be a regressive class of annelids, in which he divided tardigrades into the higher organized Echiniscidae (*Echiniscus*, *Emydium*) with a panelled carapace and a cirrus on each leg and the more degenerated Arctiscidae (*Arctiscus* (sic!), *Macrobiotus*).

Meanwhile Greeff[81] returned to anatomical studies. At the suggestion of M. Schultze, he studied the nervous system, especially “Doyéres Endhügel” (= neuromuscular endplates), using asphyctic individuals of *Arctiscon* (= *Milnesium*) *tardigradum*, and described carefully how to bring tardigrades in asphyxia, appreciating Doyére’s merits in this context (Greeff 1865). He emphasized the problems with uniting tardigrades with mites, which had been already pointed out in 1861 by Schultze, but did not suggest an alternative. Starting from Goeze, in a second article, Greeff discussed in great length taxonomy, e.g. names of taxa and species identities, described a new *Macrobiotus* species (Fig. 1.10) and added some morphological details, e.g. the pharyngeal nerve ring (Greeff 1866).

Plate[82] (1889) considerably increased knowledge of the anatomy of *Macrobiotus* spp. His comprehensive study is divided into four parts: (1) anatomy and histology (he did not prepare sections, but used asphyctic specimens), (2) biological comments, (3) systematics and (4) general remarks including relationships. Among other things he described the relations between muscles and pigmentation (the latter is absent in areas, where muscles attach); assumed that the distal epidermal thickening in the legs is related to the development of claws during the moult; proposed the body cavity cells (“blood corpuscles”) as a kind of fat body storing reserve material; reported on a channel within the stylets of *Milnesium tardigradum*; tried to explain how the buccal apparatus may work when tardigrades suck in food, suspecting that the salivary glands produce poison; thought that at least *Macrobiotus* spp. are bisexual; and noticed that the two lateral structures situated between the mid- and hindgut, which were previously considered as testes, are excretory organs that he compared with the Malpighian tubules (vasa Malpighii[83]) of mites. Further, he

---

[78]Stelechopoda (stélechos Gr. = stump; poús Gr. = foot).

[79]Linguatulida (linguátula, Lat. = small tongue) or Pentastomida (penté. Gr. = five; stóma, Gr. = mouth), tongue worms are Ecdysozoa probably related to crustaceans.

[80]Ernst Heinrich Philipp August **Haeckel** (1834–1919), extremely productive German zoologist, philosopher, physician and artist, and professor of Comparative Anatomy at the University of Jena (Germany) and impressed by Darwin’s theory; among other things known for his recapitulation theory (ontogeny recapitulates phylogeny) and the beautifully illustrated book “Kunstformen der Natur” [Art forms of Nature] from 1904.

[81]Richard **Greeff** (1829–1892), German physician, switched to zoology and became professor of zoology at the University of Marburg (Germany). He published among other things about rhizopods, annelids and echinoderms (http://www.deutsche-biographie.de/pnd116828412.html).

[82]Ludwig Hermann **Plate** (1862–1937), German zoologist and social Darwinist, professor at the Universities in Berlin and Jena; published on genetics and descent theory, student of Ernst Haeckel (1834–1919, see footnote 76); wrote his dissertation on rotifers (http://www.deutsche-biographie.de/pnd117683574.html).

[83]Named after the Italian biologist and physician Marcello **Malpighi** (1628–1694), professor of Physics at the University of Pisa, and commonly used for insect excretory organs.

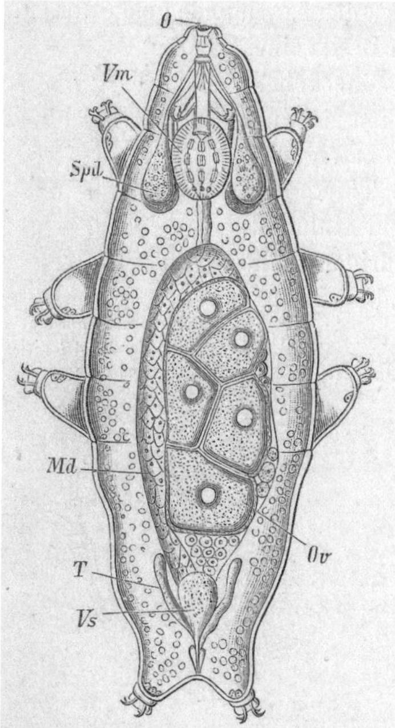

Macrobiotus Schultzei, nach Greeff. *O* Mund, *Vm* Schlundkopf, *Md* Magendarm, *Spd* Speicheldrüsen, *Ov* Ovarium, *T* Hoden, *Vs* Samenblase.

Fig. 1.10 Woodcut (after Fig. 1 on Plate 6 in Greeff 1866) of *Macrobiotus schultzei* in a textbook of zoology (Claus 1887). This figure was also used in subsequent editions (e.g. Claus and Grobben 1917). However, T (=testes) was changed in Malpighian tubules and Vs (=seminal vesicle) in accessory gland of the genital organ

added some details concerning myology and the peripheral nervous system (see Fig. 1.11).

In the third part, Plate presented determination keys of all Tardigrada known at that time, erected the genus *Echiniscoides* (see above), gave the doubtful *Lydella* the epithet *dujardini* and described specimens distinguished by rudimentary "teeth" and pear-shaped aggregations of cells surrounding the buccal tube and the absence of salivary glands, which were already observed by Doyère, as a new species, *Doyeria simplex* (Fig. 1.11 top left). Last but not least, he thoroughly discussed tardigrade relationships, concluding that they were related to arthropods,[84] based among other things on the presence of four leg pairs with claws and two Malpighian tubules and

[84]He concluded: "Die Bärthierchen sind die niedrigsten von allen bis jetzt bekannten luftathmenden Arthropoden und sind an die Spitze der Tracheaten, noch vor den Onychophoren zu setzen (...) Sie sind diejenige Thiergruppe, welche den Uebergang von den Gliederwürmern zu den luftathmenden Arthropoden am reinsten zum Ausdruck bringt und am deutlichsten erkennen lässt." [The bear-animalcules are the most primitive of all air breathing arthropods that have been identified to date and have to be set at the beginning of the Tracheata, even prior to the Onychophora...they are the animal group demonstrating most clearly the transition from annelids to air-breathing arthropods] (Plate 1889, p. 545/546).

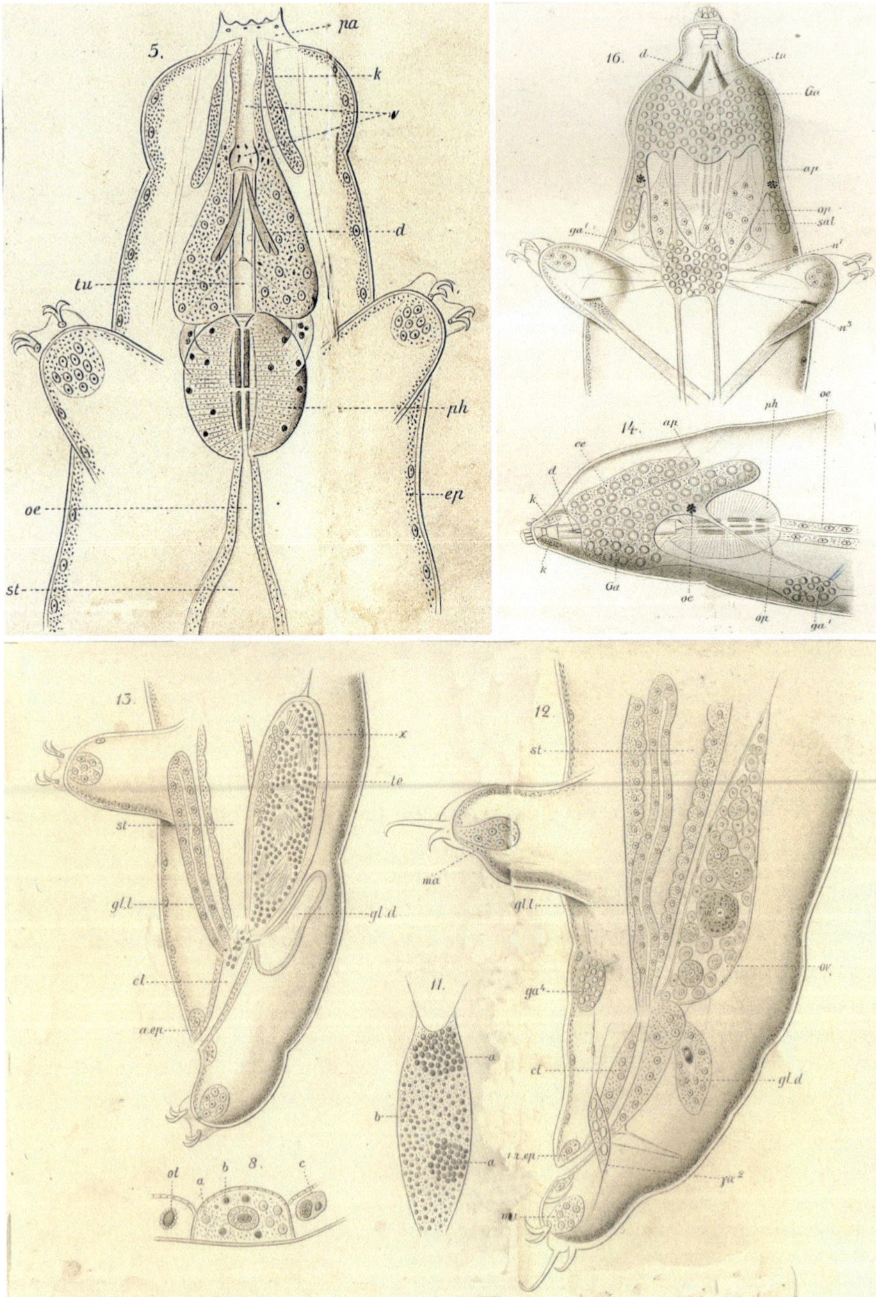

Fig. 1.11 Details from Plate 20 (top left and below) and Plate 21 (top right) from Plate (1889). Top left: head of *Doyeria simplex* n.s. Top right: brain of *Macrobiotus hufelandii*, viewed from above (1) and upper and lower oesophageal ganglion, lateral view (16). Below: sexual organs of a female *Macrobiotus oberhäuseri* (12) and a male *M. hufelandii* (13). Testis (11) and "gastric" cells of

the absence of ciliated epithelia, stating that they have a somewhat isolated position in the system.[85]

A quite sophisticated speculation concerning the relationship and descent of tardigrades should also be mentioned here. Von Kennel[86] (1891) agreed with Plate (and others) concerning the relationship to arthropods but suggested that tardigrades, in his opinion morphologically less similar to annelids than Onychophora, might be greatly reduced and simplified paedogenetic[87] higher Tracheata.[88]

Concerning anhydrobiosis some thoughts of the physiologist Preyer[89] are of interest. Obviously, he was the first to use the term anabiosis[90] to describe the process of reviviscence and not the state itself as we are doing today when using

Fig. 1.11 (continued) M. hufelandii (8). Abbreviations: an = pharyngeal nerve ring; ap = brain area with eyes; a.ep = swelling of the epidermis near the anus; cl = cloaca; ce = brain; d = tooth; ep = epidermis; $\gamma\alpha^2$ = accessory ganglion; Ga = suboesophageal ganglion; ga^1, etc., = ventral ganglions; oc = eyespot; sal = salivary gland; te = testis; gl.d = dorsal gland; gl.l = lateral gland; k = epidermal thickening at the mouth; n^3, n^4 = nerves; ma = epidermal thickening in the leg; oe = oesophagus; ov = ovary; pa = mouth papilla; ph = pharynx; st = stomach; te = testis; tu = mouth tube

[85]"Es giebt meiner Ansicht nach keine Abtheilung der Würmer oder Arthropoden, an die sich die Tardigraden direct anschließen lassen. Sie bilden eine isolirt dastehende kleine Gruppe, die sich schon ausserordentlich früh vom Stammbaum der Tracheaten abgespalten hat und daher der Urform der landbewohnenden Gliederfüssler näher steht als irgend eine andere Abtheilung". [In my opinion there is no group of worms or arthropods, to which tardigrades can be connected directly. They represent a small isolated group, which has split from the Tracheata-tree quite early; therefore they are much closer to the prototype of terrestrial arthropods than any other group] (Plate 1889, p. 547).

[86]Julius Thomas **von Kennel** (1854–1939) German zoologist and entomologist (lepidopterology, especially Microlepidoptera), professor at the University Tartu (Estonia) until 1893 Kaiserliche Universität Dorpat.

[87]von Kennel did not use the term "paedogenetic" or more neutral "paedomorphic" (pays, Gr. = child; génesis, G. = generation), i.e. the phylogenetic change that involves retention of juvenile characters by the adult, but he wrote that tardigrades might be "Arthropoden auf dem Larvenstadium, ohne Kopf, mit einem in der Segmentzahl reducirten Rumpf, der einige (secundäre) Fussstummel trägt". [Arthropods in the larval stage, without a head and with a body reduced in the number of body segments, which has some (secondary) stubs of feet] (von Kennel 1891, p. 510).

[88]Tracheata (trachýs, Gr. = rough), at that time a group within the arthropods characterized by a tracheal systems, which included myriapods and insects (see Haeckel 1896).

[89]William Thierry **Preyer** (1841–1897), British physiologist, first full professor at the University of Jena (Germany), propagated people-oriented presentations of science (http://www.deutsche-biographie.de/pnd118596411.html?anchor=adb).

[90]"Alle Organismen sind entweder lebend oder leblos. Die nicht lebenden, unlebendigen oder leblosen sind aber keineswegs todt, vielmehr sind sie entweder l e b l o s, aber lebensfähig – a n a b i o t i s c h d.h. wieder belebungsfähig, scheintodt im engeren Sinne – oder sie sind l e b l o s und l e b e n s u n fä h i g, wofür der Ausdruck todt gebräuchlich ist" [All organisms are either living or inanimate. However, the non-living, inanimate or lifeless ones are not dead, rather they are either inanimate, but viable – a n a b i o t i c, i.e. able to be revived, seemingly dead, revival capable, sham dead in the narrower sense – or they are lifeless and unlivable, for which the term 'dead' is used] (Preyer 1880, p. 28).

this term. Preyer was confident that metabolism during anabiosis ceases entirely, comparing anabiotic animals with a stopped clock[91] (see Preyer 1891). In case of anhydrobiotic animals, moistening was considered as the impetus to take up functions again.

More than 40 years after Kaufmann, von Erlanger[92] (1895a, b) also used primarily whole eggs, fixed and cleared with glycerine, to describe cleavage (v. Erlanger 1895a; preliminary note without any figures), gastrulation, germ layer formation and other developmental processes (von Erlanger 1895b). In addition, he repeatedly observed that males discharge their sperm into the female's shed cuticle during her moulting and egg laying (von Erlanger 1895a). Later researchers (e.g. von Wenck 1914; Marcus 1929a) criticized his very schematic drawings that obviously did not reflect the actual conditions (Fig. 1.12) and the careless interpretation. However, von Erlanger confirmed Kaufmann's study (and later studies) concerning the total-equal cleavage and described mesoderm formation by enterocoely, i.e. from pouches pinched off by the archenteron[93] (= primary gut) (see Fig. 1.12). However, judging from his description, Marcus doubted if von Erlanger had really seen these structures. Based on his own results (see below), Marcus noted that von Erlanger's conclusion was "das richtige Resultat seiner unrichtigen Darstellung" [the correct result of his incorrect presentation] (Marcus 1929a, p. 357). Also, von Erlanger commented on the relationships of Tardigrada and considered them (in accordance with Plate) to be allied with arthropods, but he saw them at the beginning of the arthropods and not (as Plate did) at the beginning of Tracheata, forming a transition to rotifers, nematodes (one of his arguments was the nematode-like pharynx of Tardigrada) and Nematorhyncha, i.e. gastrotrichs and kinorhynchs, which taxon was established by his teacher Otto Bütschli.[94]

Lance's[95] dissertation about the genus *Macrobiotus* (Lance 1896) again covered a wide range of topics. It is divided into three parts each considering very broadly the existing literature: (1) The anatomical part hardly brought anything new compared to Plate's study, and Lance's drawings are not of the highest artistic level (Fig. 1.13) compared to other contemporary presentations. However, he recognized that the

[91]To be more precise, he said: "Die organische Maschine stirbt also nicht jedesmal, wenn sie vollkommen still steht, so wenig wie die Uhr jedesmal zerbricht, wenn das Pendel nicht mehr schwingt" [Thus, the organic machine does not die each time it stands still completely, similar to the clock that does not break each time the pendulum does swing no longer] (Preyer 1891, p. 5).

[92]Raphael Baron Slidell **von Erlanger** (1865–1897), German zoologist, associate professor at the University of Heidelberg, research in developmental biology. As son of a banking family, he was able to acquire a laboratory from its own resources.

[93]Embryonic gut; arché Gr. = beginning; énteron Gr. = gut.

[94]Johann Adam Otto **Bütschli** (1848–1920), German zoologist, professor at the University Heidelberg, studied insect development, nematodes and Protozoa (http://www.deutsche-biographie.de/pnd117144932.html).

[95]Denis **Lance** (??–??). I did not find further bibliographic information. Lance dedicated his dissertation to "M. Milne-Edwards", i.e. Alphonse Milne-Edwards (1835–1900), who became in 1891 director of the Muséum National d'Histoire Naturelle in Paris. A. Milne-Edwards was the son of Henri Milne-Edwards, who 50 years before accepted Doyère's dissertation (see footnote 39).

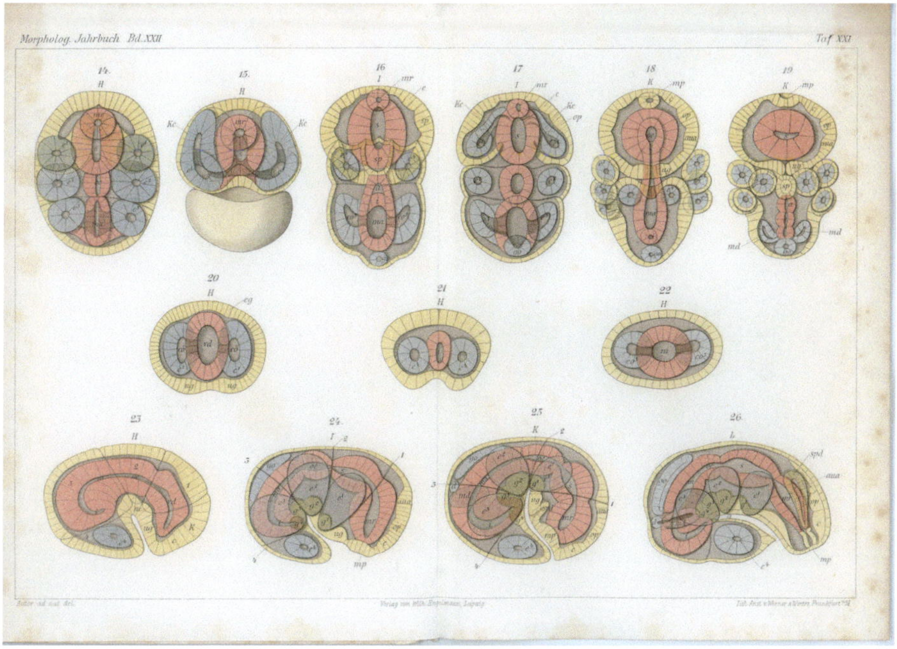

Fig. 1.12 The highly schematic drawings of developmental stages *Macrobiotus macronyx*. Plate 21 from v. Erlanger (1895b). The figures 22–24 show the supposed enterocoely

salivary glands are derived from the epidermis, described (erroneously) ciliated cells in the intestine and used (not very successfully) methylene blue to stain muscles and nerves. (2) The very detailed historical essay about "reviviscence" was followed by several chapters about "reviviscence" (from anhydrobiosis[96] and in part asphyxia) in the natural environment and the influence of various parameters such as moisture, oxygen, temperature, light, gravity, etc. and their combinations. At the end of this part, he discussed some preconditions to induce anhydrobiosis and, contrary to Doyère and Preyer (l.c.), concluded from his staining experiments that vital activities of the cytoplasm in asphyxia and in anhydrobiosis may be slowed down but may persist anaerobically. (3) This part deals with the relationships of tardigrades,

[96]This term, used by Lance, was introduced by Alfred **Giard** (1846–1908), French zoologist, professor at the faculty of sciences in Paris, to describe a variety of "hypobiotic" processes caused by loss of water including dormancy (see also Keilin 1959). He wrote "La déshydratation progressive n'augmente pas les échanges respiratoires; elle les diminue et diminue en même temps tous les phénomènes vitaux. Elle peut même aboutir à un état d'*anhydrobiose* ou vie latente par desséchement, dont le sommeil estival de nombreux animaux n'est qu'une variété remarquable" [The progressive dehydration does not increase the respiratory exchange; it reduces and diminishes in the same time all vital phenomena. It may even result in a state of anhydrobiosis or latent life by drying out, the summer sleep of many animals is a remarkable variety] (Giard 1894, p. 498).

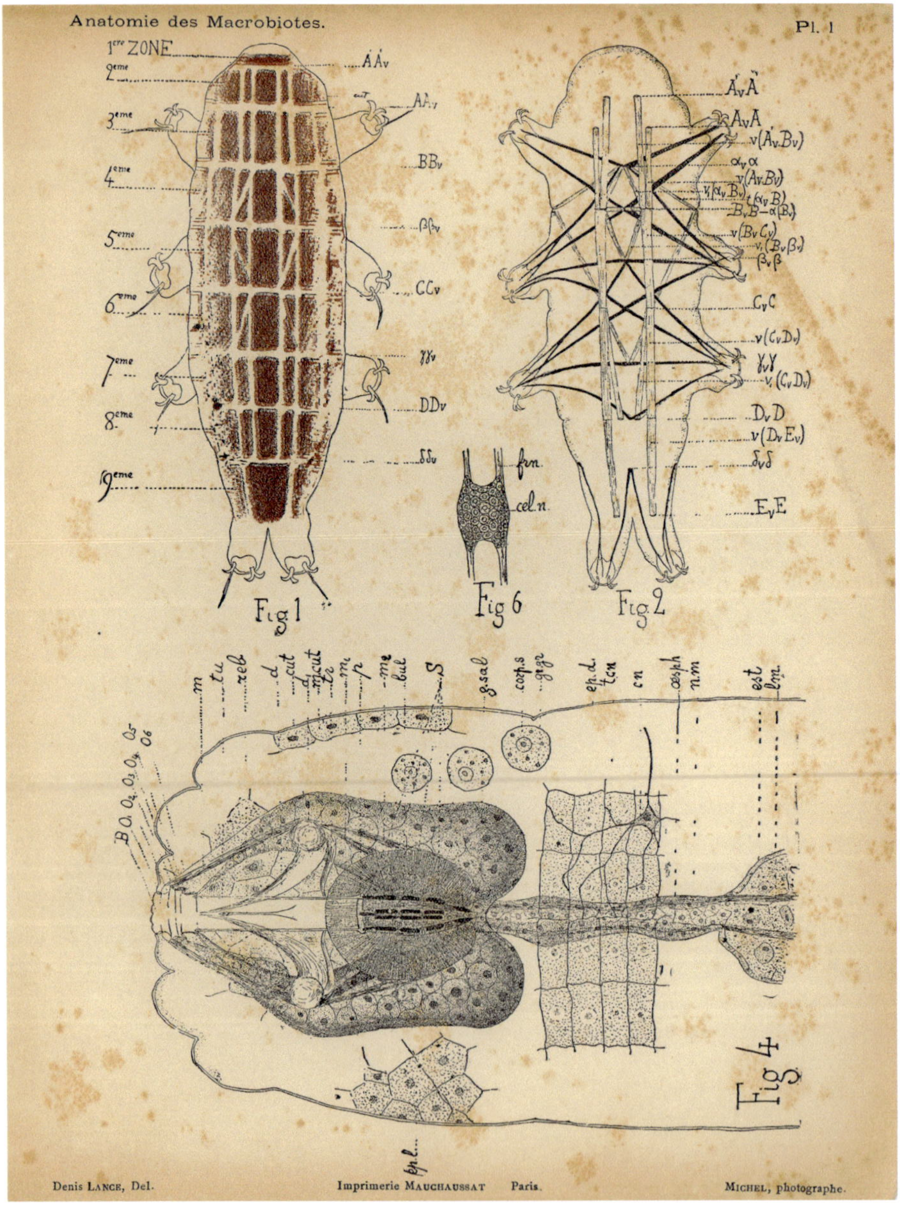

Fig. 1.13 Plate 1 from Lance's dissertation (Lance 1896)

description of *Macrobiotus* spp. and distribution of tardigrades in France, from which only the discussion about relationships is considered herein, also to demonstrate the line of arguments. Starting from Plate's view that tardigrades are the most inferior air-breathing arthropods (Tracheata), Lance argued that tardigrades (1) can hardly be distinguished from annelids because of their non-segmented cuticle, the arrangement of muscles[97] (to consider the musculature as annelid-like seems strange to us) and the presence of parapodia and that the mouthparts are not formed by limbs; on the other hand, however, tardigrades have Malpighian tubules like insects and arachnids, i.e. Tracheata; (2) are neither Tracheata (as they do not breathe by tracheae) or worms (as they have Malpighian tubules); and (3), therefore, should be positioned between worms and Tracheata. Then, he introduced the "annulated" *Peripatus*[98] (Onychophora) as a further link between annelids and Tracheata, since *Peripatus* possesses a tracheal system but apart from that has many annelid traits, and he underlined its similarity to tardigrades, e.g. *Milnesium*. Then, he carefully listed dissimilarities and similarities of both tardigrades and onychophorans and included both in the Protracheata (without wishing to imply an immediate relationship), related to annelids. Contrary to Plate (1889), who considered onychophorans to be more advanced, Lance stressed similarities of tardigrades with mites concluding that tardigrades might represent the transition from certain worms to arachnids and *Peripatus* the transition from worms to myriapods.

1.5 The First Third of the Twentieth Century

In the comparatively short time until the release of Marcus's monograph, a vast number of studies were published covering nearly all aspects of tardigradology including ecological studies, which had previously been under-represented. The number of species, which stood at 26 species at the beginning of the twentieth century, increased enormously within a little more than 10 years, namely, by the

[97]"Que ce soient des Arthropodes excessivement inférieurs cela ne fait aucun doute. En effet ils se distinguent à peine des Vers, en particulier des Annelides. Par leur cuticule à peine segmentée; par la disposition de leurs muscles longitudinaux en trois groupes (dorsaux, latéraux, ventraux) par la présence de membres à crochets, rappelant les parapodes; par l'absence de pièces buccales constituées par des membres transformés, ils se rapprochent absolument des Vers". [There is no doubt that they are extremely lower arthropods. In fact, they can hardly be distinguished from worms, especially from annelids. Be it their hardly segmented cuticle; the arrangement of their longitudinal muscles into three groups (dorsal, lateral, ventral); the presence of limbs with hooks, reminiscent of parapods; or the absence of mouthparts formed from modified limbs, it definitely places them close to worms] (Lance 1896, p. 199).

[98]perípatos, Gr. = walk, walking.

collecting activities of Richters[99] and Murray.[100] Noteworthy is the discovery of several marine species, e.g. *Halechiniscus guiteli*; the marine eutardigrade *Macrobiotus stenostomus* and *Macrobiotus appeloefi* (the spelling in the original description) (Richters 1908); and the new genera and species *Batillipes mirus* (Richters 1909a, b) and *Bathyechiniscus tetronyx* (Steiner 1926) (see Fig. 1.14).[101]

Richters began his studies on tardigrades relatively late in his career. His first publications dated from 1900, the last one, a quasi-monography, revised and posthumously published by one of the editors of the "Handbuch der Zoologie" [Handbook of Zoology][102] from 1926 (see Richters and Krumbach 1926). Richters described numerous new species (see Fig. 1.15) from different regions of the world (also to create a basis for understanding the geographical distribution of tardigrades); he used, I think for the first time, photomicrographs of tardigrades in some of his publications (e.g. Richters 1904b, 1908), corrected some errors concerning the structure of tardigrade eggs and mistook the egg shell ornamentation, especially their processes that are typical of freely deposited eggs, for adhering structures (e.g. Richters 1904a). In the same year, he considered specimens with an incomplete buccal apparatus as "parallel forms" of certain species, calling them simplex forms and thereby eliminating the genus *Doyeria* (Richters 1904b). Mainly based on the clawed legs and the various appendages of tardigrades, he revived the concept of annelid relationships[103] previously suggested by Graff and Haeckel (l.c.). He saw not only a close similarity of the clawed leg with the parapodium of

[99]Ferdinand **Richters** (1849–1914), German zoologist, curator of Crustacea at the Naturmuseum Senckenberg, Frankfurt am Main. In 1873/1874 he worked as an assistant in the Zoological Institute at Göttingen; later he found an employment at the Senckenberg Institute. In 1886 he was named vice-director of the Senckenberg Gesellschaft für Naturforschung, where, 3 years later, he was appointed first director. In his days he was one the most productive tardigradologists.

[100]James **Murray** (1865–1914), British biologist and explorer. Murray undertook both biological and bathymetric surveys, contributed to taxonomy and distribution of tardigrades and bdelloid rotifers; many of these creatures he described for the first time. In 1913, he joined a Canadian scientific expedition to the Arctic. When the ship became trapped in the ice, Murray mutinied against the captain and left the ship with three others; they were never heard of again.

[101]hals, Gr. = salt; stenós, Gr. = narrow; stóma, Gr. = mouth; batíllum, Lat. = shovel; mirus, Lat. = strange; bathýs, Gr. = deep.

[102]A multivolume manual founded in the 1920s by Willy **Kükenthal** (1861–1922), professor of Zoology and comparative anatomy at the University Breslau and Berlin and continued by Thilo **Krumbach** (1874–1949) from Breslau, who headed for some years the Zoological Station in Rovigno (Istria). The series, a treatment of the complete animal kingdom in eight volumes, started in 1925 and was restructured at the beginning in 2010 offered additionally as a database (Zoology Online), which can be easily searched and rapidly updated.

[103]"Auf alle Fälle, meine ich, haben wir volle Veranlassung, die Tardigraden nicht mehr zu den Arthropoden zu zählen, sondern sie, unter Anerkennung gewisser Anklänge an die Nematoden, die als Reminiszenzen an gemeinsame Stammformen gelten dürfen, als nächste Verwandte der Anneliden aufzufassen" [In any case, I think, we have every reason, not to include tardigrades in the arthropods, but to consider them—in recognition of a certain reminiscence of nematodes—as the nearest relatives of annelids] (Richters 1909b, S. 44).

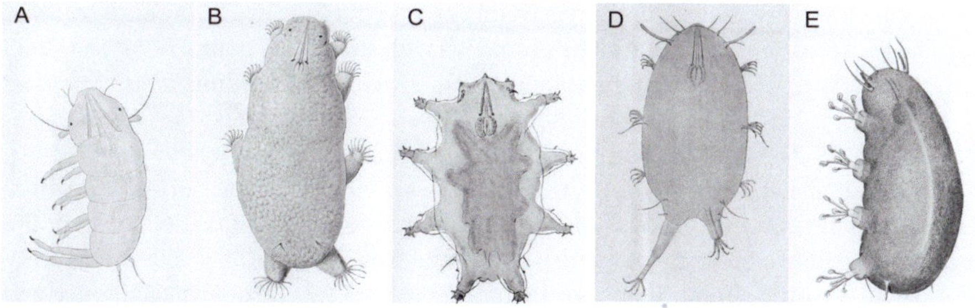

Fig. 1.14 Some marine tardigrades described before (**a**) and in the first third of the twentieth century. (**a**) *Lydella* Dujardin; (**b**) *Echiniscoides sigismundi* M. Schultze; (**c**) *Tetrakentron synaptae* Cuénot; (**d**) *Halechinicus guiteli* Richters; (**e**) *Batillipes mirus* Richters [(**a**)–(**e**) from Richters 1909a; (**a**) from Dujardin 1851]

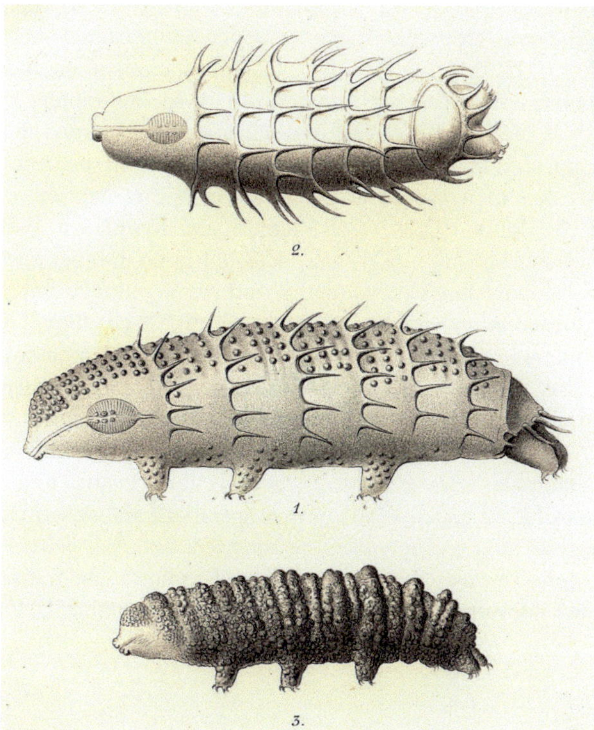

Fig. 1.15 Variations of *Macrobiotus ornatus*. From Richters (1900), detail from Plate 6

annelids—he even called them parapodia (Richters 1909a)—but also with the claws of linguatulids (Richters 1909b). In 1926 Krumbach again introduced the Stelechopoda (but this time without the myzostomids) and placed tardigrades between Annelida and Arthropoda (Richters and Krumbach 1926). Within

Tardigrada Richters distinguished between three suborders, the Prototardigrada[104] (*Batillipes*, *Halechiniscus*), the Echiniscoidea (with the single genus *Echiniscoides*) and the Eutardigrada[105] (among others with *Macrobiotus*, *Milnesium*, but also *Echiniscus*, etc.) (Richters and Krumbach 1926).

Marcus (1929a) is full of praise for the figures and diagnostic analyses of tardigrade species by Murray (e.g. 1907, 1908 and 1910 to name only a few of the numerous publications of this author). In Marcus's opinion the survey of the tardigrades collected during the British Antarctic expedition 1907–1909 (see Murray 1910) was of similar importance for taxonomy as Doyère's study for anatomy and physiology of tardigrades. Collection and description included tardigrades of the Antarctic, New Zealand, Australia and some Pacific islands and Canada resulting in 17 new species. Murray noted that the continental areas had a more diverse tardigrade fauna than the islands, but he realized that his distribution tables "indicate probably rather the amount of work done in the different regions than actual distributions" (Murray 1910, p. 121). Previously Murray (1907) and Lauterborn[106] (1906) had discovered—from all evidence independently from each other—cysts in *Macrobiotus dispar*, but Murray described the encystations in great detail, comparing cysts with pupae or mummies. He recognized the several moults involved in cyst formation and was astonished to find 1 week after encystations that most of the organs had disappeared leaving only an amorphous mass in the cyst. This observation led to the idea of a complete histolysis of the tardigrade when encysted (e.g. Heinis 1910; Rahm 1925, 1926; Richters and Krumbach 1926). In a later, more detailed article, Murray (1908) also detected cysts in other species including *Echiniscus* spp., supposed that encystment is induced by unfavourable environmental conditions, observed frequently "simplex forms" (see above) of tardigrades noting that "animals in this condition were about moult, though I doubt if this is invariably the case" (Murray 1908, p. 845) and compared encystment with similar processes in certain mites, assuming, again, relationships with them.

Basse[107] (1905) and Henneke[108] (1911) used histological techniques for the first time to study tardigrades. Basse's article (1905) is confirmatory to some extent. He examined a *Macrobiotus* species (not further specified) and described some details better than previous authors regarding the muscles, nervous system and intestinal tract. He saw, for example, the crescent-shaped cells at the transition from the oesophagus to the intestine and small crystals in the intestinal cells, and he assumed

---

[104] prótos, Gr. = first.

[105] eu, Gr. = good, right.

[106] Robert **Lauterborn** (1869–1952), German zoologist (hydrobiology), professor at the University Heidelberg, TH Karslruhe and Freiburg (Germany), studied among others things ciliates, rotifers, plankton and water quality. http://www.deutsche-biographie.de/pnd116770716.html

[107] Albert **Basse** (no further information available), doctoral student of Eugen Korschelt (1858–1946), professor of zoology and comparative Anatomy at the University of Marburg (Germany).

[108] **J. Henneke** (no further information available), doctoral student of E. Korschelt at the University of Marburg (Germany) (see Footnote 107).

that the anlage of the ovary is paired. Further, he counted 14 chromosomes during oogenesis. Concerning the systematic position, Basse followed Plate's view.

Henneke (1911), who studied *Macrobiotus macronyx*, observed the ejaculation of sperm in the female's exuvia, demonstrated zonation of the testis (i.e. spermatogenesis takes place not at the entire inner surface but in certain areas) and described spermatogenesis and the helical nucleus of mature spermatozoa and giant spermatozoa with very long heads. He distinguished nutritive cells and oocytes in the ovary and stated that not only the lateral glands of the intestine should be considered as vasa Malpighii (as Plate did) but also the third dorsal gland. It is remarkable that the author abstained from a systematic discussion.

In his comprehensive dissertation on the moss fauna in the environs of Basel (Switzerland) collected at different altitudes (up to 3000 m), Heinis[109] (1910) distinguished cosmopolitan tardigrades (most of the 35 species he identified), European species and possible glacial relicts (e.g. *Echiniscus spitzbergensis*) and discussed dissemination of tardigrades by wind, by migratory birds and several other animals found more or less by chance in mosses. In fact, he occasionally found tardigrades or their eggs on gastropods, myriapods and beetles. Later he studied cushion and rosette plants, still relatively unexplored habitats, in the Alps at altitudes of 1200–3500 m above sea level. Here he found tardigrades, in fewer numbers than in mosses and lichens, in the green and dead parts of the plants, especially in the leaf axils (Heinis 1921).

Reukauf[110] (1912a) realized that the strange tardigrades described as *Doyeria simplex* and later as "simplex" forms (see above) represented an early moulting stage ("simplex stage"), in which the cuticular parts of the ectodermal buccopharyngeal apparatus including the stylets were expelled, and wondered if the large salivary glands might be responsible for its reconstitution. In the same year, he detected a parasitic fungus (*Macrobiotophtora vimariensis*) infecting tardigrades (Reukauf 1912b). Otherwise more or less anecdotal observations on parasites (bacteria, Microsporidia, Sporozoa, fungi) and predators of tardigrades (e.g. amoebas, nematodes and conspecifics) are widely distributed in the literature and are extensively summarized in a specific chapter by Marcus (1929b).

The next study on development based on 5-μm-thick sections of exuviae of *Macrobiotus lacustris*, containing developing eggs, addressed cleavage and

---

[109]Fritz **Heinis** (1883–1979), Swiss biologist (botanist, bryologist), instructor at a secondary school, published several articles about tardigrades including his dissertation "Systematics and biology of moss-inhabiting rhizopods, rotifers and tardigrades in the environs of Basel", etc. (Heinis 1910), which was supervised by Friedrich Zschokke (1860–1936), professor of Zoology and Comparative Anatomy at the University Basel (Switzerland), known for his work in the field of hydrobiology and zoogeography.

[110]Edmund **Reukauf** (??–??), teacher at a citizen school in Weimar (Germany), botanist, published among others popular books for microscopists, e.g. in 1911 "Die mikroskopische Kleinwelt unserer Gewässer" [The microscopic small world of our water bodies]. http://zs.thulb.uni-jena.de/servlets/MCRFileNodeServlet/jportal_derivate_00170534/AB-Weimar_1920_bitunal_k.pdf

gastrulation (W. van Wenk[111] 1914; see Figs. 1.16–1.18). In addition, the article also contained some observations concerning the number of chromosomes (i.e. ten in somatic cells of the species examined), the postembryonic development, the regeneration of the stylets by the salivary glands and histological sections of cysts that appeared to disprove the putative histolysis of encysted tardigrades (see Fig. 1.17). She described very detailed successive divisions confirming previous studies that cleavage is total-equal, but noted that divisions are asynchronous leading to a morula, from which the entoderm delaminates. Without presenting relevant own results, she suggested that the mesoderm may be formed by enterocoely (see above). Finally, she rejected relationships of tardigrades to annelids, nematodes and rotifers, because she did not find any evidence of spiral cleavage, and instead she favoured connections to arachnids.

Baumann[112] (1921a, b) also used histology to broaden anatomical knowledge of several Eutardigrada (he failed to obtain useful sections from an *Echiniscus* species). Inspired by studies of Martini,[113] who demonstrated eutely in rotifers and some other organisms (e.g. Martini 1912), Baumann was also interested in the number of cells needed for the various organs. He counted 30 cells (epithelial plus muscle cells) forming the pharyngeal bulb of *Macrobiotus hufelandii* and analysed the muscular system of the buccal apparatus. He suggested that the similar pharyngeal bulb of nematodes had evolved convergently (Baumann 1921b) and gave evidence from his own studies and hints found in the literature for eutely in tardigrades (e.g. highly ordered cells in the dorsal epidermis, muscles consisting of a single cell each, lack of regeneration ability). Further, he described a seminal receptacle in female *Macrobiotus hufelandii* and—amazingly (see below)—numerous pores in the cuticle (Baumann 1921b).

Thereafter, two important studies devoted to anhydrobiosis followed in quick succession. Baumann (1922) minutely detailed the behaviour and the changes in the shape of tardigrades during the drying process free on microscope slides, with and without sand and with blotting paper, and after having wetted them. Obviously, tardigrades that survived anhydrobiosis in moss stored without additional drying substances, e.g. over calcium chloride or dried in a vacuum, were not completely deprived of water. Therefore, he concluded that the metabolism in anhydrobiotic

[111]Wanda Clara Anna **von Wenk** (1883–??), secondary school teacher, mainly in Berlin (Germany). Her dissertation published 1914 was initiated by Waldemar Schleip (1879–1948), professor at the University of Freiburg (Germany). http://bbf.dipf.de/kataloge/archivdatenbank/hans.pl?t_tunnel=idn&idn=p1611

[112]Hermann Hans Joseph Ferdinand **Baumann** (1889–1970), German pedagogue and school reformer, student of E. Korschelt (see footnote 107), worked in several educational institutions, from 1930 as a lecturer in biology and chemistry at the pedagogical academy Kassel (Germany) and later as a teacher in a secondary school (see Hesse 1995). In his last years he worked as a volunteer in the "Übersee-Museum" in Bremen (Germany).

[113]Erich Christian Wilhelm **Martini** (1880–1960), German zoologist and physician, founder of the medical entomology in Germany. Besides other places, he worked at the Bernhard Nocht Institute for Tropical Medicine in Hamburg. He introduced the term eutely to designate this concept of constancy in number and arrangement of histological units.

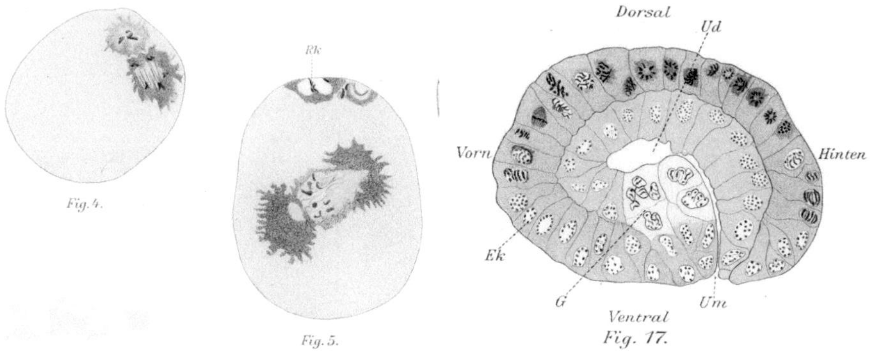

Fig. 1.16 Developmental stages of *Macrobiotus lacustris* drawn from histological sections. Left: second maturation division, metaphase (second spindle apparatus and polar body) (4), longitudinal section, metaphase of the first division. Top: first pronucleus in the prophase (combination from four sections). Right: sagittal section through the gastrula. Ek = ectoderm; G = primordial germ cells; Ud = cavity of the archenteron; Um = blastopore. From v. Wenck (1914), detail from Plates 35 and 36

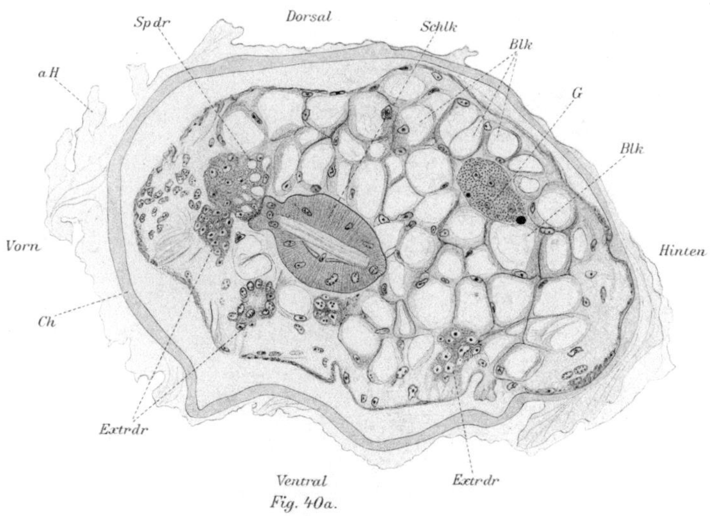

Fig. 1.17 Drawing of the front end of an older cyst of *Macrobiotus lacustris* (oblique longitudinal section) with intact tissues. aH = outer envelope of the cyst; Blk = blood corpuscle; Ch = envelope of the cyst; Extrdr = claw gland; G = primordial germ cells (sic!); Schlh = pharynx; Spdr = salivary gland. From v. Wenck (1914), detail from Plate 38

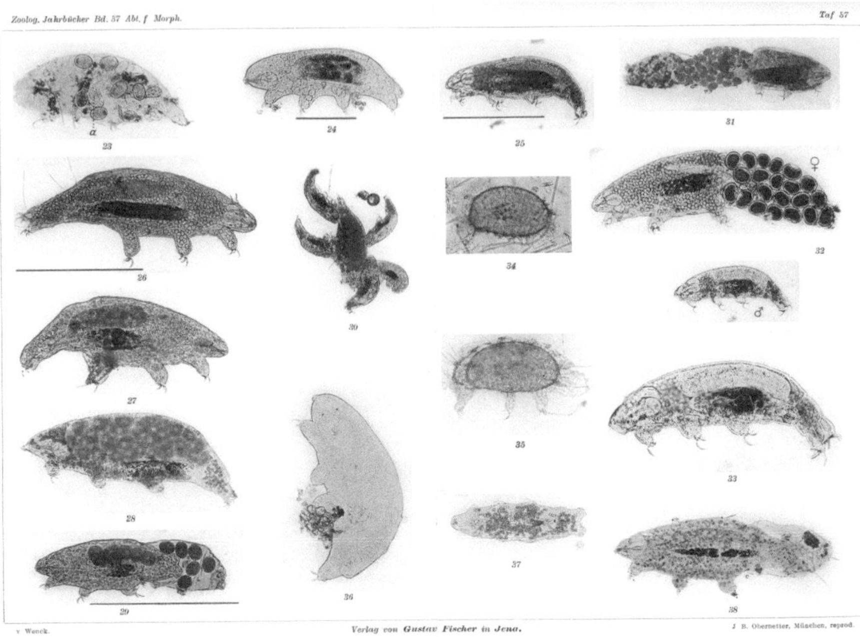

Fig. 1.18 Photomicrographs of *Macrobiotus lacustris* showing the exuvia with hatching young (23); young specimen (24); young female (25); young female with differentiating oocytes (26); young females with clearly visible oocytes in the ovary (27); mature female (28); female during deposition of eggs in the shed cuticle (29); smaller males around mature females (30); female carrying two clutches (31); female with clutch and male for comparison (32); male with characteristic claw at the first legs (Kr) and large testis (33); cyst with clearly visible organs (34); lateral view of the same cyst (35); juvenile, simplex stage (36); ditto (37); red female (38). Plate 37 from v. Wenk (1914)

tardigrades persisted,[114] albeit extremely reduced. Baumann also pointed out that secretions were discharged on the body via the above-mentioned pores to make the cuticle more impermeable.[115]

[114]"Nach allem ist es nicht mehr möglich, die Eintrocknungsfähigkeit der Tardigraden als ein Beispiel für die Unterbrechung der Kontinuität des Lebensvorganges und für Stoffwechselstillstand, kurz für das 'latente Leben' im Sinn Preyer's (. . .) anzuführen. Da in den getrockneten Körpern Wasser nachzuweisen ist, muß auch ein Stoffwechsel stattfinden, selbst wenn er noch nicht unmittelbar nachgewiesen werden kann, was bei der Kleinheit der Objekte auf Schwierigkeiten stößt". [After all, it is no longer possible to use the ability of tardigrades to dry out as an example for the interruption of the continuity of the life process and for metabolic arrest, in short for "latent life" in the sense of Preyer (...). Because there is water in the dried bodies, also metabolism must take, even though it cannot be detected directly, which will meet difficulties because of the smallness of the objects.] (Baumann 1922, p. 555).

[115]Much later, lipid extrusions were described in anhyrobiotic eutardigrades, which were assumed to be extruded from dermal glands via putative (which have not yet been proven) pore canals in the cuticle to reduce the transpiration rate (see the figure in Wright 1988, 2001; Wright et al. 1992). To my knowledge, there has not been sufficient evidence either for dermal glands or pore canals in the eutardigrade cuticle to date.

By contrast, Rahm[116] studied various external influences such as oxygen level, light, UV radiation, toxic chemicals and thermal stresses on active and anhydrobiotic tardigrades and published a relatively large number of notes and articles about this subject until 1926. Similar studies had already been done, at least in part, by former researchers, but of particular interest are Rahm's studies with high and very low temperatures. I will cite here only two. He demonstrated that anhydrobiotic tardigrades survived a 35-min exposure at +150 °C and >100 h at −190 °C and a few hours at approximately −272 °C. Initially, he assumed a "vita minima", i.e. he did not believe that the metabolism came to a complete halt, but in later publications, he argued for the complete cessation of metabolism, namely, based on his experiments at very low temperatures (e.g. Rahm 1923, 1926).

In 1925 Cuénot published a note about the cilia of the nephridia in *Peripatus* (Onychophora) and added a broad discussion about the systematic position of onychophorans, arthropods and tardigrades. He emphasized that onychophorans are not Protracheata (the term was created by Haeckel 1896); rather he argued that they are on the road to Protracheata (=the common hypothetical root of arthropods) forming a side branch with extant onychophorans as the terminal group. Adaptations to life on land (e.g. tracheae, slime glands) were independently from arthropods. The highly reduced and simplified tardigrades represent a side branch, however, closer to the Protracheata as suggested by their individual muscles, the lack of cilia and the arthropod-like nervous system (Cuénot 1925).

Between 1927 and 1929, Marcus released four substantial and broad-ranging articles (see Marcus 1927, 1928a, b, 1929a) and a synopsis of all tardigrade species previously described around the world in the series "Die Tierwelt Deutschlands"[117] (Fauna of Germany) (Marcus 1928c) and, also in 1929, as a kind of finishing touch, the above-mentioned summarizing monograph (Marcus 1929b). All these publications were produced with decisive contributions from his wife Eveline[118] (see Fig. 1.19), who obviously did not want to be named as coauthor. Naturally,

---

[116]Gilbert **Rahm** (1885–1954), German priest (Benedictine) and zoologist, obtained the doctorate at Bonn University in 1920 with the thesis "Biologische und physiologische Beiträge zur Kenntnis der Moosfauna" (supervisor Richard Hesse (1868–1944; published 1923) and qualified as a lecturer in 1925 with the thesis entitled "Beitrag zur Kenntnis der Moostierwelt der preussischen Rheinlande" (published 1925) at the University Freiburg (Switzerland). From 1929 onwards he was professor of general biology at the Catholic University Santiago (Chile), moved in 1929 to the USA, where he was engaged in studying trichinosis and covered pastoral functions (http://www.kreis-ahrweiler.de/kvar/VT/hjb1990/hjb1990.20.htm; Grothman et al. 2017).

[117]A well-introduced series founded 1925 by Friedrich **Dahl** (1856–1929), curator of arachnids at the Museum für Naturkunde in Berlin. The full title of the series is "Die Tierwelt Deutschlands und der angrenzenden Meeresteile nach ihren Merkmalen und nach ihrer Lebensweise," [The fauna of Germany and adjacent maritime zones according to their characteristics and living].

[118]Ernst **Marcus** repeatedly acknowledged the collaboration with his wife and her contributions. Already in the first article he wrote "Im Folgenden seien die Ergebnisse der von meiner Frau Eveline und mir gemeinsam angestellten Untersuchungen an zwei marinen Formen mitgeteilt,..." [In the following the results of the investigations on two marine species carried out together by my wife Eveline and me will be communicated...] (Marcus 1927, p. 487; see also the dedication in Marcus 1928c).

Fig. 1.19 Ernst Marcus and his wife Eveline (photo from the early 1950s, with kind permission of the Department of Zoology, Institute of Biosciences, University of São Paulo)

this enormous productivity led to a certain recycling of figures and text modules. However, each article covered a very broad spectrum of topics and is distinguished by the author's extensive knowledge of the relevant literature, by the critical appraisal of previous studies, and an incredible wealth of details. There is hardly anything about tardigrades on which Marcus did not make a statement. Therefore, these articles are not always easy to read (like Doyère's and Lance's theses), but are rich and stimulating sources for the reader. Since it is not possible to fully appreciate them in this brief survey, I will list only some topics (below).

In the first article, Marcus (1927) focussed on two marine tardigrades, *Echiniscoides sigismundi* and *Batillipes mirus*. He critically reviewed the marine species known at that time, called *Bathyechiniscus tetronyx* (see above) a "species dubia et inquirenda", and rejected with good reasons[119] the term "larva" for juvenile specimens used by previous authors (e.g. Heinis 1910; Thulin, 1928). He provided good tips how to handle tardigrades for histology, cultured and bred *Batillipes mirus* for months (but could not observe details due to the sandy substrate necessary for this species), and described in great detail the morphological, physiological and behavioural adaptations of both species to their habitats that differ fundamentally. Further, he suggested a uniform terminology for the head appendages; depicted the course and number of muscles; described the formation of the claws, locomotion and further morphological details; and demonstrated positive phototaxis in *Echiniscoides sigismundi*, but not in *Batillipes mirus*. Further, he presumed the presence of a

[119]"...da die aus dem Ei gekrochenen und ihre ersten Häutungen durchmachenden Tiere sich von den ausgewachsenen lediglich durch geringe Größe, geringere Krallen- und manchmal Dornenzahl und Fehlen der Geschlechtsorgane unterscheiden, nie aber provisorische oder Larvenorgane besitzen." [because the hatched specimens and those after the first moults differ from the adults only in their small size, the smaller number of claws and occasionally of spines, an in the absence of reproductive organs, however, never they have provisional or larval organs] (Marcus 1927, p. 488/489).

peritrophic membrane during defecation. Concerning tardigrade relationships, he rejected Richter's view (see above). Finally, Marcus created the Arthrotardigrada[120] as tardigrades with infoldings of the legs; included the genera *Echiniscus*, *Parechiniscus* and *Pseudechiniscus*[121] in the Echiniscoidea; and incorporated Arthrotardigrada and Echiniscoidea under the order Heterotardigrada,[122] which he compared with the Eutardigrada. This classification is valid today. Impressed by the findings of Remane,[123] he discussed similarities of tardigrades with certain aberrant gastrotrichs, but abstained from a phylogenetic assessment.

Before continuing with the next articles of Marcus, I insert here two publications by Thulin[124] (1911, 1928). Both articles captivate (in Marcus's opinion) by their excellent diagnoses for genera and species and meaningful drawings. The second article (Thulin, 1928) was published around the same time as the studies of Marcus, but Thulin considered Marcus's studies from 1927 and 1928, and Marcus, in turn, included Thulin's first article in his publications from 1928 and both in his monograph. Thulin (1928) accepted the system proposed by Marcus, but he started to subdivide the orders (which he called sections) into families, e.g. the Heterotardigrada into the Halechiniscidae and Echiniscidae and the Eutardigrada into the Macrobiotidae with *Macrobiotus*, *Hypsibius*, *Diphason* and several new genera and species and the Arctiscidae with *Arctiscon* (=*Milnesium*). Further, he split up some genera, e.g. the genus *Echiniscus* into *Echiniscus*, *Bryodelphax*[125] and *Hypechiniscus*. Most of these changes are still valid. Further he emphasized the special status of *Milnesium tardigradum* and wondered whether this species may be more closely related to Hetero- or Eutardigrada. To do this all, he used a wide range of characters such as leg morphology, cuticular plate morphology (in the Echiniscidae), the musculature, the organization of the buccopharyngeal apparatus and in great detail the cephalic sense organs and their innervations (here without specifying from where the knowledge about this issue came; he referred to a forthcoming article that, however, never was published). Finally, he constructed a tree to illustrate the phylogenetic relations among Tardigrada. Apart from that he noticed males in *Echiniscus* spp, *Pseudechiniscus* spp., and *Milnesium tardigradum*. Both, Thulin and Marcus concluded that the Heterotardigrada are more ancestral than Eutardigrada assuming an origin in the marine littoral zone (see also Marcus 1929b).

---

[120]árthron, Gr. = joint; poús, gen. podós, Gr. = Fuß.

[121]pará, Gr. = next to; pseudés, Gr. = false, mendacious.

[122]héteros, Gr. = different.

[123]Adolf **Remane** (1898–1976), German zoologist, professor at the Universities of Kiel and Halle (Germany), explored the microfauna of marine sand and refined the homology theorem in 1952 (http://www.deutsche-biographie.de/pnd11874447X.html).

[124]Gustav **Thulin** (1889–1945), Swedish zoologist, student of Sigurd Wallengren (1864–1938) and professor of Zoology the University Lund (Sweden). Thulin worked in the 1920s and 1930s in the Department of Zoology in Lund and from 1936 at the Natural History Museum in Gothenburg (Sweden). Here he studied tardigrades but also other benthic animals (www.tmbl.gu.se/libdb/taxon/.../petymol.tu.html).

[125]brýon, Gr. = Moos; délphax, Gr. = piglet; hypér, Gr. = over.

The second wide-ranging study by Marcus (1928a) is devoted to physiology and ecology, i.e. culturing and life history observations on distribution, annual cycles, reactions to light and touch stimuli, need for oxygen, way of moving (locomotion, gait rhythm and speed), feeding, defecation and digestion, excretion, moulting cycles (defecation and moulting are linked in Heterotardigrada but decoupled in Eutardigrada), anabiosis and so forth.

The third article (Marcus 1928b), again nicely illustrated by Eveline Marcus (see Figs. 1.20 and 1.21), contains a wealth of new anatomical and histological details that also relativize many findings of previous authors. Noticeable are the demonstration of relative eutely of several tardigrade species by the number of ventral and dorsal median epidermal cells, which was identical in individuals of different age, but differed slightly between species (Fig. 1.22); the intense use of the ends of the legs for systematics; the

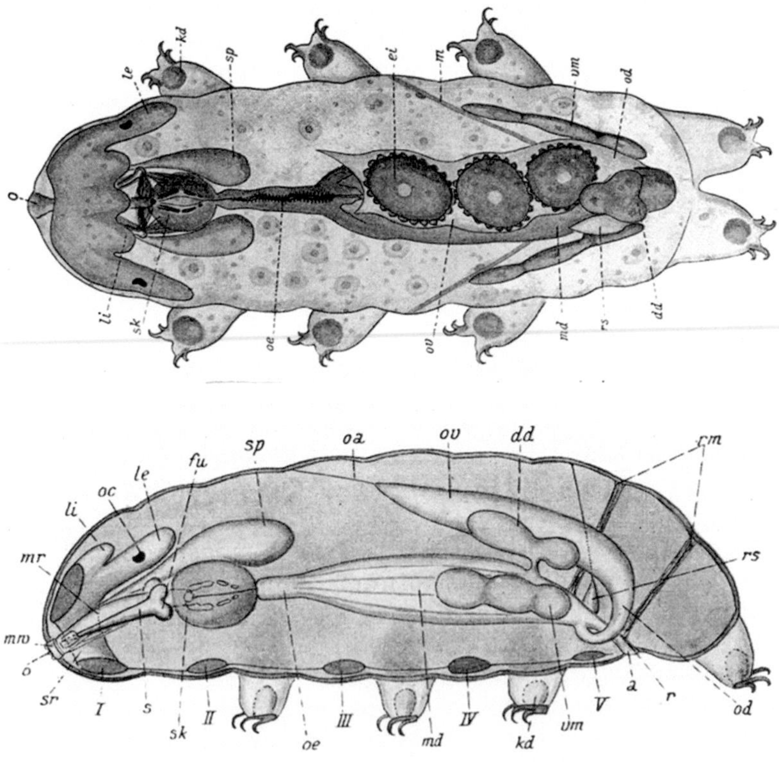

Fig. 1.20 Top: dorsal view of a female of *Macrobiotus hufelandii* (muscles omitted): Bottom: organization of a tardigrade (type, *Macrobiotus hufelandii*). The left half of the body wall and the nervous system and the left salivary gland removed; dd = dorsal gland; ei = egg; fu = furca; kd = claw gland; le = outer lobe of the brain, li = inner lobe of the brain; m = muscle; md = midgut; mr = buccal tube; o = mouth opening; oa = suspension of the ovary; oc = eye; od = oviduct; oe = oesophagus; ov = ovary; r = rectum; rm = rectal muscle; rs = receptaculum seminis; s = stylet; sk = pharyngeal bulb; sp = salivary gland; sr = pharyngeal nerve ring; vm = Malpighian tubules. I–V = ventral ganglia. Figures 1 and 2 from Marcus (1929b)

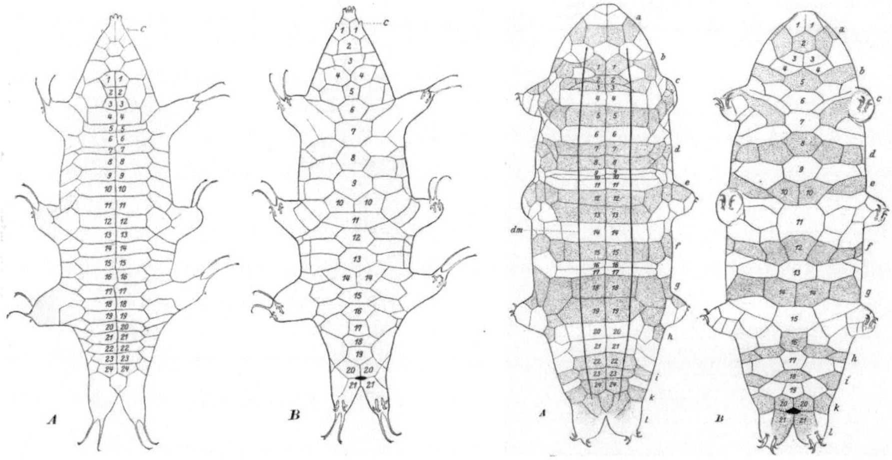

Fig. 1.21 Dorsal (**a**) and ventral view (**b**) of *Milnesium tardigradum* (left) and *Macrobiotus hufelandii* (right) after silvering of epidermal cell borders. Numbers describe the rostrocaudal sequence of epidermal cells, letters a–l the pigmented areas (dotted). Figures 8 and 9 from Marcus (1929b)

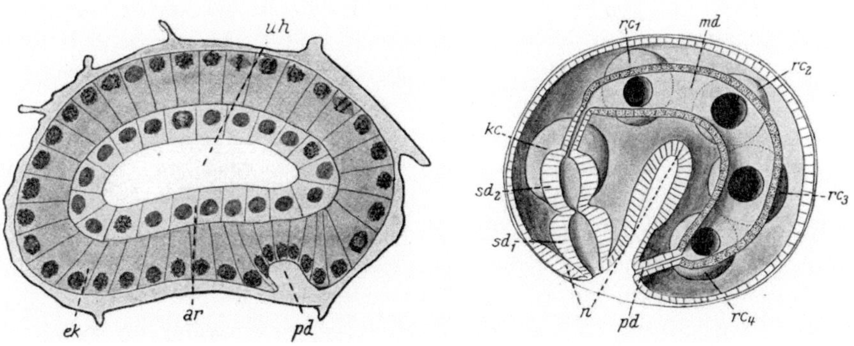

Fig. 1.22 Two stages of the development of *Hypsibius convergens*. Left: sagittal section; ar = archenteron; ek = ectoderm; pd = developing proctodaeum; uh = primary intestinal cavity. Right: right half of an embryo with colonic pouches; rc1–rc4 = colonic pouches of the trunk; kick = colonic pouch of the head; sd_1 = anlagen of the salivary gland; sd_2 = anlage of the pharynx; n = anlage of the nervous system. Figures 85 and 92 from Marcus (1929b)

evidence for transient eutely of an subepidermal epithelioid tissue of body cavity cells in *Echiniscoides sigismundi* just hatched out of the egg, which was considered as transient parietal mesoderm; the proof of mitoses and reserve storage of the body cavity cells; the detailed description of the anatomy of *Echiniscus* spp. and *Tetrakentron synaptae*, including the note that in the marine species investigated the anus and gonopore open separately, whereas eutardigrades possess a cloaca; the treatment of the intestine and its appendices, i.e. buccopharyngeal apparatus, mouth glands, Malpighian tubules, thickenings of the rectal epithelium, gonads (ovary, oogenesis, spermatozoa) and similarities and dissimilarities of the nervous system of Hetero- and Eutardigrada.

The last original article (Marcus 1929a) dealt with embryology of several Eutardigrades, primarily *Hypsibius convergens*. According to his own statement, Marcus and his wife reused for this study the histological sections made by Wanda von Wenck (see above) and prepared 2–5 μm (!) thick histological sections of approximately 5000 (!) eggs. As with the other articles, and apart from the direct issue, the article contained a lot of additional information and thoughts, i.e. notes on the formation of the chorion, chromosome numbers, copulation, outer and inner insemination and fertilization and egg deposition. In addition, Marcus speculated about parthenogenesis in *Echiniscus* and *Milnesium tardigradum* and the presence of a micropyle[126] (to my knowledge no one has seriously looked for that so far) and pointed out that tardigrades do not meet the criteria for true eutely. The main part of the study contains the very detailed description of cleavage and gastrulation, the development of the archenteron, the proctodaeum and stomodaeum, the body cavity, the gonads, the nervous system, the legs and the intestine and some notes on developmental time and hatching and finally—quasi inevitable—remarks on the systematic position of tardigrades. I will give here only an outline of the main points (see also von Wenck 1914): Eggs are isolecithal; cleavage is total-equal, asynchronous and undetermined leading to a blastula; the entoderm is formed by delamination; the proctodaeum develops after gastrulation (Fig. 1.22, left) followed by the formation of the stomodaeum, and primordial germ cells become apparent in the ventral primary entoderm; the mesoderm is formed through enterocoely (Fig. 1.22 right; see above); the primary entoderm withdraws from the ectoderm forming a schizocoel that merges with the disaggregating pouches of the archenteron; the disaggregated cells develop into muscles and body cavity cells; the fourth pair of pouches form the gonads; the buccopharyngeal apparatus, oesophagus and Malpighian tubules appear to be derived from the ectoderm.

The most puzzling result was that the mesoderm was formed by enterocoely as suggested previously, because this type of coelom formation typically occurs in deuterostomes. This finding has found its way into the textbooks of zoology but seems to have been refuted in the recent past. However, obviously this issue has not yet been fully resolved.[127]

---

[126]pyle, Gr. = door; micropyle = small hole in the egg chorion for the entry of the spermatozoon.

[127]The more recent studies using modern techniques are not free from ambiguity. J. Eibye-Jacobsen (1997), who studied *Halobiotus crispae* (Eutardigrada) and *Echiniscoides sigismundi* (Heterotardigrada) using TEM, believed she had seen traits of spiral cleavage, and she very much doubted the existence of enterocoely in tardigrades. Heijnol and Schnabel (2005a, b) studying *Thulinia stephaniae* (Eutardigrada) and *Echiniscoides sigismundi* did not detect a stereotyped cleavage pattern, but an indeterminate irregular cleavage pattern and rejected enterocoely as well, but Gabriel et al. (2007) emphasized differences in the mode of gastrulation and germ layer formation and found that the embryos of *Hypsibius dujardini* (Eutardigrada) have a stereotyped cleavage pattern with asymmetric cell divisions and that the mesoderm is formed at least in part from endomesodermal pouches (summarized in Gross et al. 2015).

In the last section of this article and even more detailed in his monograph (Marcus 1929b), Marcus applaudably and critically discussed a wide range of morphological and developmental traits of nearly all animal taxa ever suggested to be related with tardigrades over the decades. To cut this relatively long story short: Marcus concluded from the available information that as possible nearest relatives of tardigrades only Aschelminthes, Annelida and Arthropoda.[128] Among other things, he argues: Tardigrades (1) are not Aschelminthes due to their indeterminate development, the presence of enterocoely, the different organization of the nervous and excretory system and the lack of true eutely; the buccal apparatus, similar to that of nematodes, was developed by convergence; (2) tardigrades are not annelids due to their indeterminate development; the presence of a multicellular mesoderm (formed by enterocoely); the lack of a secondary body cavity (= coelom), when fully developed; the lack of a cuticulo-muscular tube; and the lack of segmental organs.

Further, he noted that all annelid traits found in tardigrades are also typical for arthropods such as the stepladder nervous system, stomodaeum and proctodaeum, the genital cavity as a remnant of the coelom and metameric segmentation and the fact that the most important primordia of organs develop ventrally.

Finally, he stated with a due degree of caution that tardigrades belong to the arthropods, but to a class of their own, placed between protracheates (i.e. onychophorans) and eutracheates (i.e. at that time the Myriapoda, Chilopoda, Apterygogenea and Insecta; see Claus and Grobben 1917), as they share characters with each of the two groups. With this he confirmed and substantiated the classification of tardigrades in a well-known and widespread textbook of zoology from this time,[129] avoiding, however, any discussion on the descent and ascent of tardigrades.[130] It is noteworthy that, apart from the more modern view that tardigrades

[128]At that time within the 'cladus' Scolecida (lower worms), the 'class' Aschelminthes (= Nemathelminthes) included the orders Rotatoria, Gastrotricha, Kinorhyncha, Nematodes, Nematomorpha, Acanthocephali, the 'cladus' Annelida (segmented woms) included the classes Archiannelida, Chaetopoda, Hirudinea, Echiuroidea and Sipuncoloidea, and the "cladus" Arthropoda comprised the classes Branchiata, Arachnoidea (including Linguatulida), Pantopoda, Protracheata (Onychophora), Tardigrada, and Eutracheata, all belonging to the 'phylum' Protostomia (see Claus and Grobben 1917).

[129]Claus and Grobben (1917); see footnote 128.

[130]"(...) eine descendenztheoretische Auseinandersetzung, die jeder plaäontologischen Basis entbehrt, muß rein hypothetisch bleiben. Da die weitere Erforschung der Tardigraden keinen Nutzen von stammesgeschichtlichen Hypothesen hat, ...so mag es genügen, wenn hier ohne Äußerung über die Descendenz und die Ascendenz der Tardigraden diese als Klasse der Arthropoden bezeichnet und zwischen den Protracheaten und den Eutracheaten eingereiht werden". [(...) a theoretical dispute on the descent with no paleontological basis must remain purely hypothetic. Since the further exploration of tardigrades does not benefit from phylogenetic hypotheses, (...), it may suffice to describe—without comments on descent and ascent—tardigrades as class of arthropods and to place them between Protracheata and Eutracheata] (Marcus 1929a, p. 143/144).

are not Articulata[131] (segmented protostomes), but belong to the clade Ecdysozoa[132] that unifies all moulting animals, and the changing nomenclature, one of the phylogenies under discussion considers Tardigrada as Panarthropoda[133] that also include Onychophora and Arthropoda. In this case, it is still a matter of debate whether tardigrades and arthropods or onychophorans and arthropods are sister groups. This issue is also discussed in modern textbooks of zoology (e.g. Westheide and Rieger 2013).

1.6 Epilogue: Ernst Marcus's Monograph (1929)

All of the above and more have been summarized in the monograph from 1929 mentioned before (Fig. 1.23). It should be added that in the year of Marcus's emigration to Brazil a further comprehensive review written in German was published in the series "Das Tierreich"[134] [The animal kingdom]. As already stated, the present survey largely follows the Chapter VII "Geschichte der Tardigradenforschung" of the monograph from 1929 published in the highly respected series "Dr. H.G. Bronn's Klassen und Ordnungen des Tier-Reichs wissenschaftlich dargestellt in Wort und Bild" (Fig. 1.23). This series was commenced by the famous geologist and palaeontologist H. G. Bronn.[135]

In the preface, Marcus stated that his monograph was not conceived as an encyclopaedia; rather it was oriented towards the requirements of university teaching (sic!) and general zoological research. Further, he emphasized that the historical

[131]Created 1812 by Jean Léopold Nicolas Frédéric **Cuvier** (1769–1832), known as Georges Cuvier, eminent French naturalist and zoologist. In his textbook in 1817 he wrote under the heading "Les animaux articulés": "Cette troisième forme générale est tout aussi caractérisée que celle des animaux vertébrés; le squelette n'est pas intérieur comme dans derniers (. . .). Les anneaux articulés qui entourent les corps et souvent les membres, en tiennent lieu (. . .) and Le sytème d'organes par lequel les animaux articulés se ressemblent le plus, c'est celui des nerfs". [This third general form is as well characterised as that of the Vertebrata; the skleton is not internal as in the latter (. . .). The articulated rings which encircle the body, and often the limbs, supply the place of it, (. . .) and the system of organs in which the Articulata particularly resemble each other, is that of the nerves] and distinguished the four classes Annelides, Crustacès, Arachnides, and Insectes (Cuvier 1817, p. 508 and 509, see also Cuvier 1812).

[132]The clade Ecdysozoa (´écdysis, Gr. = creeping out; zoon, Gr. = animal) was erected by Aguinaldo et al. (1997).

[133]pántos, Gr. = complete.

[134]This series was published on behalf of the "Preußische Akademie der Wissenschaften zu Berlin" (Prussian Academy of Sciences in Berlin), a learned society founded in 1700.

[135]Heinrich Georg **Bronn** (1800–1862), German doctor of medicine, professor of Zoology at the University of Heidelberg (Germany), geologist and palaeontologist. Prior to 1862 he wrote three volumes in the series "Die Klassen und Ordnungen des Thier-Reichs"; the work was continued by other naturalists.

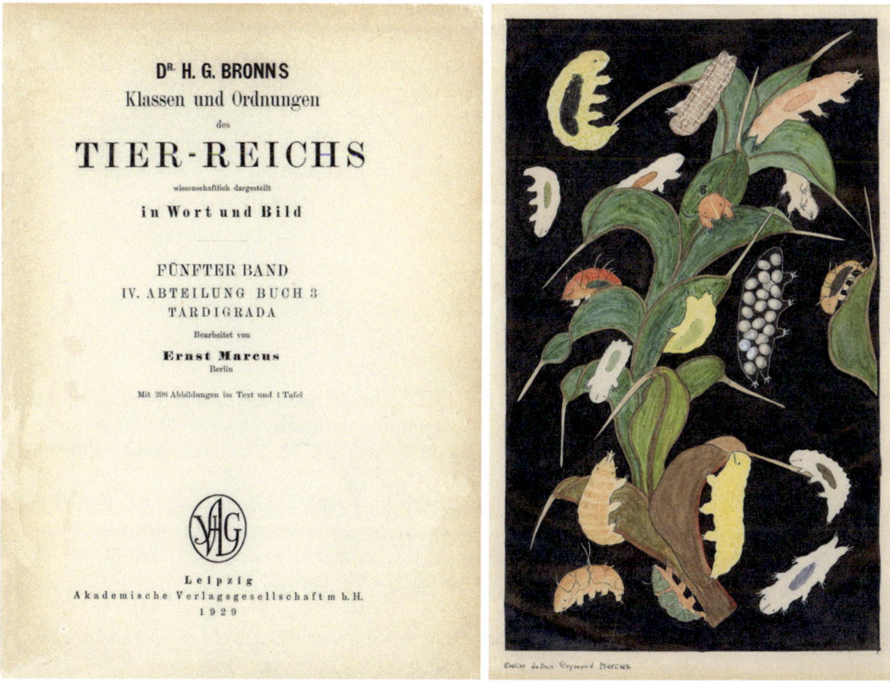
D^R. H. G. BRONNS
Klassen und Ordnungen
des
TIER-REICHS
wissenschaftlich dargestellt
in Wort und Bild

FÜNFTER BAND
IV. ABTEILUNG BUCH 3
TARDIGRADA
Bearbeitet von
Ernst Marcus
Berlin

Mit 398 Abbildungen im Text und 1 Tafel

Leipzig
Akademische Verlagsgesellschaft m. b. H.
1929

Fig. 1.23 Title page and Plate 1 from E. Marcus's monograph on Tardigrada from 1929. The plate was drawn by his wife Eveline Marcus (Original privately owned by H. Greven)

survey should be sufficient for general zoological interest.[136] From our present point of view, this is decidedly understated, as the monograph covers all aspects of tardigradology and shows the author's profound knowledge of the relevant literature. Even today it might be a wonderful reading for tardigradologists of any scientific orientation.

Acknowledgements I thank Prof Dr R Schill (Stuttgart, Germany) for the invitation to contribute to this book and for providing Doyère's dissertation; Dr H Dastych (Hamburg, Germany), who made some rare literature and illustrations available to me; Prof KI Jönsson, PhD (Kristianstad, Sweden), for a copy of Broca's report; Dr R Lacher, Museum of German Enlightenment (Halberstadt, Germany) for allowing the use of the portrait of JAE Goeze; the late Dr h.c. Eveline Marcus (1901–1990), who gave me the original of the drawing shown in Fig. 1.23 in the 1960s; Dr C Jared

[136]"Für den Grad der Ausführlichkeit der nachstehenden Naturgeschichte der Tardigraden, die keine Enzyklopädie sein will, waren mir die Bedürfnisse des Universitätsunterrichtes und der allgemein-zoologischen Forscherarbeit maßgebend (. . .). Für das allgemein-zoologische Interesse an der Geschichte der Tardigradenforschung dürfte der entsprechende Abschnitt (. . .) ausreichen". [For the level of detail of the following natural history of tardigrades, who wants to be not an encyclopedia, the needs of the university instruction and general-zoological research work had me prevail (...) for the general-zoological interest in the history of Tardigradenforschung should the relevant section (. . .) sufficient] (Marcus 1929b, p V, preface).

(São Paulo, Brazil), who supplied the contact to the Department of Zoology, Institute of Biosciences, University of São Paulo, and copied the portrait of Marcus (Fig. 1.1, bottom right), which is hanging in the amphitheatre of the University, as well as the photo of Ernst and Eveline Marcus (Fig. 1.19) from the archive of the University; Dr G Guex (Adlikon, Switzerland), who arranged the contact to the Zentralbibliothek Zürich to get the scan of Fig. 1.2 (bottom); Dr M Vakily (Collioure, France) for the help with some French texts; Prof Dr Diane Nelson (Johnson City, USA) for her careful search for linguistic imperfections; and, last but not least, all the institutions that make available old literature to the public.

References

Aguinaldo A, Turbeville JM, Linford LS, Rivera MC, Garey JR, Raff RE, Lake JA (1997) Evidence for a clade of nematodes, arthropods and other moulting animals. Nature 387:489–493

Baccetti B, Rosati F (1971) Electron microscopy on Tardigrada. III. The integument. J Ultrastruct Res 34:214–243

Bailey JW (1850) Microscopical observations made in South Carolina, Georgia and Florida. Smithsonian Contrib Knowl 2:1–48 http://www.biodiversitylibrary.org/item/100990

Basse A (1905) Beiträge zur Kenntnis des Baues der Tardigraden. Z wiss Zool 80:259–281

Baumann H (1921a) Mitteilungen zum feineren Bau der Tardigraden. Zool Anz 52:56–67 http://www.biodiversitylibrary.org/item/37571

Baumann H (1921b) Beitrag zur Kenntnis der Anatomie der Tardigraden (*Macrobiotus hufelandi*). Z wiss Zool 118:637–652 http://www.biodiversitylibrary.org/item/37608

Baumann H (1922) Die Anabiose der Tardigraden. Zool Jb Syst 45:501–556 http://www.biodiversitylibrary.org/item/38163

Bertolani R, Nelson DR (2011) Changes in scientific approaches as seen through 35 years of tardigrade symposia. J Zool Syst Evol Res 49:2–5

Broca P (1860) Rapport sur la question soumise a la Société de biologie par M.M. Pouchet, Pennetier, Tinel et Doyère au sujet de la réviviscence des animaux desséchés, lu à Société de Biologie, le 17 et le 24 mars 1860. Mem Soc Biol 3me Série II:1–139

Carus CG (1834) Über den Begriff des latenten Lebens. Arch Anat Physiol wissensch Medizin 1834:551–556 http://www.biodiversitylibrary.org/item/109154

Claus C (1887) Lehrbuch der Zoologie (4. umgearbeitete und vermehrte Auflage). Elwert'sche Verlagsbuchhandlung, Marburg

Claus C, Grobben K (1917) Lehrbuch der Zoologie (9. neubearbeitete Auflage des Lehrbuches von C. Claus). Elwert'sche Verlagsbuchhandlung, Marburg

Corti B (1774) Osservazioni microscopiche sulla Tremella e sulla circulazione del fluido in una pianta acquajuola (*Chara*). G. Rocchi, Lucca http://books.google.com

Cuénot L (1893) Commensaux et parasites des Echinodermes : Tardigrada. Rev biol N'nord de la France 5:16–19 (1 plate, no 1). http://www.biodiversitylibrary.org/item/41751

Cuénot L (1924) Tardigrades. Faune de France 24. Lechevalier, Paris. http://faunedefrance.org/bibliotheque/docs/L.CUENOT%28FdeFr24%29Tardigrades.pdf

Cuénot L (1925) L'entonnoir vibratile de la néphridie des Péripates. Ann Soc Roy Zool Belgique 56:13–17

Cuvier GB (1812) Sue un nouveau rapprochement à ètablier entre les classes qui composant le Régne Animal. Ann Mus Hist 19:73–84 (not seen)

Cuvier GB (1817) Le règne animal; distribué d'après son organisation, pour servir de base à l'histoire naturelle des animaux et d'introduction à l'anatomie comparée. Tome II. Deterville, Paris. http://www.biodiversitylibrary.org/item/18038

Davaine C (1858) Recherches sur les conditions de l'existence ou de la non-existence de la réviviscence chez des espèces appartenant au meme genre. Ann sc nat sér 4 10:335–337 http://www.biodiversitylibrary.org/bibliography/13266

de Blainville MH (1826) Sur quelques petits Animaux qui, après avoir perdu le mouvement par la desiccation, le reprennent comme auparavant quand on vient à les mettre dans l'eau. Ann sc nat 9:104–110 http://www.biodiversitylibrary.org/item/29169

Doyère ML (1840) Mémoire sur les Tardigrades (chapitre I, II, II printing error). Ann sc nat sér 2, 14(Zool):269–361 (8 plates, black and white, no 12–19). http://www.biodiversitylibrary.org/item/47974

Doyère ML (1842a) Mémoire sur l'organisation et les rapports naturels des tardigrades, et sur la propriété remarquable qu'ils possèdent de revenir a l avie après avoir été complètement desséchés. Thèses soutenues devant la Facultè des Sciences de Paris. Paul Renouard, Paris (8 plates, partly coloured)

Doyère ML (1842b) Mémoire sur les Tardigrades (chapitre III). Ann sc nat, sér 2, 17 (Zool):193–205. http://www.biodiversitylibrary.org/item/47977

Doyère ML (1842c) Mémoire sur les Tardigrades (chapitre. II). Ann sc nat sér. 2, 18(Zool):5–35. p 33–35 is an appendix). http://www.biodiversitylibrary.org/item/47978

Du Trochel (sic! = Dutrochet) (1812) Recherches sur les Rotifères. Ann mus hist nat 19:355–387 (1 plate, no 18). http://www.biodiversitylibrary.org/item/23270

Dujardin F (1838) Mémoire sur un ver parasite constituant un nouveau genre voisin des Rotifères, sur le Tardigrade et sur les Systolides ou Rotateurs en général. Ann sc nat sér 2, 10 (Zool):175–191 (1 plate, no 2). http://www.biodiversitylibrary.org/bibliography/13266

Dujardin F (1841) Histoire naturelle des zoophytes. Infusoires, comprenant la physiologie et la classification de ces animaux, et la manière de les étudier à l'aide du microscope. Roret, Paris. www.biodiversitylibrary.org/bibliography/51143

Dujardin F (1851) Sur les Tardigrades et sur une espèce à longs pieds vivant dans l'eau de mere. Observations Zoologiques II. Ann sc nat sér. 3, 15(Zool):161–166 (Pl 3). http://www.biodiversitylibrary.org/item/48051

Dutrochet RHJ (1837) Mémoires pour servir à l'histoire anatomique et physiologique des végetaux, vol 2. JB Baillière, Londres https://archive.org/details/mmoirespourser02dutr

Ehrenberg CG (1834) Über das Wiederaufleben der Moosfauna. Isis von Oken Heft 6/7:710–713 http://www.biodiversitylibrary.org/item/87984

Ehrenberg CG (1853a) Das organisch kleinste Leben über dem ewigen Schnee der höchsten Central-Alpen. Ber Bekanntm geeign Verhandl Königl Preuß Akad Wissensch Berlin 1853:315–333. http://www.biodiversitylibrary.org/item/109942

Ehrenberg CG (1853b) Diagnoses novarum formarum. Ber Bekanntm geeign Verhandl Königl Preuß Akad Wissensch Berlin 1853:526–533. http://www.biodiversitylibrary.org/item/109942

Ehrenberg CG (1854) Mikrogeologie: das Erden und Felsen schaffende Wirken des unsichtbar kleinen selbständigen Lebens auf der Erde. Leopold Voss, Leipzig http://books.google.com

Ehrenberg CG (1859) Beitrag zur Bestimmung des stationären mikroskopischen Lebens in bis 20,000 Fuß Alpenhöhe. Abhandl Kgl Akad Wissensch Berlin 1858:429–456 (Plate 1–3). http://www.biodiversitylibrary.org/item/93366

Eibye-Jacobsen J (1997) New observations on the embryology of the Tardigrada. Zool Anz 235:201–216

Eichhorn JCE (1775) Beyträge zur Natur-Geschichte der kleinsten Wasser-Thiere die mit keinem blossen Auge können gesehen werden und die sich in den Gewässern in und um Danzig befinden. Johann Emanuel Friedrich Müller, Danzig. http://books.google.com

Gabriel WN, McDuff R, Patel SK, Gregory TR, Jeck WR, Jones CD, Goldstein B (2007) The tardigrade *Hypsibius dujardini*, a new model for studying the evolution of development. Dev Biol 312:545–549

Giard A (1894) L'anhydrobiose ou ralentissement des phenomenes vitaux. CR hebd Séanc Mem Soc Biol Paris 1 10(46):497–500 http://www.biodiversitylibrary.org/item/100649

Goeze JAE (1773) 2. Beobachtung: Über den kleinen Wasserbären. In: Herrn Karl Bonnets Abhandlungen aus der Insektologie, aus dem Französischen übersetzt und mit einigen Zusätzen herausgegeben von Joh. August Ephraim Goeze, Pastor bei der St. Blasii Kirche in Quedlinburg. JJ Gebauers Wittwe und Joh Jacob Gebauer, Halle, pp 367–355. http://www.biodiversitylibrary.org/item/102801#page/7/mode/1up

Graff L (1877) Das Genus *Myzostoma* (F.S. Leuckart). Verlag von Wilhem Engelmann, Leipzig. http://www.bodleian.ox.ac.uk/dbooks

Greeff R (1865) Ueber das Nervensystem der Bärthierchen, Arctiscoida C.A.S. Schultze (Tardigraden Doyère) mit besonderer Berücksichtigung der Muskelnerven und deren Endigungen. Arch mikrosk Anat 1:101–123 (1 plate, no IV). http://www.biodiversitylibrary.org/item/49502

Greeff R (1866) Untersuchungen über den Bau und die Naturgeschichte der Bärthierchen (Arctiscoida C.A.S. Schultze). Arch mikrosk Anat 2:102–131 (2 plates, no VI and VII). http://www.biodiversitylibrary.org/item/49504

Greven H (1980) Die Bärtierchen Die Neue Brehm Bücherei 537. Ziemsen-Verlag, Wittenberg Lutherstadt

Greven H (2015) About the little water bear – A commented translation of Goeze's note „Ueber den kleinen Wasserbär" from 1773. Acta Biol Benrodis 17:1–27

Gross V, Treffkorn S, Mayer G (2015) Tardigrada. In: Wanninger A (ed) Evolutionary developmental biology of invertebrates 3: Ecdysozoa I: Non-tetraconata. Springer, Wien, pp 35–52

Grothman GT, Johansson C, Chilton G, Kagoshima H, Tsujimoto M, Suzuki AC (2017) Gilbert Rahm and the status of Mesotardigrada Rahm, 1937. Zool Sci 34:5–10

Haeckel E (1896) Systematische Phylogenie der Wirbellosen (Invertebrata). Zweiter Theil des Entwurfs einer systematischen Stammesgeschichte. Georg Reimer, Berlin. http://www.biodiversitylibrary.org/item/22322

Hay WP (1907) A bear animalcule renamed. Proc Biol Soc Washington 19(1906):46–47 http://www.biodiversitylibrary.org/item/19616

Heinis F (1910) Systematik und Biologie der moosbewohnenden Rhizopoden, Rotatorien und Tardigraden der Umgebung von Basel mit Berücksichtigung der übrigen Schweiz. Arch Hydrobiol Planktonkd 5:89–166, 217–256

Heinis F (1921) Über die Mikrofauna alpiner Polster- und Rosettenpflanzen. Festschr Feier d. 60. Geburtstages von Friedrich Zschokke 6:1–30

Hejnol A, Schnabel R (2005a) What a couple of dimensions can do for you: comparative developmental studies using 4D microscopy–examples from tardigrade development. Int Comp Biol 46:151–161 soll 2006 sein

Hejnol A, Schnabel R (2005b) The eutardigrade *Thulinia* has an indeterminate development and the potential to regulate early blastomere ablations. Development 132:1349–1361

Henneke J (1911) Beiträge zur Kenntnis der Biologie und Anatomie der Tardigraden (*Macrobiotus macronyx* Duj.). Z wiss Zool 97:721–752 (1 plate, no 33). http://www.biodiversitylibrary.org/item/37601

Hesse A (1995) Die Professoren und Dozenten der preußischen pädagogischen Akademien (1926–1933) und Hochschulen für Lehrerbildung (1933–1941). Deutscher Studien-Verlag, Weinheim

Hufeland CW (1826) Makrobiotik oder die Kunst das menschliche Leben zu verlängern, 6th edn. Macklot, Stuttgart. https://archive.org/details/makrobiotikoderd00hufeuoft

Kaufmann J (1851) Über die Entwicklung und systematische Stellung der Tardigraden. Z wiss Zool 3:220–232 (1 plate, no II). http://www.biodiversitylibrary.org/item/50073

Keilin DK (1959) The Leeuwenhoek Lecture. The problem of anabiosis or latent life: history and current concept. Proc R Soc Lond Ser B 150:149–191

Kinchin IM (1994) The biology of tardigrades. Portland Press, London

Kristensen RM (1980) Zur Biologie des marinen Heterotardigraden *Tetrakentron synaptae*. Helgoländer Meeresunters 34:165–177

Lance D (1896) Contribution á l'étude anatomique et biologique des Tardigrades (Genre Macrobiotus- Schultze), 229 pp (3 plates). Mauchassat, Paris

Lauterborn R (1906) Demonstrationen aus der Fauna des Oberrheins und seiner Umgebung. Verh Deutsch zool Ges 16:265–268 http://www.biodiversitylibrary.org/item/181877
Linné C (1790) Insecta aptera. Acarus. In: JF Gmelin (cura) Systema naturae per regna tria naturae: secundum classes, ordines, genera, species, cum characteribus, differentiis, synonymis (13 ed.) Tomus I, pars 5. Emanuel Beer, Lipsiae, p 2924. http://www.biodiversitylibrary.org/item/83109
Maire A (1892) Catalogue des Thèses de Sciences soutenues en France de 1810 a 1890 inclusivement. H. Welter, Paris
Marcus E (1927) Zur Anatomie und Ökologie mariner Tardigraden. Zool Jahrb Syst 53:487–558
Marcus E (1928a) Zur Ökologie und Physiologie der Tardigraden. Zool Jahrb Phys 44:323–370
Marcus E (1928b) Zur vergleichenden Anatomie und Histologie der Tardigraden. Zool Jahrb Phys 45:99–158 (3 plates, no 5–8) (Festschrift Hesse)
Marcus E (1928c) Spinnentiere oder Arachnoidea IV: Bärtierchen (Tardigrada). In: Dahl F (ed) Die Tierwelt Deutschlands und der angrenzenden Meeresteile. 12. Teil. Gustav Fischer, Jena, pp 1–230
Marcus E (1929a) Zur Embryologie der Tadigraden. Zool Jb Anat 50:333–384 (Pl 8)
Marcus E (1929b) Tardigrada. In: H.G. Bronn's Klassen und Ordnungen des Tierreichs 5, IV, 3. Akademische Verlagsgesellschaft, Leipzig, pp 1–608
Marcus E (1936) Tardigrada. In: Hesse R (ed) Das Tierreich 66. Lieferung. Walter de Gruyter, Berlin, pp 1–340
Martini E (1912) Studien über die Konstanz histologischer Elemente. III. *Hydatina senta*. Z wiss Zool 102:425–645 http://www.biodiversitylibrary.org/item/37603
Maucci W (1982) A history of tardigrade taxonomy. In: Proceedings of the Third International Symposium on the Tardigrada, Johnson City, Tennessee, 3–6 August 1980. East Tennessee State University Press, Johnson City, TN, pp 11–30
Morek W, Gasiorek P, Stec D, Blagden G, Michalczyk L (2016) Experimental taxonomy exposes ontogenetic variability and elucidates the taxonomic value of claw configuration in *Milnesium* Doyère, 1840 (Tardigrada: Eutardigrada: Apochela). Contrib Zool 85:173–200
Müller OF (1785) Von den Bärthierchen. Archiv der Insektengeschichte Teil I, 6. Heft, 25–31 (Pl 36). http://www.biodiversitylibrary.org/item/44317
Murray J (1907) The encystment of *Macrobiotus*. Zoologist 11:4–11
Murray J (1908) Encystment of Tardigrada. Trans R Soc Edinburgh 45 pt 4(31):837–854 (Pl I II). http://www.biodiversitylibrary.org/item/130552
Murray J (1910) British Antarctic Expedition 1907–1909. Reports on the scientific investigations, vol 1, Biology, Part V Tardigrada. William Heinemann, London, pp 81–185 (8 plates no XIV–XXI)
Needham JT (1743) Concerning chalky tubulous concretions called Malm: with some microscopical observations on the Farina of Red Lily, and of Worms discovered in Smutty Corn. Phil Trans 42:634–641 http://rstl.royalsocietypublishing.org
Perty M (1834) Einige Bemerkungen über die Familie Xenomorphidae Pty. oder die Sippe Arctiscon *Schrank* (Macrobiotus *Schultze*, Trionychicum Ehrb.) Isis von Oken 1834. Heft 12:1241–1246
Plate L (1889) Beiträge zur Naturgeschichte der Tardigraden. Zool Jahrb Anat 3:487–550 (3 plates, no XX–XII). http://www.biodiversitylibrary.org/item/41843
Preyer W (1880) Naturwissenschaftliche Thatsachen und Probleme. Populäre Vorträge. Paetel, Berlin
Preyer W (1891) Über die Anabiose. Biol Ctrbl 11:1–5. http://www.biodiversitylibrary.org/item/17522
Rahm G (1923) Biologische und physiologische Beiträge zur Kenntnis der Moosfauna. Zeitschr allg Physiol 20:1–34 (Pl 1)
Rahm G (1925) Die Cystenbildung bei den wasserlebenden Tardigraden. Verh internat Ver theor angew Limnol 3:364–371
Rahm G (1926) Die Trockenstarre (Anabiose) der Moostierwelt (Ihr Verlauf, ihre Bedeutung und ihr Unterschied von der Cystenbildung). Biol Zentralbl 46:452–477

Ramazzotti G, Maucci W (1983) Il Phylum Tardigrada (III edizione riveduta e aggiornata). Memorie dell' Istituto Italiano di Idrobiologia Dott Marco de Marchi 41:1–1012
Rebecchi L, Altiero T, Guidetti R (2007) Anhydrobiosis: the extreme limit of desiccation tolerance. ISJ 4:65–81
Reukauf E (1912a) Über das Simplexstadium und die Mundwerkzege der Macrobioten. Zool Anz 39:369–372 http://www.biodiversitylibrary.org/item/95291
Reukauf E (1912b) Ein Verderber des Wasserbären *Macrobiotus lacustris* (DUJ.), *Macrobiotophthora vimariensis* Reuk. Centralbl Bakteriol Parasitenkd Infektionskrankh 63, Abt. 1:390–393. https://archive.org/details/centralblattfr63jena
Richters F (1900) Beiträge zur Kenntnis der Fauna der Umgebung von Frankfurt a. M. Wiss Abh. Ber Senckenberg Naturf Ges 1900:21–44 (3 plates, no III–VI). http://www.biodiversitylibrary.org/item/36149
Richters F (1904a) Die Eier der Tardigraden. Ber Senckenberg Naturf Ges. II. Wissenschaftl Abhandlungen 1904:59–72 (2 plates, no IV & V). http://www.biodiversitylibrary.org/bibliography/14292
Richters F (1904b) Arktische Tardigraden. In: Römer F, Schaudinn F (eds) Fauna arctica. Eine Zusammenstellung der arktischen Tierformen. 3:493–508. http://www.biodiversitylibrary.org/item/23113
Richters F (1908) Marine Tardigraden. Zool Anz 33:77–85 http://www.biodiversitylibrary.org/item/37969
Richters F (1909a) Marine Tardigraden. Verh Deutsch Zool Ges 19:84–94 http://www.biodiversitylibrary.org/item/182001
Richters F (1909b) Tardigraden-Studien 40. Ber Senckenbergi Naturf Ges. II. Teil: Wissenschaftl Mitteilungen 40:28–44 (2 plates, no I & II). http://www.biodiversitylibrary.org/item/35680
Richters F, Krumbach T (1926) Tardigrada. In: Kükenthal W, Krumbach T (eds) Handbuch der Zoologie, vol 3. Walter de Gruyter, Berlin, pp 1–68
Schmitt O, Schultze V (1931) Wilhelm Titels Bildnisse Greifswalder Professoren. Rattsbuchhandlung L. Bamberg, Greifswald
Schultze CAS (1834a) *Macrobiotus hufelandii*, animal e crustaceorum classe novum etc. Apud Carolus Curths, Berlini (1 plate). urn:nbn:de:bvb:12-bsb10232115-1
Schultze CAS (1834b) Über *Macrobiotus Hufelandii*. Okens Isis 1834, Heft 6/7:708–710 (1 plate, no Pl 12). http://www.biodiversitylibrary.org/item/87984
Schultze CAS (1840) *Echiniscus Bellermanni*, animal crustaceum, Macrobioto Hufelandii affine. Reimer, Berolini. urn:nbn:de:bvb:12-bsb10232114-5
Schultze CAS (1861) *Echiniscus creplini*, animalculum e familia Arctiscoidum etc. Typis Frid Guil Kunike Reg Univers Typogr, Gryphiae. urn:nbn:de:bvb:12-bsb10231911-2
Schultze M (1865) *Echiniscus sigismundi*, ein Arctiscoide der Nordsee. Arch mikrosk Anat 1:1–9 (1 plate, no 26). http://www.biodiversitylibrary.org/item/49502
Senebier J (1795) Ueber die vornehmsten mikroskopischen Entdeckungen in den drey Naturreichen nebst ihrem Einfluß auf die Vervollkomnung des menschlichen Geistes. Aus dem Französischen des Herrn Senebier. Mit Anmerkungen und Zusätzen von Johann August Donndorf. Bey Heinrich Gräff, Leipzig. http://resolver.sub.uni-goettingen.de/purl?PPN647652889
Spallanzani L (1776) Opuscoli di Fisica animale, e vegetabile. Societá Tipografica, Modena. http://books.google.com
Spallanzani L (1803) Tracts of the natural history of animals and vegetables (translated from the original Italian of the Abbé Spallanzani by John Graham Dalyell with physiological illustrations), vol II, 2nd edn. Printed for William Creech and Archd Constable/TN Longman and O Rees, Edinburgh/London. https://archive.org/details/tractsonnaturalh12spal
Steiner G (1926) *Bathyechiniscus tetronyx* n. g. n. sp. ein neuer mariner Tardigrade. In: von Drygalski E (ed) Deutsche Südpolar-Expedition 1901–1903 XVIII. Band, Zoologie 10. Band. Walter de Gruyter, Berlin, pp 478–481
Sternberg K, von Krombholz JV (1838) Bericht über die 15. Versammlung deutscher Naturforscher und Ärzte in Prag im September 1837. Gottlieb Haase Söhne, Prag. http://books.google.com

Thulin G (1911) Beiträge zur Kenntnis der Tardigradenfauna Schwedens. Ark Zool 7(16):1–60. http://www.biodiversitylibrary.org/item/30233

Thulin G (1928) Über die Phylogenie und das System der Tardigraden. Hereditas 11:207–266

Usinger RL (1964) The role of Linnaeus in the advancement of entomology. Annu Rev Entomol 9 (1):16

van Leeuwenhoek A (1702) On certain animalcules found in the sediment in gutters of the roofs of houses. Letter 144. In: The select works of Antony van Leeuwenhoek, vol 2 (trans: Hoole S (1807)). London, pp 207–213. http://www.biodiversitylibrary.org/item/26600

Voigt FS (1843) Das Thierreich geordnet nach seiner Organisation. Als Grundlage der Naturgeschichte der Thiere und Einleitung in die vergleichende Anatomie vom Baron von Cuvier (nach der zweiten vermehrten Ausgabe übersetzt und durch Zusätze erweitert von F.S. Voigt, nach, vol 6 Zoophyten. Brockhaus, Leipzig, p 427–432. http://www.biodiversitylibrary.org/item/88935

von Erlanger R (1895a) Zur Morphologie und Embryologie eines Tardigraden, *Macr. macronyx*. Biol Ctrbl 15:772–777 http://www.biodiversitylibrary.org/item/17825

von Erlanger R (1895b) Beiträge zur Morphologie der Tardigraden I. Zur Embryologie eines Tardigraden: *Macrobiotus macronyx* Dujardin. Morphol Jahrb 2:491–513 (2 plates, no 20 and 21)

von Kennel J (1891) Die Verwandtschaftsbeziehungen und die Abstammung der Tardigraden. S-Ber Dorp Naturf Ges 1891:504–512

von Schrank P (1803) Fauna boica. Durchgedachte Geschichte der in Baiern einheimischen und zahmen Thieren, vol 3. Philipp Krüll, Landshut https://doi.org/10.5962/bhl.title.44923

von Siebold CT (1848) Lehrbuch der vergleichenden Anatomie der wirbellosen Thiere. Verlag von Veit & Comp, Berlin. https://ia601409.us.archive.org

von Wenck W (1914) Entwicklungsgeschichtliche Untersuchungen an Tardigraden (*Macrobiotus lacustris* Duj.). Zool Jahrb Anat 37:465–514 (4 plates, no 35–38)

Westheide W, Rieger G (2013) Spezielle Zoologie Teil1: Einzeller und Wirbellose Tiere, 3rd edn. Springer Spektrum, Heidelberg

Wright JC (1988) The tardigrade cuticle. I. Fine structure and the distribution of lipids. Tissue Cell 20:745–758

Wright JC (2001) Cryptobiosis 300 years on from van Leuwenhoek: what have we learned about tardigrades? Zool Anz 240:563–582

Wright JC, Westh P, Ramlov H (1992) Cryptobiosis in Tardigrada. Biol Rev 67:1–29

Chapter 2
Morphology and Functional Anatomy

Nadja Møbjerg, Aslak Jørgensen, Reinhardt Møbjerg Kristensen, and Ricardo C. Neves

Abstract Tardigrades are microscopic aquatic bilaterians that despite their size have a complex morphology and anatomy. These minute animals need a surrounding film of water to be active, and tardigrades residing in terrestrial environments live in moist microhabitats. The latter species are referred to as semiterrestrial, and they endure rapid dehydration by losing water over highly permeable integuments while forming a quiescent tun. Species of the class Eutardigrada are predominantly semiterrestrial and limnic with a relatively uniform morphology and anatomy. Heterotardigrades (class Heterotardigrada) comprising echiniscoideans and arthrotardigrades are much more diverse with highly variable external morphologies, ranging from clearly segmented species, over species with extremely large external sense organs and cuticular wings to species with a more "bear"-like eutardigrade body outline. In this chapter, we review the current understanding of tardigrade morphology and functional anatomy and put focus on tardigrade diversity emphasizing the often neglected marine arthrotardigrades. We give an introduction to the structure and function of major tardigrade organ systems, including the integument, body cavity, and digestive, muscular, nervous, and reproductive systems, and we highlight the diversity of tardigrade sensory appendages and overall external morphology.

N. Møbjerg (✉) · A. Jørgensen · R. C. Neves
Department of Biology, University of Copenhagen, Copenhagen, Denmark
e-mail: nmobjerg@bio.ku.dk

R. M. Kristensen
Natural History Museum of Denmark, University of Copenhagen, Copenhagen, Denmark

R. O. Schill (ed.), *Water Bears: The Biology of Tardigrades*, Zoological Monographs 2,
https://doi.org/10.1007/978-3-319-95702-9_2

2.1 Introduction

Extant tardigrades are microscopic bilaterians (approx. 50–1200 μm) with a segmented body plan composed of a head followed by a trunk that in the plesiomorphic condition bears four pairs of legs (Figs. 2.1 and 2.2). Research on this enigmatic phylum is currently at the forefront of biological research as highlighted in various editorials and book chapters across biology fields (e.g., Cross 2016; Gross et al. 2015; Maderspacher 2016; Persson and Persson 2012). Being one of a few animal phyla with extensive cryptobiotic abilities, tardigrades have historically drawn attention (Keilin 1959). Specifically, tardigrade morphology and anatomy were intensively investigated in the nineteenth century by authors such as Doyére (1840), Greeff (1865), and Plate (1889) and later by Basse (1905), Thulin (1928), and Müller (1936). Importantly, Marcus (1929) made incredible precise descriptions of tardigrade microanatomy facilitated by the transparency of the investigated species. These early studies were solely based on light microscopy, but they were, nevertheless, extremely detailed and provide a thorough background for present-day understanding of tardigrade structure. The end of the twentieth century marks the beginning of the investigation of tardigrades by electron microscopy (Greven and Kuhlmann 1972; Dewel and Clark 1973; Walz 1973, 1974, 1978, 1979; Shaw 1974; Greven and Grohé 1975; Weglarska 1975, 1987, 1989; Kristensen 1976, 1978, 1979, 1981; Dewel and Dewel 1979, 1996; Greven 1982; Wiederhöft and Greven 1996, 1999; Dewel et al. 1993, 1999; Møbjerg and Dahl 1996; Eibye-Jacobsen 1996, 1997; Rebecchi 1997, 2001; Rebecchi and Bertolani 1999; Jørgensen et al. 1999, Kristensen and Neuhaus 1999; Greven and Kristensen 2001). Data from scanning electron microscopy (SEM) and transmission electron microscopy (TEM) significantly broadened the understanding of tardigrade ultrastructure by providing high-magnification, three-dimensional resolution of external structures as well as details at the cellular level that were not previously available. To date, electron microscopy remains *state of the art* in what regards image resolution, and contemporary studies on tardigrade structure continue to build on SEM and TEM (Biserova and Kuznetsova 2012; Czerneková et al. 2017; Guidetti et al. 2013; Hyra et al. 2016a, b; Persson et al. 2014; Roszkowska et al. 2013). In addition, the advent of confocal laser scanning microscopy (CLSM)—and its combination with immunocytochemistry (ICC) and modern 3D imaging techniques—has brought a new perspective to the study of tardigrade microanatomy (Schmidt-Rhaesa and Kulessa 2007; Zantke et al. 2008; Halberg et al. 2009a, 2013a; Schulze and Schmidt-Rhaesa 2011, 2013; Persson et al. 2012, 2014; Marchioro et al. 2013; Mayer et al. 2013a, b; Schulze et al. 2014; Smith and Jockusch 2014). Future investigations into tardigrade microanatomy will likely include new techniques, such as serial block-face scanning electron microscopy (SBF-SEM; e.g., Hyra et al. 2017). Specifically, SBF-SEM combines scanning and transmission electron microscopy by mounting an ultramicrotome (used for preparing sections for TEM) inside the vacuum chamber of a SEM. The microtome cuts small, nanoscale sections of resin-embedded specimens, and the microscope subsequently prepares images of the remaining surface.

Fig. 2.1 Structural diversity of tardigrades—scanning electron microscopy of selected adult representatives of eutardigrades, echiniscoideans, and arthrotardigrades. Note the difference in overall size as well as morphology. (**a**) Ventral view of the semiterrestrial eutardigrade *Richtersius coronifer* from Öland, Sweden. Note the ventromedian muscle attachment sites, which are visible as depressions in the cuticle along the midline of the specimen. Compare to light micrograph in Fig. 2.3c revealing selected internal structures of *R. coronifer*. (**b**) Dorsal view of anhydrobiotic tun of *R. coronifer*. Depressions in the cuticle mark dorsal muscle attachment sites. (**c**) Ventrolateral view of the marine echiniscoidean *Isoechiniscoides sifae* from L'Aber de Roscoff, France. Higher

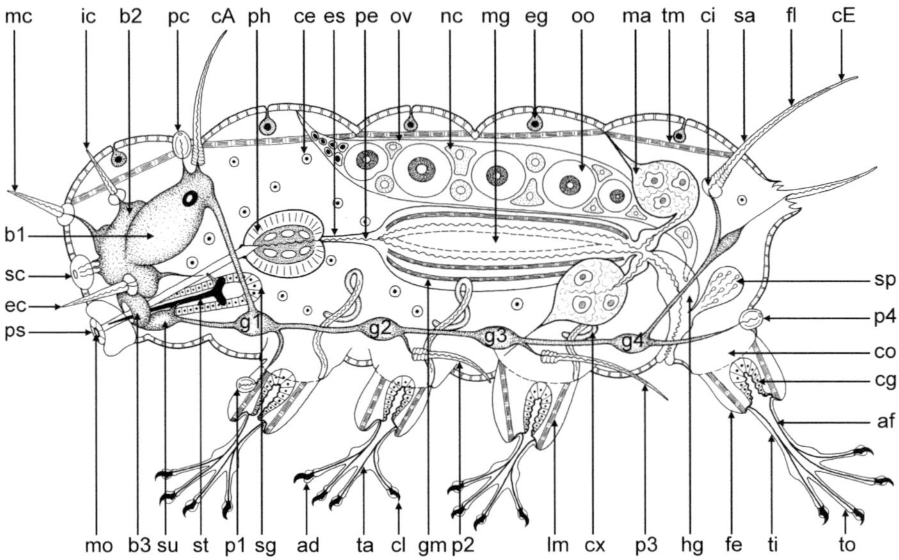

Fig. 2.2 Overview of the body plan of a hypothetical ancestral tardigrade. See the text for a detailed explanation of morphological and anatomical structures. Note that not all structures discussed in the text are presented in the drawing. ad = adhesive disc, af = cuticular attachment fibre, b1 = first brain lobe (protocerebrum), b2 = second brain lobe (deutocerebrum), b3 = third brain lobe (tritocerebrum), cA = cirrus A (lateral cirrus), cE = cirrus E, ce = coelomocyte, cg = claw gland, ci = cirrophorus, cl = claw, co = coxa, cx = coxal gland, ec = external cirrus, eg = epidermal gland, es = esophagus, fe = femur, fl = flagellum, g1–4 = trunk ganglia, gm = gut muscle, hg = hindgut, ic = internal cirrus, lm = leg muscle, ma = Malpighian tubule, mc = median cirrus, mg = midgut, mo = mouth, nc = nurse cell, oo = oocyte, ov = ovary, p1–p4 = sensory organs of legs I–IV, pc = primary clava, pe = peritrophic membrane, ph = pharyngeal bulb with placoids, ps = peribuccal sense organ, sa = scapus, sc = secondary clava, sg = stylet gland, sp = spermatheca, st = stylet, su = subpharyngeal ganglion, ta = tarsus, ti = tibia, tm = trunk muscle, to = toe

Tardigrades are aquatic organisms that need to be surrounded by water in order to be in their active, feeding, and reproducing stage. Nevertheless, numerous species have invaded terrestrial environments, where they endure periods of dehydration by entering anhydrobiosis (see Chap. 10). The latter are referred to as semiterrestrial or limno-terrestrial species. Anhydrobiosis is associated with marked alterations in morphology and anatomy (Halberg et al. 2013a; Czerneková et al. 2017). Specifically, during dehydration, semiterrestrial species contract their anterior-posterior body axis, retract their legs, and rearrange internal organs and cells while forming

---

Fig. 2.1 (continued) magnification of, respectively, the hind and head region of *I. sifae* can be found in Figs. 2.6b and 2.7b. (**d**) Ventral view of the marine arthrotardigrade *Actinarctus* cf. *doryphorus* sampled at Trezen ar Skoden off Roscoff, France. Compare to higher magnification of head region of the current specimen in Fig. 2.7c. Scale bar = 100 μm a so-called tun (Fig. 2.1b). Interestingly, marine tidal tardigrades form tuns when facing elevated external osmotic pressures, indicating that tun formation is an ancient trait, which evolved within the marine environment (Jørgensen and Møbjerg 2015; Hygum et al. 2016). In addition to tun formation, tardigrades may also form cysts and undergo cyclomorphosis. The latter processes, which are characterized by pronounced alterations in the structure of the individual animal, are dealt with in detail in Chap. 9 and will not be discussed in further detail here.

2.1.1 An Introduction to Tardigrade Structural Diversity

In order to understand tardigrade morphology and functional anatomy, one needs to assess the diversity of Tardigrada. The phylogenetic inference of morphological and molecular data supports the presence of two major evolutionary lineages within the phylum, Eutardigrada and Heterotardigrada (see Chap. 3). While eutardigrades are predominantly semiterrestrial and limnic (Fig. 2.3), heterotardigrades consist of the tidal or semiterrestrial echiniscoideans (Fig. 2.4) and the marine arthrotardigrades (Fig. 2.5). Especially the heterotardigrades are highly diverse (Figs. 2.4 and 2.5). The latter fact is unfortunately in clear contrast to the often prevailing view of tardigrades, which exclusively associates the phylum with the moss-dwelling "bear"-like eutardigrades. The focus on a few selected species (e.g., *Hypsibius exemplaris*; Fig. 2.3b) and the concurrent oversight of tardigrade diversity may introduce a clear bias in interpretations of morphological as well as molecular and physiological data. In the current review, we put focus on tardigrade diversity by reintroducing the mixotardigrade, a hypothetical ancestral tardigrade with a composite of characters from different evolutionary lineages (Fig. 2.2).

Within Heterotardigrada, arthrotardigrades are characterized by telescopic legs and a large number of well-developed cephalic sensory appendages, known as cirri and clavae (Fig. 2.5; Renaud-Mornant 1982; Hansen et al. 2012; Jørgensen et al. 2014). Body shape is extremely variable in this group, ranging from highly segmented species over species with extraordinarily enlarged sense organs and cuticular wings, the so-called alae (Fig. 2.5a, c), to species with a more traditional "bear"-like tardigrade body outline. The echiniscoideans comprise heterotardigrades with cephalic appendages that are reduced in general size and number as compared to arthrotardigrades (Fig. 2.4; Kristensen and Hallas 1980; Møbjerg et al. 2016). In eutardigrades, external evidence of cephalic sensory appendages is even more reduced (Fig. 2.3). Telescopic legs are present in a few species of echiniscoideans but are lacking in Eutardigrada. Moreover, eutardigrades are characterized by the presence of so-called Malpighian tubules (Fig. 2.2), traditionally believed to be excretory and osmoregulatory organs (Greven 1982; Kristensen 1982; Weglarska 1987; Møbjerg and Dahl 1996; Pelzer et al. 2007). In marked contrast, heterotardigrades do not have Malpighian tubules but instead possess so-called coxal or segmental glands that appear to be involved in extracellular fluid

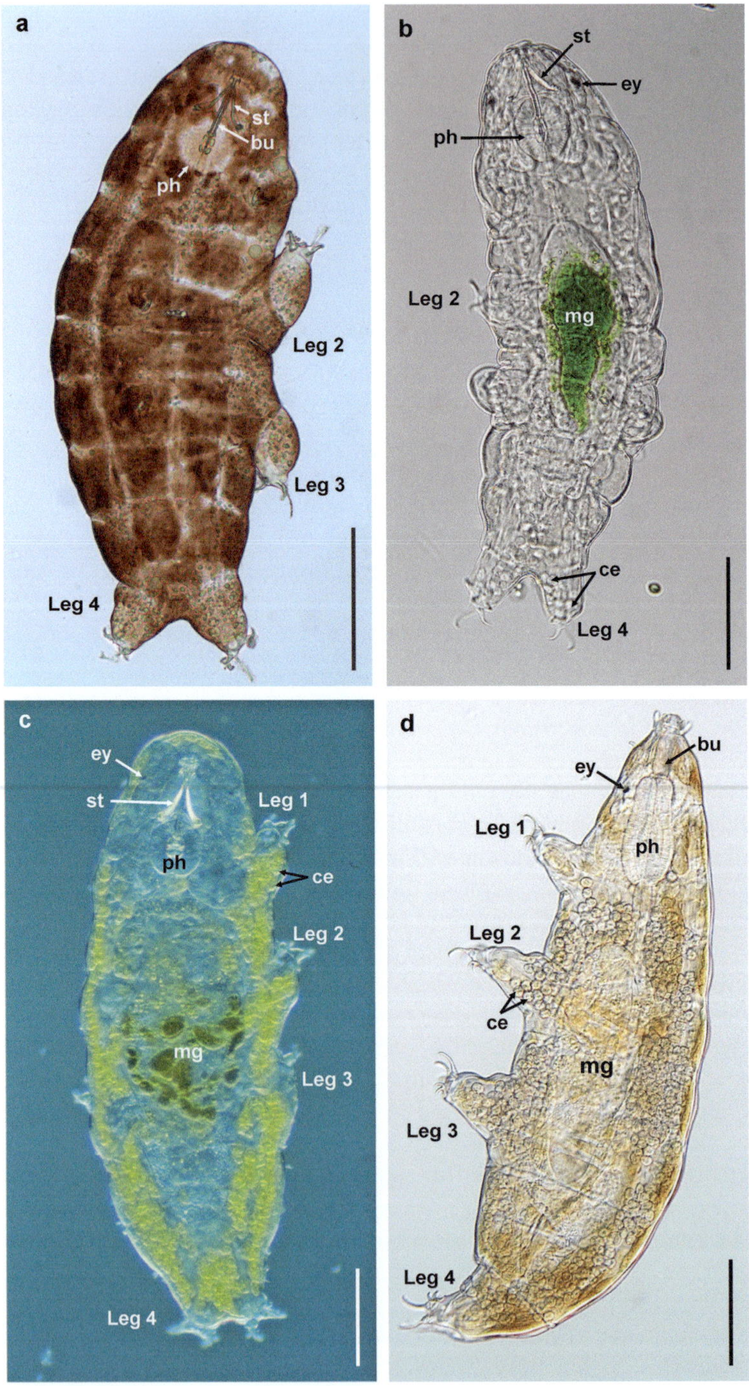

Fig. 2.3 Structural diversity of tardigrades—light microscopy of various eutardigrades with examples of species representing different families within the two eutardigrade orders Parachela

homeostasis (Fig. 2.2; Dewel et al. 1992). Importantly, in eutardigrades the gonad opens into the rectum, whereas heterotardigrades possess a distinct gonopore and anus (Figs. 2.5b and 2.6a, b; Guidetti and Bertolani 2005). In addition to the abovementioned groups, a single species, *Thermozodium esakii* Rahm, 1937, was reported from a hot spring in Japan and assigned to a distinct lineage, i.e., the mesotardigrades, characterized by well-developed sensory appendages as well as Malpighian tubules (Rahm 1937). Unfortunately, the existence of *T. esakii* has not been verified by others, and the validity of the species has consequently been questioned (Grothman et al. 2017).

Despite their small size, tardigrades are complex metazoans with highly developed organ systems. Great variations exist in cell form and size, with the largest cells represented by mature oocytes—especially large somatic cells are found within the midgut epithelium and the initial segment of the Malpighian tubules in eutardigrades. Cell counts in the eutardigrade *Halobiotus crispae* estimated that this species is composed of approx. 1060 somatic cells on average and that up to 1/3 of these cells, are concentrated in the head region (Møbjerg et al. 2011). A similar count, i.e., around 1000 cells, has been reported for the eutardigrade *Hypsibius exemplaris* (Koutsovoulos et al. 2016). Below we deal with the structure and function of major organ systems in more detail.

2.2 Integument

The tardigrade body is covered by an integument composed of a cuticle containing chitin and an underlying single-layered, cuticle secreting epidermis (Baccetti and Rosati 1971; Greven and Greven 1987). Tardigrades can have various colors. Marine species are normally transparent or white, while semiterrestrial species may be white, yellow, green, red, orange, brown, or almost black (Figs. 2.3–2.5).

Fig. 2.3 (continued) and Apochela. (**a**) Female *Ramazzottius* cf. *oberhaeuseri* (Parachela, Ramazzottiidae) from Nivå, Denmark (photo modified from Heidemann et al. 2016). *R.* cf. *oberhaeuseri* is a strong cryptobiont with an extreme tolerance toward a range of adverse conditions. The current parthenogenetic population lives on sediment within a roof gutter. Note the strong coloration of the animal—a characteristic of semiterrestrial species. (**b**) Female *Hypsibius exemplaris* (Parachela, Hypsibiidae), Sciento strain. *H. exemplaris* is a limnic (freshwater) tardigrade with moderate cryptobiotic abilities. It is easy to keep in laboratory cultures, and the genome of the parthenogenetic Sciento strain is presently available. Note the algae (green) within the midgut. (**c**) The large and yellow *Richtersius coronifer* (Parachela, Richtersiidae) is a semiterrestrial cryptobiont. The specimen is from a parthenogenetic population living within moss cushions found on limestone, Öland, Sweden. Note the coelomocytes, which gives this species its characteristic yellow color. (**d**) Female of *Milnesium tardigradum* (Apochela, Milnesiidae) from Nivå, Denmark. The large semiterrestrial *M. tardigradum* is carnivorous and a strong cryptobiont. bu = buccal tube, ce = coelomocyte, ey = eye, mg = midgut, ph = pharyngeal bulb with placoids, st = stylet. Scale bars = 100 μm

Fig. 2.4 Structural diversity of tardigrades—light microscopy of echiniscoidean heterotardigrades. (**a**) Female of the semiterrestrial cryptobiont *Echiniscus testudo* (Echiniscidae) from Nivå, Denmark. Note the orange color of the coelomocytes inside the animal. The current parthenogenetic population lives within moss cushions on the roof of houses. (**b**) Four developing *E. testudo* embryos within a shed cuticle also known as an exuvium. (**c**) *Cornechiniscus lobatus* (Echiniscidae) collected from moss in a wadi in Sinai, Egypt. Note the orange to red color of the coelomocytes inside the animal. (**d**) *Echiniscoides sigismundi* (Echiniscoididae) collected from barnacles on rocks

The coloration can be located in the integument, in various cells, or rather come from the contents of the midgut. The pronounced pigmentation of semiterrestrial species likely protects against UV radiation. Accordingly, carotenes with presumed protective abilities have been suggested and reported present in epidermal cells and within the body cavity (Greven et al. 2005; Bonifacio et al. 2012; Chap. 12). Encystment is often characterized by a darkening of the outer cuticle (Chap. 9).

Generally, tardigrades have highly permeable integuments (Crowe 1972). This is revealed when strong semiterrestrial anhydrobiotes during dehydration significantly reduce their body volume and enter the tun stage in the course of minutes (Fig. 2.1a, b; Halberg et al. 2013a). The high integument permeability is also revealed when active stage marine tardigrades upon immersion into hypo- or hyperosmotic solutions experience drastic changes in body osmolality and volume (Halberg et al. 2009b).

Notably, all members of Ecdysozoa undergo molting in order to grow, and tardigrades are characterized by regularly shedding their cuticle (ecdysis). During the process of molting, the cuticle of the integument as well as the cuticle lining internal organs, i.e., the sclerified parts of the buccal-pharyngeal apparatus and rectum, is shed. In addition, the stylets (used for piercing food) and stylet supports are reformed by stylet glands (Kristensen 1976). The molt also includes the cuticular part of the legs—the latter are reformed by so-called claw glands (Fig. 2.2).

The cuticle of the integument can be subdivided into three layers: from the outside toward the inside, the epicuticle, an inner trilaminate layer, and the basal cuticle (Kristensen and Neuhaus 1999; Weglarska 1989). In the current review, we follow the above mentioned terminology, but note that other authors use other terminologies (Greven 1984), including e.g., epicuticle, intracuticle, procuticle (Guidetti et al. 2000), and exocuticle, mesocuticle, and endocuticle (Dewel et al. 1993). The epicuticle and basal cuticle can be subdivided into various sublayers, and the epicuticle is often coated by a mucous or flocculent coat. Generally, the inner layers of the tardigrade cuticle seem to have a high amount of chitin (Kristensen and Neuhaus 1999).

Great variation is seen in cuticle structure, thickness, and ornamentation between but also within the different evolutionary lineages, as well as between the dorsal and ventral side of individual species. Notably, the epicuticle exhibits great variation and can be subdivided into an outer epicuticle that can consist of several sublayers and a highly variable inner epicuticle. In most heterotardigrades the inner epicuticle

Fig. 2.4 (continued) in the marine tidal zone at Roscoff, France. Echiniscoididae harbor the most dominant tardigrades in intertidal zones worldwide. Note the high number of claws on each leg—the multiple claws are a characteristic of the wave-exposed *Echiniscoides* and likely secure a better grip on the substrate preventing the tardigrades from being washed away by waves and tides. bu = buccal tube, cA = cirrus A (lateral cirrus), cE = cirrus E, ce = coelomocyte, cu = cuticle of exuvium, ec = external cirrus, em = embryon, ey = eye, mg = midgut, ic = internal cirrus, pc = primary clava, ph = pharyngeal bulb with placoids, sc = secondary clava, st = stylet. Scale bars = 100 μm

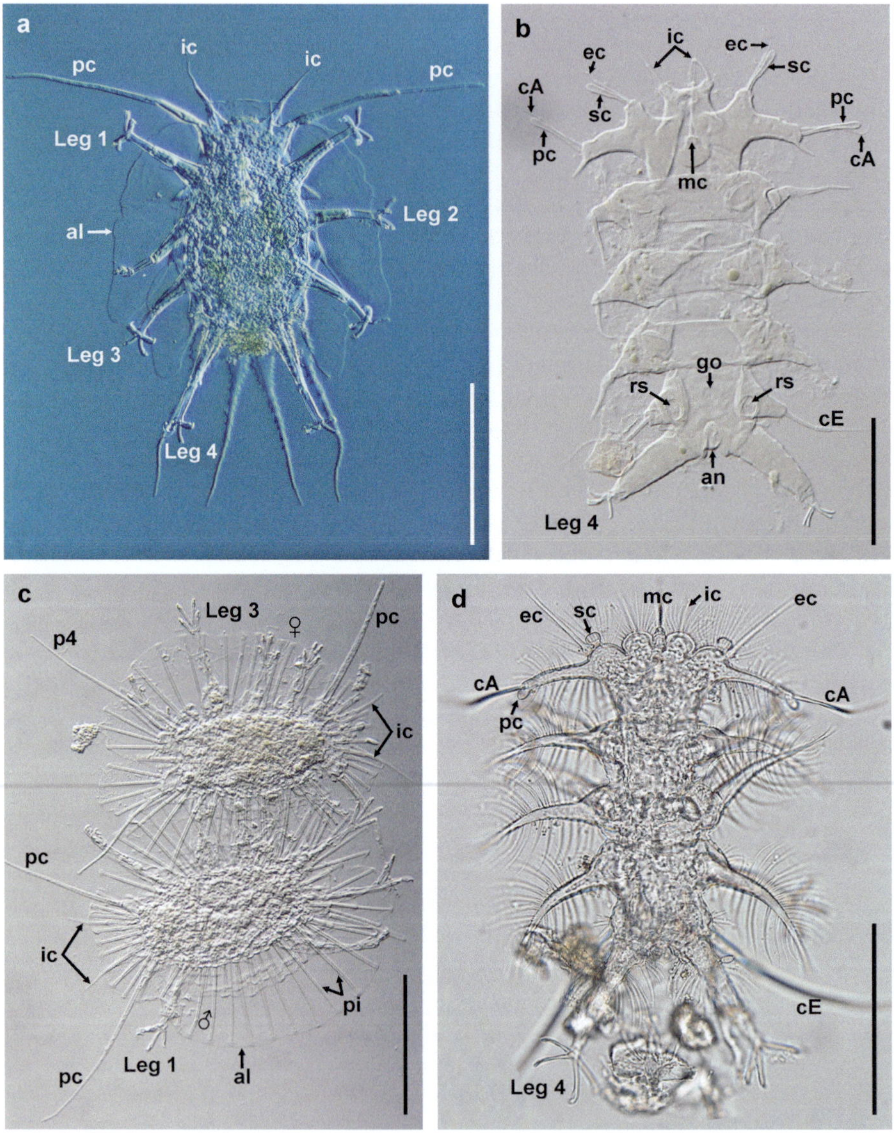

Fig. 2.5 Structural diversity of tardigrades—light microscopy of marine arthrotardigrades. (**a**) Male of *Florarctus heimi* (Halechiniscidae) sampled from coralline sand in Shark Bay, Heron Island, Australia. Note the extreme development of the alae and of the primary clavae. (**b**) Holotypic female of *Parastygarctus robustus* (Stygarctidae) from coral sand around Magdelaine Reef, Australia. Note the paired seminal receptacles and the tripartite arrangement of the head. (**c**) Female (♀) and male (♂) *Actinarctus doryphorus* sensu stricto (Tanarctidae) sampled off Roscoff, France. Note the conspicuous alae with pillars and the very large primary clavae. (**d**) Holotypic female of *Neostygarctus oceanopolis* (Neostygarctidae) sampled from the Condor Seamount in the Azores. al = alae, an = anus, go = gonopore, cA = cirrus A (lateral cirrus), cE = cirrus E, ec = external cirrus, ic = internal cirrus, mc = median cirrus, p4 = sensory organ of leg IV, pc = primary clava, pi = pillar, rs = seminal receptacle, sc = secondary clava. Scale bars = 100 μm

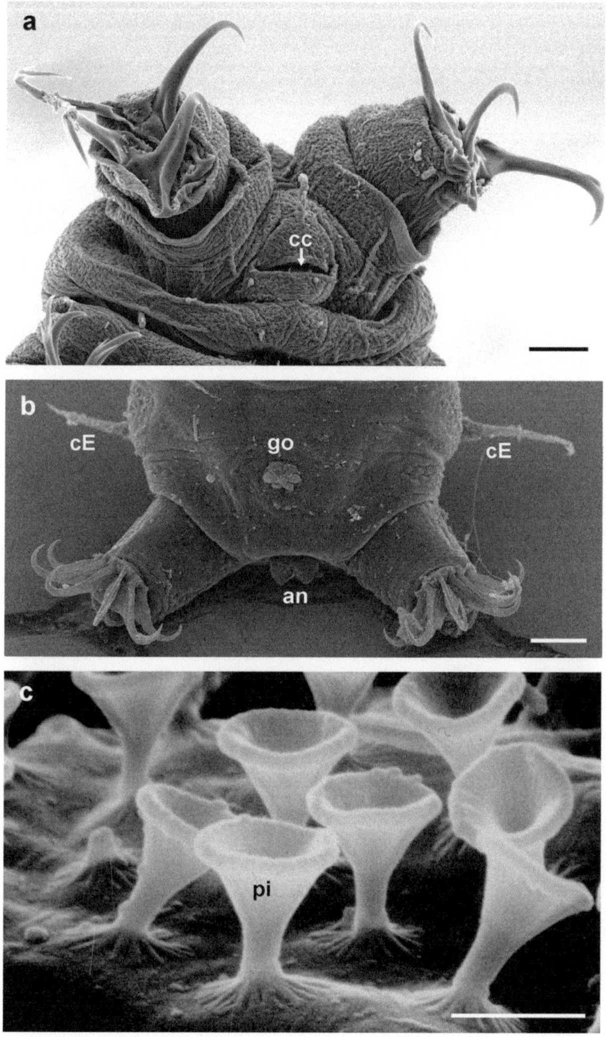

Fig. 2.6 Scanning electron microscopy revealing structural details in selected tardigrades. (**a**) Posterior region of the eutardigrade, *Halobiotis crispae* (active stage), from Saqqaq, Disko Island, Greenland, revealing the cloaca and hind legs with conspicuous double claws. (**b**) Posterior part of female specimen of the marine echiniscoidean *Isoechiniscoides sifae*. Note the conspicuous female gonopore and the six claws on the hind legs. (**c**) Pillars in the epicuticle of the marine arthrotardigrade *Batillipes mirus* from coarse sand, Florida, USA. an = anus, cc = cloaca, cE = cirrus E, go = gonopore, pi = pillar. Scale bars in (**a**) and (**b**) = 10 μm, in (**c**) = 1 μm

contains so-called pillars (Fig. 2.6c), which are covered by an outer honeycomb layer. The arthrotardigrade cuticle generally consists of—from the outside toward the inside—a membrane-like outer epicuticle, a honeycomb layer, pillars, the inner membrane-like trilaminate structure, and an amorphous basal cuticle (Kristensen and Neuhaus 1999). Within the echiniscoideans, pillars are present in most taxa but are lacking within *Echiniscoides* species (Greven and Grohe 1975; Møbjerg et al. 2016). The cuticle in eutardigrades is comparatively simple without a honeycomb pattern in the outer epicuticle and with heavily reduced or no pillars (Weglarska 1975; Guidetti et al. 2000). Nevertheless, highly complex epicuticular pillars can also be found among eutardigrades, e.g., in the freshwater species, *Murrayon dianeae*. Epicuticular pillars likely represent the plesiomorphic condition with the simpler cuticle of

Echiniscoides and most eutardigrades representing the apomorphic condition—alternatively, pillars may have evolved several times independently within the phylum (Kristensen and Neuhaus 1999).

In arthrotardigrades the cuticle may expand into conspicuous extensions, as found in species with wing- or finlike structures (alae), i.e., *Actinarctus* (Figs. 2.1d and 2.5c), *Florarctus* (Fig. 2.5a), *Raiarctus*, *Rhomboarctus*, and *Wingstrandarctus*, or arthrotardigrades with cuticular sheaths on body processes (Stygarctidae; Fig. 2.5b). Thick cuticular plates are present within the armored marine tardigrades Renaudarctidae and Stygarctidae (Hansen et al. 2012). In these arthrotardigrade taxa, the plates are formed from the basal cuticle and consist of alternating layers of electron-dense and electron-lucent amorphous material—the plates thus lack pillars. Echiniscoideans do not possess winglike cuticular extensions; however, some taxa within Echiniscidae have very thick cuticular plates, which may be ornamented and have spikes. In contrast to the arthrotardigrades, echiniscoidean plates are formed by a thickening of the epicuticle. The cuticular plates are thus analogous structures, which evolved independently in Arthrotardigrada and Echiniscoidea (Kristensen and Neuhaus 1999). Some heterotardigrades have an unornamented cuticle and a simple "bear"-like body shape, i.e., *Echiniscoides* (Fig. 2.4d), *Angursa*, *Styraconyx*, and *Coronarctus*.

Cuticle-lined sensory organs are, as outlined above, dominant structures in most heterotardigrade taxa (Figs. 2.1c, d and 2.5), whereas eutardigrades lack such appendages giving the animal a smooth and typical water "bear"-like appearance (Figs. 2.1a and 2.3). One notable exception is *Calohypsibius* Thulin, 1928, which has cuticular tubercles or spines.

2.3 Tardigrade Appendages

In the plesiomorphic condition, tardigrades have four pairs of legs and a large number of sensory organs on head and body segments, which often contain cilia (Figs. 2.2, 2.5, 2.7 and 2.8). The first three leg pairs are located ventrolaterally, pointing anteriorly, and are commonly used for locomotion, while the fourth pair is directed posteriorly and ensures retention to the substratum. In some eutardigrades the legs may, however, be reduced and the fourth pair may be lacking.

In heterotardigrades the sensory organs, cirri and clavae, may be very long and/or highly complex structures. The often long and sometimes segmented cirri (with cirrophore, scapus, and flagellum, Fig. 2.2) are traditionally believed to have both mechano- and chemoreceptive functions, whereas the shorter clavae are likely chemoreceptive (Kristensen 1981; Schulze and Persson 2016)—physiological/experimental data that can support functional assumptions are, however, lacking. Specifically, ultrastructural investigations show that clavae contain a modified cilium or cilia as well as microvilli and that they are characterized by the presence of a terminal opening in the cuticle, whereas cirri contain one or two mechanoreceptive cilia that may be strongly modified—cirri may also have a terminal pore, and

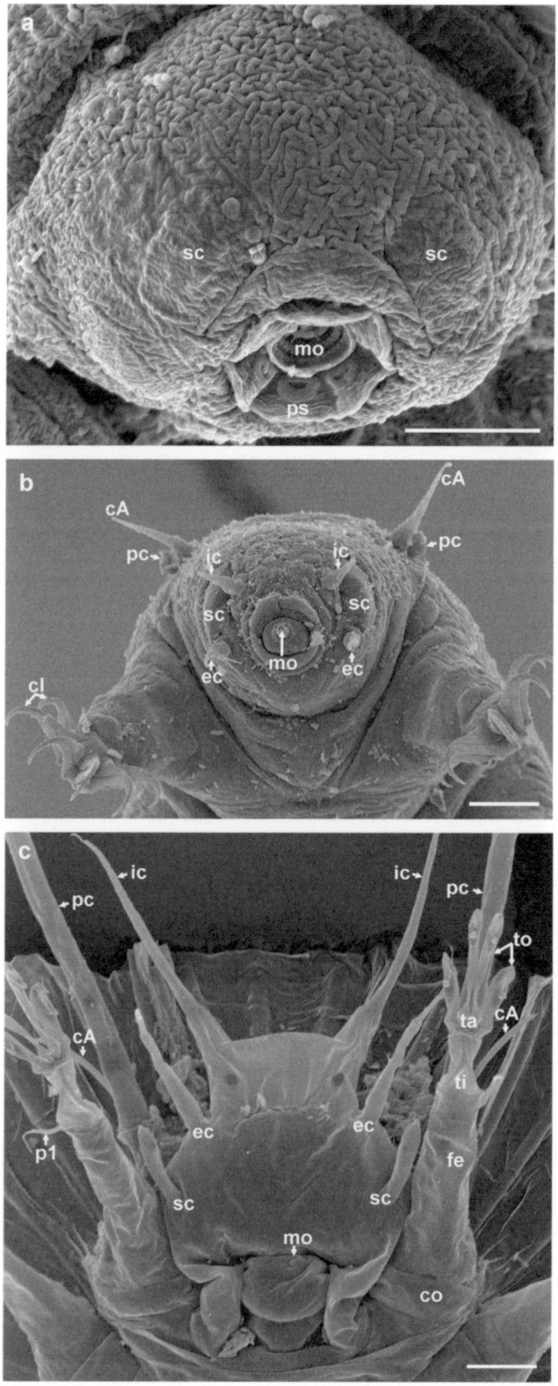

Fig. 2.7 Scanning electron microscopy of the frontal region of selected tardigrades representing the major clades. (**a**) Cephalic region of the marine eutardigrade *Halobiotus crispae* (active stage) from

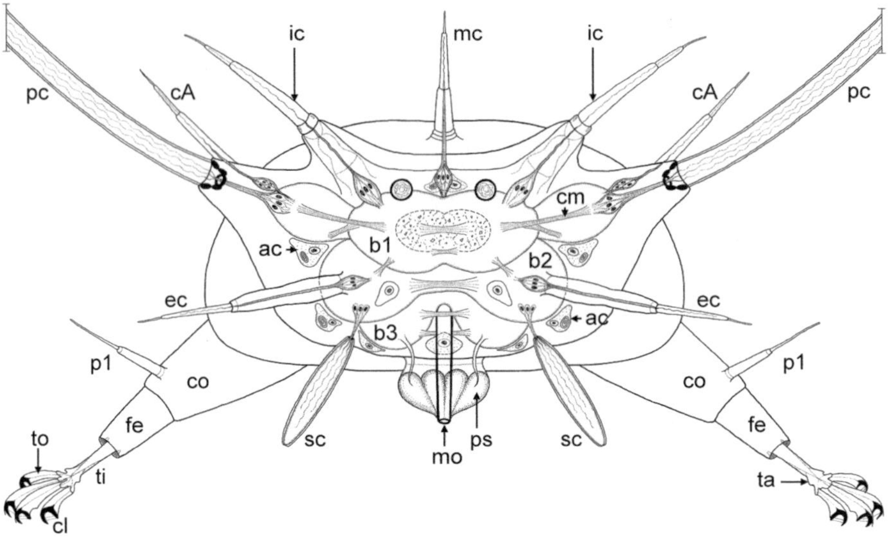

Fig. 2.8 Conceptual drawing (frontal view) of the cephalic region of the arthrotardigrade *Actinarctus doryphorus* based on light and electron microscopy. Compare to Figs. 2.1d, 2.5c, 2.7c, and 2.9d and see the text for a detailed explanation and discussion of morphological and anatomical structures. ac = amoebocyte (specialized coelomocytes found attached to the brain and subpharyngeal ganglia in *Actinarctus* and *Tanarctus*), b1 = first brain lobe (protocerebrum), b2 = second brain lobe (deutocerebrum), b3 = third brain lobe (tritocerebrum), cA = cirrus A (lateral cirrus), cl = claw, cm = commissure, co = coxa, ec = external cirrus, ic = internal cirrus, mc = median cirrus, mo = mouth, p1= sensory organ of first leg, pc = primary clava, ps = peribuccal sense organ, sc = secondary clava, ta = tarsus, ti = tibia, to = toe

may thus also be involved in chemosensing (Kristensen 1981). The sensory organs of eutardigrades are much reduced as compared to the conspicuous cirri and clavae of heterotardigrades (Figs. 2.7 and 2.8). Nevertheless, Walz (1975a, 1978, 1979) provided evidence of sensory cells with characteristics of, respectively, mechanoreceptors and chemoreceptors in the cephalic sensory fields of the eutardigrade

---

Fig. 2.7 (continued) Saqqaq, Disko Island, Greenland. In eutardigrades of the order Parachela, cephalic sensory organs are greatly reduced in external appearance and appear as sensory regions. In *H. crispae* the frontal sensory region is known as cephalic papillae (secondary clavae). (**b**) Frontal region of female specimen of the marine echiniscoidean *Isoechiniscoides sifae*. The head has two anterior pairs of cirri, i.e., internal and external cirri, as well as a more posterior pair of lateral cirri (cirri A). Paired primary and secondary clavae are present in connecting with, respectively, cirri A and the external cirri. Note the six claws on the first pair of legs. The presence of six claws on each leg is a characteristic of the genus *Isoechiniscoides*. Compare to Fig. 2.6b. (**c**) Ventral view of frontal region of the marine arthrotardigrade *Actinarctus* cf. *doryphorus* with highly developed cephalic sensory appendages. Compare to Fig. 2.1d. Note the conspicuous primary clavae and the first pair of legs. cA = cirrus A (lateral cirrus), cl = claw, co = coxa, ec = external cirrus, fe = femur, ic = internal cirrus, mo = mouth, p1= sensory organ of first leg, pc = primary clava, ps = peribuccal sense organ, sc = secondary clava, ta = tarsus, ti = tibia, to = toe. Scale bars = 10 μm

Macrobiotus hufelandi (reviewed in Dewel et al. 1993). Sensory organs are innervated by the nervous system—simple pigment cup eyes located in the first outer lobe of the brain are present in many eutardigrades and in echiniscoideans (Figs. 2.2, 2.3b–d, and 2.4a, c, d; Greven 2007). The sensory structures within the pigment cups comprise microvilli as well as a single cilium (Kristensen 1982).

2.3.1 The Tardigrade Leg

A light microscopic comparative investigation of tardigrade legs may be confusing, and the number of legs (i.e., four pairs) may initially appear as the only commonality between families. The legs of arthrotardigrades are clearly telescopic/retractable (Figs. 2.5c and 2.7c), whereas similar retractability is less evident in echiniscoideans and eutardigrades.

Van der Land (1968) introduced an arthropod-derived terminology to describe tardigrade leg structure, with up to four distinguishable divisions—in proximal to distal direction—coxa, femur, tibia, and tarsus (Figs. 2.2 and 2.7c). This terminology has been retained in newer literature, and it is currently used within descriptions of new arthrotardigrade species (Hansen et al. 2012; Jørgensen et al. 2014). In arthrotardigrades, the tarsus may either bear claws of highly variable type and number or cuticular "toes" (digits) that terminate in claws and/or adhesive disks. An electron microscopic investigation by Kristensen (1976) suggested that the proximal part of arthrotardigrade legs is fundamentally built after the same pattern and that the major variation between taxa is in the distal cuticular parts, formed in the claw glands following a molt (Fig. 2.9a, b). The proximal part of the leg (coxa and femur) contains epidermal cells, nerves, muscles, a sensory seta, and the claw gland (Figs. 2.2, 2.7c, and 2.9a). During each molt the distal noncellular parts of the extremities, i.e., tibia, tarsus, toes, and claws (or adhesive disks), are reformed by these glands, in the same manner as the body cuticle is formed by epidermal cells of the integument. Thus, the claw gland epithelium likely represents a derivation of the epidermis, and these glands are likely homologous to the stylet glands situated alongside the pharyngeal apparatus (see below and Fig. 2.9c). Specifically, the syntheses of claws and stylets by these glands seem to occur in a similar way (Kristensen 1976). Notably, in arthrotardigrades the tibia, tarsus, toes, and claws (or adhesive disks) are cuticular specializations, which are shed during a molt—these structures are thus analogous to regions of the arthropod appendage of the same name. Furthermore, these specializations (with exception of the claws) are only found in arthrotardigrades and may be considered as an adaptation to the interstitial way of life. In eutardigrades and echiniscoideans, the molting process is relatively simple—only the claws are reformed. Electron and confocal laser scanning microscopic investigations have verified that tardigrade legs are innervated by leg ganglia, which connect to four ventral trunk ganglia. Leg muscles seem to originate from the dorsal, lateral, and ventral musculature (see below).

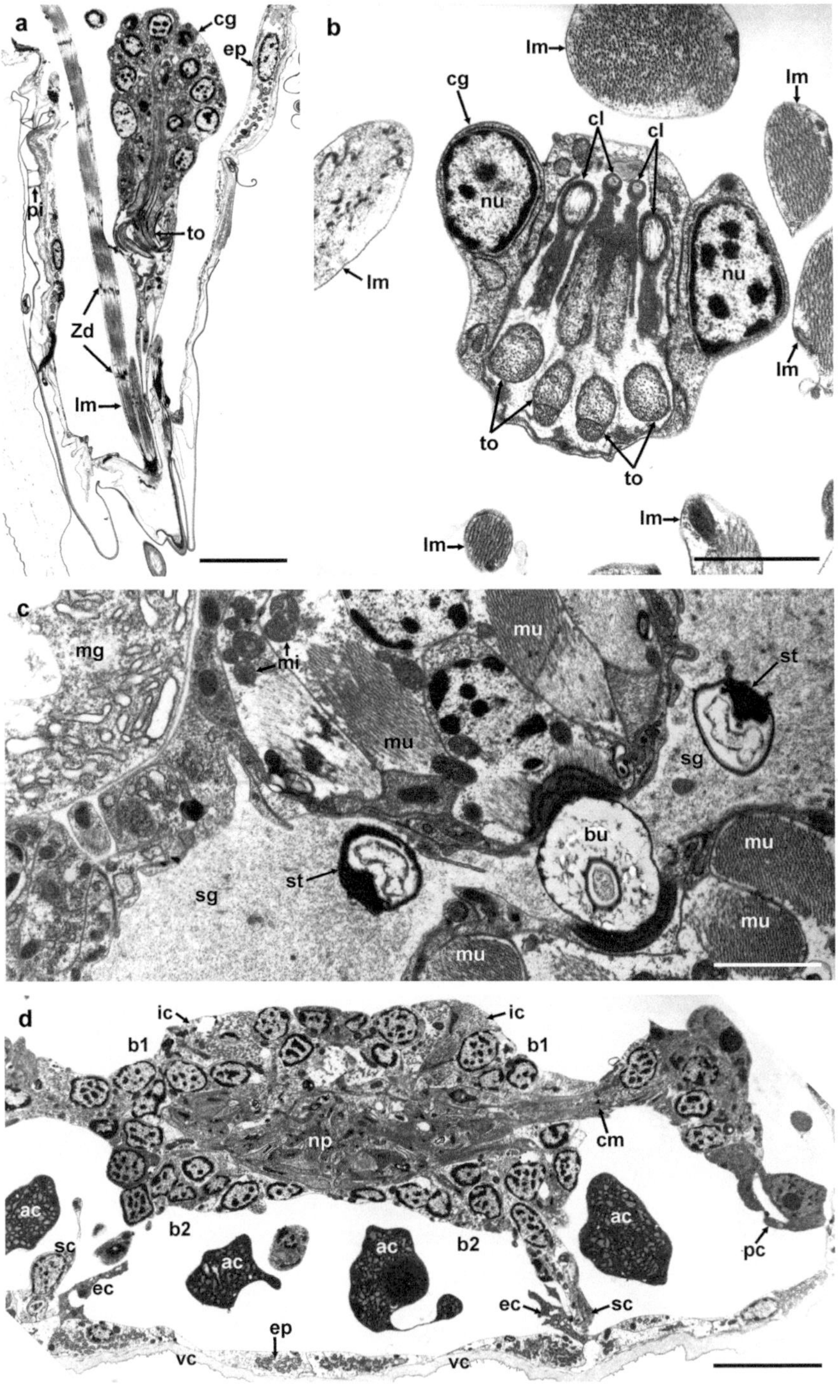

Fig. 2.9 Transmission electron microscopy revealing ultrastructural details of the arthrotardigrade *Actinarctus doryphorus* sensu stricto collected off Hirsholmene, Denmark. Compare to Figs. 2.1d,

2.3.2 Overview of Tardigrade Claw Morphology

In most tardigrade taxa, each leg terminates in cuticular claws inserted directly on the leg or, alternatively, on digits (Fig. 2.2). Generally, eutardigrades have 4, and echiniscoideans 4–13, claws inserted directly on each leg (Figs. 2.3, 2.4, and 2.6a, b). Notably, the multiple claws of the wave-exposed *Echiniscoides* species probably secure a better grip on the substrate and prevent the animals from being washed away by wave action (Fig. 2.4d). Eutardigrades generally have two double claws on each leg (Fig. 2.6a). In apochelans, the primary and secondary branches of the double claws attach directly to the leg, whereas, in parachelans, the primary and secondary branches fuse before the basal tract attaches to the leg. Generally, the parachelan claw is composed of accessory spines on the primary branch, a basal tract, and a basal lunule. The primary and secondary branches are fused at the distal end of the shaft. The edge of the lunule might be smooth or dentate. Claw morphology is widely used within eutardigrade taxonomy, and major parachelan claw types can be recognized, i.e., macrobiotid, hypsibid, isohypsibid, and eohypsibiid, along with subcategories within these (Marley et al. 2011). In some eutardigrades the number or size of the legs has been reduced (i.e., in *Apodibius* and *Hexapodibius*). Notably, *Apodibius* Dastych, 1983, is clawless.

Many arthrotardigrades also have four claws inserted directly on the leg. However, the arthrotardigrade families Halechiniscidae, Tanarctidae, and Renaudarctidae, and the genera, *Neostygarctus* and *Prostygarctus*, have legs with four digits each bearing a single claw (Figs. 2.7c and 2.8). Arthrotardigrade claws are very variable with regard to shape. Simple crescent-shaped claws are the most common; however, in the family Styraconyxidae, each claw is composed of three or four hooks that are partly retractable into a claw sheath (Kristensen and Higgins 1984a). In Renaudarctidae and Stygarctidae, long accessory spines or terminal filaments are present. Minute dorsal hooks on the claws are often present—

Fig. 2.9 (continued) 2.5c, 2.7c, 2.8, and 2.10b. (**a**) Longitudinal section through the femur of the second leg revealing the cross-striated pattern of the leg musculature. Note the claw gland, which is in the process of forming new toes and claws. Scale bar = 5 μm. (**b**) Cross section of the claw gland within the third leg from the same specimen as in (**a**). Note the cuticular toes, internal and external claws, which are being formed within the gland. Scale bar = 0.5 μm. (**c**) Cross section of the stylet glands, which are in the final process of forming stylets and new cuticular parts of the buccal tube (anterior part). Scale bar = 1 μm. (**d**) Cross section through the anterior part of the brain, revealing the conspicuous central neuropile of the first brain lobes, the region of overlap between the first and second lobes, and the connection between the brain and cephalic sense organs, i.e., primary and secondary clavae and external and internal cirri (modified from Persson et al. 2014). Note the commissure, which extends from the lateral nerve cluster (ganglion of the primary clava and cirrus A) toward the center of the first lobe. Scale bar = 5 μm. ac = amoebocyte, b1 = first brain lobe (protocerebrum), b2 = second brain lobe (deutocerebrum), bu = buccal tube, cg = claw gland, cl = claw, cm = commissure, ep = epidermis, lm = leg muscle, mi = mitochondria, mg = midgut, mu = muscle, nu = nucleus, np = neuropil, pc = primary clava, pi = pillar, sg = stylet gland, st = stylet, to = toe, vc = ventral cuticle, Zd = Z-disk sometimes these are only visible using scanning electron microscopy. An internal hook-like structure, the calcar, is often present and is the part of the claw where the cuticular fibers fasten, so the claw can be withdrawn inside the digit or claw sheath. In adults of the arthrotardigrade family Batillipedidae, six digits each terminate in a disk or shovel, the so-called suction disk (*Batillipes*, shovel feet). In the subfamily Orzeliscinae (Halechiniscidae), the species have four digits that end in adhesive pads with or without claws. Arthrotardigrade disks and pads likely take advantage of capillary adhesion, electrostatic or van der Waals attraction forces. As noted above, the legs of heterotardigrades often bear a sense organ (Figs. 2.2, 2.7c, and 2.8).

2.3.3 Tardigrade Sensory Appendages

The cephalic region of heterotardigrades is characterized by numerous well-developed, serially organized cirri and clavae (Figs. 2.2, 2.4, 2.5, 2.7b, c, and 2.8). The two most anterior pairs of head cirri are referred to as internal and external cirri, while the generally more posterior pair is called lateral cirri or cirri A (Figs. 2.2, 2.7c, and 2.8). In addition, a single unpaired median cirrus may be present as well as primary, secondary, and tertiary clavae. Thus, extant tardigrades (e.g., the arthrotardigrade *Neoarctus primigenius*) have up to 13 cephalic appendages comprising primary, secondary, and tertiary clavae and internal, external, and lateral cirri, in addition to the single unpaired median cirrus. In most arthrotardigrades the tertiary clavae are missing, and in echiniscoideans external evidence of both tertiary clavae and the median cirrus are lacking (Fig. 2.7b). In eutardigrades cephalic sensory organs are greatly reduced in external appearance, and within the variety of species of the eutardigrade order Parachela, they now appear as sensory regions composed of papillae and sensory fields (Fig. 2.7a). Members of the eutardigrade order Apochela have distinct oral and lateral cephalic papillae (Wiederhöft and Greven 1999). In eutardigrades, internal evidence of a reduced median cirrus may be present in the form of retained innervation to this area (Zantke et al. 2008; Person et al. 2012; Schulze and Persson 2016). Several studies have shown the presence of cilia within cirri, clava, and sensory fields/papilla, and it would thus seem that modified cilia are a characteristic of tardigrade sensory organs (Walz 1975a; Kristensen 1981; Wiederhöft and Greven 1999; Biserova and Kuznetsova 2012).

In heterotardigrades each leg normally bears a sense organ, and trunk segments can have paired lateral cirri, named cirri B, C, D, and E, respectively, with the latter referring to the two often enlarged cirri of the most posterior segment (Figs. 2.2, 2.5b–d, and 2.6b). Trunk cirri B, C, and D are rare in arthrotardigrades except in *Coronarctus*. Eutardigrades lack these trunk sensory appendages giving the animal a smooth and typical water "bear"-like appearance (Fig. 2.1a). The leg sense organs of many heterotardigrades may have the shape of spines and bulbs or be greatly modified structures, such as those located on the fourth pair of legs in many tanarctids (Jørgensen and Kristensen 2001). Additionally, heterotardigrades may have either paired or unpaired dorsal cirri located at the center of the trunk segments

(Fujimoto and Miyazaki 2013; Kristensen et al. 2015). In summary, tardigrade trunk segments may have up to five sensory appendages as seen in the caudal segment of *Neostygarctus lovedeluxe*. These include a leg sense organ on each leg, a dorsolateral cirrus on each side of the trunk, and a central dorsal cirrus.

2.3.4 Evolutionary Origin of the Tardigrade Head

The evolutionary origin of the tardigrade head has in recent literature been highly debated, and the status of tardigrade segmentation remains controversial. Tardigrades are, according to the traditional view, composed of seven to eight segments with the head contributing three or four segments and the trunk being formed by four segments (reviewed in Kristensen 1987; Dewel and Dewel 1996, 1997). Dewel and Dewel (1997) further suggest that the gonad and its ducts could represent a limbless vestigial segment, in which case the trunk would be composed of five segments. Recent investigations seem to agree that the trunk is composed of four segments; however, researchers working with Hox gene expression and immunostainings of the nervous system in eutardigrades conclude that the tardigrade head is composed of a single segment (Mayer et al. 2013a; Gross and Mayer 2015; Smith et al. 2016), whereas other (Persson et al. 2012, 2014) comparative neuroanatomical investigations support the view that the head represents at least three segments.

As highlighted above, tardigrades are diverse and eutardigrades are, among other traits, characterized by much reduced sensory structures, whereas the marine arthrotardigrades carry a large number of sensory appendages. The greatly reduced condition found in eutardigrades therefore calls for extra caution when drawing general conclusions at the phylum level—preferably the latter should only be done after consulting heterotardigrades and if possible representatives of the marine arthrotardigrades.

It should be noted that the paired cirri and clavae of arthrotardigrades may have a common base (a cirrophore). Interestingly, in some arthrotardigrades (e.g., *Neoarctus*, *Actinarctus*, *Neostygarctus*), the posteriorly situated primary clava and lateral cirrus (cA) clearly bear the outer resemblance of a modified leg with its leg sense organ (Fig. 2.7c). Arthrotardigrades may also have secondary and tertiary clavae with associated external and internal cirri, underlining the complexity of the tardigrade head and supporting the hypothesis that the head is composed of at least three segments. Interestingly, two pairs of stylet glands (likely modified claw glands) are present in the buccal-pharyngeal area of some eutardigrades (Hansen et al. 2017), which indicates the presence of at least two segments and stresses the fact that comparative morphology/anatomy does not support the assumption of a one-segmented tardigrade head.

2.4 Body Cavity

Like other small animals, tardigrades lack circulatory and respiratory systems. The integument delineates a fluid filled body cavity, which bathes the internal organs and extends into the legs (Fig. 2.2). The fluid within this cavity assists in bulk movement of gas, nutrients, water, and electrolytes, the exchange of which otherwise occurs by diffusion alone or, as regards solutes, through facilitated or active transport by membrane proteins (Halberg and Møbjerg 2012). Tardigrades hyper-regulate (Halberg et al. 2009b, 2013b; Møbjerg et al. 2011), which ensure a constant influx of water into the body cavity with excess water likely being expelled by osmoregulatory epithelia. Somatic muscles attach to the cuticle (via the epidermis), and locomotor activity relies on the antagonistic hydrostatic pressure of the body cavity fluid, generated by this regulated hyperosmolarity.

Body cavity volume is generally larger in eutardigrades than in heterotardigrades as well as in well-nourished as compared to starved individuals (Dewel et al. 1993). Notably, the cavity fluid contains body cavity cells, also known as coelomocytes or storage cells (Figs. 2.2, 2.3c, d, and 2.4c). In eutardigrades and echiniscoideans, coelomocytes move passively around in the fluid or attach to the basement membranes (extracellular matrix) of the epidermis and internal organs, whereas they always seem to be attached in arthrotardigrades (Marcus 1929; Dewel et al. 1993; Kinchin 1994). In eutardigrades of the genus *Ramazzottius*, they are found in numbers of approximately 200 (Kinchin 1994). Body cavity cells are thought to function as storage cells, as they accumulate polysaccharides, proteins, and lipids (Hyra et al. 2016a). Their numbers seem to decline during starvation and thus vary with the nutritional state of the animal (Weglarska 1975). Accordingly, it has been shown that mitosis occurs in storage cells (as well as in other cell types; Bertolani 1970a, b; Czerneková and Jönsson 2016; Hyra et al. 2016b), which strongly questions whether tardigrades are truly eutelic as sometimes reported. Interestingly, storage cells may additionally play a defensive role, phagocytizing bacteria (Greven 1980).

2.5 Digestive System

At the phylum level, tardigrades appear omnivorous with the food source being either bacteria or of plant, fungal, animal, or other eucaryote origin. Yet, little is known of general tardigrade biology, and we thus lack in depth knowledge of what most tardigrades eat. When investigating tardigrade microanatomy, however, it becomes apparent that tardigrades possess a complex feeding apparatus and associated alimentary canal (Figs. 2.2, 2.3, and 2.4a, d).

The tardigrade digestive system can be divided into a foregut, a midgut, and a hindgut (Dewel and Clark 1973). Cuticle extends into and covers foregut and hindgut—the latter are separated by the large midgut. Major differences between evolutionary lineages are represented by the presence of Malpighian tubules and a cloaca in eutardigrades, whereas heterotardigrades lack these organs and have a separate anus and gonopore (Fig. 2.6a, b).

The tardigrade foregut includes the mouth, a complex buccal-pharyngeal apparatus, and an esophagus (Fig. 2.2). Moreover, stylets and associated stylet glands represent foregut-associated structures. The mouth is located terminally or ventrally and is surrounded by sense organs in eutardigrades (Fig. 2.7a). These organs are referred to as peribuccal or circumoral and infrabuccal structures, e.g., peribuccal lamellae, ciliary organs, papillae, mucrones, and infrabuccal ridges (Dewel et al. 1993).

Notably, the mouth ring is telescopic. The mouth opens into a buccal tube that can be rigid or flexible. A flexible buccal tube and esophagus is likely an advantage for anhydrobiotic species, as this allows for a better packing of internal structures during tun formation (Halberg et al. 2013a). The paired stylets used for piercing food items enter the anterior part of the buccal tube and can be protruded through the mouth opening. Cuticular extensions of the buccal tube, the stylet sheaths, may surround the anterior part of the stylets guiding their entrance into the tube, whereas stylet supports, when present, anchor the posterior part of the stylets to the tube (Dewel and Clark 1973; Eibye-Jacobsen 2001a). Stylet protractor and retractor muscles attach to the tube and to the muscular pharynx (Dewel and Clark 1973).

The buccal tube opens into the muscular pharynx (Figs. 2.2, 2.3, and 2.4a, d), also known as the pharyngeal bulb, composed of myoepithelial elements. The bulb provides a suction action responsible for ingestion of food, such as contents from organisms and cells pierced by the stylets or even whole animals, e.g., rotifers, small nematodes, or other tardigrades (Roszkowska et al. 2016). The food source apparently depends on and varies between species and their specific microhabitats as well as between different developmental stages of a single species. The luminal side of the pharyngeal bulb is characterized by cuticular thickenings with conspicuous supportive structures, so-called placoids. These placoids serve as attachment sites for pharyngeal muscles in echiniscoideans and eutardigrades of the order Parachela and as a rigid support of the pharyngeal lumen in Arthrotardigrada (Eibye-Jacobsen 2001b). Placoids are missing in eutardigrades of the order Apochela (and reduced or absent in species of the parachelan genus *Itaquascon*). In Parachela, placoids are composed of a string of macro- and microplacoids, and in the Heterotardigrada they are fused encrusted bars—the placoids might not be homologous structures throughout Tardigrada (Eibye-Jacobsen 2001b). A comparative examination of the tardigrade pharyngeal apparatus is provided by Guidetti et al. (2013).

Stylet glands (Fig. 2.9c), also known as "salivary glands," are present in the buccal-pharyngeal area—there is no direct evidence for the assumption that these structures produce digestive secretions. These glands are responsible for production of new stylets and stylet supports during the molting process and are likely homologous with the claw glands that are responsible for reforming the cuticular parts of the legs following a molt (Fig. 2.9a–c; Kristensen 1976). Accordingly, the stylet glands are indicated as transformed claw glands and the stylets and the stylet supports as modified claws implicating that the stylet apparatus is a derivate of a pair of internalized legs (Halberg et al. 2009a; Nielsen 2001). As holds for the formation of claws, by the claw glands, the first product synthetized in the stylet glands is a thin cuticular membrane. Late in the molt, $CaCO_3$ incrustations supposedly stabilize the stylets and the stylet supports (Marcus 1929). Notably, two pairs of stylet glands are found in the rather aberrant semiterrestrial eutardigrades *Itaquascon*, *Bertolanius*, and *Eohypsibius* (Hansen et al. 2017). The pharyngeal bulb opens into the cuticle-lined esophagus composed of small cuboidal epithelial cells. In eutardigrades, the esophagus produces a peritrophic membrane—a cuticular film that extends into the midgut (Dewel et al. 1993). In heterotardigrades the cuticle of the very short esophagus does not extend into the midgut, which accordingly is without a peritrophic membrane.

The esophagus leads to a large midgut (Figs. 2.2, 2.3b–d, and 2.4d) lined by a single layer of large epithelial cells (Dewel et al. 1993). The midgut likely represents the key site for nutrient uptake, overall solute and water exchange with the body cavity (and its cellular elements), and it may serve excretory functions (Halberg and Møbjerg 2012; Hyra et al. 2016b). In heterotardigrades the midgut has pronounced lateral diverticula that sometimes extend into the legs, whereas the midgut is a straight tube-like structure with conspicuous apical microvilli in eutardigrades.

In eutardigrades, three Malpighian tubules empty into the transition zone between the midgut and the cuticular lined hindgut (Greven 1982; Weglarska 1987; Møbjerg and Dahl 1996; Pelzer et al. 2007). Each of the tubules can be divided into an initial segment, a distal and a proximal part—with the latter providing the opening into the gut (Møbjerg and Dahl 1996). Based on the position and on ultrastructure, these organs have generally been assumed to be involved in urine production with isoosmotic primary urine being produced by secretion across the basal labyrinth of the large initial segment cells. The urine would then be transported and perhaps modified in the distal and proximal parts of the organs before being passed into the hindgut for further modification through selective reabsorption of solutes and perhaps water (Dewel and Dewel 1979; Greven 1982; Halberg et al. 2009b). Although experimental data have provided evidence of a high water turnover and production of dilute urine in eutardigrades, evidence directly relating the Malpighian tubules to an osmoregulatory role is still lacking (Halberg et al. 2009b, 2013b; Halberg and Møbjerg 2012).

In eutardigrades, the hindgut ends in a cloaca, which receives the contents from the midgut, the Malpighian tubules, as well as gametes from the reproductive tract (Fig. 2.6a). Conversely, in heterotardigrades the hindgut ends in an anus, which is closed with a cuticular valve (Fig. 2.6b).

2.6 Muscle Systems

Tardigrades possess a highly complex musculature that can be subdivided into a number of somatic and visceral components (Fig. 2.10a, b). Marcus (1929) made incredibly precise descriptions of the musculature based entirely on light microscopy. The first ultrastructural investigations were performed in the 1970s and provided insights into, e.g., the contractile elements and the structure of muscle attachment sites (Dewel and Clark 1973; Walz 1973, 1974, 1975b; Shaw 1974; Greven and Grohé 1975; Kristensen 1978; Dewel and Dewel 1979). The somatic muscles of tardigrades are composed of structurally independent muscle fibers containing sarcomeres. In arthrotardigrades the sarcomeres give the musculature a clear cross-striated pattern, whereas the pattern has been described as intermediate between smooth and obliquely striated in eutardigrades (Walz 1974; Kristensen 1978; Dewel et al. 1993). Detailed ultrastructural investigations of muscles in echiniscoideans are still lacking. In arthrotardigrades, Z-disks providing anchorage for actin, actin-rich I-bands, and myosin- and actin-containing A-bands are clearly visible, whereas H-zones are less conspicuous (Figs. 2.9a and 2.10b; Kristensen 1976; Dewel et al. 1993). In addition, the sarcoplasmic reticulum and the T-system appear well developed in arthrotardigrades (Kristensen 1976). The musculature of the pharyngeal bulb, stylets, and legs seems to be cross-striated in both heterotardigrades and eutardigrades (Kristensen 1978; Eibye-Jacobsen 1996; Halberg et al. 2009a). More recently, an approach using fluorescent-coupled phalloidin, an F-actin-binding toxin, and CLSM imaging has been used to assess the 3D myoanatomy of tardigrades (Fig. 2.10a, b; Schmidt-Rhaesa and Kulessa 2007; Halberg et al. 2009a, 2013a; Schulze and Schmidt-Rhaesa 2011; Marchioro et al. 2013; Smith and Jockusch 2014). In the following, we give an account of somatic and visceral components of the tardigrade musculature and discuss the functional significance of the muscle groups.

2.6.1 *Somatic Musculature, Locomotion, and Tun Formation*

The tardigrade somatic musculature can be subdivided into several distinct muscle groups, i.e., dorsal, lateral (including ventrolateral), ventral, dorsoventral (or transverse), and leg muscles (Schmidt-Rhaesa and Kulessa 2007; Halberg et al. 2009a, 2013a; Schulze and Schmidt-Rhaesa 2011; Marchioro et al. 2013). The muscles traverse the tardigrade body attaching to the cuticle via epithelial cells of the integument (Shaw 1974). The dorsal and lateral muscle groups are situated beneath the epidermis; the ventral muscles are located dorsal to the ventral nerve cord, whereas leg muscles originate from dorsal, lateral, as well as ventral attachment sites. The dorsoventral muscles cross the tardigrade body connecting the dorsal and ventral side. While the leg musculature provides the means for locomotion and anchorage to the substrate, the remaining somatic muscle groups enable delicate flexing and bending movements (species with thick cuticular plates are generally less flexible) involved in, e.g., foraging and mating behavior. The somatic muscles furthermore provide the basis for body contraction involved in cryptobiotic tun formation—tardigrades generally respond to unfavorable conditions by contracting. The latter response is well known to most tardigradologists as specimens tend to contract during fixation, making ultrastructural investigations especially difficult

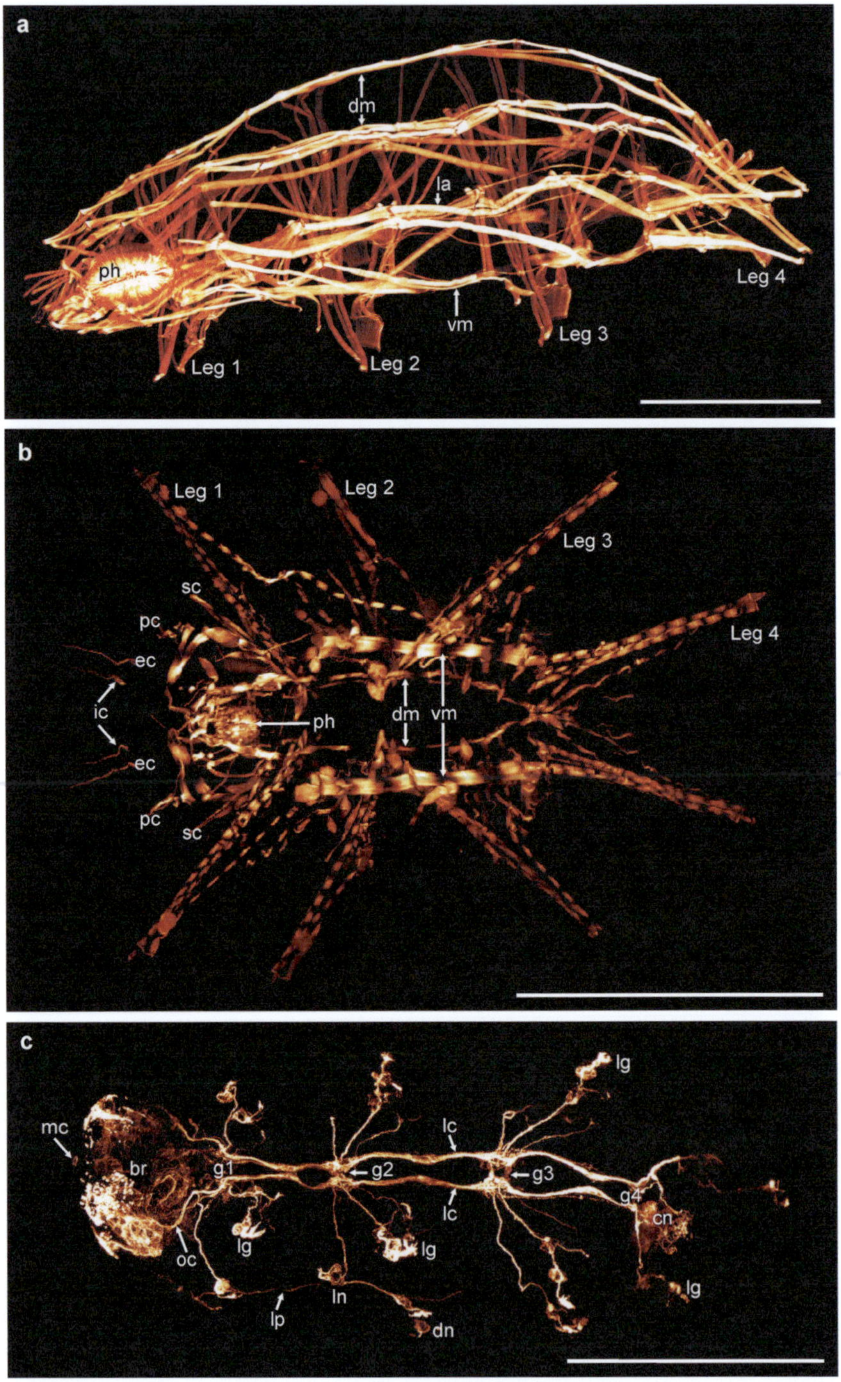

Fig. 2.10 Tardigrade anatomy as visualized by (immuno) cytochemistry and confocal laser scanning microscopy of paraformaldehyde fixed whole mounts. (**a**) and (**b**) Fluorescent-coupled

(e.g., in regard to the musculature, individual sarcomere zones and bands are often hard to recognize).

Attachment sites, anchoring muscle fibers to the cuticle via epithelial cells of the integument, are present all over the tardigrade body, including the head and legs, and can often be recognized as intrusions on the cuticle in SEM preparations. Most prominent are perhaps the seven paired, but partly or entirely fused ventromedian attachment sites, which are clearly visible on the external surface of the cuticle as seven depressions along the midline of the body (Fig. 2.1a). The ventromedian attachment sites exhibit an alternating pattern in which every other site connects to similar muscles within the specific segments. The arrangement of attachment sites and their associated fibers has been highlighted as giving tardigrade myoanatomy a metameric pattern (Marchioro et al. 2013; Smith and Jockush 2014). Notably, attachment sites are also found in connection with internal organs, such as the pharyngeal bulb.

Although the overall architecture of the somatic musculature is similar within Tardigrada, myoanatomy appears more complex in eutardigrades than in heterotardigrades, and within the heterotardigrades, echiniscoideans seem to have fewer and thicker fibers (Marchioro et al. 2013). Generally, the dorsal musculature is composed of two parallel strands of longitudinal muscles on each side of the animal—the inner and outer strands—that span the length of the body attaching to the dorsal cuticle via numerous attachment sites. Depending on the species, the two strands may, however, have different lengths and can be composed of a different number of fibers, and they may or may not interconnect at selected attachment sites (Schmidt-Rhaesa and Kulessa 2007; Halberg et al. 2009a, 2013a; Schulze and Schmidt-Rhaesa 2011; Marchioro et al. 2013). The lateral musculature is also characterized by two parallel bands of paired muscles that span the length of the body, attaching to the cuticle through lateral attachment sites (Halberg et al. 2009a; Schmidt-Rhaesa and Kulessa 2007; Schulze and Schmidt-Rhaesa 2011; Marchioro et al. 2013). However, the echiniscoidean, *Echiniscus testudo*, has a single thick

Fig. 2.10 (continued) phalloidin (F-actin) staining and CLSM imaging, maximum projection. (**a**) Lateral view of the musculature of the eutardigrade *Halobiotus crispae* (modified from Halberg et al. 2009a). The eutardigrade somatic musculature has been described as an intermediate between smooth and obliquely striated. (**b**) Ventral view of the musculature of the arthrotardigrade *Actinarctus doryphorus* sensu stricto. Note the clear cross-striated pattern of the somatic musculature (e.g., leg and ventral longitudinal musculature) as well as the anterior actin staining within the cephalic sense organs. (**c**) Ventral view of the nervous system in *Halobiotus crispae* as visualized by immunoreactivity against antiacetylated α-tubulin in combination with CLSM imaging, maximum projection (modified from Persson et al. 2012). The staining provides an overview of the CNS revealing the brain, longitudinal nerve cords and ventral ganglia as well as parts of the PNS. br = brain, cn = cloacal neurons, dm = dorsal longitudinal musculature, dn = dorsal neuron, ln = lateral neuron, ec = external cirrus, g1–4 = ventral ganglion 1–4, ic = internal cirrus, la = lateral longitudinal musculature, lc = longitudinal nerve cord, lg = leg ganglion, lp = lateral projection, mc = nerve corresponding to nerve of the median cirrus found in heterotardigrades, oc = outer connective, pc = primary clava, ph = pharyngeal bulb, sc = secondary clava, vm = ventral longitudinal musculature. Scale bars = 100 μm muscle with a characteristic zig-zag arrangement (Marchioro et al. 2013). In eutardigrades, the two lateral muscles are composed of four and six fibers, respectively, that cross one another (Marchioro et al. 2013). The ventral musculature seems highly conserved across tardigrades with two parallel longitudinal muscles on each side of the animal spanning the anterior-posterior axis (Halberg et al. 2009a, 2013a; Schmidt-Rhaesa and Kulessa 2007; Schulze and Schmidt-Rhaesa 2011; Marchioro et al. 2013). The two muscle strands connect to the ventral cuticle via ventrolateral attachment sites, and they may, depending on species, be partially or almost entirely fused.

The three muscle groups described above have been interpreted as functionally independent structures and the dorsal subsystem regarded as the most isolated of the three groups (Marchioro et al. 2013). Indeed, the three muscle groups only connect at the most anteroventral and posterodorsal attachment points, as well as through the legs (see below). In addition, dorsoventral muscles have been described in several species (Halberg et al. 2009a; Schmidt-Rhaesa and Kulessa 2007; Schulze and Schmidt-Rhaesa 2011). These muscles originate from either a ventromedian or ventrolateral attachment point and extend toward the dorsal or dorsolateral region of the body (Halberg et al. 2009a; Schulze and Schmidt-Rhaesa 2011). The leg musculature is composed of fibers that originate from dorsal, lateral, as well as the prominent ventromedian attachment sites. The number of muscles involved in controlling movement of the most anterior three leg pairs appears relatively similar, whereas the posterior leg pair is characterized by significantly fewer muscle fibers (Marchioro et al. 2013), reflecting that the fourth leg pair is used for retention to the substratum rather than locomotion. For example, in the eutardigrade *Halobiotus crispae*, the most anterior three legs of each side have 13, 10, and 9 muscles, respectively, while each of the most posterior legs only has 3 (Halberg et al. 2009a).

While the different somatic muscle groups and their associated cuticular attachment sites provide the underlying machinery for tardigrade locomotion, controlled movement of the body and paired appendages also relies on a regulated body volume and osmolyte concentration. Specifically, coordinated movement in tardigrades seems to rely on a balance between the osmotic pressure of the body cavity and the force generated by the somatic muscle tissue. Tardigrades are rendered immobile if they are subjected to and inflated by hypoosmotic shock. Specifically, it is well known—within the meiofauna literature—that marine tardigrades can be retrieved from their sediment through freshwater shocking, i.e., by applying freshwater to the marine sediment. Due to the high integument permeability, this technique will make marine tardigrades swell almost instantly, thereby releasing them from the sediment, which they are no longer capable of holding on to. The sudden influx of water during hypoosmotic shock likely renders the tardigrades immobile due to an excessive pressure on the cuticle, which no longer has the flexibility to bend. In addition, dilution of tardigrade body fluids, and thereby ions involved in electrical signals, likely poses a problem for locomotion through direct interference with the function of excitable muscle and nervous tissues. Notably, while incapacitated for shorter or longer periods in fresh—or even ultrapure water—cryptobiotic marine tardigrades (*Echiniscoides*) readily regain activity following retransfer to seawater, underlining the incredible plasticity of tardigrade physiology. In agreement with their hyper-regulatory strategy, dehydration and loss of water appears to be less of a problem to muscle movement, as seen in dehydrating tardigrades, which keep moving until tun formation and complete desiccation has occurred (Fig. 2.1b). Importantly, the musculature facilitates the structural reorganization that is associated with desiccation and tun formation in semiterrestrial tardigrades (Halberg et al. 2013a). Specifically, the dorsal, ventral, and lateral muscle groups are involved in contracting the body and generating the compact body shape of the tun, which in the eutardigrade *Richterisus coronifer* involves an 87% reduction of body volume (Halberg et al. 2013a). Tun formation is moreover characterized by withdrawal of the legs into the body cavity mediated via the leg muscles. Visceral muscles (see below) are likely also involved in both entrance into and exit out of the cryptobiotic tun state, which involves a rearrangement of the internal organs.

2.6.2 Visceral Musculature Including Stylet and Pharyngeal Musculature

In the current review, we define the visceral musculature as the muscles associated with internal organs, i.e., the digestive and reproductive systems, but note that the stylet musculature could just as well be included under the somatic musculature due to its structural resemblance to the leg musculature. Generally, the internal organs do not lie free within the body cavity but are rather suspended from ligaments and muscle cells that attach to the integument or perhaps to other parts of the digestive system (Fig. 2.2). The visceral musculature, thus, includes muscles that suspend organs—e.g., the gonad and associated ducts, Malpighian tubules, and the hindgut—within the body cavity, as well as muscles that are involved in opening of the eutardigrade cloaca or heterotardigrade gonopore (Dewel and Dewel 1979; Møbjerg and Dahl 1996; Halberg et al. 2009a).

Visceral muscles also include a group of fine longitudinal muscles radially arranged, spanning the length of the midgut in eutardigrades (Figs. 2.2 and 2.10a). These muscles seem to be absent in heterotardigrades, but they are, nonetheless, designated as the only true visceral muscles by some authors (e.g., Dewel et al. 1993; Marchioro et al. 2013). The longitudinal gut muscles originate at the junction between the esophagus and the midgut in eutardigrades and extend posteriorly into the region of the hindgut (Halberg et al. 2009a). The muscles are characterized by the presence of multiple adhesion sites or attachment plaques, anchoring fibers to the basal side of the gut epithelium (Walz 1975a; Dewel et al. 1993). The number of longitudinal gut muscles appears to vary among eutardigrade species: only six were found in *Halobiotus crispae*, and eight were found in *Milnesium tardigradum* and *Hypsibius* sp., whereas *Ramazzottius oberhauseri* has been reported to possess 13 of these longitudinal muscles (Schmidt-Rhaesa and Kulessa 2007; Halberg et al. 2009a).

According to the current definition, visceral muscles also include the muscles involved in food acquisition, i.e., the musculature of the buccal-pharyngeal apparatus. As discussed above (see Sect. 2.5), the pharyngeal bulb is a muscular pump—specifically, it is an oval organ composed of radially arranged monosarcomeric myoepithelial cells (Fig. 2.10a, b; Eibye-Jacobsen 1996, 1997; Halberg et al. 2009a; Schmidt-Rhaesa and Kulessa 2007). Rows of placoids situated in the center of the bulb serve as muscle attachments for the myoepithelial cells in echiniscoideans and parachelan eutardigrades. For example, in *Halobiotus crispae* each row contains three macroplacoids and one microplacoid (Kristensen 1982). In the eutardigrades investigated so far, the stylet musculature consists of six pairs of muscles arranged in a cone-shaped manner on either side the buccal tube (Fig. 2.9c; Schmidt-Rhaesa and Kulessa 2007; Halberg et al. 2009a). Stylet muscles emerge from the most posterior part of the stylet, the so-called furca, and extend onto either the buccal tube or the pharyngeal bulb (Dewel et al. 1993; Schmidt-Rhaesa and Kulessa 2007; Halberg et al. 2009a).

2.7 The Tardigrade Nervous System

The nervous system of tardigrades may be subdivided into peripheral (PNS) and central (CNS) parts with the latter comprising a large anterior lobate brain and a ventral ganglionated nervous system that extends into the posterior region of the body (Figs. 2.2 and 2.10c; Marcus 1929; Dewel and Dewel 1996). The peripheral system comprises nervous structures of the trunk that lay outside the central nerve cord, such as leg ganglia and lateral and dorsal neurons (Fig. 2.10c; Dewel et al. 1993; Zantke et al. 2008; Persson et al. 2012; Mayer et al. 2013b). Glial cells have been observed but generally seem to be present in a very small portion of the total nervous system (Greven and Kuhlmann 1972; Dewel et al. 1993). Notably, the tardigrade nervous system lies free in the body cavity only surrounded by a simple basement membrane (Greven and Kuhlmann 1972; Dewel et al. 1993). We thus hypothesize that the nervous system, besides rapid communication via electrochemical signaling, could serve endocrine functions releasing hormones to the body cavity, thereby providing the basis for more long-term adjustments of body functions, e.g., controlling reproduction. Generally, very little is known of signal transduction within the tardigrade nervous system—however, immunohistochemical studies suggest that neurotransmitters include serotonin, RFamide-like peptides, allatostatin, and perisulfakinin (summarized in Schulze and Persson 2016).

The microanatomy of the tardigrade nervous system has in recent years been highly debated—a debate, which has its base in divergent observations obtained with different techniques and which to a large extent is fueled by divergent opinions on the phylogenetic affinities of the phylum. Most of the more recent investigations into the tardigrade nervous system have mainly been approached using ICC in combination with CLSM, employing commercially available primary antibodies against tubulin and various neurotransmitters (Zantke et al. 2008; Persson et al. 2012, 2014; Mayer et al. 2013a, b; Schulze and Schmidt-Rhaesa 2013; Schulze et al. 2014). This technique provides insightful images that significantly help researchers make three-dimensional interpretations of tardigrade anatomy. However, this approach is not flawless as anatomical details are only revealed if the antibody recognizes the specific epitope within the tardigrade tissue. Here, we will not go into details on the contradicting hypothesis regarding tardigrade segmentation and the evolutionary origin of the nervous system (addressed in e.g. Dewel et al. 1999; Mayer et al. 2013a; Persson et al. 2014; Schulze and Persson 2016; Smith and Goldstein 2017). As the current chapter deals with morphology and anatomy, we will, however, urge authors engaging in studies on tardigrades to consult morphological and anatomical investigations. Below, we give a more detailed account of the tardigrade nervous system.

2.7.1 The Tardigrade Brain

The tardigrade brain occupies a large region of the head, which in the eutardigrade *Halobiotus crispae* accounts for approximately one-quarter to one-third of the total number of cells (Møbjerg et al. 2011; Persson et al. 2012). Extending from the dorsal to ventral region of the head, the brain is traversed by the buccal tube and the stylets (Fig. 2.2). The ganglionate or lobate organization of the brain was already recognized in the detailed descriptions of authors from the nineteenth and early twentieth centuries (reviewed in Dewel and Dewel 1996; Zantke et al. 2008; Schulze and Persson 2016). While recent publications generally confirm this lobate organization, it is currently debated whether the presence of lobes reflects a metameric organization of the brain (reviewed in Smith and Goldstein 2017). The brain is bilateral symmetric and can be described as composed of three paired lobes with each lobe in a pair constituted by one or two nerve cell clusters (Fig. 2.9d; Persson et al. 2012, 2014; Schulze and Schmidt-Rhaesa 2013; Smith and Goldstein 2017). The two lobes of each pair connect through transverse tubulin-rich commissures (Figs. 2.8 and 2.9d; reviewed in Schulze and Persson 2016; Smith and Goldstein 2017). The suprabuccal region of the brain is constituted by two of these paired lobes, referred to as the outer/first and inner/second lobes, or following the arthropod-derived terminology reintroduced by Van der Land (1968; see Sect. 3), tentatively as protocerebrum and deutocerebrum, respectively (Kristensen 1982). Notably, simple pigment cup eyes are found in the outer/first brain lobe (protocerebrum) in many eutardigrades and in echiniscoideans (Figs. 2.2, 2.3b–d, and 2.4a, c, d; reviewed in Greven 2007). The large first brain lobe further receives sensory input from anterodorsal cephalic sense organs (Fig. 2.9d). In arthrotardigrades, this includes the primary clava and cirrus A and in eutardigrades, the large dorsal sensory area sometimes referred to as the temporalia. In eutardigrades, the first lobes span into the most posterior region of the head being significantly larger than the second lobes, whereas the first lobes appear short but wide in heterotardigrades (reviewed in Dewel and Dewel 1996).

The first and second brain lobes may appear more or less fused, but a tract seems to propagate from the second through the third brain lobes connecting the brain to the ventral part of the central nervous system (Persson et al. 2012, 2014). Interestingly, the first brain lobes also connect directly to the first ventral trunk ganglia via so-called paired outer connectives (Figs. 2.2 and 2.10c). The second brain lobes seem to receive the major part of the sensory input from the frontal sensory field or papilla cephalica (secondary clava) of eutardigrades and from the external cirrus and secondary clava of arthrotardigrades, whereas the internal cirri seem to interconnect with the first brain lobe (Fig. 2.9d; Persson et al. 2014). Detailed comparative investigations into the association between the brain and the cephalic sensory appendages are, however, lacking—as are studies on the brain of, e.g., neoarctid arthrotardigrades, which clearly have three pairs of cephalic cirri and clavae. Interestingly, a small unpaired median lobe or ganglion is present near the center of the brain in both heterotardigrades and eutardigrades. This ganglion receives input from a conspicuous nerve that connects to the median cirrus in heterotardigrades and from a corresponding nerve that connects to the dorsal epidermis in eutardigrades (Fig. 2.9d; Marcus 1929; Kristensen and Higgins 1984a; Zantke et al. 2008; Persson et al. 2012, 2014; Mayer et al. 2013a).

The third, posteroventrally situated brain lobes flank the buccal tube. These lobes receive sensory input from the peribuccal lamellae surrounding the mouth opening, and they innervate the musculature of the stylet apparatus (Persson et al. 2012, 2014; Mayer et al. 2013a; Schulze and Schmidt-Rhaesa 2013). The organization of this ventral circumbuccal region of the brain, which tentatively has been referred to as tritocerebrum (Kristensen 1982, 1987; Kristensen and Higgins 1984a, b), has been highly debated (Dewel and Dewel 1996; Dewel et al. 1999; Zantke et al. 2008; Persson et al. 2012; Mayer et al. 2013a). Transmission electron microscopy and some immunohistochemical investigations confirm older light microscopic studies revealing that the ventral part of the brain connects to a subpharyngeal ganglion (Dewel and Dewel 1996; Wiederhöft and Greven 1996; Dewel et al. 1999; Persson et al. 2012, 2014; Gross and Mayer 2015: referred to as AV cells). Other immunohistochemical studies have not identified this ventral sub-buccal structure (Zantke et al. 2008; Mayer et al. 2013a). Specifically, the subpharyngeal ganglion connects the brain to the most anterior pair of trunk ganglia. In addition, paired inner connectives connect the third brain lobes directly to the first ventral trunk ganglia.

2.7.2 The Nervous System of the Trunk

It is generally accepted that the central nervous system of the tardigrade trunk appears ladder-like with a clear segmented pattern constituted by four paired ventral ganglia connected through paired longitudinal nerve cords (Fig. 2.10c; Persson et al. 2012; Mayer et al. 2013b; Schulze and Schmidt-Rhaesa 2013; Schulze et al. 2014). The two ganglia of each pair are connected through transverse commissures, whereas the four ganglia on each side of the body are connected by one of the longitudinal nerve cords. Anteriorly the nerve cords connect to the subpharyngeal ganglion and the brain (see above). Extra-ganglionic commissures connecting the two nerve cords have been reported in both eutardigrades and heterotardigrades (Mayer et al. 2013b; Schulze et al. 2014). Two neurites emerge from each ventral ganglion and extend laterally into the associated leg (Persson et al. 2012; Schulze et al. 2014; Mayer et al. 2013b). One of these neurites connects with a leg ganglion located in the proximal region of the leg (Fig. 2.10c), whereas the second neurite extends into the most distal region of the leg. In addition, a third neurite emerges from each trunk ganglion, the so-called lateral nerve. This neurite connects the trunk ganglion to a lateral neuron, which in turn extends dorsally and connects to a dorsal neuron (Fig. 2.10c). Thus, each side of the body has a total of four lateral neurons connected through lateral longitudinal projections as well as four dorsal neurons. Recently, a stomatogastric ganglion was reported in the eutardigrade *Mesobiotus* cf. *harmsworthi* (Mayer et al. 2013a). This ganglion is composed of at least four cells and seems to be located at the junction between the esophagus and the midgut. Three posteriorly oriented neurites emerge from the ganglion projecting along the wall of the midgut, while a single neurite was reported to span the esophagus in an anterior direction ending blindly behind the pharynx.

2.8 Reproductive System

Tardigrades have a single dorsal sac-like gonad, suspended from the body wall by ligaments (Fig. 2.2, Kristensen 1979; Ramazzotti and Maucci 1983; Dewel et al. 1993; Bertolani 2001; Rebecchi and Bertolani 1994). The gonad terminates in a single gonoduct or paired sperm ducts that open into a cloaca (eutardigrades) or a gonopore (heterotardigrades). Tardigrades are generally gonochoristic but may also be hermaphroditic. Noticeably, parthenogenesis is widespread among semiterrestrial species. As a full chapter (Chap. 8) within this book is devoted to reproduction, development, and life history, we will only briefly discuss the organization of the tardigrade reproductive system in the following paragraphs.

The ovary of female tardigrades ends posteriorly in a single oviduct, which in eutardigrades opens into a cloaca and in heterotardigrades into a conspicuous six-leaved rosette-like gonopore comprising six myoepithelial cells (Fig. 2.6a, b). The number of oocytes maturing in the large ovaries of mature females varies between 1 and 20. Oocytes can either produce yolk autosynthetically or acquire it from associated nurse cells or through micropinocytosis. In males, the unpaired dorsal testis connects via paired sperm ducts to the cloaca in eutardigrades or to the round or oval-shaped gonopore of heterotardigrades through two small pores. In some males of both heterotardigrades and eutardigrades, mature spermatozoa are stored in seminal vesicles formed by the distal part of the sperm ducts. Tardigrade spermatozoa terminate in a motile flagellum with a classical 9+2 axoneme microtubule structure. In heterotardigrades spermatozoa generally appear round-headed with two free mitochondria, whereas they generally are threadlike and have a tufted tail in eutardigrades.

Fertilization in arthrotardigrades is external. Spermatozoa are often deposited in cuticular pockets—seminal receptacles, which are found lateral to the female gonopore of many arthrotardigrades (Fig. 2.5b). These receptacles store the sperm from the male(s) prior to fertilization of the eggs. The morphology and position of the receptacles can be highly variable even within a family, as seen in Stygarctidae (Hansen et al. 2012). Some female arthrotardigrades, from a few species within Megastygarctidinae and Styraconyxinae, are characterized by external genital structures formed from extensions of the ducts of the receptacles—these extensions may be involved in copulation and/or insemination of the eggs (Hansen and Kristensen 2006). In eutardigrades, fertilization may be external via ejaculation into the exuvia or internal involving copulation. In the latter case, a spermatheca may be found in the female hindgut. Postcopulatory modifications of spermatozoa have been reported from both heterotardigrades and eutardigrades (Rebecchi 1997; Jørgensen et al. 1999). Recently, a study on mating behavior in the eutardigrade *Isohypsibius dastychi* was published revealing a complex process involving mutual stimulation of the male as well as female (Bingemer et al. 2016). Mating behavior involving stimulation of the female has also been reported in marine heterotardigrades (Kristensen 1979).

Tardigrade eggs may be laid freely in the substratum or in exuvia during molting (Fig. 2.4b). They can be laid singly or numerously, even in clusters from multiple females as seen in *Echiniscoides*. The eggshells may be smooth and unornamented or highly ornamented as in many eutardigrades, where eggs represent important taxonomic characters. It has been hypothesized that ornamentation of the eggs could function as anchors to the substratum, protection against predation, a buffer against rapid desiccation, or as an increased surface for gas exchange (Kinchin 1994). In eutardigrades development is direct with juveniles having two double claws and a cloaca just as the adults. Larval/juvenile morphology in heterotardigrades differs from adult morphology in other aspects than merely size—the situation is, however, complex with varying claw numbers in the juveniles of different families. Generally, the juvenile that leaves the egg is two-clawed, with aberrant looking cephalic sense organs and without anus and gonopore (Bertolani et al. 1984). In a subsequent stage (at the latest, in the final juvenile stage), the anus emerges and the gonopore appears as a little oval dot.

Acknowledgments Funding came from the Carlsberg Foundation and the Danish Council for Independent Research (grant no. DFF – 4090-00145).

References

Baccetti B, Rosati F (1971) Electron microscopy on tardigrades. III. The integument. J Ultrastruc Res 34:214–243

Basse A (1905) Beiträge zur Kenntnis des Baues der Tardigraden. Z Wiss Zool 80:259–280

Bertolani R (1970a) Mitosi somatiche e costanza cellular numerica nei Tardigradi. Atti Accad naz Lincei Rc Ser 8a 48:739–742

Bertolani R (1970b) Variabilita numerica cellulare in alcuni tessuti di Tardigradi. Atti Accad naz Lincei Rc Ser 8a 49:442–446

Bertolani R (2001) Evolution of the reproductive mechanisms in tardigrades – a review. Zool Anz 240:247–252

Bertolani R, Grimaldi de Zio S, D'Addabbo Gallo M et al (1984) Postembryonic development in heterotardigrades. Monit Zool Ital 18:307–320

Bingemer J, Hohberg K, Schill RO (2016) First detailed observations on tardigrade mating behaviour and some aspects of the life history of *Isohypsibius dastychi* Pilato, Bertolani & Binda 1982 (Tardigrada, Isohypsibiidae). Zool J Linn Soc 178(4):856–862

Biserova NM, Kuznetsova KG (2012) Head sensory organs of *Halobiotus stenostomus* (Eutardigrada, Hypsibiidae). Biol Bull 39:579–589

Bonifacio A, Guidetti R, Altiero T et al (2012) Nature, source and function of pigments in tardigrades: *in vivo* Raman imaging of carotenoids in *Echiniscus blumi*. PLoS One 7(11): e50162. https://doi.org/10.1371/journal.pone.0050162

Cross R (2016) Water bear genomes start to reveal hardy critter's secrets. Chem Eng News 94 (44):20–21

Crowe H (1972) Evaporative water loss by tardigrades under controlled relative humidities. Biol Bull 142:407–416

Czerneková M, Jönsson KI (2016) Mitosis in storage cells of the eutardigrade *Richtersius coronifer*. Zool J Linn Soc 178(4):888–896

Czerneková M, Jönsson KI, Chajec L et al (2017) The structure of the desiccated *Richtersius coronifer* (Richters, 1903). Protoplasma 254:1367. https://doi.org/10.1007/s00709-016-1027-2

Dewel RA, Clark WH (1973) Studies on the tardigrades. II. Fine structure of the pharynx of *Milnesium tardigradum* Doyère, 1840. Tissue Cell 5:147–159

Dewel RA, Dewel WC (1979) Studies on the Tardigrades. IV. Fine structure of the hindgut of *Milnesium tardigradum* Doyère. J Morph 161:79–110

Dewel RA, Dewel WC (1996) The brain of *Echiniscus viridissimus* Peterfi, 1956 (Heterotardigrada): a key to understanding the phylogenetic position of tardigrades and the evolution of the arthropod head. Zool J Linn Soc 116:35–49

Dewel RA, Dewel WC (1997) The place of tardigrades in arthropod evolution. In: Fortey R, Thomas RH (eds) Arthropod relationships. Chapman and Hall, London, pp 109–123

Dewel RA, Dewel WC, Roush BG (1992) Unusual cuticle associated organs in the heterotardigrade *Echiniscus viridissimus*. J Morph 212:123–140

Dewel RA, Nelson DR, Dewel WC (1993) Tardigrada. In: Harrison FH, Rice ME (eds) Microscopic anatomy of invertebrates, Onychophora, Chilopoda, and Lesser Protostomata, vol 12. Wiley-Liss, New York, pp 143–183

Dewel RA, Dewel WC, Budd GE, Castano DF et al (1999) The organization of the subeosphageal nervous system in tardigrades: insights into the evolution of the arthropod hypostome and tritocerebrum. Zool Anz 238:191–203

Doyére M (1840) Mémoire sur les Tardigrades. Ann Sci Nat 2(Série 14):269–361

Eibye-Jacobsen J (1996) On the nature of pharyngeal muscle cells in the Tardigrada. Zool J Linn Soc 116:123–138

Eibye-Jacobsen J (1997) Development, ultrastructure and function of the pharynx of *Halobiotus crispae* Kristensen, 1982 (Eutardigrada). Acta Zool 78:329–347

Eibye-Jacobsen J (2001a) A new method for making SEM preparations of the tardigrade buccopharyngeal apparatus. Zool Anz 240:309–319

Eibye-Jacobsen J (2001b) Are the supportive structures of the tardigrade pharynx homologous throughout the entire group? J Zool Syst Evol Res 39:1–11

Fujimoto S, Miyazaki K (2013) *Neostygarctus lovedeluxe* n. sp. from the Miyako Islands, Japan: the first record of Neostygarctidae from (Heterotardigrada: Arthrotardigrada) from the Pacific. Zool Sci 30:414–419

Greeff R (1865) Über das Nervensystem der Bärthierchen, Arctiscoida C.A.S. Schultze (Tardigraden Doyère) mit besonderer Berücksichtigung der Muskelnerven und deren Endigungen. Arch Mikrosk Anat 1:101–122

Greven H (1980) Die Bärtierchen. Die Neue Brehm-Bucherei, vol 537. Ziemsen Verlag, Wittenberg-Lutherstadt, p 101

Greven H (1982) Homologues or analogues? A survey of some structural patterns in Tardigrada. In: Nelson D (ed) Proceedings of the third international symposium on Tardigrada, Johnson City, Tennessee, 1980. East Tennessee State University Press, Johnson City, TN, pp 55–76

Greven H (1984) Tardigrada. In: Bereiter-Hahn J, Matolsty AG, Richards KS (eds) Biology of the integument, Invertebrates, vol 1. Springer, Berlin, pp 714–727

Greven H (2007) Comments on the eyes of tardigrades. Arthropod Struct Dev 36:401–407

Greven H, Greven W (1987) Observations on the permeability of the tardigrade cuticle using lead as an ionic tracer. In: Bertolani R (ed) Biology of tardigrades, selected symposia and monographs, U.Z.I. 1. Mucchi, Modena, pp 35–43

Greven H, Grohé G (1975) Die Feinstruktur des Integuments und der Muskelansatzscellen von *Echiniscoides sigismundi*. Helgol Wiss Meeres 27:450–460

Greven H, Kristensen RM (2001) Some aspects of spermiogenesis and spermatozoan ultrastructure in *Echiniscoides sigismundi* (Heterotardigrada: Echiniscoididae). Zool Anz 240:331–339

Greven H, Kuhlmann D (1972) Die Struktur des Nervengewebes von *Macrobiotus hufelandi* C.A.S. Schultze (Tardigrada). Z Zellforsch Mikrosk Anat 132:131–146

Greven H, Dastych H, Kraus H (2005) Notes on the integument of the glacier-dwelling tardigrade *Hypsibius klebelsbergi* Mihelcic, 1959 (Tardigrada). Mitt hamb zool Mus Inst 102:11–20

Gross V, Mayer M (2015) Neural development in the tardigrade *Hypsibius dujardini* based on anti-acetylated α-tubulin immunolabeling. EvoDevo 6:12. https://doi.org/10.1186/s13227-015-0008-4

Gross V, Treffkorn S, Mayer G (2015) Tardigrada. In: Wanninger A (ed) Evolutionary developmental biology of invertebrates 3. Ecdysozoa I: Non-tetraconata. Springer, Wien, pp 35–52

Grothman GT, Johansson C, Chilton G et al (2017) Gilbert Rahm and the status of Mesotardigrada Rahm, 1937. Zool Sci 34(1):5–10

Guidetti R, Bertolani R (2005) Tardigrade taxonomy: an updated check list of the taxa and a list of characters for their identification. Zootaxa 845:1–46

Guidetti R, Rebecchi L, Bertolani R (2000) Cuticle structure and systematics of the Macrobiotidae (Tardigrada, Eutardigrada). Acta Zool 81(1):27–36

Guidetti R, Bertolani R, Rebecchi L (2013) Comparative analysis of the tardigrade feeding apparatus: adaptive convergence and evolutionary pattern of the piercing stylet system. J Limnol 72:24–35

Halberg K, Møbjerg N (2012) First evidence of epithelial transport in tardigrades: a comparative investigation of organic anion transport. J Exp Biol 215:497–507

Halberg KA, Persson D, Møbjerg N et al (2009a) Myoanatomy of the marine tardigrade *Halobiotus crispae* (Eutardigrada: Hypsibiidae). J Morph 270:996–1013

Halberg KA, Persson D, Ramløv H et al (2009b) Cyclomorphosis in Tardigrada: adaptation to environmental constraints. J Exp Biol 212:2803–2811

Halberg KA, Jørgensen A, Møbjerg N (2013a) Desiccation tolerance in the tardigrade *Richtersius coronifer* relies on muscle mediated structural reorganization. PLoS One 8(12):e85091

Halberg KA, Larsen KW, Jørgensen A et al (2013b) Inorganic ion composition in Tardigrada: cryptobionts contain large fraction of unidentified organic solutes. J Exp Biol 216:1235–1243

Hansen JG, Kristensen RM (2006) The 'hyena female' of tardigrades and descriptions of two new species of *Megastygarctides* (Arthrotardigrada: Stygarctidae) from Saudi Arabia. Hydrobiologia 558:81–101

Hansen JG, Kristensen RM, Jørgensen A (2012) The armoured marine tardigrades Stygarctidae (Arthrotardigrada, Tardigrada). Kgl Dan Vidensk Selsk Biol Skr 2:1–91

Hansen JG, Kristensen RM, Bertolani R et al (2017) Comparative analyses of *Bertolanius* species (Eohypsibiidae; Eutardigrada) with the description of *Bertolanius birnae* sp. nov. from northern polar regions. Polar Biol 40:123–140

Heidemann N, Smith D, Hygum TL et al (2016) Osmotic stress tolerance in semi-terrestrial tardigrades. Zool J Linn Soc 178(4):912–918. https://doi.org/10.1111/zoj.12502

Hygum TL, Clausen LKB, Halberg KA et al (2016) Tun formation is not a prerequisite for desiccation tolerance in the marine tidal tardigrade *Echiniscoides sigismundi*. Zool J Linn Soc 178(4):907–911. https://doi.org/10.1111/zoj.12444

Hyra M, Rost-Roszkowska MM, Student S et al (2016a) Body cavity cells of Parachela during their active life. Zool J Linn Soc 178(4):878–887

Hyra M, Poprawa I, Włodarczyk A et al (2016b) Ultrastructural changes in the midgut epithelium of *Hypsibius dujardini* (Doyère, 1840) (Tardigrada, Eutardigrada, Hypsibiidae) in relation to oogenesis. Zool J Linn Soc 178(4):1096–3642

Hyra M, Czernekova M, Student S et al (2017) Traditional and modern methods in tardigrade analysis. The 16th European Microscopy Congress, Lyon, France. http://emc-proceedings.com/abstract/traditional-and-modern-methods-in-tardigrade-analysis/

Jørgensen A, Kristensen RM (2001) A new tanarctid arthrotardigrade with buoyant bodies. Zool Anz 240:425–439

Jørgensen A, Møbjerg N (2015) Notes on the cryptobiotic capability of the marine arthrotardigrades *Styraconyx haploceros* (Halechiniscidae) and *Batillipes pennaki* (Batillipedidae) from the tidal zone in Roscoff, France. Mar Biol Res 11:214–217

Jørgensen A, Møbjerg N, Kristensen RM (1999) Ultrastructural studies on spermiogenesis and postcopulatory modifications of spermatozoa of *Actinarctus doryphorus* Schulz, 1935 (Arthrotardigrada: Halechiniscidae). Zool Anz 238:235–257

Jørgensen A, Boesgaard TM, Møbjerg N et al (2014) The tardigrade fauna of Australian marine caves: with descriptions of nine new species of Arthrotardigrada. Zootaxa 28:401–443. https://doi.org/10.11646/zootaxa.3802.4.1

Keilin D (1959) The problem of anabiosis or latent life: history and current concept. Proc R Soc Lond, B 150:149–191

Kinchin IM (1994) The biology of tardigrades. Portland Press, London. 186 pp

Koutsovoulos G, Kumar S, Laetsch DR et al (2016) No evidence for extensive horizontal gene transfer in the genome of the tardigrade *Hypsibius dujardini*. PNAS 113(18):5053–5058. https://doi.org/10.1073/pnas.1600338113

Kristensen RM (1976) On the fine structure of *Batillipes noerrevangi* Kristensen, 1976. 1. Tegument and molting cycle. Zool Anz 197:129–150

Kristensen RM (1978) On the fine structure of *Batillipes noerrevangi* Kristensen, 1978. 2. The muscle attachments and the true cross-striated muscles. Zool Anz 200:173–184

Kristensen RM (1979) On the fine structure of *Batillipes noerrevangi* Kristensen, 1978 (Heterotardigrada). 3. Spermiogenesis. Zesz Nau Uni Jagiel Prace Zool 25:97–105

Kristensen RM (1981) Sense organs of two marine arthrotardigrades (Heterotardigrada, Tardigrada). Acta Zool 62:27–41

Kristensen RM (1982) The first record of cyclomorphosis in Tardigrada based on a new genus and species from Arctic meiobenthos. Z Zool Syst Evol 20:249–270

Kristensen RM (1987) Generic revision of the Echiniscidae (Heterotardigrada), with a discussion of the origin of the family. In: Bertolani R (ed) Biology of tardigrades, selected symposia and monographs, U.Z.I. 1. Mucchi, Modena, pp 261–335

Kristensen RM, Hallas TE (1980) The tidal genus *Echiniscoides* and its variability, with erection of Echiniscoididae fam. n. (Tardigrada). Zool Scr 9:113–127

Kristensen RM, Higgins RP (1984a) Revision of *Styraconyx* (Tardigrada: Halechiniscidae), with descriptions of two new species from Disko Bay, West Greenland. Smithson Contr Zool 391:1–40

Kristensen RM, Higgins RP (1984b) A new family of Arthrotardigrada (Tardigrada: Heterotardigrada) from the Atlantic coast of Florida, U.S.A. Trans Am Microsc Soc 103 (3):295–311

Kristensen RM, Neuhaus B (1999) The ultrastructure of the tardigrade cuticle with special attention to marine species. Zool Anz 238:261–281

Kristensen RM, Sørensen MV, Hansen JG et al (2015) A new species of *Neostygarctus* (Arthrotardigrada) from the Condor Seamount in the Azores, Northeast Atlantic. Mar Biodiv 45(3):453–467. https://doi.org/10.1007/s12526-015-0323-y

Maderspacher F (2016) Zoology: the walking heads. Cur Biol 26(5):R194–R197

Marchioro T, Rebecchi L, Cesari M et al (2013) Somatic musculature of Tardigrada: phylogenetic signal and metameric patterns. Zool J Linn Soc 169:580–603

Marcus E (1929) Tardigrada. In: Bronn HG (ed) Klassen und Ordnungen des Tierreichs, vol 5 sec 4 part 3. Akademische Verlagsgesellschaft, Leipzig, p 609

Marley NJ, McInnes SJ, Sands CJ (2011) Phylum Tardigrada: a re-evaluation of the Parachela. Zootaxa 2819:51–64

Mayer G, Kauschke S, Rüdiger J et al (2013a) Neural markers reveal a one-segmented head in tardigrades (water bears). PLoS One 8(3):e59090

Mayer G, Martin C, Rüdiger J et al (2013b) Selective neuronal staining in tardigrades and onychophorans provides insights into the evolution of segmental ganglia in panarthropods. BMC Evol Biol 13:230

Møbjerg N, Dahl C (1996) Studies on the morphology and ultrastructure of the Malpighian tubules of *Halobiotus crispae* Kristensen, 1982 (Eutardigrada). Zool J Linn Soc 116:85–99

Møbjerg N, Halberg KA, Jørgensen A et al (2011) Survival in extreme environments – on the current knowledge of adaptations in tardigrades. Acta Physiol 202:409–420

Møbjerg N, Kristensen RM, Jørgensen A (2016) Data from new taxa infer *Isoechiniscoides* gen. nov. and increase the phylogenetic and evolutionary understanding of echiniscoidid tardigrades (Echiniscoidea: Tardigrada). Zool J Linn Soc 178:804–818

Müller J (1936) Zur vergleichenden Myologie der Tardigraden. Z wiss Zool 147:171–204

Nielsen C (2001) Animal evolution. interrelationships of the living phyla, 2nd edn. Oxford University Press, London, pp 226–231

Pelzer B, Dastych H, Greven H (2007) The osmoregulatory/excretory organs of the glacier-dwelling eutardigrade *Hypsibius klebelsbergi* Mihelcic, 1959 (Tardigrada). Mitt hamb Zool Mus Inst 104:61–72

Persson AB, Persson PB (2012) Getting a kick out of thermoregulation. Acta Physiol 204:291–293. https://doi.org/10.1111/j.1748-1716.2011.02400.x

Persson DK, Halberg KA, Jørgensen A et al (2012) Neuroanatomy of *Halobiotus crispae* (Eutardigrada: Hypsibiidae): tardigrade brain structure supports the clade Panarthropoda. J Morph 273:1227–1245

Persson D, Halberg KA, Jørgensen A et al (2014) Brain anatomy of the marine tardigrade *Actinarctus doryphorus* (Arthrotardigrada). J Morph 275(2):173–190

Plate L (1889) Beiträge zur Naturgeschichte der Tardigraden. Zool Jahrb Abt Anat Ontog Tiere 3:487–550

Rahm G (1937) Eine neue Tardigraden-Ordnung aus den heißen Quellen von Unzen, Insel Kyushu, Japan. Zool Anz 120:65–71

Ramazzotti G, Maucci W (1983) Il phylum Tardigrada, 3rd edn. Mem Ist Ital Idrobiol 41:1–1012

Rebecchi L (1997) Ultrastructural study of spermiogenesis and the testicular and spermathecal spermatozoon of the gonochoristic tardigrade *Xerobiotus pseudohufelandi* (Eutardigrada, Macrobiotidae). J Morphol 234(1):11–24

Rebecchi L (2001) The spermatozoon in tardigrades: evolution and relationships with the environment. Zool Anz 240:525–533

Rebecchi L, Bertolani R (1994) Maturative pattern of ovary and testis in eutardigrades of freshwater and terrestrial habitats. Invertebr Reprod 26:107–117
Rebecchi L, Bertolani R (1999) Spermatozoon morphology of three species of Hypsibiidae (Tardigrada, Eutardigrada) and phylogenetic evaluation. Zool Anz 238:319–328
Renaud-Mornant J (1982) Species diversity in marine Tardigrada. In: Nelson DR (ed) Proceedings of the third international symposium on the tardigrada, Johnson City, Tennessee. East Tennessee State University Press, Johnson City, TN, pp 149–177
Roszkowska MM, Proprawa I, Hyra M et al (2013) The fine structure of the midgut epithelium in *Xerobiotus pseudohufelandi* (Iharos, 1966) (Tardigrada, Eutardigrada, Macrobiotidae). J Limnol 72(s1):54–61
Roszkowska MM, Bartels PJ, Gołdyn B et al (2016) Is the gut content of *Milnesium* (Eutardigrada) related to buccal tube size? Zool J Linn Soc 178:794–803. https://doi.org/10.1111/zoj.12459
Schmidt-Rhaesa A, Kulessa J (2007) Muscular architecture of *Milnesium tardigradum* and *Hypsibius* sp. (Eutardigrada, Tardigrada) with some data on *Ramazottius oberhaeuseri*. Zoomorphology 126:265–281
Schulze C, Persson D (2016) Tardigrada. In: Schmidt-Rhaesa A, Harsch S, Purschke G (eds) Structure and evolution of invertebrate nervous systems. Oxford University Press, Oxford, pp 383–389
Schulze C, Schmidt-Rhaesa A (2011) Organisation of the musculature of *Batillipes pennaki* (Arthrotardigrada, Tardigrada). Meiofauna March:195–207
Schulze C, Schmidt-Rhaesa A (2013) The architecture of the nervous system of *Echiniscus testudo* (Echiniscoidea, Heterotardigrda). J Limnol 72(s1):44–53
Schulze C, Neves RC, Schmidt-Rhaesa A (2014) Comparative immunohistochemical investigation on the nervous system of two species of Arthrotardigrada (Heterotardigrada, Tardigrada). Zool Anz 253:225–235
Shaw K (1974) The fine structure of muscle cells and their attachments in the tardigrade *Macrobiotus hufelandi*. Tissue Cell 6:431–445
Smith FW, Goldstein B (2017) Segmentation in Tardigrada and diversification of segmental patterns in Panarthropoda. Arthropod Struct Dev 46:328–340
Smith FW, Jockusch EL (2014) The metameric pattern of *Hypsibius dujardini* (Eutardigrada) and its relationship to that of other panarthropods. Front Zool 11:66
Smith FW, Boothby TC, Giovannini I et al (2016) The compact body plan of tardigrades evolved by the loss of a large body region. Cur Biol 26:224–229
Thulin G (1928) Über die Phylogenie und das System der Tardigraden. Hereditas 11:207–266
Van der Land J (1968) *Florarctus antillensis*, a new tardigrade from the coral sand of Curaçao. Stud Fauna Curaçao Caribb Isl 25:140–146
Walz B (1973) Zur Feinstruktur der Muskelzellen des Pharynx-Bulbus von Tardigraden. Z Zellforsch Mikrosk Anat 140:389–399
Walz B (1974) The fine structure of somatic muscles of Tardigrada. Cell Tissue Res 149:81–89
Walz B (1975a) Modified ciliary structures in receptor cells of *Macrobiotus hufelandi* (Tardigrada). Cytobiologie 11:181–185
Walz B (1975b) Ultrastructure of muscle cells in *Macrobiotus hufelandi*. Mem Ist Ital Idrobiol 32:425–443
Walz B (1978) Electron microscopic investigation of cephalic sense organs of the tardigrade *Macrobiotus hufelandi* C.A.S. Schulze. Zoomorphologie 89:1–19
Walz B (1979) Cephalic sense organs of Tardigrada: current results and problems. Zesz Nauk Uniw Jagiell Pr Zool 62:27–41
Weglarska B (1975) Studies on the morphology of *Macrobiotus richtersi* Murray, 1911. In: Higgins RP (ed) Proceedings of the first international symposium on the tardigrades. Mem Ist Ital Idrobiol 32:445–464
Weglarska B (1987) Studies on the excretory system of *Isohypsibius granulifer* Thulin (Eutardigrada). In: Bertolani R (ed) Biology of tardigrades, selected symposia and monographs, U.Z.I. Mucchi, Modena, pp 15–24

Weglarska B (1989) Ultrastructure of tardigrade *Dactylobiotus dispar* (Murray, 1907) cuticle. Acta Biol Cracoviensia Ser Zool 31:57–62

Wiederhöft H, Greven H (1996) The cerebral ganglia of *Milnesium tardigradum* Doyère (Apochela, Tardigrada): three dimensional reconstruction and notes on their ultrastructure. Zool J Linn Soc 116:71–84

Wiederhöft H, Greven H (1999) Notes on head sensory organs of *Milnesium tardigradum* Doyere, 1840 (Apochela, Eutardigrada). Zool Anz 238:338–346

Zantke J, Wolff C, Scholtz G (2008) Three-dimensional reconstruction of the central nervous system of *Macrobiotus hufelandi* (Eutardigrada, Parachela): implications for the phylogenetic position of Tardigrada. Zoomorphology 127:21–36

Chapter 3
Phylogeny and Integrative Taxonomy of Tardigrada

Aslak Jørgensen, Reinhardt M. Kristensen, and Nadja Møbjerg

Abstract Tardigrade phylogeny is currently the object of intense investigations driven by an increasing amount of molecular data from a broad taxonomic range of tardigrades. New information obtained from these investigations plays a crucial role in establishing a reliable systematic framework of Tardigrada. Importantly, contemporary phylogenetic investigations readily incorporate and reinvestigate morphological characters and character systems. Most of these newer investigations are, however, primarily based on a few conservative nuclear genes (in addition to the mitochondrial COI). Emerging transcriptomic and genomic data sets provide promising new information for future phylogenetic studies.

Currently the traditional major taxa of Tardigrada are still accepted, with Eutardigrada constituted by Apochela and Parachela and Heterotardigrada by Echiniscoidea and Arthrotardigrada. However, the arthrotardigrades seem to be a paraphyletic assemblage of distantly related taxa. During the last decade, major rearrangements have occurred in the parachelan taxa Eohypsibioidea, Hypsibioidea, Isohypsibioidea and Macrobiotoidea. Likewise, progress has been made within Arthrotardigrada and Echiniscoidea, but stable phylogenies have yet to be reached within these major taxa.

At the species level, an integrative approach to taxonomy has recently been implemented. Thus, molecular methods in support of morphological investigations are used to facilitate the identification and characterization of different tardigrade species. Phylogenetic inference methods have been combined with DNA barcoding approaches, and the ITS2 secondary structure has been applied as a marker for taxonomy.

In this review we present the current knowledge of tardigrade phylogeny. We emphasise that major changes likely will occur within the phylogenies of arthrotardigrades and echiniscoideans as additional species and genes are sampled

A. Jørgensen (✉) · N. Møbjerg
Department of Biology, University of Copenhagen, Copenhagen, Denmark
e-mail: nmobjerg@bio.ku.dk

R. M. Kristensen
Natural History Museum, University of Copenhagen, Copenhagen, Denmark

R. O. Schill (ed.), *Water Bears: The Biology of Tardigrades*, Zoological Monographs 2,
https://doi.org/10.1007/978-3-319-95702-9_3

for use in molecular phylogenies. We also foresee that integrative approaches to species identification and description will become more widespread securing a firm basis for future investigations in tardigrade taxonomy and systematics.

3.1 Position of Tardigrada in the Animal Kingdom

Tardigrades likely originated and developed in the marine environment (Kristensen 1981; Renaud-Mornant 1982) before becoming successful inhabitants of various terrestrial and limnic habitats. Both morphological and molecular data show that tardigrades are ecdysozoans, i.e. protostome animals that have lost their epithelial cilia and shed their cuticle in a process called ecdysis (Edgecombe 2010; Telford et al. 2008; Fig. 3.1). This was already evident from the first use of tardigrade molecular data (Garey et al. 1996; Giribet et al. 1996; Moon and Kim 1996), which predate the establishment of Ecdysozoa (Aguinaldo et al. 1997). Before the establishment of the Ecdysozoa, some authors considered tardigrades as closely related to arthropods, while others considered them as members of the Aschelminthes, an obsolete phylum of pseudocoelomates and various other small invertebrates (see, e.g. Kinchin 1994). Tardigrades were primarily included in the polyphyletic Aschelminthes due to their similarities with nematodes (triradiate pharynx, buccal stylets and anhydrobiosis). It is now clear that both nematodes and tardigrades belong to Ecdysozoa (Aguinaldo et al. 1997; Dunn et al. 2008; Borner et al. 2014).

Recently, the phylogenetic position of tardigrades within Ecdysozoa has been debated, i.e. whether tardigrades are cycloneuralians (nematodes, nematomorphs, priapulids, kinorhynchs and loriciferans) or panarthropods (euarthropods and onychophorans). Traditionally panarthropods were included in Articulata together with annelids and molluscs and not in Ecdysozoa. Articulata was established by Cuvier in 1812 and was composed of animals with segmented bodies (Annelida, Crustacea, Arachnida, Insecta) (Cuvier 1812; Nielsen 1995). Phylogenies inferred from molecular data have placed tardigrades as sister group to Nematoida (Nematoda and Nematomorpha) (Dunn et al. 2008; Lartillot and Philippe 2008; Borner et al. 2014). However, the sister group relationship with nematodes is likely a long-branch attraction artefact, and a panarthropod relationship is inferred when using slowly evolving genes and including the analysis of additional microRNA data (Campbell et al. 2011). Other recent studies based on mitochondrial genomes, microRNA and genomic data also infer a panarthropod relationship (Rota-Stabelli et al. 2010, 2013). The current textbook consensus is that tardigrades are panarthropods, a clade that shares segmentation, appendages with hooked tips, moulting in order to grow, a cuticle containing chitin and protein and panarthropod sensilla (Brusca and Brusca 2003; Nielsen 2012; Ruppert et al. 2003; see also Chap. 2 on morphology). Compared to arthropods the absence of mixocoel (fusion of the coelomic cavities and the primary body cavity/embryonic blastocoels), heart with openings to the hemocoel and nephridia might be due to miniaturization of

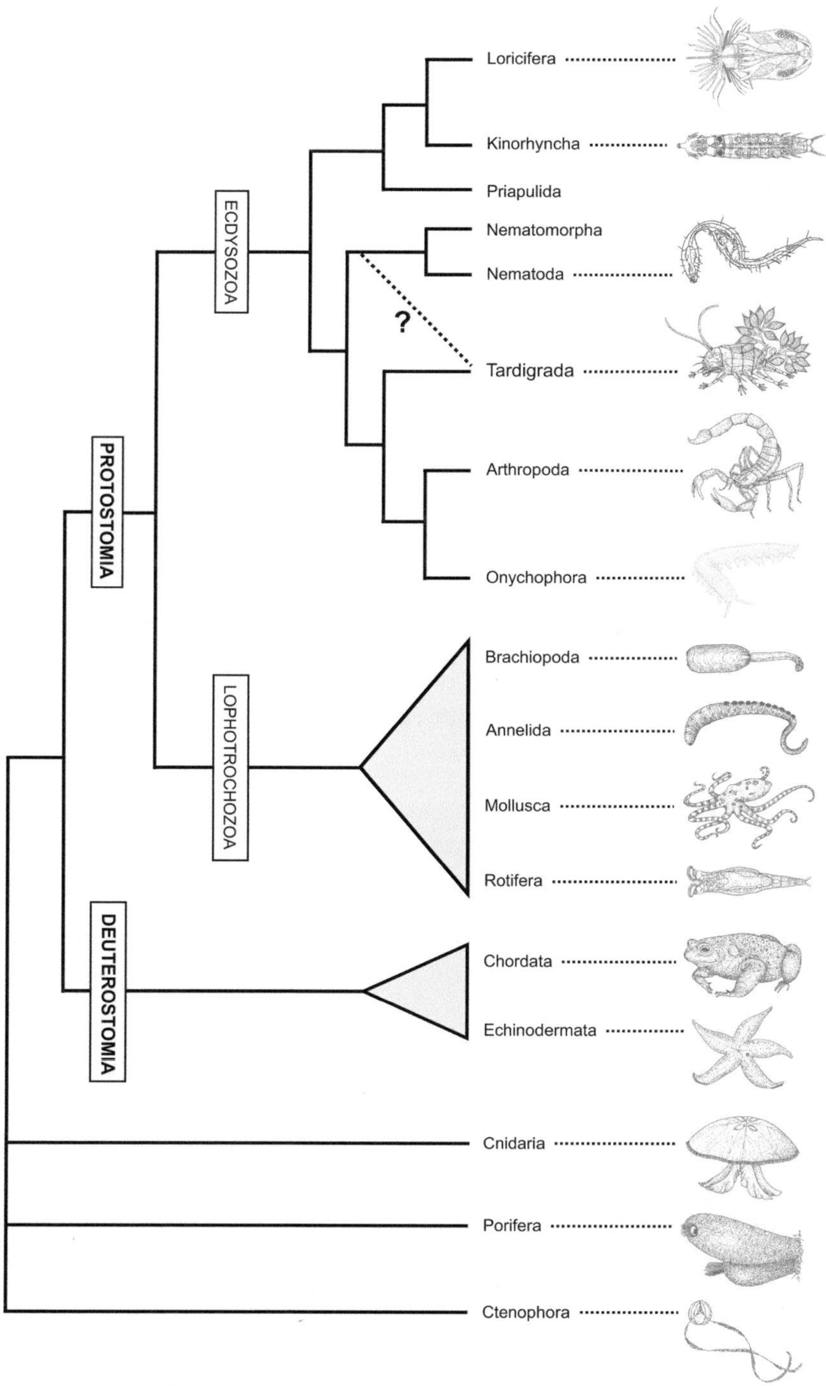

Fig. 3.1 Ecdysozoan relationships. The stippled line and question mark indicate the recent debate regarding the phylogenetic position of Tardigrada, i.e. sister group to cycloneuralians or panarthropods. Most data point towards a Panarthropoda relationship

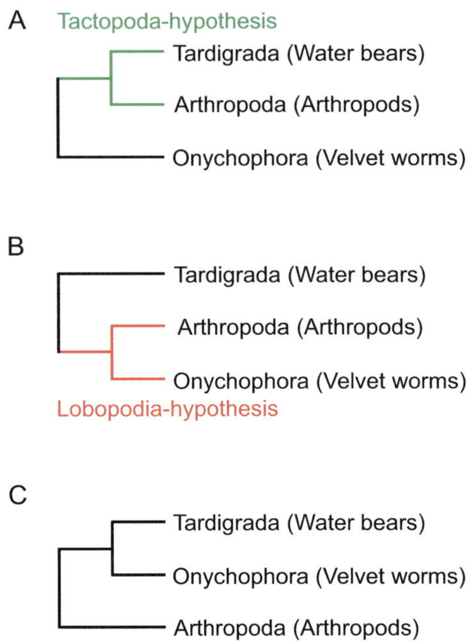

Fig. 3.2 Panarthropoda relationships. (**a**)–(**c**) The three possible panarthropod relationships. The Tactopoda (**a**) and Lobopodia (**b**) hypotheses are currently receiving most support

tardigrades (Hejnol and Schnabel 2005; Schmidt-Rhaesa 2001). However, Marcus (1929) reported a mixocoel in eutardigrades.

With tardigrades being panarthropods, the relationships as either the sister group to arthropods (Tactopoda hypothesis), arthropods plus onychophorans (Lobopodia hypothesis) or onychophorans are currently given attention, and a consensus has not yet been reached (Fig. 3.2). Currently a sister group relationship between Tardigrada and Onychophora has received little support but has been inferred in a few analyses of molecular data (Campbell et al. 2011 supplementary material; Rota-Stabelli et al. 2010). The Lobopodia is an informal group of legged vermiform animals (Onychophora + many fossils), most of which are placed in the stem group of arthropods. The Lobopodia hypothesis, which is preferred in Nielsen (2012), places tardigrades as sister group to a clade composed of Arthropoda and Onychophora, and this relationship has received recent support from many analyses of molecular data (Hejnol et al. 2009; Campbell et al. 2011; Rota-Stabelli et al. 2013). If this hypothesis is true, then the character losses due to miniaturization, as mentioned above, never happened as these structures did not evolve in tardigrades. The Tactopoda hypothesis (Budd 2001) suggests that Arthropoda and Tardigrada are sister groups. This hypothesis requires secondary loss of characters due to miniaturization; however, the similarities of cuticle, ganglionated ventral nerve cord and limbs in tardigrades and Euarthropoda support this sister group relationship (Telford et al. 2008), as does the presence of a three-parted brain (Persson et al. 2012, 2014).

Presumed tardigrade fossils have been used to place tardigrades as most closely related to arthropods (Budd 2001) or various fossil lobopods and onychophorans

(Lui et al. 2011). Lui and Dunlop (2014) regard it as likely that the fossil lobopodians are a paraphyletic grade of legged worms that might have given rise to onychophorans and arthropods. However, all fossil lobopods were marine, and all recent onychophorans are terrestrial, so perhaps the small micro-lobopodians were more closely related to marine tardigrades (Maas et al. 2007). Tardigrade fossils are extremely rare, and the earliest known presumed tardigrade fossil is found in Middle Cambrian rock from Siberia (Maas and Walossek 2001; Maas et al. 2007; Walossek et al. 1994) and shares several similarities with recent tardigrades, e.g. sensory structures, claw structure and cuticular pillars. Much more recent tardigrade fossils from Cretaceous amber are *Milnesium swolenskyi* Bertolani & Grimaldi, 2000 and the 10–15 million years younger *Beorn leggi* Cooper, 1964 (Bertolani and Grimaldi 2000; Cooper 1964).

3.2 Phylogeny of Tardigrada

3.2.1 Phylogenetic Analyses and Morphological and Molecular Characters

Prior to the turn of the century, a number of phylogenetic investigations of tardigrades at various taxonomic levels were published (Kristensen and Renaud-Mornant 1983; Kristensen and Higgins 1984a, b; Renaud-Mornant 1984, 1987; Kristensen 1987; Pollock 1995; Bello and Grimaldi de Zio 1998). Despite the scientific value of these contributions, one must realize that the depicted trees or cladograms were manually derived and not necessarily follow recent standards of phylogenetic reasoning. Often only few key characters were used in non-reproducible analyses or simply stated as facts without pointing out how the taxa were character coded. In the following these manually derived phylogenies are referred to as informal, and often they reflect the evidence present in various authors favourite character set, e.g. claw/foot structures, cephalic appendages, bucco-pharyngeal apparatus or sclerotized plates.

Tardigrade morphology and functional anatomy are described in more detail in Chap. 2. The morphology-based data can and should be used together with molecular data in phylogenetic analyses and constitute an important part of formally defining and describing the inferred clades.

Beside the multigene phylogenomic (Dunn et al. 2008) and transcriptomic (Borner et al. 2014) approaches, relatively few genes have been used for phylogenetic inference in tardigrades. Most commonly the highly conserved nuclear ribosomal genes 18S and 28S have been used with huge success in inferring major evolutionary lineages. Also the nuclear protein coding genes elongation factor 1α, elongation factor 2 and the large subunit of RNA polymerase II have been used for phylogenetic inference (Regier et al. 2004). The mitochondrial gene cytochrome c oxidase subunit I (COI) has also been used in phylogenetic inference; however, this gene might be best suited for DNA barcoding approaches as it evolves very fast. The nuclear internal transcribed spacer 2 (ITS2) has been used in phylogeographic analysis and species discrimination.

3.2.2 *Eutardigrada, Mesotardigrada and Heterotardigrada*

Tardigrades are currently composed of three classes, i.e. the Eu-, Meso- and Heterotardigrada (Ramazzotti and Maucci 1983). Eutardigrada and Heterotardigrada were established by Marcus (1929) and the Mesotardigrada by Rahm (1937) (Fig. 3.3). Mesotardigrada is monotypic and established based on the description of *Thermozodium esakii* Rahm, 1937 from a Japanese hot spring. However, the species is widely considered as being dubious as the type material probably no longer exists. The presence of a mixture of heterotardigrades and eutardigrade characters, i.e. cephalic appendages (including well-developed lateral cirri) and a cloaca, made Rahm (1937) suggest a new tardigrade class. Binda and Kristensen (1986) suggested that *Thermozodium* Rahm, 1937 could be related to the limnic *Carphania* (Carphanidae, Echiniscoidea) due to the presence of a cloaca in both taxa but warned that the similarity might be caused by convergent evolution. Despite the fascinating position as a taxon of intermediate position between the eutardigrades and heterotardigrades, the mesotardigrades will not receive any further discussion here. Grothman et al. (2017) consider *T. esakii* and Mesotardigrada for nomen dubium until physical evidence is found. The monophyly of the two classes Eutardigrada and Heterotardigrada has been supported by molecular studies focusing on the higher levels of tardigrades (Jørgensen and Kristensen 2004; Nichols et al. 2006; Jørgensen et al. 2010) and studies including comprehensive taxon and gene sampling (Sands et al. 2008; Guil and Giribet 2012; Guil et al. 2013). Interestingly in

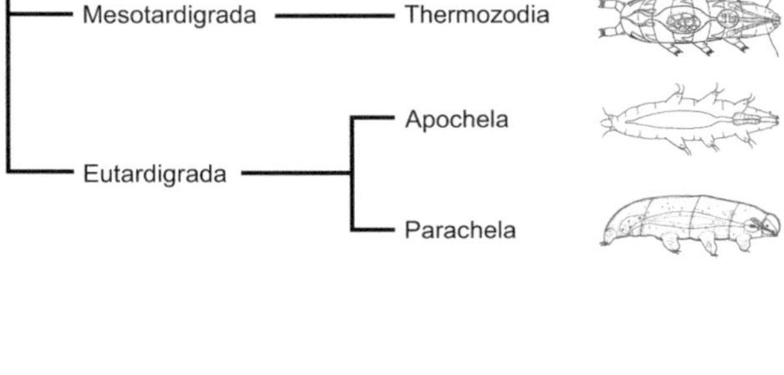

Fig. 3.3 Classical Higher-level tardigrade relationships. Mesotardigrada might be considered as a nomen dubium due to the missing type material and destroyed type locality

some outgroup combinations, the study by Guil and Giribet (2012) inferred a sister group relationship between Heterotardigrada and the eutardigrade order Apochela. This intriguing phylogenetic position could indicate a close relationship between *Milnesium* Doyère, 1840, and *Thermozodium* suggesting that Apochela are the real tardigrades in the middle, i.e. the Mesotardigrada.

3.2.2.1 Phylogeny of Eutardigrada

Eutardigrada predominantly inhabits limno-terrestrial habitats and is the class harbouring the majority of the currently described species (Degma et al. 2016). The eutardigrades are composed of the two orders Apochela Schuster, Nelson, Grigarick & Christenberry, 1980 and Parachela Schuster, Nelson, Grigarick & Christenberry, 1980. The monophyly of Eutardigrada has been repeatedly inferred from molecular data. Eutardigrada is morphologically characterized by the presence of a cloaca, a straight midgut, presence of excretory organs (Malpighian tubules), two double claws on each leg and the general absence of the numerous cephalic, body and leg sensory organs known from the heterotardigrades (see Chap. 2).

Phylogeny of Apochela

Apochela consists of the single family Milnesiidae Ramazzotti, 1962 currently composed of four genera, i.e. *Bergtrollus* Dastych, 2011, *Limmenius* Horning, Schuster & Grigarick, 1978, *Milnesioides* Claxton, 1999 and *Milnesium*. The phylogenetic relationship between these genera is unknown, and molecular data only exists for *Milnesium*. Apochela is morphologically characterized by the presence of lateral cephalic papillae and primary and secondary branches of the double claws that attach directly to the leg. Recently, Guil et al. (2019) suggested that Apochela should be elevated to the rank of class and given the name Apotardigrada.

Phylogeny of Parachela

Prior to the use of modern phylogenetic principles and methods, the Parachela was composed of five families, i.e. Calohypsibiidae Pilato, 1969, Eohypsbiidae Bertolani & Kristensen, 1987, Hypsibiidae Pilato, 1969, Macrobiotidae Thulin, 1928 and Necopinatidae Ramazzotti & Maucci, 1983 (Kinchin 1994). Figure 3.4a shows the traditional classification of Parachela. The taxon-rich Hypsibiidae included the three subfamilies Diphasconinae Dastych, 1992, Hypsibiinae Pilato, 1969 and Itaquasconinae Rudescu, 1964. The phylogeny of Macrobiotidae has received some attention (Guidetti and Bertolani 2001; Guidetti et al. 2005); however, the phylogeny at the parachelan family level was largely unknown until the use of molecular data (Kiehl et al. 2007; Sands et al. 2008). Parachela is morphologically characterized by the absence of external cephalic sensory appendages (except

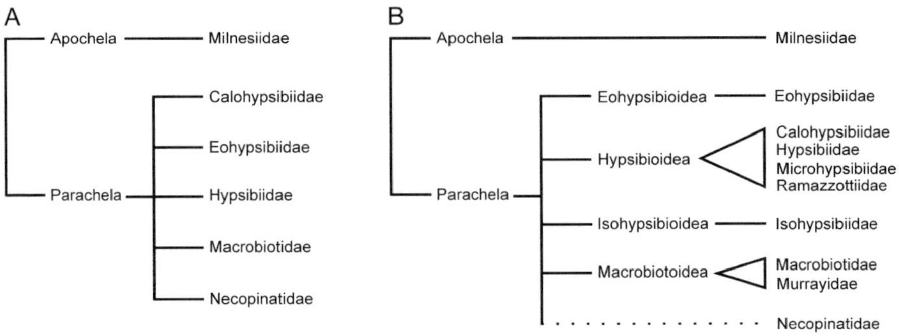

Fig. 3.4 Eutardigrade relationships. (**a**) The traditional classification of Eutardigrada. (**b**) The current phylogeny showing at least four major evolutionary lineages within Parachela. The sister group relationships between the major evolutionary lineages are currently unknown. Necopinatidae has not yet been included in the molecular analyses

peribuccal lamellae and circumoral sensory fields) and the presence of fused primary and secondary branches of the double claws and pharyngeal cuticular placoids. Secondarily evolved variations are present in a few species with regard to the presence/absence or structure of claws and placoids (see also Chap. 9 on cyclomorphosis and encystment).

The most recent studies using 18S (Marley et al. 2011) and combined 18S and 28S molecular data (Guil and Giribet 2012; Bertolani et al. 2014) inferred the existence of four major evolutionary lineages within Parachela (Fig. 3.4b), i.e. Eohypsibioidea Bertolani & Kristensen, 1987 in Marley et al. 2011, Hypsibioidea Pilato, 1969 in Marley et al. 2011, Isohypsibioidea Sands, McInnes, Marley, Goodall-Copestake, Convey & Linse, 2008, Macrobiotoidea Thulin, 1928 in Marley et al. 2011. These inferred groups were only in partial agreement with the classical morphology-based classification, but supporting morphological evidence has recently been provided by Marley et al. (2011) and Bertolani et al. (2014).

Currently several evolutionary lineages have been inferred within the four parachelan superfamilies (Bertolani et al. 2014; Fig. 3.4b). Eohypsibioidea consists of a single family Eohypsibidae Bertolani & Kristensen, 1987. Hypsibioidea consists of two main evolutionary lineages, i.e. a Ramazzotiidae clade and a much more diverse assemblage of taxa traditionally classified in Calohypsibiidae, Diphasconinae, Hypsibiinae and Itaquasconinae. Isohypsibioidea consists of several evolutionary lineages currently constituting a single family Isohypsibiidae. However, within Isohypsibioidea the phylogenetic relationships are at present unclear, and its genera come from the formerly known Calohypsibiidae and Hypsibiinae. The enigmatic clawless *Apodibius* Dastych, 1983 have recently been inferred within Isohypsibiidae (Dabert et al. 2014). Macrobiotoidea also consists of several evolutionary lineages forming major clades of unresolved taxonomic status. The macrobiotic family Murrayidae is supported by both morphological and molecular data, whereas the macrobiotic family, Macrobiotidae currently is regarded as polyphyletic and should be reconsidered in future studies (Bertolani et al. 2014). A new family, Richtersiidae, was recently established within Macrobiotoidea comprising the genera *Richtersius*, *Diaforobiotus* and *Adorybiotus* (Guidetti et al. 2016).

Despite the recent inference of the major parachelan evolutionary lineages by the nuclear genes 18S and 28S, the resolution of the phylogeny is poorly supported or incongruent between phylogenetic methods. A consensus of the phylogenetic relationships between the major parachelan clades has not yet been established, and Necopinatidae has not yet been included in the molecular analyses. Phylogenomic approaches will likely increase the phylogenetic resolution of the parachelan lineages, and with the increased accessibility of genomic and transcriptomic sequencing, data from several species are now feasible.

3.2.2.2 Phylogeny of Heterotardigrada

Heterotardigrada has traditionally and is currently divided into the predominantly marine Arthotardigrada (Marcus, 1927) and Echiniscoidea (Richters, 1926) with tidal, semiterrestrial and limnic species (Fig. 3.5). Whereas Echiniscoidea almost always appears monophyletic [an exception is Nichols et al. (2006)], the Arthrotardigrada has been shown to be paraphyletic based on phylogenetic inference of molecular data (Jørgensen et al. 2010; Guil and Giribet 2012). Guil et al. (2013) inferred common ancestry of most arthrotardigrades except *Tanarctus*, which is represented in the analyses by an aberrant 28S sequence. Future analyses with the addition of DNA sequences from a broader gene and taxon sampling of arthrotardigrades will give a much needed resolution of the relationships within the paraphyletic Arthrotardigrada. Heterotardigrada is morphologically characterized by the midgut having lateral diverticula, the presence of distinct anus and gonopore and numerous cephalic sensory organs.

Phylogeny of Arthrotardigrada

Arthrotardigrada currently consists of seven to eight families, but the taxonomic status of some taxa is still being debated, and there is currently no well-established phylogeny of the group. Arthrotardigrada is morphologically characterized by the presence of a median cirrus, three pairs of cephalic cirri, two to three pairs of clavae and a cirrus E with cirrophore (see Chap. 2).

The phylogeny of Arthrotardigrada was investigated by Kristensen and Higgins (1984b) in an informal analysis in which they plotted key apomorphic morphological characters on the cladogram defining each branch (Fig. 3.5a). Halechiniscidae Thulin, 1927 is the most taxon-rich arthrotardigrade family, and its phylogenetic relationships have traditionally been investigated using morphological data (Grimaldi de Zio and D'Addabbo Gallo 1987; Pollock 1995; Renaud-Mornant 1984). Based on phylogenetic inference from molecular data, it seems probable

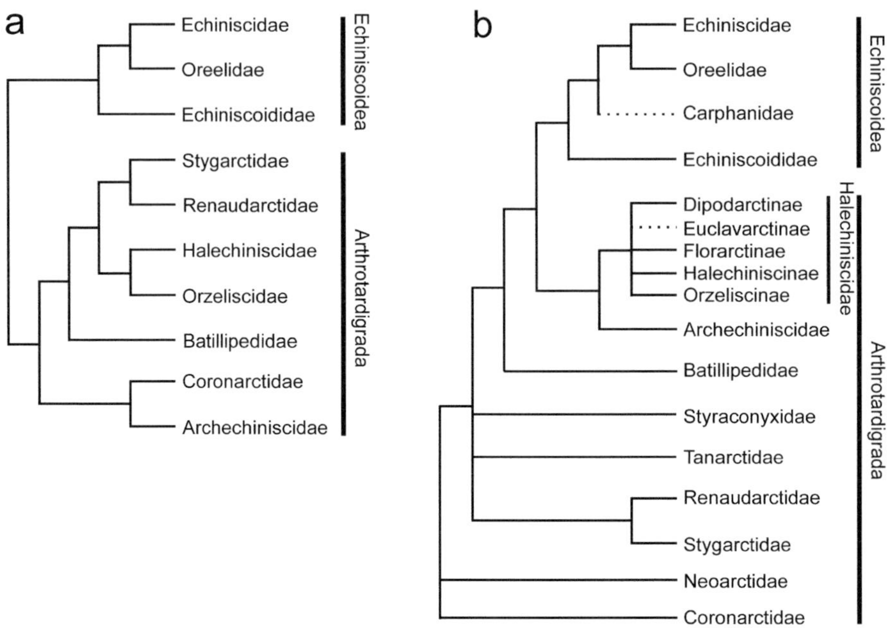

Fig. 3.5 Arthrotardigrade relationship. (**a**) The proposed phylogeny by Kristensen and Higgins (1984b) based on key morphological characters. (**b**) A preliminary consensus phylogeny based primarily on the molecular phylogeny inferred by Fujimoto et al. (2016). Taxa without data and uncertain positions are indicated by punctuated lines. Note the unresolved polytomies and paraphyletic Arthrotardigrada. It is still uncertain, which taxon represents the ancestral condition (i.e. is the sister group to all the other arthrotardigrades)

that Halechiniscidae is a "ragbag" of different evolutionary lineages that have been lumped together (Jørgensen et al. 2010). When a more complex data set recently was generated by Fujimoto et al. (2016), the analyses inferred four families that previously were included in Halechiniscidae, i.e. Archechiniscidae, Halechiniscidae, Styraconyxidae and Tanarctidae. An informal phylogenetic analysis by Kristensen and Renaud-Mornant (1983) suggests that the taxonomic status of some genera in the Styraconyxinae Kristensen & Renaud-Mornant, 1983 is highly problematic. The phylogeny of the armoured marine tardigrades in the family Stygarctidae Schulz, 1951 has furthermore been investigated using morphological data (Bello and Grimaldi de Zio 1998; Hansen et al. 2012). In the comprehensive treatment of the taxonomy and phylogeny of Stygarctidae, Hansen et al. (2012) introduced the seminal receptacles as a new character system in arthrotardigrade phylogeny. The analyses suggested that Neostygarctidae de Zio Grimaldi, D'Addabbo Gallo & De Lucia Morone, 1987 should be included in Stygarctidae and that the many peculiar characteristics of *Neostygarctus* Grimaldi de Zio, D'Addabbo Gallo & Morone De Lucia, 1982 were autapomorphies for the genus. It was also evident that the

Megastygarctidinae Bello & de Zio Grimaldi 1998 is very different from the other stygarctids.

The major obstacles that currently prevent the phylogeny of arthrotardigrades (Fig. 3.5b) to be as well-founded as that of Eutardigrada is the lack of molecular data (especially 18S) with sequences from more than a single specimen and the absence of a comprehensive morphological dataset. Notably, the current knowledge of the genetic and morphological diversity of arthrotardigrades might simply represent a scattered picture of the full, currently unknown, arthrotardigrade diversity. Hopefully more attention will be given to the diversity and phylogeny of arthrotardigrades in the next decades.

Phylogeny of Echiniscoidea

Echiniscoidea currently consists of four families, i.e. Echiniscidae Thulin, 1928, Oreellidae Ramazzotti, 1963, Echiniscoididae Kristensen & Hallas, 1980 and Carphanidae Binda & Kristensen, 1986. However, the position of *Carphania* Binda, 1978 is questionable. When Richters (1926) established Echiniscoidea, it only contained *Echiniscoides* [*Echiniscoides sigismundi* M. Schultze, 1865]. Oreellidae (*Oreella mollis* Murray, 1910) and Echiniscidae (various species of *Echiniscus* C.A.S. Schultze, 1840) were regarded as eutardigrades. Echiniscoidea is morphologically characterized by reduced cephalic sensory organs as compared to arthrotardigrades, i.e. the median cirrus is missing, and other cephalic sensory organs are reduced except for the secondary clava (buccal papilla). The leg sensory organs are also reduced or absent.

Absence of the median cirrus in some arthrotardigrades has historically caused taxonomic problems. Schulz (1953) described the species *Archechiniscus marci* Schulz, 1953 from intertidal barnacles and placed it in the order Echiniscoidea (family Oreellidae) because it lacks the median cirrus. Subsequently, Binda (1978) transferred the species to the Arthrotardigrada and established the family Archechiniscidae. The archechiniscids were moved to the Halechiniscidae (subfamily Archechniscinae) when the new species *Archechiniscus minutus* Grimaldi de Zio & d'Addabbo Gallo, 1987 from subtidal sand was described. Recently *Archechiniscus* specimens with a median cirrus were collected subtidally at the Solomon Islands (Kristensen unpubl.). The genus *Archechiniscus* may lack the median cirrus (intertidal) or have a small median cirrus (interstitial in sand). So the key character for the difference between Echiniscoidea and Arthrotardigrada, i.e. the presence or absence of the median cirrus, can be misleading if additional evidence is not taken into consideration. The genus *Anisonyches* (currently Echiniscoididae) could play a major role in the morphological definition of the two orders: Arthrotardigrada and Echiniscoidea. Like in *Archechiniscus*, the median cirrus may be present as found in intertidal *Anisonyches* specimens from Saudi Arabia (Kristensen unpubl.).

Members of the family Echiniscidae are predominantly found in terrestrial habitats as mosses and lichens, where they can be numerous. The phylogeny of

Echiniscidae was first investigated by Kristensen (1987) who presented a manually derived phylogenetic tree constructed using cladistic principles. Jørgensen (2000) used the morphological characters from Kristensen (1987) to create a data matrix that was used to infer the phylogeny of Echiniscidae. This work was subsequently expanded by the addition of DNA sequence data (Jørgensen et al. 2011) that inferred a clade consisting of *Antechiniscus* Kristensen, 1987, *Cornechiniscus* Maucci & Ramazzotti, 1981, *Proechiniscus* Kristensen, 1987 and *Pseudechiniscus islandicus* (Richters, 1904) resembling the '*Pseudechiniscus* lineage' proposed by Kristensen (1987), but without *Mopsechiniscus* du Bois-Reymond Marcus, 1944. Furthermore, this study showed that the genus *Pseudechiniscus* Thulin, 1911 is not monophyletic comprising at least two genera. The genus *Multipseudechiniscus* Schulte & Miller, 2011 was established by transferring *Pseudechiniscus raneyi* (Grigarick, Mihelčič & Schuster, 1964) to the new genus (Schulte and Miller 2011). Further phylogenetic analysis of Echiniscidae were conducted by Vicente et al. (2013), who, in an integrative taxonomy approach, established the new genus *Diploechiniscus* Vicente, Fontoura, Cesari, Rebecchi, Guidetti, Serrano & Bertolani, 2013 based on *Echiniscus oihonnae* (Richters, 1903). Also Guil et al. (2013) investigated the phylogeny of Echiniscidae and added a substantial amount of DNA sequences extracted from single specimens to the phylogeny. This study elucidated both congruences and conflicts between the traditional morphological data based on cuticular design and the molecular data. As in Jørgensen et al. (2011), Guil et al. (2013) inferred the modified '*Pseudechiniscus* lineage' but also a clade including *Hypechiniscus* Thulin, 1928, *Testechiniscus* Kristensen, 1987, *Diploechiniscus* and *Echiniscus* C.A.S. Schultze, 1840 resembling Kristensen's (1987) '*Echiniscus* lineage' but excluding *Bryodelphax* Thulin, 1928 and *Bryochoerus* Marcus, 1936. Although the basal topology of the Echiniscidae is still largely unresolved, the use of molecular data resulted in phylogenies with many similarities to the phylogenies derived from morphological characters. A consensus phylogeny of Echiniscidae is presented in Fig. 3.6.

Echiniscoididae is predominantly marine with most of the 17 currently described taxa reported from the intertidal zone, where they often use barnacles as substrate. Some of the tardigrades tolerating the most extreme environmental conditions can be found in this family (Møbjerg et al. 2014). The phylogeny of Echiniscoididae was recently investigated using 28S and COI sequences, and a new subfamily, Isoechiniscoidinae Møbjerg, Kristensen & Jørgensen, 2016, was established to accommodate a new genus *Isoechiniscoides* Møbjerg, Kristensen & Jørgensen, 2016 comprising species with isonych claws, pillars in the epicuticle and exceptionally long sensory appendages (Møbjerg et al. 2016). Based on the molecular analyses, Isoechiniscoidinae was inferred as sister group to all other Echiniscoididae taxa. Echiniscoididae currently comprises two other genera, i.e. *Anisonyches* Pollock, 1975 and *Echiniscoides* Plate, 1888. From studies on the genetic diversity of *Echiniscoides*, it has become very clear that this genus comprises cryptic complexes with several clades of *Echiniscoides* taxa inferred from molecular analyses (Faurby et al. 2011, 2012). Currently, however, concomitant descriptions of morphotypes are lacking. Moreover, the phylogeny of Echiniscoidea needs an improvement of the

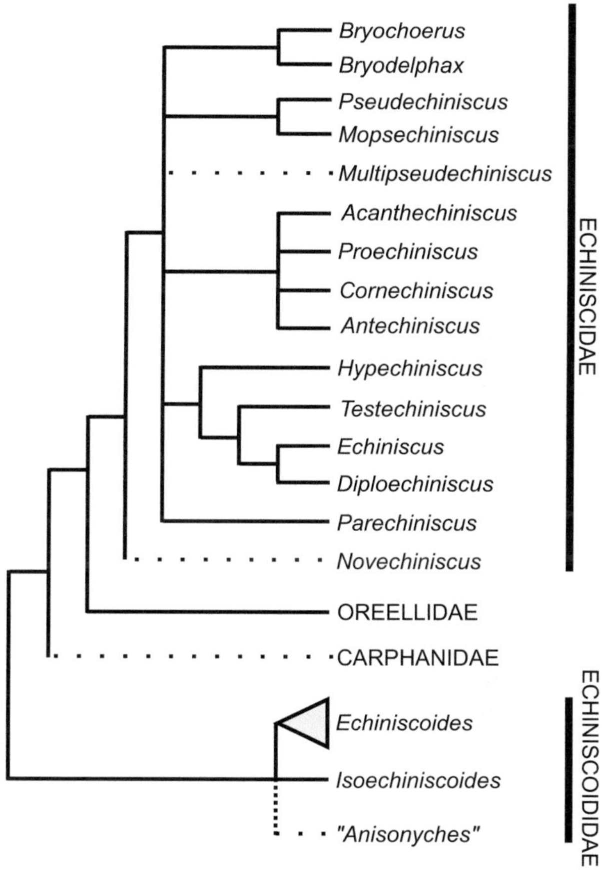

Fig. 3.6 Echiniscoidean relationships. Consensus phylogeny based on both morphological and molecular evidence (Guil et al. 2013; Jørgensen et al. 2011; Møbjerg et al. 2016; Vecchi et al. 2016). Taxa without molecular data have been indicated with punctuated lines. Note that *Echiniscoides* is currently an unresolved group of taxa at various taxonomical levels and that *Anisonyches*, although currently placed in Echiniscoidea, is more closely related to arthrotardigrade taxa than to the other echiniscoidean taxa

taxon and gene sampling from Oreellidae and Carphanidae, and also sequences from more basal *Echiniscoides* species and *Anisonyches*. A better understanding of the phylogenetic relationships within Echiniscoididae, a major tardigrade lineage comprising the tardigrades of intertidal zones worldwide, would elucidate the colonization of terrestrial environments by echiniscoideans and shed light on the evolution of cuticular sclerotization, claw number and cryptobiosis.

3.3 Integrative Taxonomy

Taxonomy encompasses the description, identification, nomenclature and classification of organisms. Traditional taxonomy was mostly morphology and anatomy based, and it is important to note that detailed morphological descriptions are still the foundation for species descriptions; however, integrative taxonomy seeks to implement all methods and data to improve species discovery and description (Padial et al.

2010). Especially the use of molecular methods in support of morphological investigations has attracted much attention with DNA barcoding (Hebert et al. 2003) as the main avenue of investigation. In the last decade, the identification and characterization of different tardigrade species have been facilitated using DNA barcoding approaches and ITS2 secondary structure as a marker for taxonomy.

It is important to obtain voucher specimen information related to the acquired DNA sequences before the microscopic tardigrade specimens are destroyed during DNA extraction (Cesari et al. 2011; Guil et al. 2013). So far different kinds of substitute voucher specimens have been used or proposed, i.e. the additional slide preparations of similar/identical specimens from the same sample, digital images, eggs or progeny from the DNA extracted specimen. The most convenient method is currently to use high magnification digital images as voucher specimens. Though it is possible to amplify genes from single cells, this technology has not yet been optimised for tardigrades, but could in the future be implemented on, e.g. storage cells isolated from voucher specimens.

3.3.1 DNA Barcoding

In animal DNA barcoding an app. 650 bp fragment of the mitochondrial gene, cytochrome oxidase c subunit I (COI), has been chosen as the standard barcode (Hebert et al. 2003). Note that other genes are also used as DNA barcodes by various authors depending on the objectives of the studies. The COI gene is amplifiable with universal primers and provides interspecific variation and a relatively small amount of intraspecific variation (taxonomic gap between species). The major shortcoming of DNA barcoding is the lack of sequence variation between sibling species. A DNA barcoding approach has the strongest potential in designation of species already within the database and also with regard to the emphasis/discovery of hidden DNA sequence variation in cryptic complexes. DNA sequences without a species label can be clustered into molecular operational taxonomic units (MOTUs) an approach used by Blaxter and co-workers (Blaxter et al. 2004, 2005). The generation of MOTUs can be used when the taxonomy of the specimens is unknown and will allow non-tardigrade researchers to readily include tardigrades in their studies. However, linking MOTUs and DNA barcodes to species and vouchers should be the preferred approach placing anonymous sequences firmly within the taxonomic framework.

DNA barcoding approaches have been applied in both eutardigrades (Cesari et al. 2009, 2011, 2013; Bertolani et al. 2010, 2011a, b) and echiniscoideans (Vicente et al. 2013). Faurby et al. (2012), although not a taxonomic study, used the barcoding terminology and various cut-off criterions to define phylogenetically inferred clusters within *Echiniscoides*. They found 43 distinct clusters at the 3% COI sequence variation cut-off criterion (may represent 43 distinct species) and 34–35 clusters at the 10–15% COI cut-off criterion (these clusters could represent undescribed genera). As discussed above, the genetic diversity within *Echiniscoides* is of a magnitude to warrant the erection of several new species and even genera. Jørgensen et al.

(2011) reported up to 50% COI sequence variation within Echiniscidae, and Vicente et al. (2013) reported 20% COI variation between the echiniscids *Testechiniscus* and *Diploechiniscus*. As noted above taxonomic work on *Echiniscoides* is difficult due to very similar morphology of the taxa, and an integrative taxonomic approach is needed to resolve the cryptic species complexes. *Echiniscoides* can be regarded as a 'Pandora's box' containing intriguing mysteries rather than the evils of the world. We expect that investigations into the various aspects of echiniscoidid biology will have far-reaching consequences for our understanding of tardigrade biology, an understanding that is currently dominated by our knowledge of eutardigrades.

Only relatively few tardigrade COI barcodes is currently available for referencing in the 'Barcode of Life Data Systems' or 'Tardigrade Barcoding Project' databases, and tardigrade researchers are strongly encouraged to supply more sequences.

3.3.2 Compensatory Base Changes in ITS2 Secondary Structure

The nuclear internal transcribed spacer 2 (ITS2) is flanked by the ribosomal genes 5.8S and 28S, and it is presumed to have a regulatory function in the transcription of ribosomal subunits (Van Nues et al. 1995). Generally, ITS fold in a secondary structure composed of conserved stems and variable loops (Fig. 3.7a). ITS2 has four main helix domains (DI–DIV), each composed of a stem and a loop. The DII and DIII helix domains are conserved, while DI and DIV are variable with substantial interspecific sequence variation both with regard to nucleotide composition and with insertions or deletions of nucleotides (Coleman 2009).

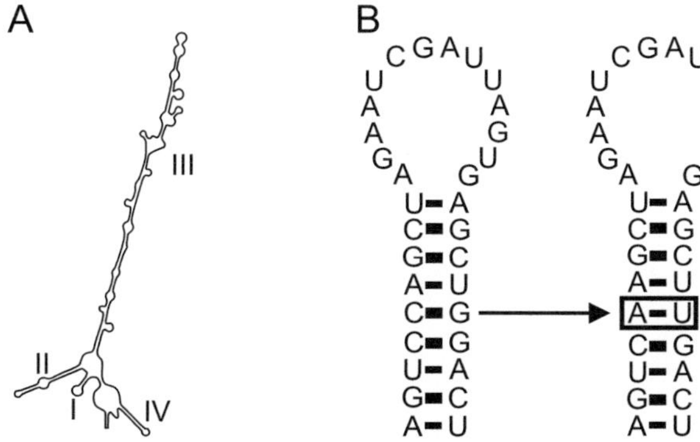

Fig. 3.7 Secondary structure of internal transcribed spacer 2 (ITS2). (**a**) Outline of the secondary structure of ITS2 in *Milnesium* (based on Wełnicz et al. 2011). The four main helix domains are indicated (I–IV). (**b**) Simultaneous nucleotide substitutions at both of the two pairing positions in the stem region of a primary RNA transcript are known as a compensatory base change

Simultaneous nucleotide substitutions at both of the two pairing positions in a double-stranded helix (the paired region of a primary RNA transcript) are known as compensatory base changes (CBCs; Fig. 3.7b). A CBC maintains the conserved secondary structure because the nucleotides in the stem are still able to pair. A relationship exists between the interbreeding ability of closely related taxa and CBCs (Coleman 2000, 2009). It has been hypothesized that taxa differing by a single CBC are incapable of intercrossing (Coleman and Vacquier 2002). Müller et al. (2007) found that a single CBC was able to discriminate between two closely related species with 93% confidence in a large dataset.

When a secondary structure template for ITS2 is known, an automated CBC estimation has been developed (Schultz and Wolf 2009). CBCs have been used with promising results in a study of *Paramacrobiotus* Guidetti, Schill, Bertolani, Dandekar and Wolf, 2009 in which the CBC approach was able to distinguish between six closely related species and identify three undescribed species (Schill et al. 2010).

Acknowledgements We are grateful to scientific illustrator Stine Elle for her drawings of animals in the cladograms. Funding came from the Carlsberg Foundation and the Danish Council for Independent Research (grant no. DFF – 4090-00145).

References

Aguinaldo AA, Turbeville JM, Linford LS et al (1997) Evidence for a clade of nematodes, arthropods and other moulting animals. Nature 387:489–493

Bello G, de Zio Grimaldi S (1998) Phylogeny of the genera of the Stygarctidae and related families (Tardigrada: Heterotardigrada). Zool Anz 237:171–183

Bertolani R, Grimaldi D (2000) A new eutardigrade (Tardigrada: Milnesiidae) in amber from Upper Cretaceous (Turonian) of New Jersey. In: Grimaldi D (ed) Studies on fossils in amber, with particular reference to the cretaceous of New Jersey. Backhuys Publishers, Leiden, pp 103–110

Bertolani R, Rebecchi L, Cesari M (2010) A model study for tardigrade identification. In: Nimis PL, Vignes Lebbe R (eds) Tools for identifying biodiversity: progress and problems. EUT-Edizioni Università di Trieste, Trieste, pp 333–339

Bertolani R, Rebecchi L, Giovannini I et al (2011a) DNA barcoding an integrative taxonomy of *Macrobiotus hufelandi* C.A.S. Schultze 1834, the first tardigrade species to be described, and some related species. Zootaxa 2997:19–36

Bertolani R, Biserov VI, Rebecchi L et al (2011b) Taxonomy and biogeography of tardigrades using an integrated approach: new results on species of the *Macrobiotus hufelandi* group. Inv Zool 8:23–36

Bertolani R, Guidetti R, Marchioro T et al (2014) Phylogeny of Eutardigrada: new molecular data and their morphological support lead to the identification of new evolutionary lineages. Mol Phylogenet Evol 76:110–126

Binda BG (1978) Risistemazione di alcuni Tardigrada con l'istituzione di un nuova genere di Oreellidae e della nuova famiglia Archechiniscidae. Animalia 5:307–314

Binda MG, Kristensen RM (1986) Notes on the genus *Oreella* (Oreellidae) and the systematic position of *Carphania fluviatilis* Binda, 1978 (Carphanidae fam. nov., Heterotardigrada). Animalia 13:9–20

Blaxter M, Elsworth B, Daub J (2004) DNA taxonomy of a neglected animal phylum: an unexpected diversity of tardigrades. Proc R Soc Lond B 271:S189–S192

Blaxter M, Mann J, Chapman T et al (2005) Defining operational taxonomic units using DNA barcode data. Phil Trans R Soc Lond B Biol Sci 360(1462):1935–1943

Borner J, Rehm P, Schill RO et al (2014) A transcription approach to ecdysoan phylogeny. Mol Phylogenet Evol 80:79–87

Brusca RC, Brusca GJ (2003) Invertebrates, 2nd edn. Sinauer Associates, Sunderland, MA, p 895

Budd GE (2001) Tardigrades as 'stem-group arthropods': the evidence from the Cambrian fauna. Zool Anz 240:265–279

Campbell LI, Rota-Stabelli O, Edgecombe GD et al (2011) MicroRNAs and phylogenomics resolve the relationships of Tardigrada and suggest that velvet worms are the sister group of Arthropoda. PNAS 108:15920–15924

Cesari M, Bertolani R, Rebecchi L et al (2009) DNA barcoding in Tardigrada: the first case study on *Macrobiotus macrocalix* Bertolani & Rebecchi 1993 (Eutardigrada, Macrobiotidae). Mol Ecol Res 9:699–706

Cesari M, Giovannini I, Bertolani R et al (2011) An example of problems associated with DNA barcoding in tardigrades: a novel method for obtaining voucher specimens. Zootaxa 3104:42–51

Cesari M, Guidetti R, Rebecchi L et al (2013) A DNA barcoding approach in the study of tardigrades. J Limnol 72:182–198

Coleman AW (2000) The significance of a coincidence between evolutionary landmarks found in mating affinity and a DNA sequence. Protist 151:1–9

Coleman AW (2009) Is there a molecular key to the level of biological species in eukaryotes? A DNA guide. Mol Phylogenet Evol 50:197–203

Coleman AW, Vacquier VD (2002) Exploring the phylogenetic utility of ITS sequences for animals: a test case for abalone (*Haliotis*). J Mol Evol 54:246–257

Cooper KW (1964) The first fossil tardigrade: *Beorn leggi*, from Cretaceous Amber. Psyche (Camb) 71(2):41–48

Cuvier GB (1812) Sur un nouveau rapprochement à établir entre les classes qui composant le Règne Animal. Ann Mus Hist 19:73–84

Dabert M, Dastych H, Hohberg K et al (2014) Phylogenetic position of the enigmatic clawless eutardigrade genus *Apodibius* Dastych, 1983 (Tardigrada), based on 18S and 28S rRNA sequence data from its type species *A. confusus*. Mol Phylogenet Evol 70:70–75

Degma P, Bertolani R, Guidetti R (2009–2016) Actual checklist of Tardigrada species. Ver. 31: 15-12-2016. http://www.tardigrada.modena.unimo.it/miscellanea/Actual%20checklist%20of%20Tardigrada.pdf. Accessed 20 Dec 2016

Dunn CW, Hejnol A, Matus DQ et al (2008) Broad phylogenomic sampling improves resolution of the animal tree of life. Nature 452:665–780

Edgecombe GD (2010) Arthropod phylogeny: an overview from the perspectives of morphology, molecular data and the fossil record. Arthropod Struct Dev 39:74–87

Faurby S, Jørgensen A, Kristensen RM et al (2011) Phylogeography of North Atlantic tidal tardigrades: refugia, cryptic speciation and the history of the Mid Atlantic islands. J Biogeogr 38:1613–1624

Faurby S, Jørgensen A, Kristensen RM et al (2012) Distribution and speciation in marine tidal tardigrades: a test of the roles of climate and geographic isolation. J Biogeogr 39(9):1596–1607

Fujimoto S, Jørgensen A, Hansen JG (2016) A molecular approach to arthrotardigrade phylogeny (Heterotardigrada, Tardigrada). Zool Scr 46(4):496–505

Garey JR, Krotec M, Nelson DR et al (1996) Molecular analysis supports a tardigrade arthropod association. Invert Biol 115:79–88

Giribet G, Carranza S, Baguna J et al (1996) First molecular evidence for the existence of a Tardigrada + Arthropoda clade. Mol Biol Evol 13(1):76–84

Grimaldi de Zio S, d'Addabbo Gallo M (1987) *Archechiniscus minutus* n. sp. and its systematic position within Arthrotardigrada (Tardigrada: Heterotardigrada). In: Bertolani R (ed) Biology of Tardigrades. Selected symposia and monographs U.Z.I. 1. Mucchi, Modena, pp 253–260

Grothman GT, Johansson C, Chilton G et al (2017) Gilbert Rahm and the status of Mesotardigrada Rahm, 1937. Zool Sci 34(1):5–10

Guidetti R, Bertolani R (2001) Phylogenetic relationships in the Macrobiotidae (Tardigrada: Eutardigrada: Parachela). Zool Anz 240:371–376

Guidetti R, Gandolfi A, Rossi V et al (2005) Phylogenetic analysis in Macrobiotidae (Eutardigrada, Parachela): a combined morphological and molecular approach. Zool Scr 34:235–244

Guidetti R, Rebecchi L, Bertolani R et al (2016) Morphological and molecular analyses on *Richtersius* (Eutardigrada) diversity reveal its new systematic position and lead to the establishment of a new genus and a new family within Macrobiotoidea. Zool J Linn Soc 178(4):834–845

Guil N, Giribet G (2012) A comprehensive molecular phylogeny of tardigrades–adding genes and taxa to a poorly resolved phylum-level phylogeny. Cladistics 28:21–49

Guil N, Jørgensen A, Giribet G et al (2013) Congruence between molecular phylogeny and cuticular design in Echiniscoidea (Tardigrada, Heterotardigrada). Zool J Linn Soc 169 (4):713–736

Guil N, Jørgensen A, Kristensen R (2019) An upgraded comprehensive multilocus phylogeny of the Tardigrada tree of life. Zool Scripta 48(1):120–137

Hansen JG, Kristensen RM, Jørgensen A (2012) The armoured marine tardigrades (Arthrotardigrada, Tardigrada). Scientia Danica, series B Biologica 2, The Royal Danish Academy of Sciences and Letters, p 91

Hebert PDN, Cywinska A, Ball SL et al (2003) Biological identifications through DNA barcodes. Proc R Soc Lond (Biol) 270:313–321

Hejnol A, Schnabel R (2005) The eutardigrade *Thulinia stephaniae* has an indeterminate development and the potential to regulate early blastomere ablations. Development 132:1349–1361

Hejnol A, Obst M, Stamatakis A et al (2009) Assessing the root of bilaterian animals with scalable phylogenomic methods. Proc R Soc Lond (Biol) 276:4261–4270

Jørgensen A (2000) Cladistic analysis of the Echiniscidae Thulin, 1928 (Tardigrada: Heterotardigrada: Echiniscoidea). Steenstrupia 25:11–23

Jørgensen A, Kristensen RM (2004) Molecular phylogeny of Tardigrada – investigation of the monophyly of Heterotardigrada. Mol Phylogenet Evol 32:666–670

Jørgensen A, Faurby S, Hansen JG et al (2010) Molecular phylogeny of Arthrotardigrada (Tardigrada). Mol Phylogenet Evol 54:1006–1015

Jørgensen A, Møbjerg N, Kristensen RM (2011) Phylogeny and evolution of the Echiniscidae (Echiniscoidea, Tardigrada) – an investigation of the congruence between molecules and morphology. J Zool Syst Evol Res 49(Suppl 1):6–16

Kiehl E, Dastych H, D'Haese H et al (2007) The 18S rDNA sequences support polyphyly of the Hypsibiidae (Eutardigrada). J Limnol 66(Suppl 1):21–25

Kinchin IM (1994) Origins and systematics. In: Kinchin IM (ed) The biology of Tardigrades. Portland Press, London, pp 7–13

Kristensen RM (1981) Sense organs of two marine arthrotardigrades (Heterotardigrada, Tardigrada). Acta Zool 62:27–41

Kristensen RM (1987) Generic revision of the Echiniscidae (Heterotardigrada), with a discussion of the origin of the family. In: Bertolani R (ed) Biology of Tardigrades. Selected symposia and monographs U.Z.I. 1. Mucchi, Modena, pp 261–335

Kristensen RM, Higgins RP (1984a) Revision of *Styraconyx* (Tardigrada: Halechiniscidae), with description of two new species from Disko Bay, West Greenland. Smithson Contr Zool 391:1–40

Kristensen RM, Higgins RP (1984b) A new family of Arthrotardigrada (Heterotardigrada: Tardigrada) from the Atlantic coast of Florida, USA. Trans Am Microsc Soc 103:295–311

Kristensen RM, Renaud-Mornant J (1983) Existence d'Arthrotardigrades semi-benthiques de genres nouveaux de la sous-famille des Styraconyxinae subfam. nov. Cah Biol Mar 24:337–353

Lartillot N, Philippe H (2008) Improvement of molecular phylogenetic inference and the phylogeny of Bilateria. Phil Trans R Soc B 363:1463–1472

Liu J, Dunlop JA (2014) Cambrian lobopodians: a review of recent progress in our understanding of their morphology and evolution. Palaeogeogr Palaeocl 398:4–15

Liu J, Steiner M, Dunlop JA et al (2011) An armoured Cambrian lobopodian from China with arthropod-like appendages. Nature 470:526–530

Maas A, Waloszek D (2001) Cambrian derivatives of the early arthropod stem lineage, pentastomids, tardigrades and lobopodians – an "Orsten" perspective. Zool Anz 240 (3/4):451–459

Maas A, Mayer G, Kristensen RM et al (2007) A Cambrian micro-lobodian and the evolution of arthropod locomotion and reproduction. Chin Sci Bull 52(24):3385–3392

Marcus E (1929) Tardigrada. In: Bronn HG (ed) Klassen und Ordnungen des Tier-reichs vol 5 sec 4 part 3, pp 1–609

Marley NJ, McInnes SJ, Sands CJ (2011) Phylum Tardigrada: a re-evaluation of the Parachela. Zootaxa 2819:51–64

Møbjerg N, Andersen K, Hygum L et al (2014) Extreme stress tolerance of the intertidal tardigrade *Echiniscoides sigismundi*. In: The physiologist, APS intersociety meeting: comparative approaches to grand challenges in physiology, San Diego, California, 5–8 October 2014. Program number 39.3

Møbjerg N, Kristensen RM, Jørgensen A (2016) Data from new taxa infer *Isoechiniscoides* gen. nov. and increase the phylogenetic and evolutionary understanding of echiniscoidid tardigrades (Echiniscoidea: Tardigrada). Zool J Linn Soc 178:804–818

Moon SY, Kim W (1996) Phylogenetic position of the Tardigrada based on the 18S ribosomal RNA gene sequences. In: McInnes SJ, Norman DB (eds) Tardigrade biology. Zool J Linn Soc 116 (1/2):61–69

Müller T, Philippi N, Dandekar T et al (2007) Distinguishing species. RNA 13:1469–1472

Nichols PB, Nelson DR, Garey JR (2006) A family level analysis of tardigrade phylogeny. Hydrobiologia 558:53–60

Nielsen C (1995) Animal evolution: interrelationships of the living phyla. Oxford University Press, Oxford, p 467

Nielsen C (2012) Animal evolution: interrelationships of the living phyla, 3rd edn. Oxford University Press, Oxford, p 402

Padial JM, Miralles A, De la Riva I et al (2010) The integrative future of taxonomy. Front Zool 7:16

Persson D, Halberg KA, Jørgensen A et al (2012) Neuroanatomy of *Halobiotus crispae* (Eutardigrada: Hypsibiidae): Tardigrade brain structure supports the clade Panarthropoda. J Morph 273:1227–1245

Persson DK, Halberg KA, Jørgensen A et al (2014) Brain anatomy of the marine tardigrade *Actinarctus doryphorus* (Arthrotardigrada). J Morph 275(2):173–190

Pollock LW (1995) New marine tardigrades from Hawaiian beach sand and phylogeny of the family Halechiniscidae. Invertebr Biol 114:220–235

Rahm G (1937) Eine neue Tardigraden-Ordnung aus den heißen Quellen von Unzen, Insel Kyushu, Japan. Zool Anz 120:65–71

Ramazzotti G, Maucci W (1983) Il Phylum Tardigrada. Terza edizione riveduta e corretta. Memorie dell'Isituto Italiano di Idrobiologia Dott. Marco De Marchi 1–1012:41 (English translation by Clark W. Beasley)

Regier JC, Shultz JW, Kambic RE et al (2004) Robust support for tardigrade clades and their ages from three protein-coding nuclear genes. Invertebr Biol 123:93–100

Renaud-Mornant J (1982) Species diversity in marine Tardigrada. In: Nelson D (ed) Proceedings of the Third International Symposium on Tardigrada, Johnson City, Tennessee, 1980. East Tennessee State University Press, Tennessee, pp 149–178

Renaud-Mornant J (1984) Halechiniscidae (Heterotardigrada) de la campagne Benthedi, canal du Mozambique. Bull Mus natn Hist nat Paris 6(1):67–88

Renaud-Mornant J (1987) Halechiniscidae nouveaux de sables coralliens tropicaux (Tardigrada, Arthrotardigrada). Bull Mus natn Hist nat Paris 9(2):353–373

Richters F (1926) Tardigrada. In: Kükenthal W, Krumbach T (eds) Handbuch der Zoologie III. Walter de Gruyter, Berlin, pp 58–61

Rota-Stabelli O, Kayal E, Gleeson D et al (2010) Ecdysozoan mitogenomics: evidence for a common origin of the legged invertebrates, the Panarthropoda. Genome Biol Evol 2:425–440

Rota-Stabelli O, Daley AC, Pisani D (2013) Molecular timetrees reveal a Cambrian colonization of land and a new scenario for ecdysozoan evolution. Curr Biol 23:392–398

Ruppert EE, Fox RS, Barnes RD (2003) Invertebrate zoology: a functional evolutionary approach, 7th edn. Cengage Learning, Hampshire, p 1008

Sands CJ, McInnes SJ, Marley NJ et al (2008) Phylum Tardigrada: an "individual" approach. Cladistics 24:1–11

Schill RO, Förster F, Dandekar T et al (2010) Using compensatory base change analysis of internal transcribed spacer 2 secondary structures to identify three new species in *Paramacrobiotus* (Tardigrada). Org Divers Evol 10:286–296

Schmidt-Rhaesa A (2001) Tardigrades – are they really miniaturized dwarfs? Zool Anz 240:549–555

Schulte R, Miller WR (2011) Tardigrades of North America: the elevation of a new genus from California. In: Abstracts of the 143rd annual meeting of the Kansas Academy of Science, Baker University, Baldwin City, Kansas, 8–9 April 2011. Transactions of the Kansas Academy of Science 114(1/2):177

Schultz J, Wolf M (2009) ITS2 sequence-structure analysis in phylogenetics. A how to manual for molecular systematics. Mol Phylogenet Evol 52:520–523

Schulz E (1953) Eine neue Tardigradengattung von der pazifischen Küste. Zool Anz 151:306–310

Telford MJ, Bourlat SJ, Economou A et al (2008) The evolution of the Ecdysozoa. Phil Trans R Soc Lond B Biol Sci 363:1529–1537

van Nues RW, Rientjes MJ, Morré SA et al (1995) Evolutionarily conserved structural elements are critical for processing of Internal Transcribed Spacer 2 from *Saccharomyces cerevisiae* precursor ribosomal RNA. J Mol Biol 250:24–36

Vecchi M, Cesari M, Bertolani R et al (2016) Integrative systematic studies on tardigrades from Antarctica identify new genera and new species within Macrobiotoidea and Echiniscoidea. Invertebr Syst 30:303–322

Vicente F, Fontoura P, Cesari M et al (2013) Integrative taxonomy allows the identification of synonymous species and the erection of a new genus of Echiniscidae (Tardigrada, Heterotardigrada). Zootaxa 3613(6):557–572

Walossek D, Müller KJ, Kristensen RM (1994) A more than half a billion years old stem-group tardigrade from Sibiria. In: Sixth International Symposium on Tardigrada, Selwyn College, Cambridge, 22–26 August 1994

Wełnicz W, Grohme MA, Kaczmarek Ł et al (2011) ITS-2 and 18S rRNA data from *Macrobiotus polonicus* and *Milnesium tardigradum* (Eutardigrada, Tardigrada). J Zool Syst Evol Res 49 (1):34–39

Chapter 4
Tardigrade Biogeography

S. J. McInnes and P. J. A. Pugh

Abstract We report on the history and the progress made towards understanding tardigrade biogeography. We have updated our data and provided new analyses of both limno-terrestrial and marine tardigrade faunas to show the limitations of the earlier techniques, demonstrated newer ideas and highlighted the regions requiring further study.

4.1 Introduction

Biogeography, the study of the geographic distribution of biological organisms, has a long history; indeed, Aristotle asked the question 'How are organisms distributed?' (Aristotle *Historia Animālium* Book VIII). The fundamental principle of biogeography is based on the works of Georges-Louis Leclerc, Comte de Buffon (Buffon 1749–1767), whose observation that 'environmentally similar but isolated regions have distinct assemblages of mammals and birds' led to 'Buffon's Law' and 'migration as causal explanation for biotic distribution'. These ideas helped formulate the principles of biological evolution by Charles Darwin and Alfred Wallace. The tenets of evolutionary biology that 'biotic distributions are a result of dispersal away from a centre of origin' have usurped the original comparative biological study of biogeography into the modern phylogenetic biogeography and evolutionary biological research, where individual lineages are traced back to a common origin (Ebach et al. 2003). Much of this early comparative and phylogenetic study, which extends into the 1960s, assumed a static earth. The idea of continental drift, suggested as early as 1596 (Ortelius 1596), was slow to take hold (Wegener 1912), until the mechanism of plate movement was proven (Isacks et al. 1968). This has turned Buffon's Law of 'phylogeny explains biotic distribution' into

S. J. McInnes (✉)
British Antarctic Survey, Cambridge, UK
e-mail: smcinnes@bas.ac.uk

P. J. A. Pugh
Department of Biology, Anglia Ruskin University, Cambridge, UK

R. O. Schill (ed.), *Water Bears: The Biology of Tardigrades*, Zoological Monographs 2,
https://doi.org/10.1007/978-3-319-95702-9_4

'biogeographic distributions may explain phylogeny' (Ebach et al. 2003), though some care is needed in the interpretation of results.

The tardigrade is a microscopic metazoan, with adult sizes ranging from 150 to 1500 μm. Multicellular organisms requiring free water to metabolise tardigrades are found in all habitats from the poles to the tropics, high altitudes through coastal to abyssal depths (see Chap. 7). Parthenogenesis, the reproductive strategy for many limno-terrestrial tardigrades, allows a single female arriving at a suitable habitat to give rise to a new population (see Chap. 8). Limno-terrestrial tardigrades are also capable of surviving extreme conditions by cryptobiosis—the reversible cessation of metabolism under extreme environmental conditions (see Chaps. 10–13). Under cryptobiosis the tardigrade reduces free body water and collapses into the 'tun', a resistant stage named for its similarity to a barrel. Tuns are dehydrated tardigrades that have 'shrunk' to ca. 30% of their original volume (Wright et al. 1992).

Tardigrades are micro-invertebrates and so limited by their size to localised movement often within small niches, e.g. a moss cushion, and must rely on passive dispersal for wider distribution. Aerial dispersal favours small and light objects such as terrestrial tardigrades (e.g. Wright et al. 1992; Kinchin 1994), but while adults are small enough to be carried in the 'aerial plankton', only eggs and tuns are both light and resistant enough to form viable airborne propagules (e.g. Kinchin 1994). Desiccation, the principal feature of tun formation, also reduces density to yield a very light propagule which is not only suitable for aerial transport but also physiologically adapted to withstand the rigorous (dry and cold) conditions in high-altitude winds (e.g. Kristensen 1987; Nelson and Higgins 1990; Wright et al. 1992). Indeed the genus *Echiniscus* is reported as common in raindrops and 'aerial plankton' after föhn storms in Greenland (Kristensen 1987). However, low wind velocity (2 m/s) is not sufficient to disperse tardigrades (Sudzuki 1972).

There is little information on the modes of dispersal used by freshwater and marine tardigrades. Freshwater tardigrades form cysts rather than tuns, which do not appear to be as tolerant to desiccation (Kinchen 1994; McInnes pers. obs.), while eggs, at early and embryonic stage, are also sensitive to desiccation (McInnes et al. 1997; Schill and Fritz 2008). It is possible to speculate that aquatic species are distributed via rapid moving floodwaters or storm surges altering the coastal currents, though this does not explain the presence of freshwater species in remote habitats.

Such traits as aerial dispersal and cryptobiosis predispose the tardigrade to the potential of ubiquity. The data, however, suggest otherwise indeed extant limno-terrestrial tardigrades display unequivocal geological scale signals (Bertolani and Rebecchi 1993; McInnes and Pugh 1998, 2007; Pilato and Binda 2001), while marine tardigrade global diversity and biogeography remain largely unknown (Kaczmarek et al. 2015a). The current checklist (Degma et al. 2014) reports a total of ca.1200 tardigrade species, of which 1012 are limno-terrestrial and 190 marine. This extreme disparity does not reflect actual numbers of marine species but rather the number of marine studies (i.e. 1000+ 'limno-terrestrial' versus ca. 500 'marine' papers and 200+ terrestrial 'countries' versus 13 'coastal/oceanic' regions).

4.2 Tardigrade Biogeography: The Rationale

In this chapter we have opted to differ from the normal overview of a subject, which are largely reviews of the current tardigrade literature (e.g. Meyer 2013; Kaczmarek et al. 2014, 2015b, 2016). Instead of reviewing tardigrade biogeography per se, we review the methods used and the potential pitfalls and future of biogeography as modelled by the phylum *Tardigrada*. Biogeography, and in particular analytical biogeography, is defined by two elements of data: (1) the tardigrade data, which appears to effectively double every 10–12 years, and (2) the computer-intensive analytical methods, which have a 'shelf life' of the same period or generally less.

Pilato and Binda (2001) used the data on tardigrade distribution relatively simply, identifying species considered cosmopolitan and the number of taxa for a given region. They determined that the results indicated: '(a) passive dispersal of animals adapted to a limited microenvironmental niche would be restricted by the availability of suitable habitats. Thus explaining "insular" distribution;' and '(b) the fact that extant tardigrade distribution patterns still reflect palaeogeographical events, indicating the slowness of tardigrade evolution'. They concluded that 'few species enlarged their geographic distribution while most, despite passive dispersal, have remained localised'. This explained the paucity of cosmopolitan species relative to parochial and endemic species (Pilato and Binda 2001).

Our initial foray (McInnes and Pugh 1998) was the first ever global biogeographic analysis of a phylum, albeit small phylum and incorporating all the non-marine taxa. The method, using a simple 'weighted mean pair-group analysis' (WMPGA) of taxa present or non-recorded (Sokal and Sneath 1963), was adapted from a contemporary study of the spatial structure of the much smaller Antarctic acarofauna (Marshall and Pugh 1996). Faced with a very sparse data set dominated by rare taxa with few regionally shared species or even genera, we were forced to use what would now be considered an unusual clustering technique. Despite the poor data quality and, with hindsight, the problem of too many (17–23) necessarily ad hoc prescribed regions or 'operational geographic units' (OGUs—Farris 1970), the story was one of global tardigrades having a Laurasian origin which then colonised Gondwana and the subsequent fragments of both supercontinents as they disintegrated. This did, and indeed still does, appear entirely sensible even though we were only able to 'suggest' in terms of best fit rather than prove or verify the directions taken by tardigrades.

In our second paper (McInnes and Pugh 2007), we applied a more conventional (presence/no record) Sørensen clustering algorithm and confirmed the groupings via an independent principal component analysis (PCA). Although both (cluster and principal components) analyses use the same input data, they are independent measures. Cluster analysis extracts a signal from a spreadsheet of row/OGU or column (specific/generic or even familial taxa or 'operational taxonomic units' (OTUs—Farris 1970), while PCA extracts its signal from the complementary variance. But both cluster and principal components 'multivariate' analyses are purely scalar; they cluster data into groups based on inherent (high) similarity or (low) dissimilarity. Despite claims to the contrary, neither can impart a directional vector component and effectively suggest that our animals 'went that way'. However, it did match the results from the earlier paper (McInnes and Pugh 1998) and provided analysis at a more exacting species level.

However, there are tools out there that can impart a vector quality to simple binary data, indeed parsimony analysis or 'cladistics', the mainstay of evolutionary biology since the early 1990s can designate related organisms as being 'ancestral' or 'derived' in terms of shared characters which are generally coded as '0' or '1' for 'either' and 'or'. Brian Rosen (1988) adapted this technique to fit biogeographic data by simply transposing the traditional '1/0' presence/absence distribution record spreadsheet, so that rows represent regional OGUs and columns the taxonomic OTUs, and then importing this spreadsheet as an input matrix to cladistics software. The area cladogram output shows 'area clades' of OGUs which share groups of OTUs and has a true vector quality imparted from an 'uninhabited' theoretical outlier fixed to the bottom of the area cladogram.

Cladistics has some biogeographically useful tools. For example, Wagner or Fitch parsimony, which assumes equitable forward (i.e. $0 \rightarrow 1$) and reverse ($1 \rightarrow 0$) character transitions, forms a good dispersal proxy allowing taxa to immigrate (i.e. $0 \rightarrow 1$) or emigrate ($1 \rightarrow 0$) from any prescribed OGU. But Wagner parsimony is problematic in that it simply allows any changes even ones that should not reverse. Dollo parsimony was implemented to correct this problem by allowing free reversal but very restricted forward change. This prevents complex structures from 're-evolving' and dramatically reduces homoplasy or data conflict. The catch is that characters must be polarised a priori, which is often a problem in evolutionary biology where it may not be clear which character state is ancestral and which derived. Not so in biogeography where '0' and '1' are clearly defined and so Dollo parsimony will restrict ($0 \rightarrow 1$) immigration and speciation but not ($1 \rightarrow 0$) local extirpation.

We had high hopes and applied both Wagner and Dollo parsimonies to the updated tardigrade distribution data using the method of De Grave (2001). However, the results (McInnes and Pugh 2007) provided an interpretation which was completely contrary to the previous paper (McInnes and Pugh 1998), with tardigrades originating in the Southern Hemisphere. Convincing enough to publish, but nonetheless perplexing. Indeed sufficiently perplexing that we added caveats to the apparent impasse, suggesting where we may have 'taken a wrong turn'.

Also in this, from the 1990s to the current period, the use of integrated taxonomy (combining molecular techniques with alpha taxonomy) has infiltrated the tardigrade world (e.g. Bertolani et al. 2011). This has paved the way for biogeographic studies, usually at the finer, population level dynamics, and local speciation events using haplotype networks. Haplotype networks represent the relationships between haploid genotypes and pool identical sequences into a single node. Haplotype networks are based on a suite of techniques (Tang et al. 2014) and have been applied to the Antarctic tardigrade *Acutuncus antarcticus* (Cesari et al. 2016).

Another larger-scale study examined the potential links between the Americas, including the Great American Biotic Interchange (GABI) (Kaczmarek et al. 2016).

Here the known tardigrade taxa for the Americas were analysed using Biodiversity 1.0 (Laffan et al. 2010), a tool for the spatial analysis of diversity using indices based on taxonomic, phylogenetic and matrix (e.g. genetic dissimilarity) relationships. These authors concluded that some tardigrades from either side of the Isthmus of Panama supported the predictions of GABI; however, Central America had a unique fauna with more endemism than the connection of two continents would otherwise predict.

4.3 The Way Forward

The last decade has seen a major upheaval in evolutionary biology with cladistic/parsimony techniques being increasingly replaced by Bayesian mechanics. Some regard the latter as simply more powerful or 'better', but this is not the case, and this underlies a simple misconception. Bayesian mechanics are much more efficient at analysing larger data sets, and evolutionary biology is continually addressing bigger data in terms of longer gene sequences, multiple gene sequences and 50+ taxa. Nevertheless, as our previous effort showed, trying to address large numbers of areas/OGUs is counterproductive, especially where taxa are poorly reported and so inflate regional endemism. Here, for the limno-terrestrial tardigrades, we have pared the world back to only 11 OGUs and, here for the first time, analysed the marine tardigrades with 9 OGUs.

4.4 Methods

4.4.1 Limno-Terrestrial Tardigrada

We abstracted all records of non-marine *Eutardigrada* and *Heterotardigrada* from the current literature (Degma et al. 2014) and stripped these down to their essential systematics and distribution attributes. We reconciled the systematics to 1065 species (and subspecies), 73 genera and 16 families and the distribution to 32 primary OGUs in MS Excel where, as in our previous study, we used a combination of cluster analysis and principal component analysis to define larger groups within the data. The specific data showed excess endemism and the familial data predominantly widely distributed to cosmopolitan, so we focused on the 49 broadly distributed but non-cosmopolitan genera and families. We reduced the original OGUs to a working set of 14 and, after preliminary analyses, to only 11 by removing 3 groups ((1) Central Pacific, (2) Madagascar and Macaserene, (3) South Atlantic and South Indian Ocean subantarctic islands) showing particular high levels of endemism, which we attribute to low-level collecting. This left only 11 primary OGUs of North America, Central America (and Caribbean), South America, Europe, Eurasia, South-East Asia,

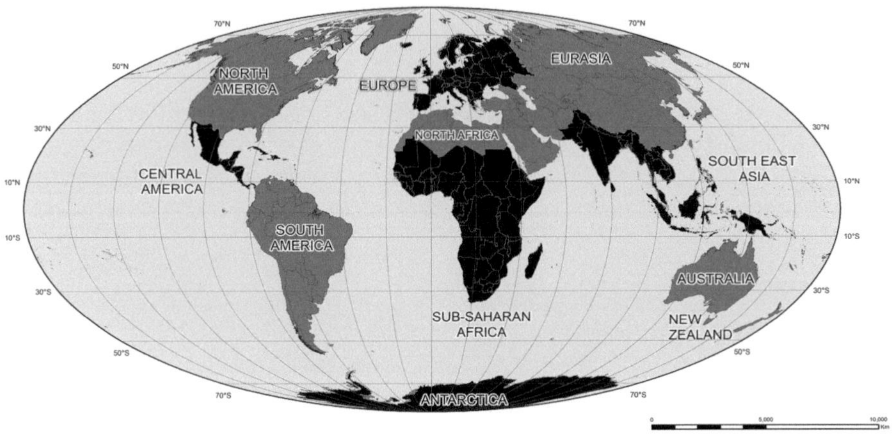

Fig. 4.1 Map of the world displaying operational geographic units used in this analysis

North Africa (and Mediterranean islands), sub-Saharan Africa, Australia, New Zealand (and South Pacific islands) and Antarctica (Fig. 4.1).

As before we used PAE as our primary analytical tool though we added a few 'recent twists', for example, running the data with either an all-zero (0000000) or all-one (1111111) (Cano and Gurrea 2003; Ribichich 2005) root with forward (i.e. '1/0') or reverse ('0/1') presence/absence coding (Pugh manuscript). These tools allow us to adopt a more flexible approach, particularly modelling dispersal from a refugium or towards a taxon sink (i.e. 'from' or 'to' an area supporting a high tardigrade diversity). We used both PHYLIP (Felsenstein 2007) and PAUP (Swofford 2002) to perform parsimony analysis using 100 or 1000 bootstrap pseudoreplicates to compare and verify the outputs. Instead of outputting the results as conventional rectangular cladograms, we adopted a less conventional but more appropriate approach of using more free-form 'splits networks'. Splits networks (assembled using 'Dendroscope' (Huson and Scornavacca 2012) replace weakly supported branches with alternative 'splits' to form loops. We must however add a word of caution and make it clear that these loops represent signal conflict or uncertainty rather than 'alternative dispersal routes', hence 'splits networks'.

We produced eight splits networks representing all (zero/full root vs. forward/reverse polarity vs. Wagner/Dollo parsimony) combinations but only six different outputs (full root/forward polarity/Wagner parsimony were identical to zero/reverse/Wagner and full/reverse/Wagner to zero/forward/Wagner). We rejected suboptimal scenarios identified via 'long' conventional area cladogram length or a 'spider web' splits network. This left us two possible scenarios of zero/reverse/Wagner and zero/reverse/Dollo; both represent (outward) centrifugal dispersal from a refugium with nominal (Wagner) (Fig. 4.2) or elevated (Dollo) levels of speciation (Fig. 4.3).

Both imply a Palaearctic (Europe/Eurasia) origin with subsequent colonisation of Gondwana via New World (Nearctic and Neotropical) in the Wagner scenario or via several routes in the Dollo scenario. Both have their issues. The Wagner model is

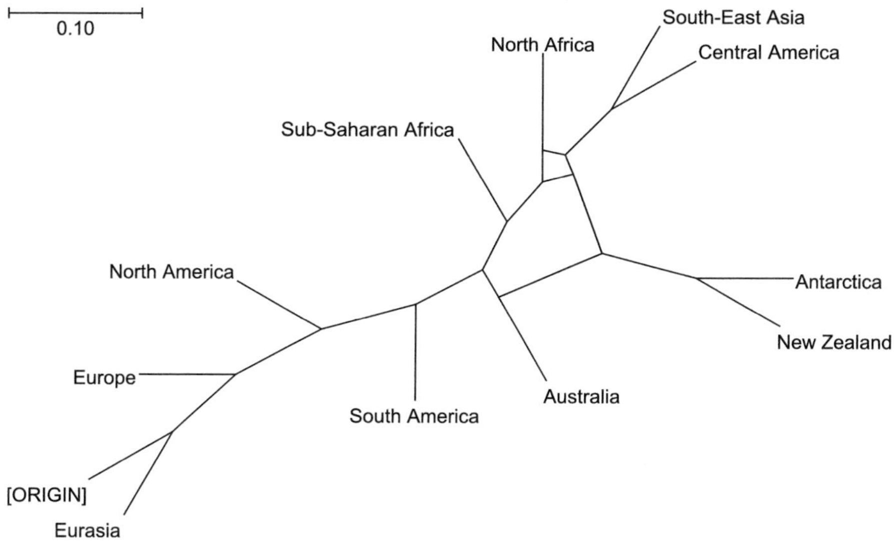

Fig. 4.2 Reticulate 'splits tree' network representing zero/reverse/Wagner with outward centrifugal dispersal from a refugium and nominal speciation/extinction

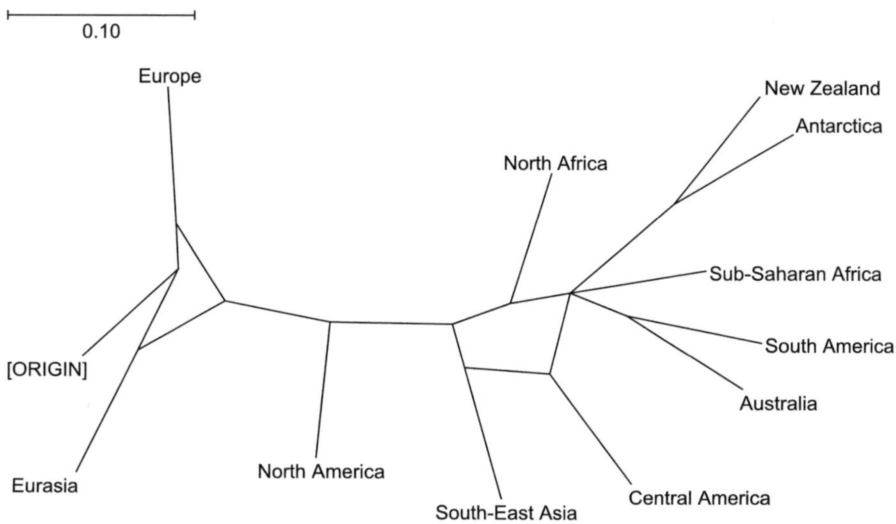

Fig. 4.3 Reticulate 'splits tree' network representing zero/reverse/Dollo with enhanced speciation/ supressed extinction

better supported (length, 128; consistency index, 0.453; retention index, 0.584) than the Dollo (length, 145; consistency index, 0.400; retention index, 0.627). This could be a reflection of an 'easier' mathematical scenario, though the reticulation/uncertainty is limited to Gondwana where the (Australia, New Zealand, Antarctica and

North Africa and sub-Saharan Africa) groupings are more plausible. The adjoining positions of SE Asia and Central America are anomalous in both (indeed all other) scenarios, though this most likely reflects the paucity of data from these two regions.

This new model clearly appears to corroborate our first attempt (McInnes and Pugh 1998) rather than the second (McInnes and Pugh 2007).

4.4.2 *Marine* Tardigrada

We abstracted all 237 species, 28 genera and 9 (sub-)familial records of marine *Heterotardigrada*, *Echiniscoidea* and Isohypsibioidea (*Eutardigrada*) from a literature-based database (McInnes—unpublished data). We reduced an initial set of 13 primary ocean basin OGUs to only nine of West Atlantic, East Atlantic, Caribbean, Mediterranean, Indian Ocean, NW Pacific, NE Pacific, SW Pacific and SE Pacific harbouring at least 16 species (Fig. 4.4). The Southern Ocean and other 'missing' regions were deleted as having too few records. In contrast to the terrestrial fauna, the generic and familial data showed high levels of shared endemism, so we focused on the 97 regionally shared species.

The aforementioned selection criteria again left us with two possible scenarios of zero/reverse/Wagner and zero/reverse/Dollo, that is (outward), centrifugal dispersal from a refugium with nominal (Wagner) (Fig. 4.5) or elevated (Dollo) levels of speciation (Fig. 4.6). The two models are very similar though the Wagner model has marginally better support (length, 196 vs. 212; consistency index, 0.495 vs. 0.475; retention index, 0.505 vs. 542). Both imply an Atlantic origin with subsequent colonisation of the Indian and Pacific basins via the Panama Gap which existed

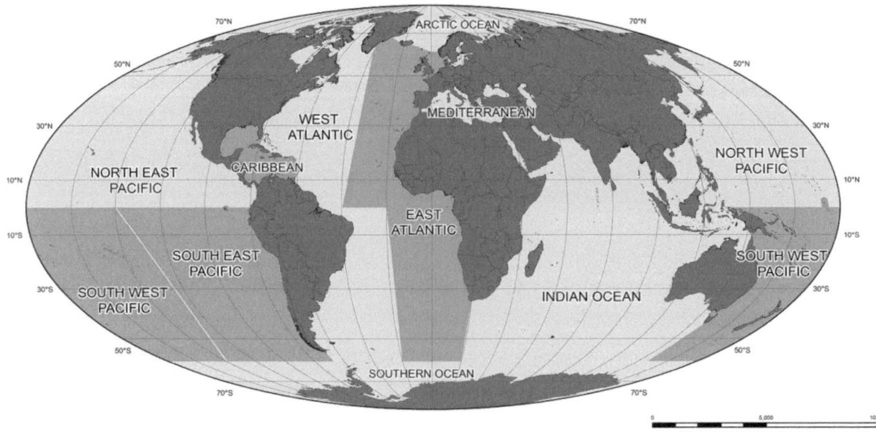

Fig. 4.4 Map of the world displaying oceanic operational geographic units (OGUs) used in this analysis [OGU data from SeaVox gazetteer (Claus et al. 2014)]

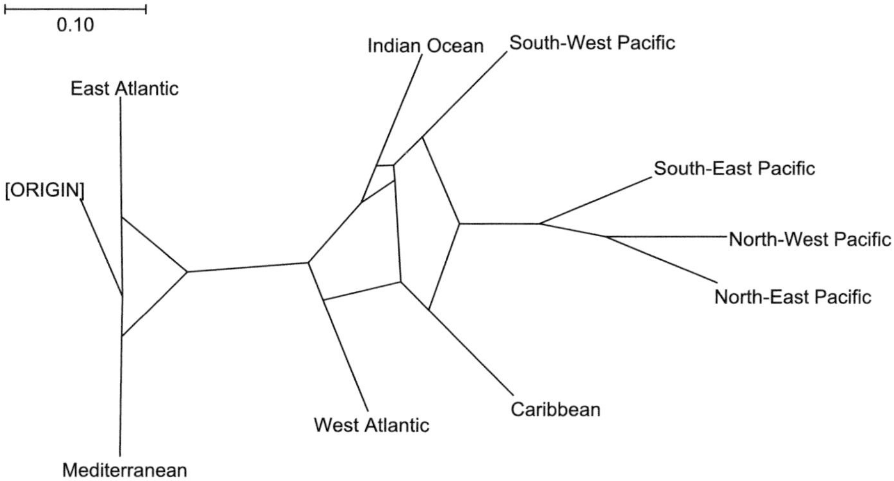

Fig. 4.5 Reticulate 'splits tree' network representing zero/reverse Wagner with nominal speciation/extinction

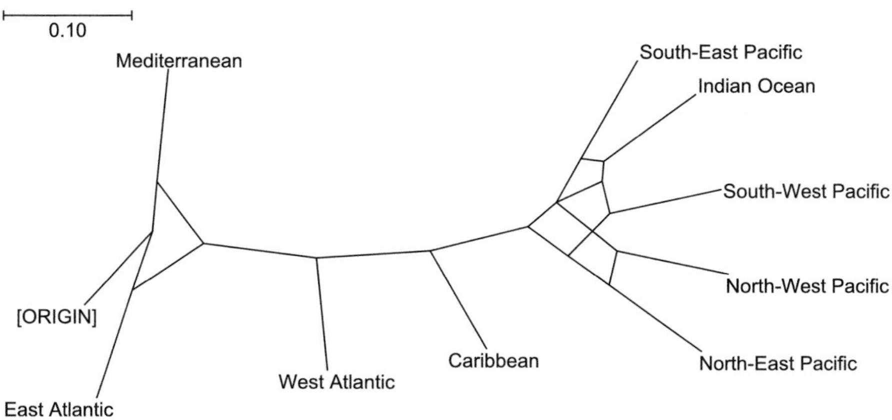

Fig. 4.6 Reticulate 'splits tree' network representing zero/reverse Dollo with enhanced speciation/suppressed extinction

prior to the recent establishment of the current Isthmus of Panama. We do however urge caution as this apparent Atlantic origin is very likely an artefact resulting from a huge record bias with the Atlantic (including Caribbean and Mediterranean) harbouring 69–141 species compared with 73 in the Indian Ocean and 16–43 in the Pacific.

4.5 So . . . Does This Explain Global Tardigrade Biogeography?

Clearly the answer is 'no'. This story will be retold repeatedly (hopefully by others) as we obtain more data and develop new techniques. While the application of splits networks to the global tardigrade data is the latest development, it is still a rather blunt tool, and future work will, indeed, focus on smaller areas.

For any of these techniques, 'the devil is in the detail'. The core problem is that biogeographic data are fluid, and boundaries continually move with advances in tardigrade taxonomy. Tardigrades challenge the taxonomist as they have a limited suit of morphological characters upon which to base alpha-taxonomic studies. Classical morphology of this group has increasingly incorporated more characteristics, allometry and updated techniques (Bartels et al. 2011; Cesari et al. 2013; Pilato 2013). Original descriptions were sparse in detail, e.g. *Macrobiotus hufelandi* original in Schultze (1834) and redescribed in Bertolani and Rebecchi (1993). However, even early researchers were amazed at the variety of species to be found in the *Tardigrada* causing Murray (1907) to comment that '. . .till quite recently few supposed that there were numerous species of Tardigrada'. For many years tardigrade species had been considered largely ubiquitous and cosmopolitan, but recent research is questioning this assumption. Early alpha taxonomy using a restricted set of morphological characters led to a number of species groups or species complexes, e.g. *Echiniscus arctomys*, *Macrobiotus hufelandi*, *Milnesium tardigradum*, etc. Modern integrative taxonomy combining morphology and molecular techniques is showing that many of these species groups comprise previously cryptic species (e.g. Bertolani et al. 2010; Guidetti et al. 2005, 2013). This approach is gathering pace, filling in the species details and revealing that the tardigrade species discovery rate is still on the exponential curve even in well-reported regions such as Europe (Fontaine et al. 2012).

Marine tardigrades have more limited research data, and our analysis probably indicates more bias towards regional study effort than true biogeography. The study of regional biodiversity is only slowly expanding known biogeographic limits for individuals and genera (e.g. research in the Seychelles and Maldives expanded the ranges for *Neostygarctus acanthophorus*, *Archechiniscus minutus*, *Halechiniscus paratuleari*, *Styraconyx tyrrhenus* and *S. sardiniae* from Mediterranean reports, *Tanarctus velatus* and *Florarctus heimi* from Pacific Ocean reports and *Renaudarctus psammocriptus* from Atlantic and Pacific Oceans (Gallo et al. 2007; Balsamo et al. 2008)). Gallo et al. (2007) hypothesised that, to some extent, plate tectonics explain the existence of the rather high numbers of cosmopolitan species in the intertidal and subtidal sediments of the Seychelles and Maldives. There is also a curious pattern to the genus *Parmursa* that, while restricted to bathyal depths, has a northern circum-global distribution (North Atlantic Ocean, Indian Ocean and Indo-Pacific Ocean) (Hansen 2007).

A study limited to the Northern Atlantic (Faurby et al. 2011) explored the relationships of intertidal *Echiniscoides* species (associated with barnacles on rocky shores). This work, based on a single mitochondrial marker, highlighted the problems of potential cryptic species (Faurby et al. 2011) and indicated the potential for strong geographical differentiation, with both northern and southern species recovered from either sides of the Atlantic but only the northern species from the mid-Atlantic Islands. The trends were interpreted as long-term isolation of the east and west populations of the northern species despite the potential for mid-Atlantic islands acting as stepping stones, with indications of postglacial recolonisation from several refugia. The southern species signals also suggested the potential for secondary contact between formerly isolated groups (Faurby et al. 2011).

Our foray into tardigrade biogeography was prompted by looking at the origins of Antarctic and Arctic faunas. We wanted to know whether the geological events of glacial maxima would have erased all tardigrades from the poles or if the tardigrades survived in refugia. To the north we found that nearly three quarters of the Arctic tardigrades occur in the temperate Holarctic, with many 'American' species occurring throughout the Arctic, while 'European' species are largely confined to the Palaearctic (Pugh and McInnes 1998). We hypothesised that the prevailing winds during the Holocene were, and continue to be, circumpolar from the west, and therefore light airborne propagules (eggs or adults in resistant stage 'tuns') were, and still are, dispersed into deglaciated Arctic terrain, in particular from Northern America (Pugh and McInnes 1998). The southern fauna is proving more challenging to explain in terms of global biogeography but does suggest tardigrades have survived glacial maxima in situ (McInnes and Pugh 1998, 2007; Czechowski et al. 2012; Guidetti et al. 2014).

4.6 Where Next?

Some of the problems associated with analytical biogeography we cannot answer (e.g. tardigrades do not make good fossils, so it is unlikely that we will be able to include accurate geological timescales within the analysis). However, with time more data will become available regarding species diversity and distribution. But this is not the end of the story.

Network analysis is used in the social sciences to investigate flow, diffusion (shortest), path length, friendships, genealogy, group cohesion, affiliation and rank in business, familial, government and other organisational structures (Newman 2010; De Nooy et al. 2011). This technique is being employed by biologists, for example, it has been used to explain the relationships between hydrothermal vent communities (Moalic et al. 2012). Here we use simply the technique to display the relationships within tardigrade distribution data which present no obvious hierarchical structure and, in particular, highlight which areas appear to be 'well described/known' and which are less so.

We condensed records of non-marine tardigrades to numbers of species within the 13 families in each of the 11 primary regional OGUs and formatted the data as a two-mode matrix (de Nooy et al. 2011), with separate 'family' and 'OGU' partitions.

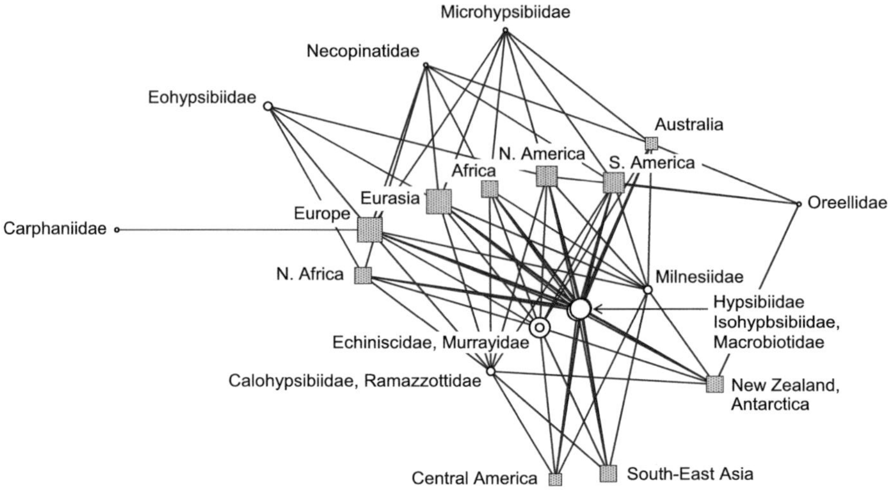

Fig. 4.7 This network shows the principal OGUs (circles) forming two distinct groups of 'restricted distribution' Carphaniidae, Eohypsibiidae, Necopinatidae, Microhypsibiidae and Oreellidae versus the more central and overlapping groups of 'cosmopolitan' Calohypsibiidae, Ramazzottidae, Echiniscidae, Murrayidae, Isohypsibiidae, Macrobiotidae, Hypsibiidae and Milnesiidae

We scaled the symbols representing both OGUs (boxes) and tardigrade families (circles) to represent number of included tardigrade species as sizes 1–6 with boundaries at 10, 50, 100, 200 and 300 species. We uploaded the plain text network file to Pajek version 3.12 (de Nooy et al. 2011), where we applied sequential Kamada-Kawai (1989) and 2-D Fruchterman Reingold (1991) algorithms to yield optimal vertex (OGU and OTU) separation (Fig. 4.7) but moved any vertices (or more often their labels) clear of the screen/network space edge.

There are two groups of both OGUs/locations (shaded squares) and tardigrade families (open circles). The 'lower' series of OGUs (Central America, South-East Asia, Antarctica and New Zealand) harbour only the 'core groups' of cosmopolitan, tardigrade families (Calohypsibiidae, Ramazzottidae, Echiniscidae, Murrayidae, Isohypsibiidae, Macrobiotidae, Hypsibiidae, Milnesiidae and Oreellidae). The remaining upper OGU series harbours both these cosmopolitan families and five others (Carphaniidae, Eohypsibiidae, Necopinatidae, Microhypsibiidae and Oreellidae) with more restricted ranges. The three concentric groupings of Calohypsibiidae/Ramazzottidae, Echiniscidae/Murrayidae and Hypsibiidae/Isohypsibiidae/Macrobiotidae suggest three identical ranges. The anomalous positions of both South-East Asia and Central America in the preceding area splits networks could be explained by them sharing only these relatively few but widespread families. The addition of the Oreellidae alone is suffice to shift Antarctica and New Zealand into a distinct second group.

4.7 Conclusion

In this chapter we endeavour to show the history and development of tardigrade biogeography. The story is not a finished work but provides a stepping stone to the next generation of investigations. We hope that by showing the limitations of the techniques and demonstrating the regions requiring further study, this will enthuse others to explore the diversity and distribution of this enigmatic and fascinating wee beastie—the tardigrade.

Acknowledgements We thank Oliva Martin-Sanchez (BAS, Cambridge) for providing the maps for Figs. 4.1 and 4.4.

References

Balsamo M, Semprucci F, Sbrocca C, Colantoni P, Baldelli G, Sandulli R, de Leonardis C, D'Addabbo R, Gallo M (2008) Biodiversità della meiofauna nelle Isole Maldive. In: Bianchi CN, Gnone G, Lasagna R, Morri C (eds) Ricerche italiane sugli ecosistemi marini tropicali. CEMT (Centro di Ecologia Marina Tropicale), Genoa, p 20

Bartels PJ, Nelson DR, Exline RP (2011) Allometry and the removal of body size effects in the morphometric analysis of tardigrades. J Zool Sys Evol Res 49:17–25. https://doi.org/10.1111/j1439-0469201000593x

Bertolani R, Rebecchi L (1993) A revision of the *Macrobiotus hufelandi* group (Tardigrada, Macrobiotidae), with some observations on the taxonomic characters of eutardigrades. Zool Scr 22:127–152

Bertolani R, Rebecchi L, Cesari M (2010) A model study for tardigrade identification. In: Nimis PL, Vignes Lebbe R (eds) Tools for identifying biodiversity: progress and problems. Edizioni Università di Trieste, Trieste, pp 333–339

Bertolani R, Biserov V, Rebecchi L, Cesari M (2011) Taxonomy and biogeography of tardigrades using an integrated approach: new results on species of the *Macrobiotus hufelandi* group. Invert Zool 8:23–36

Cano JM, Gurrea P (2003) La distribución de las zigenas (Lepidoptera, Zygaenidae) Ibéricas: una consecuencia del efecto peninsula. Graellsia 59:273–285

Cesari M, Guidetti R, Rebecchi L, Giovannini I, Bertolani R (2013) A DNA barcoding approach in the study of tardigrades. J Limnol 72(s1):182–198. https://doi.org/10.4081/jlimnol2013s1e23

Cesari M, McInnes SJ, Bertolani R, Rebecchi L, Guidetti R (2016) Genetic diversity and biogeography of the south polar water bear *Acutuncus antarcticus* (Eutardigrada: Hypsibiidae)–evidence that it is a truly pan-Antarctic species. Invertebr Syst 30(6):635–649

Claus S, De Hauwere N, Vanhoorne B, Souza Dias F, Hernandez F, Mees J (Flanders Marine Institute) (2014) MarineRegions.org. http://www.marineregions.org. Accessed 29 August 2014

Czechowski P, Sands CJ, Adams BJ, D'Haese CA, Gibson JAE, McInnes SJ, Stevens MI (2012) Antarctic Tardigrada: a first step in understanding molecular operational taxonomic units (MOTUs) and biogeography of cryptic meiofauna. Invertebr Syst 26:526–538

de Buffon GLLC (1749–1767) Histoire Naturelle, générale et particulière, avec la description du Cabinet du Roi, vol 1–15. De L'Imprimerie royale, Paris

De Grave S (2001) Biogeography of indo-Pacific Pontoniinae (Crustacea, Decapoda): a PAE analysis. J Biogeogr 28:1239–1253

De Nooy W, Mrvara A, Batagelj V (2011) Exploratory social network analysis with Pajek, Structural analysis in the social sciences no 34. Cambridge University Press, Cambridge

Degma P, Bertolani R, Guidettii R (2014) Actual checklist of Tardigrada species (2009–2014, Ver. 25: 10-05-2014). http://www.tardigrada.modena.unimo.it/miscellanea/Actual%20checklist%20of%20Tardigrada.pdf

Ebach MC, Humphries CJ, Williams DM (2003) Phylogenetic biogeography deconstructed. J Biogeogr 30:1285–1296. https://doi.org/10.1046/j1365-2699200300928x

Farris JS (1970) Methods for computing Wagner trees. Syst Zool 19:83–92

Faurby S, Jørgensen A, Kristensen RM, Funch P (2011) Phylogeography of North Atlantic intertidal tardigrades: refugia, cryptic speciation and the history of the mid-Atlantic Islands. J Biogeogr 38:1613–1624. https://doi.org/10.1111/j.1365-2699.2011.02533.x

Felsenstein J (2007) PHYLIP (phylogeny inference package) version 367. Distributed by the author, Department of Genome Sciences, University of Washington, Seattle

Fontaine B, van AK, Alonso-Zarazaga MA et al (2012) New species in the old world: Europe as a frontier in biodiversity exploration, a test bed for 21st century taxonomy. PLoS One 7(5): e36881. https://doi.org/10.1371/journal.pone.0036881

Fruchterman TMJ, Reingold EM (1991) Graph drawing by force-directed placement. Softw Pract Exper 21:1129–1164. https://doi.org/10.1002/spe4380211102

Gallo M, D'addabbo R, De Leonardis C, Sandulli R, De Zio Grimaldi S (2007) The diversity of Indian Ocean Heterotardigrada. In: Pilato G, Rebecchi L (Guest eds) Proceedings of the tenth international symposium on Tardigrada. J Limnol 66(s1):60–64

Guidetti R, Gandolfi A, Rossi V, Bertolani R (2005) Phylogenetic analysis of Macrobiotidae (Eutardigrada, Parachela): a combined morphological and molecular approach. Zool Scr 34:235–244

Guidetti R, Peluffo JR, Rocha AM, Cesari M, Moly de Peluffo MC (2013) The morphological and molecular analyses of a new south American urban tardigrade offer new insights on the biological meaning of the *Macrobiotus hufelandi* group of species (Tardigrada: Macrobiotidae). J Nat Hist 47(37–38):2409–2426

Guidetti R, Rebecchi L, Cesari M, McInnes SJ (2014) Mopsechiniscus franciscae, a new species of a rare genus of Tardigrada from continental Antarctica. Polar Biol 37:1221–1233. https://doi.org/10.1007/s00300-014-1514-x

Hansen JG (2007) The deep sea elements of the Faroe Bank tardigrade fauna with a description of two new species. In: Pilato G, Rebecchi L (Guest Eds) Proceedings of the tenth international symposium on Tardigrada. J Limnol 66(s1):12–20

Huson DH, Scornavacca C (2012) Dendroscope 3: an interactive tool of rooted phylogenetic trees and networks. www.dendroscope.org

Isacks B, Oliver J, Sykes LR (1968) Seismology and the new global tectonics. J Geophys Res 73 (18):5855–5899. https://doi.org/10.1029/JB073i018p05855

Kaczmarek Ł, Michalczyk Ł, McInnes SJ (2014) Annotated zoogeography of non-marine Tardigrada. Part I: Central America. Zootaxa 3763:1–62

Kaczmarek Ł, Bartels PJ, Roszkowska M, Nelson DR (2015a) The zoogeography of marine Tardigrada. Zootaxa 4037:1–189

Kaczmarek Ł, Michalczyk Ł, McInnes SJ (2015b) Annotated zoogeography of non-marine Tardigrada. Part II: South America. Zootaxa 3923:1–107

Kaczmarek Ł, Michalczyk Ł, McInnes SJ (2016) Annotated zoogeography of non-marine Tardigrada. Part III: North America. Zootaxa 4203:1–249

Kamada T, Kawai S (1989) An algorithm for drawing general undirected graphs. Inform Process Lett 31:7–15. https://doi.org/10.1016/0020-0190(89)90102-6

Kinchin IM (1994) The biology of tardigrades. Blackwell, London

Kristensen RM (1987) Generic revision of the Echiniscidae (Heterotardigrada), with a discussion of the origin of the family. In: Bertolani R (ed) Biology of Tardigrades, Selected symposia and monographs U.Z.I., vol 1. Mucchi, Modena, pp 261–335

Laffan SW, Lubarsky E, Rosauer DF (2010) Biodiverse, a tool for the spatial analysis of biological and related diversity. Ecography 33:643–647

Marshall DJ, Pugh PJA (1996) Origin of the inland Acari of continental Antarctica, with particular reference to Dronning Maud land. Zool J Linn Soc Lond 118:101–118
McInnes SJ, Pugh PJA (1998) Biogeography of limno–terrestrial Tardigrada, with particular reference to the Antarctic fauna. J Biogeogr 25:31–36
McInnes SJ, Pugh PJA (2007) An attempt to revisit the global biogeography of limno-terrestrial. In: Pilato G, Rebecchi L (Guest eds) Proceedings of the tenth international symposium on Tardigrada. J Limnol 66(s1):90–96
McInnes SJ, Pugh PJA, Robinson K (1997) Application of low temperature SEM: how eggs of a tardigrade respond to humidity changes. Microsc Anal 59:38–39
Meyer HA (2013) Terrestrial and freshwater Tardigrada of the Americas. Zootaxa 3747:1–71
Moalic Y, Desbruyères D, Duarte CM, Rozenfeld AF, Bachraty C, Arnaud-Haond S (2012) Biogeography revisited with network theory: retracing the history of hydrothermal vent communities. Syst Biol 61:127–137
Murray J (1907) XXXI. - Encystment of Tardigrada. Trans Roy Soc Edin 45:837–857
Nelson DR, Higgins RP (1990) Tardigrada. In: Dindal DL (ed) Soil biology guide. Wiley, New York, pp 393–419
Newman MEJ (2010) Networks: an introduction. Oxford University Press, Newman
Ortelius A (1596) Thesaurus geographicus, recognitus et actus. Ex officina Plantiniana, Antwerp
Pilato G (2013) The past, the present and the future of eutardigrade taxonomy. J Limnol 72(s1):1–7. https://doi.org/10.4081/jlimnol2013s1e1
Pilato G, Binda MG (2001) Biogeography and limno-terrestrial tardigrades: are they truly incompatible binomials? Zool Anz 240:511–516
Pugh PJA, McInnes SJ (1998) The origin of Arctic terrestrial and freshwater tardigrades. Polar Biol 19:177–182
Ribichich AM (2005) From null community to non-randomly structured actual plant assemblages: parsimony analysis of species co-occurrences. Ecography 28:88–98
Rosen BR (1988) From fossils to earth history: applied historical biogeography. In: Myers AA, Giller PS (eds) Analytical biogeography: an integrated approach to the study of animal and plant distributions. Chapman & Hall, London, pp 437–481
Schill RO, Fritz GB (2008) Desiccation tolerance in embryonic stages of the tardigrade. J Zool 276:103–107
Schultze CAS (1834) *Macrobiotus hufelandii* animal e crustaceorum classe novum, reviviscendi post diuturnam asphixiam et aridiatem potens. Apud Carolum Curths, Berolini, pp 1–7
Sokal RR, Sneath PHA (1963) Numerical taxonomy. WH Freeman, San Fransisco
Sudzuki M (1972) An analysis of colonization in freshwater micro-organisms. II. Two simple experiments on the dispersal by wind. Jpn J Ecol 22(5):222–225
Swofford DL (2002) PAUP*: phylogenetic analysis using parsimony (* and other methods), version. Sinaur Associates, Sunderland, MA, p 4
Tang CQ, Humphreys AM, Fontaneto D, Barraclough TG (2014) Effects of phylogenetic reconstruction on the robustness of species delimitation methods using single-locus data. Methods Ecol Evol 5:1086–1094
Wegener A (1912) Die Herausbildung der Grossformen der Erdrinde (Kontinente und Ozeane), auf geophysikalischer Grundlage (in German) Petermann Geogr Mitt 63:185–195, 253–256, 305–309. Presented at the annual meeting of the German Geological Society, Frankfurt am Main, 6 Jan
Wright JC, Westh P, Ramløv H (1992) Cryptobiosis in Tardigrada. Biol Rev 67:1–29

Chapter 5
Paleontology and Molecular Dating

Roberto Guidetti and Roberto Bertolani

Abstract Few fossil tardigrade records are known, probably due to their minute size and absence of consistent hard body parts. A possible but not certain tardigrade ancestor was found as an "Orsten"-type secondarily phosphatized fossil from the Middle Cambrian limestone. It is characterized by only three pairs of legs as a possible consequence of a plesiomorphic anameric ontogeny that led to four leg segments in the extant tardigrades. Its other characters, related to claw and cuticle morphologies, resemble those of heterotardigrades. The first findings of sure tardigrades come from Cretaceous amber, in which the species *Beorn leggi* and *Milnesium swolenskyi* were found. These species, and especially the latter, evidence that more than 90 million years ago, "modern" tardigrades already existed, and only few morphological differences, mainly related to morphometric tracts, occur with respect to known modern species. Quaternary subfossils of tardigrades have been found with palynological studies during paleoecology researches. They have been recorded mainly in polar regions and classified within the non-pollen palynomorphs category. Cuticular remains of animals and eggs can persist in the substrates for very long time, so tardigrades can be used as paleoclimatic bioindicators, but despite their potentiality, a few studies have focused to find these organisms in microfossil samples and in sediments for paleoecological studies.

For estimating the dates of phylogenetic events related to the origin of tardigrades and/or their evolutionary lineages, molecular clock analyses have been used. Molecular data indicate that the origin of the phylum should be posed during or before the Cambrian period, placing the origin of Tardigrada in marine environment and their terrestrialization not before that of other ecdysozoans.

R. Guidetti (✉) · R. Bertolani
Department of Life Sciences, University of Modena and Reggio Emilia, Modena, Italy
e-mail: roberto.guidetti@unimore.it; roberto.bertolani@unimore.it

R. O. Schill (ed.), *Water Bears: The Biology of Tardigrades*, Zoological Monographs 2,
https://doi.org/10.1007/978-3-319-95702-9_5

5.1 Introduction

The peculiar position of tardigrades among the tree of life places this phylum in a key position to unveil new information on *Panarthropoda* and therefore *Arthropoda* evolution. In particular, the presence among tardigrades of taxa colonizing the most diverse habitats (pertaining marine and continental environments) and showing peculiar cryptobiotic capabilities can add further comprehension to the major events that led to the emergence of continental life.

The tempo and mode of the evolution of cryptobiosis within Tardigrada are still unknown, but it is clear that comprehending this process is key to better understand the evolutionary history and ecology of this phylum more broadly. In addition, given the ubiquitous continental distribution of tardigrades, understanding the tardigrade evolution would allow for a better understanding of the process of their terrestrialization (i.e., colonization of the land masses and their freshwater environments).

5.2 Paleontology: Fossil and Subfossil Records

Few fossil tardigrade records are known (Cooper 1964; Bertolani and Grimaldi 2000). The scarcity of tardigrade fossil records could be due, other than to their minute size (<1 mm), to the absence of consistent hard body parts. Soft-bodied microinvertebrates have extremely low fossilization potential, and their absence from the fossils of the earliest terrestrial biota is unsurprising.

Tardigrada, together with *Arthropoda* and *Onychophora*, form the *Panarthropoda*. They are considered to evolve from lobopodians, a paraphyletic group of Paleozoic vermiform animals with soft legs (e.g., Chen and Zhou 1997; Waloszek 2003; Mass et al. 2007; Lui et al. 2011; Caron and Aria 2017). Within this group, the Tardipolypodia (Chen and Zhou 1997), extinct metazoans characterized by an elongate trunk supported by clawed uniramous lobopodia, are considered closely related to *Tardigrada*. Tardipolypods genera (e.g., *Hallucigenia*, *Aysheaia*, *Microdictyon*, *Onychodictyon*, *Paucipodia*, *Cardiodictyon*) lack anterior antennae, have a terminal mouth and head not greatly differentiated from the rest of the body, and posteriorly present a pair of terminal legs as tardigrades. A possible but not certain tardigrade ancestor has been found in a mid-Cambrian limestone (Müller et al. 1995; Maas and Waloszek 2001). Paleontological records related to true tardigrades are limited to records from Cretaceous amber (Cooper 1964; Bertolani and Grimaldi 2000) with a large gap up to subfossil records from Pleistocene (Durante and Maucci 1972; Jankovská 1992, 2008; Cromer et al. 2006, 2008; Bernardová and Košnar 2012).

5.2.1 Fossil Records

5.2.1.1 Possible Stem Tardigrade

"Orsten"-type secondarily phosphatized fossil from the Middle Cambrian Kuonamka Formation (Olenek Uplift, West Siberia) contained microfossils considered larvae stages of a possible stem group of tardigrades (Müller et al. 1995; Maas and Waloszek 2001). "Orsten"-type preservation is the phosphatization of cuticular surfaces without any further deformation and has yielded completely three-dimensional fossils, mainly arthropods at scale of 0.1–2.0 mm. Records of such exceptional fossils are now reported from several continents and from the Early Cambrian to the Early Cretaceous, approximately between 520 and 100 million year ago (mya) (Maas and Waloszek 2001). The four animals, dated approximately 514 ± 1.0 mya (Waloszek et al. in preparation), are 250–350 μm long and have a soft integument and three pairs of parapodial legs with two claws at the extremities. The presence of two claws in each of the six legs led to consider these animals as possible tardigrades in larval stage, because the first developmental stages (called larvae) of heterotardigrades have two claws instead of four at the leg extremities.

According to Maas and Waloszek (2001), the sister group relationship of these fossil forms to extant tardigrade is apparent from their segmentation starting with only three pairs of legs, thus demonstrating plesiomorphic anameric ontogeny (anameric = addition of one more somite to the body at each molting during development), while the number of legs in living tardigrades is four. To support their idea, Maas and Waloszek (2001) interpreted the expanded posterior area of one of the fossils as a primordium (anlagen) of a further pair of legs. According to the authors, besides size and general shape, the design of the inner cuticle supports the relationships of these fossils to tardigrades, due to the presence of pillar-like structures resembling those of many marine (and some terrestrial) tardigrade species (Kristensen and Neuhaus 1999; Guidetti et al. 2000). Furthermore, the anteroventral mouth of the fossils is surrounded by fine vela and sensorial structures similar to the clavae and cirri of heterotardigrades. The legs of the fossils point onward (while in tardigrades they point backward) and have two claws each. The claws in the posterior legs are unforked, while those in the first two legs are three-forked similar to those found in some species of *Styraconyx* (Heterotardigrada, Arthrotardigrada).

Maas and Waloszek (2001) did not describe a new species for these fossils, but they included them into *Tardigrada* to embrace the fossil and all living representatives of the phylum. We consider that, due to the morphological differences present (e.g., number and orientation of legs) and the lack of internal anatomical features (e.g., presence of piercing stylets considered an apomorphy of the phylum), for now it is more parsimonious to consider these fossils as belonging to a possible stem group of Tardigrada.

5.2.1.2 Cretaceous Fossils

Even though amber is a matrix renowned for exquisite preservation of minute, soft-bodied organisms, even of their internal organs, to date only in a few amber sample tardigrades have been identified.

Beorn leggi Cooper, 1964

In summer 1940, William M. Legg made a small but valuable collection of chemawinite, or Canadian amber, from sparse secondary deposits along beaches not far from the entrance of the Saskatchewan River into Cedar Lake, southeast of The Pas, Manitoba (Canada), containing Cretaceous fauna of some 60–80 mya.

One specimen of tardigrade, attributed to a new species, *Beorn leggi* Cooper, 1964, and to the new family Beornidae, has been identified (Cooper 1964). In the same chemawinite, a heterotardigrade (probably an Echiniscidae) was observed, but it was not enough preserved for a better identification. These specimens are in the Legg collection of Canadian amber fossils (Dept. of Entomology, Museum of Comparative Zoology, Harvard). Subsequently, *Beorn leggi* has been considered similar to the present eutardigrades by Pilato (1979), having very long claws, an isohypsibiid, or a murrayid by Bertolani and Grimaldi (2000), i.e., in any case attributed to a currently existing family.

Milnesium swolenskyi Bertolani and Grimaldi, 2000

In a sample of amber collected in Sayreville (New Jersey, USA), a well-preserved specimen of tardigrade has been found (Bertolani and Grimaldi 2000). It has been attributed to a new species of the existing genus *Milnesium*, namely, *Milnesium swolenskyi* Bertolani and Grimaldi, 2000 (Fig. 5.1). It is worthy of note that it is closely related to the actual species of the genus, existing only morphometric differences, in particular in the claw length. These differences are similar to those found among some actual species of the genus, which up to some years it was considered monospecific and currently composed by 38 species. This fossil tardigrade from Turonian New Jersey amber (89.3 mya) provides a minimum of 89 mya for the split of the Milnesiidae lineages.

5.2.1.3 Quaternary Subfossils: Non-pollen Palynomorphs

During palynological studies, related to paleoecology, microfossils of non-pollen origin can be found. These fossils are called non-pollen palynomorphs (NPP). Fossil tardigrades classified as non-pollen palynomorphs are not commonly used in paleoecological studies, mainly because they have been reported only rarely (Durante and

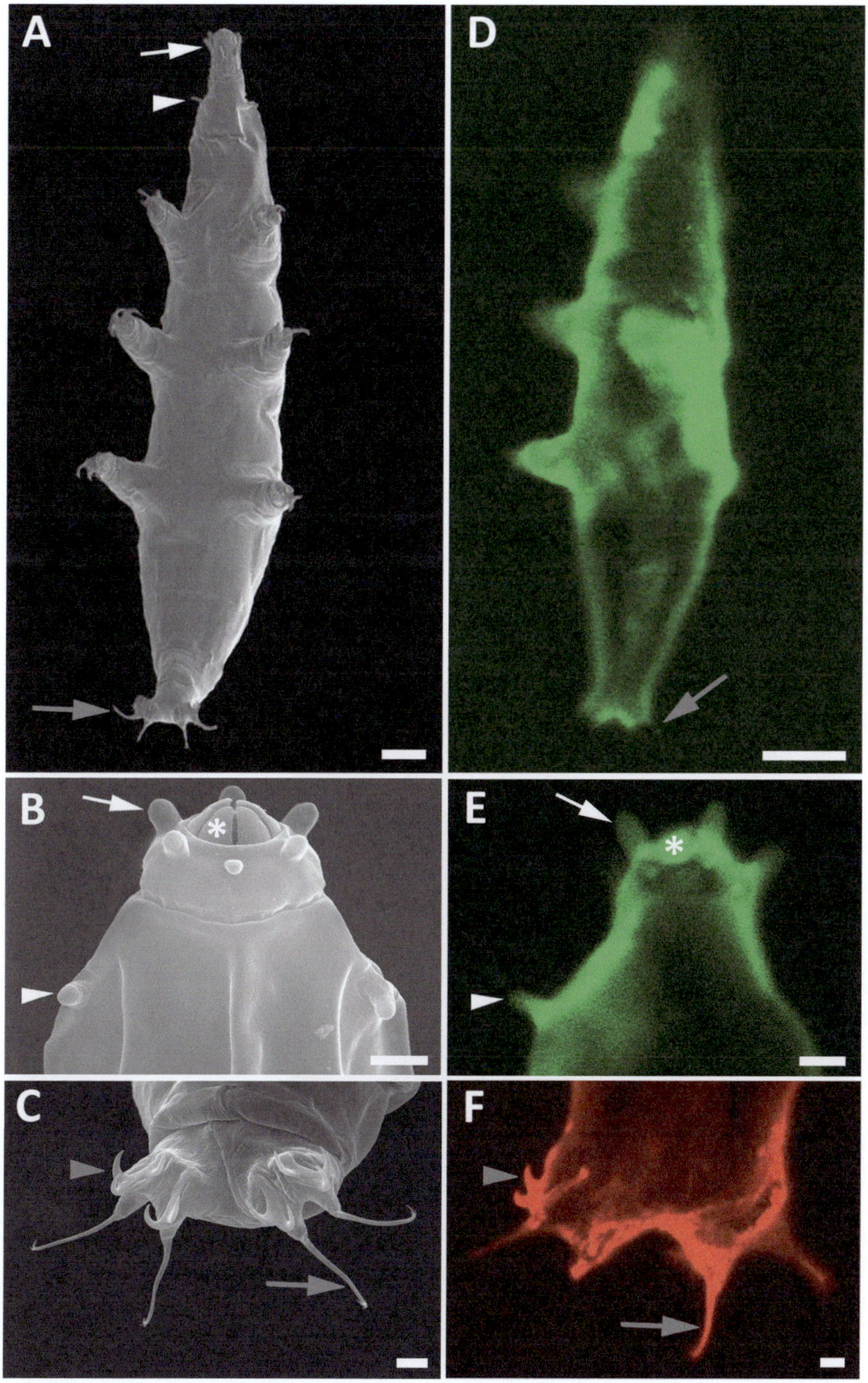

Fig. 5.1 *Milnesium* cf. *tardigradum* (**a–c**; SEM) and *Milnesium swolenskyi* Bertolani and Grimaldi, 2000 (**d–f**; CLSM). (**a**, **d**) Animals in toto. (**b**, **e**) Frontal end of the animal. (**c**, **f**) Hind

Maucci 1972; Jankovská 1992, 2008; Miller and Heatwole 2003; Jankovská et al. 2016). According to a recent study, tardigrade remains (e.g., cuticular sclerified structures such as buccal tube, claw, exuvia, eggshell) are frequent NPP elements of pollen analyses, at least in polar regions; therefore, they can be considered indicators in paleoecological studies (Jankovská et al. 2016).

Pleistocene

The oldest subfossils have been found in a quaternary travertine from Alti Lessini (Verona), Italian Alps (Durante and Maucci 1972). Two eggs originally cited as *Macrobiotus hufelandi*, but belonging to *Macrobiotus macrocalix* (see Bertolani and Grimaldi 2000), have been found together with pollens of *Tsuga pattoniana*, allowing to date them as pre-Rissian, i.e., older than 200,000 years. The species is still existing; therefore, those eggs should be considered subfossils. This finding is a further sign of a slow morphological change occurring during tardigrade evolution.

Holocene

Tardigrade exuviae and eggs have also been found on pollen slides. Information exists on high arctic (Spitsbergen, Norway, and Devon Island, Canada) sediments (Jankovská 1992, 2008). The names of the species are here referred according to the current nomenclature: *Dactylobiotus ambiguus*, *Murrayon hibernicus*, *Paramacrobiotus areolatus*, *Paramacrobiotus richtersi*, *Richtersius coronifer*, *Diaforobiotus islandicus*, *Macrobiotus hufelandi*, *Macrobiotus echinogenitus*, *Mesobiotus harmsworthi*, and *Minibiotus intermedius*. Moreover, in Spitsbergen, in a sediment profile retrieved from periodic lake with a basal radiocarbon dated from 580 ± 25 to 5080 ± 30 years ago, corresponding to 590–5900 calibrated years ago, tardigrade remains were found in high abundance (Bernardová and Košnar 2012). Six species were identified: *Mesobiotus* cf. *harmsworthi* dominated the whole core, *Murrayon* cf. *hastatus* and *Richtersius* cf. *coronifer* were also abundant, and *Macrobiotus* cf. *hufelandi* and *Macrobiotus* cf. *echinogenitus* were found only rarely, while exuviae of *Echiniscus* cf. *blumi* were very abundant in the uppermost layer.

Holocene subfossils have also been found in sediments of Antarctic lakes, with an age ranging from 280 ± 160 years to 11,210 ± 330 years (Cromer et al. 2006, 2008; Gibson et al. 2007): eggs of *Dactylobiotus* cf. *ambiguus*, *Mesobiotus furciger*,

---

Fig. 5.1 (continued) pair of legs. Asterisk, buccal lamellae around the mouth opening; gray arrow, main branch of the claw; gray arrow head, secondary branch of the claw; white arrow, papillae around the mouth opening; white arrow head, lateral papillae of the head. Images **d–f** were obtained from an animal embedded in amber, detecting by CLSM equipped with a HeNe laser (543 nm/1.2 mW) the autofluorescence of chitin in the cuticular parts of the tardigrade. Scale bars: **a**, **d** , 100 μm; **b**, **c**, **d**, **e**, 5 μm

Mesobiotus blocki, *Minibiotus weinerorum*, and *Acutuncus antarcticus* and exuviae of *Acutuncus antarcticus*, *Isohypsibius* sp., and *Diphascon* sp.

5.3 Neontology: Molecular Information

Much more information on the tardigrade origin and their phylogenetic relationships with other ecdysozoan phyla and within the phylum has been obtained using a molecular approach on current tardigrade species, even though the obtained results not always, or not yet, are in agreement among them.

5.3.1 The Molecular Clock and Tardigrades

The molecular clock (a.k.a. molecular dating or molecular palaeobiology) is a powerful technique in molecular evolution that uses fossil constraints and rates of molecular change for estimating the dates of phylogenetic events related to the origin of taxa and/or evolutionary lineages. Although molecular estimates often significantly predate the fossil record, these analyses can identify important trends in the early evolutionary history of tardigrades and related phyla. Regier et al. (2004) first applied this technique to tardigrades. Their analyses were based on three nuclear coding sequences (elongation factor-1α, elongation factor-2, RNA polymerase II) from six tardigrade species: one heterotardigrade Echiniscidae and five eutardigrades belonging both to Apochela and Parachela (Fig. 5.2). Divergence times have been estimated from amino acid sequence data using an empirical Bayesian statistical approach. They considered Tardigrada and Arthropoda to be sister taxa relative to Onychophora; therefore, they added to the analyses two Arthropoda and two Onychophora. With Onychophora as out-group, they could then treat Arthropoda as part of the in-group and apply temporal constraints on that portion of the in-group topology, since they considered tardigrade fossils too recent to represent useful constraints.

They constrained the node separating the ostracod *Cypridopsis* and the xiphosuran *Limulus* between 601 and 1017 mya (lower and upper bounds were needed to constrain the standard deviation of the estimated time). They estimated divergence times throughout the *Tardigrada* using a program developed by Thorne and Kishino (2002) that combines evolutionary information separately inferred for multiple genes within an analytical framework that allows violation of the strict molecular clock assumption. Key features of the program were that rates at adjoining nodes were assumed to be autocorrelated and that each gene had a separate autocorrelation parameter.

Based on the tree topology that they obtained and on the esteems of the rate of evolution of the different branch length, Regier et al. (2004) were able to date the tardigrade cladogenic events.

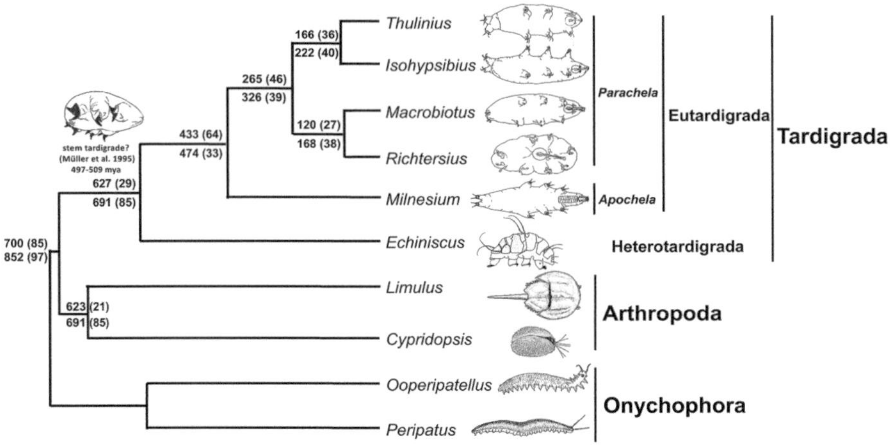

Fig. 5.2 Molecular clock analyses showing the divergence time estimates for tardigrade clades based on Regier et al. (2004). The time estimates in millions of years ago (mya) are showed above (minimum) and below (maximum) each nodes with their standard deviations in parentheses. The topology of the tree is not linearized with respect to time

According to their results (Fig. 5.2), the origin of the tardigrade stem (the split between the evolutionary lineages that lead to the extant arthropods and tardigrades) was dated back in a pre-Ediacaran period during Proterozoic (852–700 mya), so well before the Cambrian explosion. The origin of the two tardigrade classes took place 100 my later (691–627 mya), while the origin of the two orders of eutardigrades was after the Cambrian period during the Early Ordovician (474–433 mya). The split between the parachelan superfamilies Isohypsibioidea and Macrobiotoidea took place 326–265 mya when tardigrades already colonized the land; the divergence estimates place the split between genera within the superfamilies between Triassic and Cretaceous (222–120 mya).

The most problematic assumption when performing molecular clock analyses is tree topology and branch length tree, because divergence times depend on these factors.

The results obtained by Regier et al. (2004) could be affected by some methodological problems. The phylogeny on which they inferred their conclusions was obtained without the new evolutionary model that could have allowed a better esteem of the branch lengths (i.e., the number of substitutions per site per branch). Moreover, only one calibration point was used outside the phylum. In addition, to calibrate the Chelicerata-Mandibulata split, Regier et al. (2004) used a secondary calibration interval (1017–610 mya) that most likely overestimated the true divergence time among the considered taxa (e.g., Peterson et al. 2008; Erwin et al. 2011; Rota-Stabelli et al. 2013). They consider tardigrades as sister group of arthropods, but this assumption could be incorrect (see Chap. 3). Lastly, the taxon sampling used by Regier et al. (2004) was very reduced, and the estimation of branch-specific substitution rates could be affected (Blouin et al. 2005; Nabhan and Sarkar 2012).

A time scale of ecdysozoan evolution based on multiple molecular data sets, a large set of fossil calibrations, and thorough series of validation analyses have been performed by Rota-Stabelli et al. (2013). A total of 300 genes from 158 taxa were analyzed along with 67 ecdysozoan fossil calibrations. Evolutionary relationships were reconstructed using Bayesian inference and the heterogeneous CAT phylogenetic model. Relaxed molecular clock analyses were performed for five data sets using on average 30 calibration points (among the 67 available for the study). The calibrations defined soft bounded intervals, where estimated divergence times were allowed to fall outside the proposed fossil-defined intervals, with a given probability. This approach used multiple, lineage-specific substitution rates to infer divergence times and minimize errors associated with the fossil record, such as possible misidentifications and inaccurate age estimations. Relaxed molecular clock analyses were performed under the best-fitting molecular clock model for each considered data set, which was the autocorrelated CIR model for all data sets but RNA-related genes. Not all data sets recovered the same set of relationships, highlighting a discrepancy among data sets. However, most phylogenies confirmed monophyly of *Panarthropoda*, with *Tardigrada* as sister group of *Arthropoda* plus *Onychophora* (Fig. 5.3).

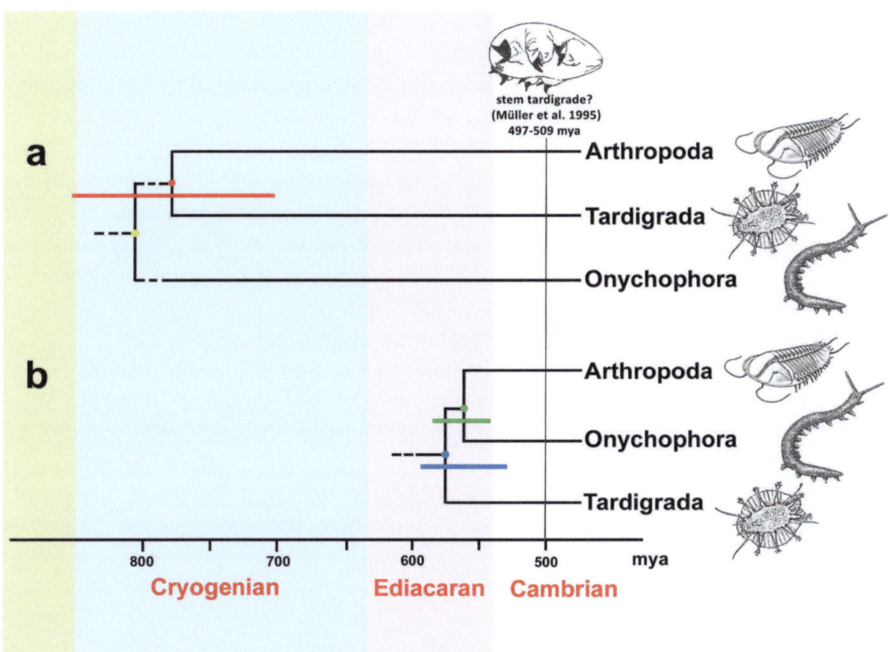

Fig. 5.3 Comparison of the molecular clock analyses for the time estimates in million years ago (mya) of the tardigrade origin between (**a**) the results of Regier et al. (2004) and (**b**) the results of Rota-Stabelli et al. (2013). The colored bars represent the minimum-maximum time estimate range for each node

The origin of ecdysozoan body plans was esteemed during Ediacaran (587–543 mya) and diversified into their constituent phyla by the beginning of the Cambrian (Rota-Stabelli et al. 2013). Therefore, the split between tardigrade stem group and lobopodia (Arthropoda + Onychophora) took place during Ediacaran, but the radiation of extant lineages has been relatively recent (442 mya), such as those of nematodes and onychophorans. The extremely long stem (ghost) lineages of these three phyla, with rare fossils being relatively young in age, induce to consider the inferences about their evolutionary patterns with caution (Rota-Stabelli et al. 2013). Probably, it is not a case that the authors reported that the only significant disagreement between the data sets was the timing of Eutardigrada diversification; they impute this result to the fact that Tardigrada is a fast-evolving lineage (Telford et al. 2008). Another problem in their analyses was the use of only one calibration point related to tardigrades. This fossil calibration was placed at 581–502 mya and corresponded to the Middle Cambrian fossil from Siberia (Müller et al. 1995). Rota-Stabelli et al. considered this fossil as "The earliest tardigrade from Siberia," but the real tardigrade nature of this fossil is not established (see above) and could lead to wrong assumptions. Considering this animal as a tardigrade and not as a representative of a possible extinct sister group of tardigrades or belonging to a stem group of tardigrades could move backward in time the radiation point of this phylum.

A recent study explored possible evolutionary scenarios for the origin of a component of the Antarctic meiofauna (i.e., the heterotardigrade *Mopsechiniscus franciscae*) using molecular dating analyses and historical biogeography (Guidetti et al. 2014, 2017). This study indicates that the molecular dating of tardigrades using genes coding for ribosomal RNA (18S and 28S rDNA) is a complex task, revealed by the very wide range of posterior densities obtained and by the relative difficulty in discriminating between competing models. The authors of this study calibrated the clock using two distinct calibration sets based on replacement rates and posterior estimates for the split of the Eutardigrada derived from Regier et al. (2004) and Rota-Stabelli et al. (2013). According to their findings, the split between the two tardigrade classes predate 550 mya; therefore, the origin of the phylum has to be placed during the Ediacaran or even before (Guidetti et al. 2017).

The results obtained by all the molecular clock analyses performed so far on tardigrades (Regier et al. 2004; Rota-Stabelli et al. 2013; Guidetti et al. 2017) are in agreement in placing the origin of the phylum well before Cambrian (Figs. 5.2, 5.3), at least during Ediacaran or before. These results placed tardigrade cladogenic event in the Proterozoic, consistently with the divergence of basal metazoan group estimates based on multigene analyses (Feng et al. 1997; Wang et al. 1999; Nei et al. 2001).

5.4 Conclusions

Even though we do not have very old fossils certainly attributable to tardigrades, but only probably attributable to the stem group of the phylum, molecular data strictly indicate that the origin of the phylum should be posed during or before the Cambrian period. Being arthropods monophyletic (Nielsen 2012) and originated within the sea around 550 mya (Rota-Stabelli et al. 2013), whether tardigrades are the sister group of *Arthropoda+Onychophora* or whether they are the sister group of *Arthropoda*, in both cases, the origin of the phylum *Tardigrada* has to be considered marine. Current marine tardigrades belong to *Arthrotardigrada* (*Halobiotus*, an eutardigrade, is secondarily marine) that however seems to be a paraphyletic taxon (Jørgensen et al. 2010). Tardigrade terrestrialization probably does not predate that of other ecdysozoans. The terrestrializations of Onychophora, Nematoda, and (Eu) tardigrada (late Silurian-Early Devonian) resulted more recent than that of arthropods (Ordovician) and mirrored the establishment of vascular plants or forests and the radiation of insects (Rota-Stabelli et al. 2013).

The first findings of sure tardigrades in the Cretaceous amber evidence that more than 90 million years ago, "modern" tardigrades already existed. It means that a very low change in morphology happened during tardigrade evolution, confirmed by the existence of many currently existing species in the subfossil findings (old even more than 200,000 years). Despite the potentiality of tardigrades as paleoclimatic bioindicators, a few studies have focused to find and to use these organisms in sediments for paleoecological studies, and more attention deserves their research in microfossil samples.

Acknowledgments We thank Ralph Schill for inviting us to write this chapter and for his patience and the anonymous referees for their valuable suggestions.

References

Bernardová A, Košnar J (2012) What do Holocene sediments in Petuniabukta, Spitsbergen reveal? Pol Polar Res 33:329–345

Bertolani R, Grimaldi D (2000) A new Eutardigrade (Tardigrada: Milnesiidae) in amber from the upper cretaceous (Turonian) of New Jersey. In: Grimaldi D (ed) Studies on fossils in amber, with particular reference to the cretaceous of New Jersey. Backhuys, Leiden, pp 103–110

Blouin C, Butt D, Roger AJ (2005) Impact of taxon sampling on the estimation of rates of evolution at sites. Mol Biol Evol 22(3):784–791

Caron JB, Aria C (2017) Cambrian suspension-feeding lobopodians and the early radiation of panarthropods. BMC Evol Biol 17(1):29

Chen JY, Zhou GQ (1997) Biology of the Chengjiang fauna. Bull Natl Mus Nat Sci 10:11–106

Cooper KW (1964) The first fossil tardigrade: *Beorn leggi* Cooper, from cretaceous amber. Psyche 71:41–48

Cromer L, Gibson JAE, Swadling KM, Hodgson DA (2006) Evidence for a lacustrine faunal refuge in the Larsemann Hills, East Antarctica, during the last glacial maximum. J Biogeogr 33:1314–1323

Cromer L, Gibson JAE, McInnes SJ, Agius JT (2008) Tardigrade remains from lake sediments. J Paleolimnol 39:143–150

Durante MV, Maucci W (1972) Descrizione di *Hypsibius* (*Isohypsibius*) *basalovoi* sp. nov. e altre notizie su Tardigradi del veronese. Mem Mus Civ St Nat Verona 20:275–281

Erwin DH, Laflamme M, Tweedt SM, Sperling EA, Pisani D, Peterson KJ (2011) The Cambrian conundrum: early divergence and later ecological success in the early history of animals. Science 334:1091–1097

Feng DE, Cho G, Doolittle RF (1997) Determining divergence times with a protein clock: update and reevaluation. PNAS 94:13028–13033

Gibson JA, Cromer L, Agius JT, McInnes SJ, Marley NJ (2007) Tardigrade eggs and exuviae in Antarctic lake sediments: insights into Holocene dynamics and origins of the fauna. J Limnol 66(Suppl 1):65–71

Guidetti R, Rebecchi L, Bertolani R (2000) Cuticle structure and systematics of the Macrobiotidae (Tardigrada, Eutardigrada). Acta Zool 81(1):27–36

Guidetti R, Rebecchi L, Cesari M, McInnes SJ (2014) *Mopsechiniscus franciscae*, a new species of a rare genus of Tardigrada from continental Antarctica. Polar Biol 37(9):1221–1233

Guidetti R, McInnes SJ, Cesari M, Rebecchi L, Rota-Stabelli O (2017) Evolutionary scenarios for the origin of an Antarctic tardigrade species based on molecular clock analyses and biogeographic data. Contrib Zool 86(2):97–100

Kristensen RM, Neuhaus B (1999) The ultrastructure of the tardigrade cuticle with special attention to marine species. Zool Anz 238:261–282

Jankovská V (1992) Unbekannte Objekte in Pollenpräparaten – Tardigrada. In: J Kovar-Eder (ed) Palaeovegetational development in Europe and regions relevant to its palaeofloristic evolution. Proceedings of the Pan-European Palaeobotanical Conference, Museum of Natural History, Vienna, pp 19–23, 19–23 Sept 1991

Jankovská V (2008) Non pollen Palynomorphs in Pollen slides: their determination and utilization in palaeoecology. Новости палеонтологии и стратиграфии: Приложение к журналу 10–11:488–491

Jankovská V, Roszkowska M, Kaczmarek Ł (2016) Remains of non-pollen-palynomorphs–tardigrades from Spitsbergen found during pollen analyses. Polar Rec 52(04):450–463

Jørgensen A, Faurby S, Hansen JG, Møbjerg N, Kristensen RM (2010) Molecular phylogeny of Arthrotardigrada (Tardigrada). Mol Phyl Evol 54(3):1006–1015

Liu J, Steiner M, Dunlop JA, Keupp H, Shu D, Ou Q, Han J, Zhang Z, Zhang X (2011) An armoured Cambrian lobopodian from China with arthropod-like appendages. Nature 470 (7335):526–530

Maas A, Waloszek D (2001) Cambrian derivatives of the early arthropod stem lineage, pentastomids. tardigrades and lobopodians Zool Anz 240:451–459

Maas A, Mayer G, Kristensen RM, Waloszek D (2007) A Cambrian micro-lobopodian and the evolution of arthropod locomotion and reproduction. Chin Sci Bull 52(24):3385–3392

Miller WR, Heatwole HF (2003) Tardigrades of the sub-Antarctic: 5000 year old eggs from Marion Island. Abstract. 9th International Symposium on Tardigrada, Florida, USA

Müller KJ, Walossek D, Zakharov A (1995) "Ortsen" type phosphatized soft-integument preservation and a new record from the middle Cambrian Kuonamka formation in Siberia. N Jb Geol Paleont Abh 197:101–118

Nabhan AR, Sarkar IN (2012) The impact of taxon sampling on phylogenetic inference: a review of two decades of controversy. Brief Bioinform 13(1):122–134

Nei M, Xu P, Glazko G (2001) Estimation of divergence times from multiprotein sequences for a few mammalian species and several distantly related organisms. PNAS 98(5):2497–2502

Nielsen C (2012) Animal evolution: interrelationships of the living phyla, 3rd edn. Oxford University Press, New York

Peterson KJ, Cotton JA, Gehling JG, Pisani D (2008) The Ediacaran emergence of bilaterians: congruence between the genetic and geologic fossil records. Philos T Roy Soc B 363:435–1443

Pilato G (1979) Correlations between cryptobiosis and other biological characteristics in some soil animals. Boll Zool 46:319–332

Regier JC, Shultz JW, Kambic RE, Nelson DR (2004) Robust support for tardigrade clades and their ages from three protein-coding nuclear genes. Invertebr Biol 123:93–100

Rota-Stabelli O, Daley AC, Pisani D (2013) Molecular timetrees reveal a Cambrian colonization of land and a new scenario for ecdysozoan evolution. Curr Biol 23(5):392–398

Telford MJ, Bourlat SJ, Economou A, Papillon D, Rota-Stabelli O (2008) The evolution of the Ecdysozoa. Philos T Roy Soc B 363:1529–1537

Thorne JL, Kishino H (2002) Divergence time and evolutionary rate estimation with multilocus data. Syst Biol 51(5):689–702

Waloszek D (2003) The 'Orsten'window—a three-dimensionally preserved upper Cambrian meiofauna and its contribution to our understanding of the evolution of Arthropoda. Paleont Res 7 (1):71–88

Wang DY-C, Kumar S, Hedges SB (1999) Divergence time estimates for the early history of animal phyla and the origin of plants, animals and fungi. Proc R Soc London B 266:163–171

Chapter 6
Cytology and Cytogenetics

Roberto Bertolani and Lorena Rebecchi

Abstract Several cytological aspects have been considered in tardigrades. Firstly, the cell constancy which is not a true eutely being several mitoses present even after hatching, even though some organs, such epidermis and nervous ganglia, have the same cell number in juveniles and adults. The total number of these cells is species-specific. Then the ultrastructure of cuticle, epidermis, feeding and digestive apparatus, excretory and osmoregulatory organs, muscles, nerve cells, sensory cells and storage cells has been considered. Instead, the ultrastructure of the germ cells has been considered in the chapter on reproduction. With regard to chromosome number and shape, it has been observed that generally there is little difference among the species ($n = 5$ or $n = 6$), but several cases of polyploid populations exist, often very similar to diploid populations from a morphological point of view. In most cases the polyploid populations do not have males and reproduce by apomixis. Studies on the genome size have confirmed the presence of polyploid populations, as well as the presence of nuclei with multiple amounts of DNA within the same specimen. The genome size of the tardigrades is always relatively small and does not seem related to phylogenetic lineages. Studies on tardigrade genomes have placed this phylum at the centre of discussions on the evolution of Metazoa and have considered the role of horizontal gene transfer in animal evolution with contrasting results.

R. Bertolani (✉)
Department of Education and Humanities, University of Modena and Reggio Emilia, Modena, Italy

Civic Museum of Natural History, Verona, Italy
e-mail: roberto.bertolani@unimore.it

L. Rebecchi
Department of Life Sciences, University of Modena and Reggio Emilia, Modena, Italy
e-mail: lorena.rebecchi@unimore.it

R. O. Schill (ed.), *Water Bears: The Biology of Tardigrades*, Zoological Monographs 2,
https://doi.org/10.1007/978-3-319-95702-9_6

6.1 Introduction

Tardigrades are always microscopic, and their reduced body size is an advantage in colonizing interstices of several types of substrates, not only in marine and freshwater habitats but also and especially in terrestrial habitats. Such a small body size is necessarily formed by a limited number of cells, with some organs formed by a constant cell number. Nonetheless, there is no true eutely, as in nematodes, in which mitoses do not occur after hatching. In some tardigrade organs or systems, such as in the nervous system or epidermis, the cells do not increase in number with growth, whereas in other cases (e.g. midgut, storage cells) they clearly vary in number. Other than the gonad (ovary, testis and ovotestis), in which mitoses and meioses frequently occur, mitoses have been frequently observed in storage cells (coelomocytes) and in the so-called transient cells in the beginning of the midgut (Bertolani 1970; Czernekova and Jönsson 2016). Increase in storage cell mitoses has been related to moulting and the late developmental stage of eggs (Czernekova and Jönsson 2016). Variation in storage cell size and number has also been related to anhydrobiosis (Jönsson and Rebecchi 2002).

6.2 Cytology

Ultrastructural morphology of different types of cells of tardigrades has been studied since the 1970s. In this chapter, we will describe the general cell morphology in various organs and systems of tardigrades, excluding those of gonads and related germ cells, which are discussed in another chapter (Chap. 8).

6.2.1 Cuticle and Epidermis

The epidermal cells form a monolayer of square, large, flat cells covered by a chitinous cuticle (Bussers and Jeuniaux 1973; Greven and Peters 1986; Kristensen and Neuhaus 1999). The cuticle, from the outside to the inside, is formed by the epicuticle (outer and inner epicuticle), intracuticle (trilaminar layer) and procuticle. According to Greven (1984), the epidermal cells contain the usual cytoplasmic organelles (nucleus, mitochondria, RER, free ribosomes, etc.) and numerous membrane-bound pigment granules and occasionally lipid droplets. In reality, epidermal pigment is evident only in some species or genera, for example, in the eutardigrade *Ramazzottius*, but it is absent in clearly pigmented genera, such as the heterotardigrade *Echiniscus blumi*, in which Raman microspectroscopy on living specimens identified carotenoids (beta-carotene) within the animal body and eggs, but pigments are absent in the cuticle and epidermis (Bonifacio et al. 2012). The epidermal cells are connected to each other by tight and intermediate junctions, desmosomes or septate junctions (Baccetti and Rosati 1971; Kristensen 1976; Greven 1984). According to Marcus (1929) the epidermal cells are constant in number, but mitoses have been observed (Bertolani 1970).

6.2.2 Cells of the Feeding and Digestive Apparatus

The anterior part of the feeding apparatus is composed of non-cellular, cuticular structures (mouth, buccal tube, stylet supports); only the stylets are mainly composed of calcium carbonate. On the sides of the buccal tube or the pharynx, there are two large salivary glands, connected with the mouth and containing heterogeneous mucus. According to Avdonina et al. (2007), these glands contain scattered areas without limiting membranes and with electron-lucent material, and each gland is surrounded by thin muscle processes and neuromuscular contacts. According to Dewel and Clark (1973a, b), the heterogeneity of the mucus could be either a result of different stages of mucous secretion or an artefact of fixation.

The buccal tube penetrates into the pharynx, a myoepithelial structure whose cells contain several mitochondria concentrated under the cuticle lining the cavity of the pharynx (Avdonina et al. 2007). Cellular parts of the digestive apparatus have been examined in several species of eu- and heterotardigrades. In the oesophagus of *Ramazzottius tribulosus*, Avdonina et al. (2007) have observed epithelial cells characterized by a flat basal surface, some large vacuoles and an abundance of rough endoplasmic reticulum (RER). The wall of the midgut is formed by a basal membrane with one layer of large epithelial cells and is surrounded by visceral muscles. Epithelial cells exhibit one large nucleus and one large nucleolus, a Golgi apparatus with numerous dictyosomes, many small and large vacuoles, electron-dense inclusions and well-developed RER, which do not form concentric whorls (Avdonina et al. 2007). The surface of the midgut is enlarged by the presence of a high number of folds (first-order and second-order folds) extending the length of the midgut. The apical surface of the epithelial cells exhibits many microvilli (determined in *R. tribulosus* to be 111 per μm^2 and about 0.5 μm in length) forming a regular brush border and surrounded by a well-developed glycocalyx. In cross section, the microvilli form very compact hexagonal, honeycomb-shaped cells.

According to Dewel et al. (1989), the midgut cells of the heterotardigrade *Echiniscus viridissimus* differ from those of the eutardigrades in a number of ways. The nucleus is located apically rather than centrally. Much of the RER is found as a single cisterna paralleling the cell periphery or surrounding membrane-bound inclusions. Golgi bodies appear to be less numerous, and apical secretory (digestive) granules and large multivesicular or granular bodies are absent. According to Avdonina et al. (2007), microvilli density of the midgut epithelium is higher in the phytophagous *R. tribulosus* than in the zoophagous *Paramacrobiotus spatialis* and *Isohypsibius prosostomus*, whereas according to Hyra et al. (2016b) the number of microvilli decreases towards the hindgut. These authors also identified two types of cells in the midgut epithelium: digestive cells (principal cells) and regenerative cells. The regenerative cells are gathered at the anterior part of the midgut and in the posterior end. The anterior ones were already known as transient cells ("Übergangszellen"), often with the presence of mitoses (Bertolani 1970). Studies on *Hypsibius dujardini* verified that the cytoplasm of the regenerative cells is poor in organelles, and primarily has sporadic mitochondria and cisterns of endoplasmic reticulum (Hyra et al. 2016b). These authors observed that the regenerative cells do not modify their shape during the various stages of oogenesis, whereas evident changes occur in the digestive cells. Just after oviposition, when the animals resume feeding, the digestive cells are cuboidal in shape. Their cytoplasm contains ribosomes, mitochondria, cisterns of rough endoplasmic reticulum and Golgi complexes. Then the amount of reserve material increases, such as the number of mitochondria, free ribosomes, cisterns of the rough endoplasmic reticulum and spheres having content with a different electron density. Subsequently the midgut epithelial cells begin to become cylindrical in shape, and the epithelium creates folds projecting into the midgut lumen. The number of large and small spheres with electron-dense content increases significantly. During egg shell formation, large, irregularly shaped vacuoles with a medium electron-dense content are observed in the digestive cells. Lastly, the amount of reserve material that had earlier accumulated in the cytoplasm of the digestive cells decreases significantly. The digestive cells change their shape into a cuboidal form; their cytoplasm is rich in organelles such as ribosomes, mitochondria and cisterns of the rough endoplasmic reticulum.

In *Isohypsibius granulifer*, using a histochemical and ultrastructural approach, Rost-Roszkowska et al. (2011) showed that in the midgut epithelium cells, accumulation of glycogen granules, lipid droplets and structures of various electron densities occurs gradually. According to these authors, the midgut epithelium takes part in yolk precursor synthesis. Consequently, during vitellogenesis and choriogenesis, in the cytoplasm of the midgut cells, they observed an increasing number of organelles responsible for intensive synthesis of lipids, proteins and saccharides, such as cisterns of endoplasmic reticulum and Golgi complexes. At the end of oogenesis, autophagy also intensifies in midgut epithelial cells, probably caused by the great amount of reserve material.

Halberg and Møbjerg (2012), using Chlorophenol Red, evidenced that the tardigrade midgut is involved in organic anion transport, which is active and transporter-mediated.

With regard to the hindgut, electron microscopic observations on *Milnesium* by Dewel and Dewel (1979) demonstrated that the anterior hindgut is a specialized transporting epithelium. The luminal surface is covered by a thin layer of cuticle, which penetrates into channel-like invaginations. Numerous mitochondria are concentrated apically. The basal and lateral surfaces are also folded. The cells are joined apically by deep tight junctions and a simple basal lamina lines the entire hindgut. Avdonina et al. (2007) observed that in *R. tribulosus* the cell surface has many indentations of plasmalemma, which form channels penetrating the cuticle of the rectum lumen. According to Dewel and Dewel (1979), the anterior hindgut could be involved in the response to a hypoosmotic habitat, evaporative water loss during the induction of anhydrobiosis and low oxygen tension.

6.2.3 Cells of the Excretory and Osmoregulatory Organs

The two lateral and the dorsal Malpighian tubules have been studied from an ultrastructural point of view in several species of eutardigrades by several authors (Dewel and Dewel 1979; Greven 1979; Węglarska 1987a, b, c; Dewel et al. 1993; Pelzer et al. 2007). Møbjerg and Dahl (1996) described the ultrastructure of the cells of these organs in the marine eutardigrade *Halobiotus crispae*. These tubules consist of an initial segment with three large cells, a thin transitional distal part lacking a nucleus and a proximal part with 9–12 nuclei. The initial segment possesses deep basal infoldings and interdigitating, finger-shaped processes of the plasma membrane, large mitochondria and giant nuclei. The distal part is a short section supporting the initial segment. Cellular offshoots from the succeeding proximal part constitute the distal part. The distal and proximal parts contain intercellular canals with concretions of variable size. The exit of the proximal part into the digestive tract is characterized by the presence of microvilli. Observing the ultrastructure of the Malpighian tubules of *Hypsibius klebelsbergi*, Pelzer et al. (2007) noted that the cells of the distal part are characterized by a large nucleus, a fair number of mitochondria, a basal labyrinth, interdigitating plasma membranes and an irregular surface. They also noted that apical spaces extend into the middle part, which is in large part without nuclei and characterized by basal infolding and sparse mitochondria. Interwoven cell projections and microvilli give this part and the proximal part a vacuolated appearance.

Heterotardigrades do not have Malpighian organs, but several echiniscids have unusual organs lying above the ventromedial body wall approximately at the level of the second and third pair of legs (Dewel et al. 1992, 1993). Each organ is composed of one medial and two lateral cells and possesses a pair of convoluted ducts. The first one is characterized by a high density of mitochondria interwoven among the tubular infolds. The latter have a dorsolateral surface with an elevated basement membrane whose filaments can be transferred into the body cavity. The lateral cell extension forms "tight" zonula adherens junctions with the walls of the cleft in the medial cell, thus isolating the duct of the organ from the body cavity.

6.2.4 Muscle Fibres

Four types of muscles are known in tardigrades: somatic muscles, pharyngeal muscles, stylet muscles and visceral muscles.

The somatic muscles are formed by isolated elongated muscle strands arranged parallel to the longitudinal axis of the animal (Marchioro et al. 2013). Each is enveloped by a basement membrane and contains thin and thick filaments and a peripheral nucleus surrounded by numerous clumped mitochondria (Dewel et al. 1993). In eutardigrades, the thin filaments enter staggered Z bodies, and the filaments are arranged in irregular clumps: "ribbon-like" rows in cross section and in diagonal bands of I and A zones in longitudinal section (Shaw 1974; Walz 1974, 1975; Dewel et al. 1993). Cross-striated muscles have been observed by Halberg et al. (2009) in *Richtersius coronifer*. The somatic muscles of marine arthrotardigrades are cross striated (Kristensen 1978) with two to five sarcomeres per cell (in *Batillipes noerrevangi*). No H zone is visible in longitudinal sections. In somatic muscle, the sarcoplasmic reticulum is formed by smooth, flattened cisternae lying between the contractile elements and the sarcolemma and within the I zones of the contractile apparatus (Dewel et al. 1993).

The contractile cells of the pharynx contain a monosarcomeric contractile apparatus (Dewel and Clark 1973b; Walz 1973; Shaw 1974; Kristensen 1978). The thin filaments attach to the apical and basal plasmalemma by hemidesmosomes. The thick filaments are found in a distinct A zone, which is displaced basally and has a broad apical zone (Dewel et al. 1993). An H zone is not visible.

The stylet muscles are cross striated both in eu- and heterotardigrades in all examined species (Walz 1974; Kristensen 1978). They insert on hemidesmosomes when attached on the buccal tube or pharyngeal apical cells. Z bodies are slightly out of alignment with one another. Here too, H zones are not visible.

The visceral muscles are small cells with multiple adhesion sites along the sides of the cell. These multiple adhesion sites have a structure similar to hemidesmosomes and have been considered as the functional equivalents of the absent Z bodies and the insertion sites for loosely organized myofibrils (Shaw 1974; Walz 1975). Near the attachment plaques I zones are clearly evident; H zones can be seen in cross sections (Walz 1975).

6.2.5 Ultrastructure of the Nerve Cells

The ultrastructure of nerve cells of the cerebral ganglion has been studied in *Milnesium tardigradum* and well-described by Wiederhoft and Greven (1996). The entire ganglion is surrounded by a thin neural (basal) lamella. Cell bodies of the ganglia are about 5 μm in diameter, irregularly shaped and with large nuclei. The cytoplasm, even if scarce, contains numerous vesicles, RNA-particles, heterogeneous cytosomes, mitochondria, small amounts of rough endoplasmic reticulum and dictyosomes. The neuropil consists of unsheathed nerve fibres containing fine filaments, mitochondria, a few microtubules and numerous vesicles. These vesicles can be distinguished based on their size and electron density. Small electron-lucent vesicles are concentrated on the presynaptic side of the synaptic junctions and probably correspond to synaptic vesicles. Larger electron-lucent vesicles are likely to be cholinergic vesicles. The other vesicles resemble dense-core bodies which contain biogenic amines. Nerve fibres of the neuropil can be differentiated based on their diameter and the number of vesicles. Type I has a diameter of about 250 nm and contains microtubules, mitochondria and electron-lucent vesicles. Type II has a more irregular outline, is more electron dense than type I and contains large electron-lucent vesicles and vesicles with a dense core. Type III is fairly electron-lucent and flat; it is located largely between the inner and outer lobes at the underside of the cerebral mass near the neural lamella. Type IV is relatively electron-dense and has a large diameter and vesicles of all types. Membranes of adjacent axons are connected by septate junctions. Synaptic junctions are numerous. Synapses with dense bar-like material incorporated into the membrane of the postsynaptic terminal are present when axon II is adjacent to axon I. Another type of synapsis is characterized by dense material incorporated into the membrane of the presynaptic terminal between two axons of type I. A third type of synapsis is present when three nerve fibres of type I share a common presynaptic terminal. Dense material can be observed in the zones of contact.

The neuropils contain peripheral glial cell bodies and enveloping processes (Greven and Kuhlmann 1972; Dewel et al. 1993). The glial cells, recognizable by tightly packed ribosomes, do not cover the perikarya of most neurons or form a sheath around the ganglia (Greven and Kuhlmann 1972).

Precerebral ganglia are recognizable by a more heterogeneous structure (Wiederhoft and Greven 1996). Many cells are fairly electron-lucent. They are located preferentially at the side of the ganglion which faces the cerebral ganglion. They contain vesicles of different types, mitochondria and rough endoplasmic reticulum. Other cells are more electron-dense and similar to the cell bodies of the cerebral ganglia. Investigations on the brain and neuroanatomy of the eutardigrade *Halobiotus crispae* reveal pronounced immunoreactivity against acetylated α-tubulin, particularly in the brain as well as the longitudinal nerve cords and ventral ganglia (Persson et al. 2012). These studies have evidenced the presence of transverse commissures in the ventral ganglia and suggest that the brain is partitioned into at least three lobes. In the arthrotardigrade *Actinarctus doryphorus* too, the brain is partitioned into three paired lobes, and these lobes exhibit a more pronounced separation as compared to that of eutardigrades (Persson et al. 2014).

6.2.6 Sensory Cells

Sensory receptors have often different shapes in hetero- and eutardigrades and sometimes are positioned differently. Sensory neurons innervating the sensory structures are always bipolar, can be two or more in number and can be distinguished on the basis of the structure of their outer dendritic segments.

According to Kristensen (1981), the different appendages of heterotardigrades have the same basic pattern: one or two sensory cells, each bearing one modified cilium surrounded by microvilli-like projections. A second ("trichogen") cell envelopes both the ciliary and the inner cell. A third ("tormogen") cell externally surrounds the base of the sensillum. Cephalic cirri are considered as mechanoreceptive. Nonetheless, the external cirri of *Halechiniscus* and *Batillipes* have two sensory cells, one of them with a cilium terminating in a pore, which suggests even a chemoreceptive function. Clavae contain numerous both cilia and microvilli and have a large terminal pore; therefore, they are presumably chemoreceptors.

In the Parachela, which lack cephalic appendages, Walz (1978, 1979) identified two sensory fields in the anterior head region of *Macrobiotus hufelandi*: an anterior-lateral sensory field and a circumoral sensory field. He also identified three types of dendrites in the circumoral sensory organs, each with a highly modified ciliary region, without a centriole, and with a highly modified basal body. He also identified receptor cells having cilia with 9 + 0, 11 + 0, 12 + 0, 15 + 0, 17 + 0 and 31 + 0 patterns. The Type 1 sensory cell might be chemoreceptive. Type 2 cells are provided with highly ordered microtubules and filaments in the outer dendritic segment, resembling a mechanoreceptor. Type 3 are branched sensory cells with unknown function. The anterior sensory field is innervated by three pairs of fibre bundles, each divided into an outer brush-like and an inner dendritic segment by the ciliary region (Walz 1978). Approximately 40 dendrites terminate under the cuticle. There is also a pharyngeal organ, represented by four cuticular pockets, located at the transition between mouth cavity and mouth tube and innervated by 3–5 dendrites each. The cuticular pockets enclose considerable receptor lymph cavities which communicate with the lumen of the anterior foregut over multiple pores. Sensory fields also occur in the foregut.

6.2.7 Storage Cells

The storage cells, or body cavity cells, have an ameboid shape and move freely within the body cavity. The size of the storage cells is positively correlated with the amount of accumulated nutrient reserves (Marcus 1929). The reserves consist of polysaccharides and lipids (Węglarska 1975). Although typical glycogen granules have not been found in cells prepared for electron microscopy, these polysaccharides have been located there histochemically (Węglarska 1975). In *Milnesium tardigradum* lipid droplets are often surrounded by a second, densely staining inclusion. Occasionally, this material appears layered around the periphery of lipid droplets. Large phagocytic vacuoles (hetero- or autophagic or a combination of both) are common whether lipid is present or not. Heterophagosomes may be prevalent when lipid reserves are adequate if the cells are functioning as general scavengers, and autophagosomes may be plentiful during starvation. A decrease in the cell size of these cells has been observed over a period of anhydrobiosis (Jönsson and Rebecchi 2002). Storage cells are also involved in vitellogenesis (Jönsson and Rebecchi 2002) and in the synthesis of vitellogenins (Hyra et al. 2016b). Mitoses frequently occur in storage cells (Bertolani 1970; Czernekova and Jönsson 2016).

According to Hyra et al. (2016a), the centre of each storage cell is occupied by a large, irregular nucleus with a large non-homogeneous nucleolus surrounded by clumps of heterochromatin. The internal part of the nucleolus has a lower electron density than the external part. The cytoplasm contains cisterns of rough endoplasmic reticulum (RER), Golgi complexes, free ribosomes and mitochondria. Spheres of reserve material with different sizes and electron densities also occur. Up to four types of spheres are distinguishable: with a high electron density (small, large), with a medium electron density and with a lower electron density. The ultrastructure of the storage cells and the number of spheres of reserve material can change according to the stage of oogenesis. At the beginning of vitellogenesis, there is a slight reduction in the number of the cisterns of the RER and a numeric increase of mitochondria and spheres. Accumulation of the reserve material in the cytoplasm of the storage cells occurs until the middle stage of oogenesis. During late vitellogenesis and the beginning of choriogenesis, there is again an increase in the number of cisterns of the RER, ribosomes and mitochondria, in comparison with some autophagosomes. At this stage the amount of the reserve material decreases significantly. At the end of choriogenesis, the storage cells contain cisterns of the RER, ribosomes, mitochondria, single Golgi complexes and autophagosomes, whereas the amount of the reserve material consistently decreases until the end of oviposition.

6.3 Karyology and Cytotaxonomy

Until the 1960s, there was only a vague (and not always correct) estimate of the chromosome number of tardigrades, obtained from histological cross sections. The first clear datum on the karyotype of a tardigrade species was obtained by Ammermann (1962, 1967) who, during a study on parthenogenesis in *Hypsibius dujardini*, evidenced the presence of a diploid number of ten chromosomes (Fig. 6.1). Some years later, Bertolani conducted a series of karyological studies on several species of eutardigrades (for review see Bertolani 1975, 1982). These studies were mainly performed on oocytes, which showed larger chromosomes compared to those observed during mitoses in gametogonia or somatic cells, or in meioses in the male germ cells (Fig. 6.2). In an all-female population at that time attributed to *Paramacrobiotus richtersi*, currently attributable to *Paramacrobiotus fairbanski*, he discovered the first case of polyploidy, namely, triploidy (17–18 chromosomes), which was tied to apomictic (ameiotic, with univalents; Fig. 6.3a) parthenogenesis (Bertolani 1971, 1972). In a population attributable to the same morphospecies, but found in a different locality, Bertolani (1971) identified males and females (i.e. an amphimictic population) with mitoses with 12 chromosomes and male and female meioses with 6 bivalents (i.e. a diploid population) (Fig. 6.3b–d). The presence of polyploidy (often triploidy, but also tetraploidy; Fig. 6.4) has been subsequently found in several other species of different eutardigrade genera (*Macrobiotus*, *Paramacrobiotus*, *Xerobiotus*, *Pseudobiotus*, *Eremobiotus*, *Ramazzottius*; for review see Bertolani 1975, 1982), in most cases bound to apomixis. Only in one case, tetraploidy was found in a hermaphroditic population (Bertolani 1975, 1982). The species was firstly attributed to *Isohypsibius*

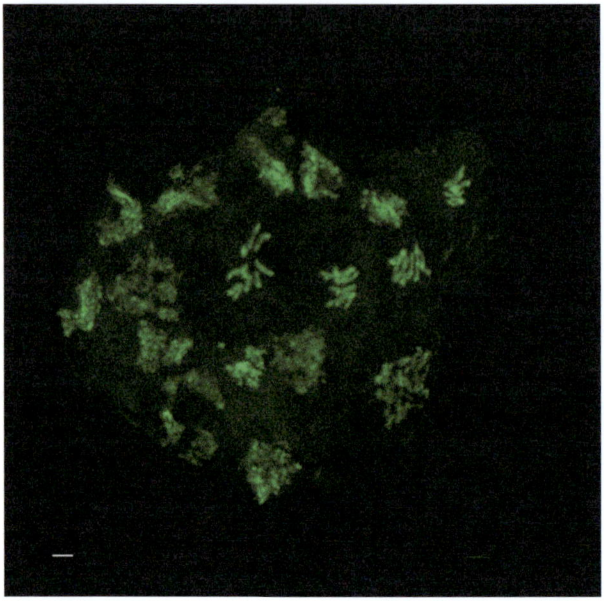

Fig. 6.1 Mitoses in young embryos of *Hypsibius dujardini*. Confocal microscopy; bar = 5 μm

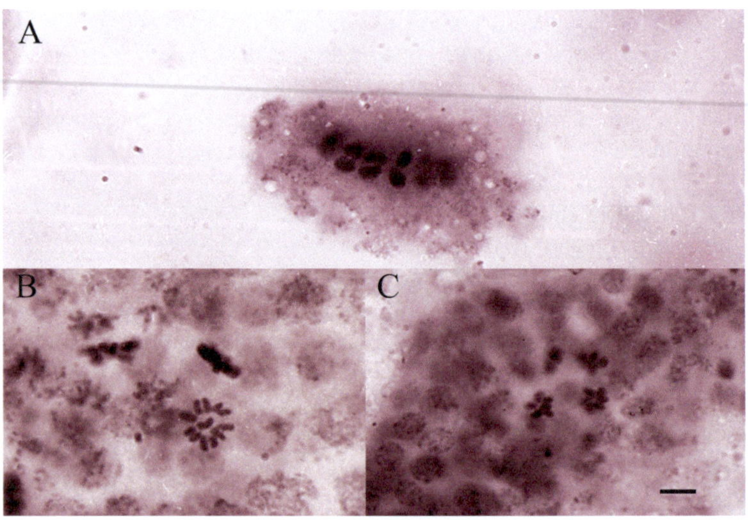

Fig. 6.2 Cell divisions in *Macrobiotus* cf. *hufelandi*. (**a**) Oocyte metaphase, lateral view; (**b**) mitotic metaphase; (**c**) spermatocyte metaphases. Orcein; bar = 5 μm

cf. *granulifer* (Bertolani 1975) and then described as a new species, *Isohypsibius marii* (Bertolani 1981). The presence of diploid populations, normally with males, within morphospecies also having triploid and sometimes even tetraploid populations, is relatively frequent. Recent molecular studies using *cox1* as a marker

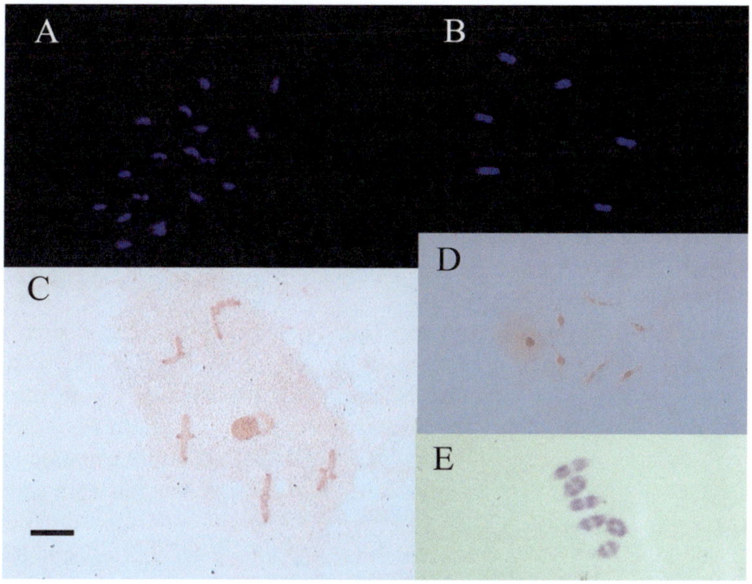

Fig. 6.3 Cell divisions in oocytes of *Paramacrobiotus*. (**a**) Triploid metaphase with 17 univalents in *Paramacrobiotus fairbanski*. (**b–e**): Diploid cell divisions in *Paramacrobiotus spatialis*: (**b**) metaphase with 6 bivalents; (**c**) diplotene with residual nucleolus and 6 bivalents with chiasmata and terminal silver positive bands on both extremities. (**d**) Metaphase with 6 bivalents. (**e**) Metaphase with 6 bivalents. (**a–b**) DAPI, (**c–d**) silver staining, (**e**) Giemsa. Bar = 5 μm

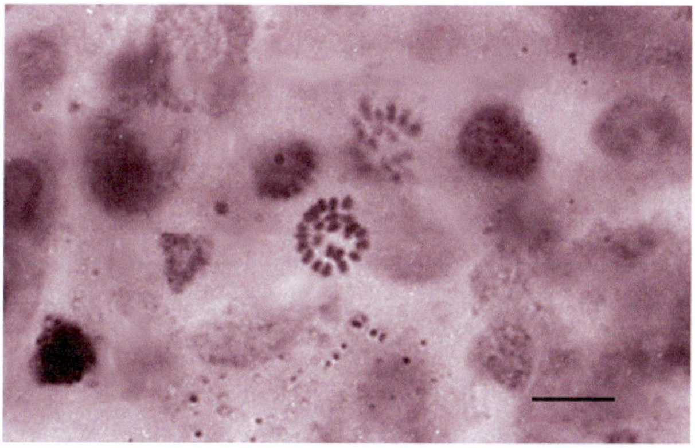

Fig. 6.4 Mitotic metaphase in a tetraploid specimen of *Ramazzottius*. Orcein; bar = 5 μm

(Guidetti et al. 2019; Bertolani unpublished) reveal that the populations with different ploidy should be attributed to different species, in spite to their very close morphological resemblance. A similar conclusion can be deduced by studies using allozymes. For example, two kinds of diploid populations attributable to the morphospecies *Richtersius coronifer* (one population with females and males and another one with only females) greatly differed in the mutation number of some allozymes (Rebecchi et al. 2003). Guidetti et al. (2016) confirmed the presence of cryptic species in this genus.

Cytology of egg maturation in diploid parthenogenetic populations of tardigrades always evidenced meiotic (automictic) parthenogenesis with pairing of the homologous chromosomes (Ammermann 1967; Bertolani 1975, 1982; Rebecchi et al. 2003) and, except *Hypsibius dujardini*, with chiasmata during oocyte prophase. Therefore, after several automictic generations, complete homozygosity occurs. This has been clearly shown in the morphospecies *Richtersius coronifer* using allozymes (Rebecchi et al. 2003).

Bertolani (1975, 1982) observed very little variation in the chromosome number and shape among tardigrade species, with the exception of polyploidy. The basic number was normally 5 or 6, and there was little difference in size among the various chromosomes. Using banding techniques, Rebecchi et al. (2002) confirmed the presence of diploid and polyploid populations but also added information on chromosome shape and behaviour. In most cases, these chromosomes were acrocentric (Fig. 6.5), and, even though they always paired during meiosis, crossing over occurred only in oocytes (Fig. 6.2c) and not in spermatocytes (Altiero and Rebecchi 2003), as in many insects. Studies on Feulgen-DNA content and genome size have confirmed the presence of diploid and polyploid populations, as well as the presence of nuclei with multiple amounts of DNA within the same specimen (Bertolani et al. 1987, 1994; Redi and Garagna 1987).

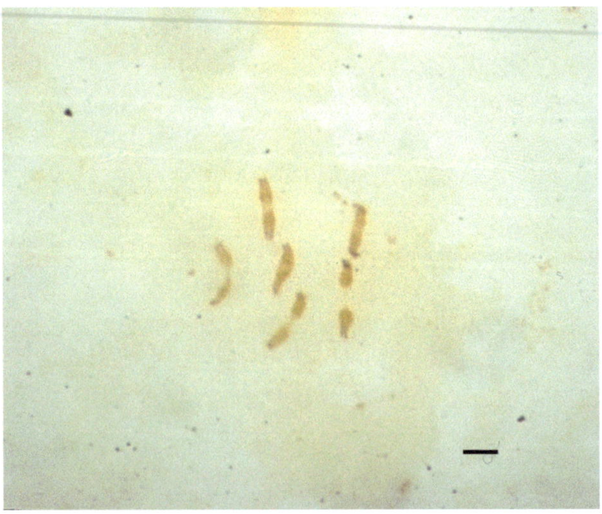

Fig. 6.5 Meiotic metaphase in a diploid specimen of *Paramacrobiotus spatialis*; note the positive staining of the telomeres. Silver staining; bar = 5 μm

6.4 Cytogenetics

The genome size of some species of tardigrades, evaluated measuring the Feulgen-DNA content of spermatids and spermatozoa in arbitrary units (a.u.) and then eventually transformed in pg (1 a.u. = 0.31 pg), is always relatively small, even though it ranges widely from 0.08 pg of *Isohypsibius lunulatus* to 0.73 pg of *Bertolanius volubilis* (Bertolani et al. 1994). The genome size values do not seem related to phylogenetic lineages: for instance, family Eohypsibiidae comprises the highest value in *Bertolanius volubilis* (2.34 a.u. in haploid cells) and one of the lowest values, found in *Eohypsibius nadjae* (0.56 a.u. at least in diploid cells). Within Macrobiotidae, *Paramacrobiotus spatialis* and *Xerobiotus pseudohufelandi* have similar values (in male gametes 0.84 and 0.76 a.u., respectively), while in the same kind of cells *Macrobiotus furcatus* has 0.44 a.u. The lowest value (in a.u.) has been found in spermatozoa of *Ramazzottius tribulosus* (0.31), but relatively low values have been found even in *Parhexapodibius ramazzottii*, *Necopinatum mirabile* and *Milnesium tardigradum* (0.70, 0.60, 0.84 a.u. in diploid cells, respectively). Higher values have been found in two echiniscid heterotardigrades: *Cornechiniscus lobatus* and *Echiniscus blumi* (1.36 and 1.09 a.u., respectively), in which the degree of ploidy is unknown.

A common finding in almost all diploid populations and in some with unknown ploidy is the presence of four doublings of the diploid, or presumed diploid content (2c) of DNA, independently from the genome size of the species (Bertolani et al. 1987, 1994). The number of doublings is lower in specimens of polyploid populations, reaching the same maximum Feulgen-DNA content (8.7–8.9 pg) as in the diploid ones. Notably in organs in which physiological cell renewal is rapid (such as gonads, midgut and storage cells complexes), DNA content is a multiple of the 2c ploidy, whereas in organs in which cell divisions are rare or absent (in particular the nervous system and epidermis), the ploidy is only 2c.

6.5 Cytogenetics, Animal Phylogeny and Gene Expression

Studies on tardigrade genomes have placed this phylum at the centre of discussions on the evolution of Metazoa, the biology of survival in extreme environments and the role of horizontal gene transfer in animal evolution (see Chap. 10). Three species of eutardigrades have been considered: *Hypsibius dujardini*, *Ramazzottius varieornatus* and *Paramacrobiotus richtersi*. Arakawa et al. (2016) have developed an ultra-low input library sequencing protocol to enable the genome sequencing of a single *H. dujardini* individual, a prevalently limnic species with low ability to desiccate. Hashimoto et al. (2016) have determined a high-quality genome sequence of *R. varieornatus*, one of the most stress-tolerant tardigrade species. These studies have revealed the presence of a small proportion (1.2% or less) of putative foreign genes, loss of gene pathways that promote stress damage, expansion of gene families related to ameliorating damage and evolution and high expression of novel tardigrade-unique proteins. Other studies of genomics and metagenomics on *H. dujardini* have shown that 3.8–7.1% of the genome of that species is due to horizontal gene transfer from bacteria, Archaea, plants and fungi and that in general a large part of an animal genome can be derived from foreign sources (Boothby et al. 2015; Boothby and Goldstein 2016). Other authors suspected that contamination was responsible for this result (Bemm et al. 2016; Arakawa 2016; Koutsovoulos et al. 2016), but Boothby and Goldstein (2016) confirmed the validity of their data, although they revised their first evaluation of approximately one-sixth (16.67%) of the genome from foreign origin. Yoshida et al. (2017) identified few horizontally transferred genes, some of them involved in entering anhydrobiosis. Tardigrade-specific intrinsically disordered proteins, called TDPs, essential for desiccation tolerance, have been very recently identified both in *H. dujardini* and in *P. spatialis* by using transcriptomics (Boothby et al. 2017).

Using human cultured cells, Hashimoto et al. (2016) have demonstrated that a tardigrade-unique DNA-associating protein suppresses X-ray-induced DNA damage by ~40% and improves radiotolerance. Yoshida et al. (2017) re-sequenced the genomes of *H. dujardini* and *R. varieornatus*, showing that the two species have contrasting gene expression responses to anhydrobiosis, with major transcriptional change in *H. dujardini* but limited regulation in *R. varieornatus*.

Other information based on tardigrade nuclear and mitochondrial gene sequences has been used for taxonomy (DNA barcoding), phylogenetics and metabolomics and is discussed in other chapters.

References

Altiero T, Rebecchi L (2003) First evidence of achiasmatic male meiosis in the water bears *Richtersius coronifer* and *Macrobiotus richtersi* (Eutardigrada, Macrobiotidae). Hereditas 139:116–120

Ammermann D (1962) Parthenogenese bei dem Tardigraden *Hypsibius dujardini* (Doy). Naturwissenschaften 49:1–2

Ammermann D (1967) Die Cytologie der Parthenogenese bei dem Tardigraden *Hypsibius dujardini*. Chromosoma 23:203–213

Arakawa K (2016) No evidence for extensive horizontal gene transfer from the draft genome of a tardigrade. Proc Natl Acad Sci USA 113(22):E3057

Arakawa K, Yoshida Y, Tomita M (2016) Genome sequencing of a single tardigrade *Hypsibius dujardini* individual. Sci Data 3:160063. https://doi.org/10.1038/sdata.2016.63

Avdonina AM, Biserova NM, Bertolani R, Rebecchi L (2007) Ultrastructure of the digestive system of *Ramazzottius tribulosus* and *Macrobiotus richtersi* (Eutardigrada) in relationship with diet. J Limnol 66(Suppl 1):5–11

Baccetti B, Rosati F (1971) Electron microscopy on tardigrades. III. The integument. J Ultrastruct Res 34:214–243

Bemm F, Weiß CL, Schultz J, Förster F (2016) Genome of a tardigrade: horizontal gene transfer or bacterial contamination? Proc Natl Acad Sci U S A 113:E3054. https://doi.org/10.1073/pnas.1525116113

Bertolani R (1970) Mitosi somatiche e costanza cellulare numerica nei Tardigradi. Atti Accad Naz Lincei Rend Ser 8ª 49:739–742
Bertolani R (1971) Partenogenesi geografica triploide in un Tardigrado (*Macrobiotus richtersi*). Atti Accad Naz Lincei Rend Ser 8ª 50:487–489
Bertolani R (1972) Sex ratio and geographic parthenogenesis in *Macrobiotus* (Tardigrada). Experientia 28:94
Bertolani R (1975) Cytology and systematics in Tardigrada. Mem Ist Ital Idrobiol 32(Suppl):17–35
Bertolani R (1981) A new genus and five new species of Italian fresh-water tardigrades. Boll Mus Civ St Nat Verona 8:249–254
Bertolani R (1982) Cytology and reproductive mechanisms in tardigrades. In: Nelson DR (ed) Proceedings of third international symposium on the Tardigrada. East Tennessee State Univ Press, Johnson City, Tennessee, pp 93–114
Bertolani R, Garagna S, Manicardi GC, Redi CA (1987) *Macrobiotus pseudohufelandi* Iharos as a model for cytotaxonomic study in populations of eutardigrades. Experientia 43:210–213
Bertolani R, Garagna S, Manicardi GC, Rebecchi L, Redi CA (1994) New data on the nuclear DNA content in some species of tardigrades. In: Argano A, Cirotto C, Grassi Milano E, Mastrolia L (eds) Contributions to animal biology. Halocynthia, Palermo, pp 103–109
Bonifacio A, Guidetti R, Altiero T, Sergo V, Rebecchi L (2012) Nature, source and function of pigments in tardigrades: *in vivo* Raman imaging of carotenoids in *Echiniscus blumi*. PLoS One 7 (e50162):1–8
Boothby TC, Goldstein B (2016) Reply to Bemm et al. and Arakawa: identifying foreign genes in independent *Hypsibius dujardini* genome assemblies. Proc Natl Acad Sci USA 113(22):E3058. https://doi.org/10.1073/pnas.1601149113
Boothby TC, Tenlen JR, Smith FW, Wang JR, Patanella KA, Nishimura EO, Tintori SC, Li Q, Jones CD, Yandell M, Messina DN, Glasscock J, Goldstein B (2015) Evidence for extensive horizontal gene transfer from the draft genome of a Tardigrade. Proc Natl Acad Sci USA 112 (52):15976–15981
Boothby TC, Tapia H, Brozena AH, Piszkiewicz S, Smith AE, Giovannini I, Rebecchi L, Pielak GJ, Koshland D, Goldstein B (2017) Tardigrades use intrinsically disordered proteins to survive desiccation. Mol Cell 65:975–984
Bussers JC, Jeuniaux C (1973) Chitinous cuticle and systematic position of Tardigrada. Biochem Syst 1:77–78
Czernekova M, Jönsson KI (2016) Mitosis in storage cells of the eutardigrades *Richtersius coronifer*. Zool J Linn Soc 178:888–896
Dewel RA, Clark WH (1973a) Studies on the tardigrades. II. Fine structure of the pharynx of *Milnesium tardigradum* Doyère. Tissue Cell 5:147–159
Dewel RA, Clark WH (1973b) Studies on the tardigrades. III. Fine structure of the esophagus of *Milnesium tardigradum* Doyère. Tissue Cell 5:161–169
Dewel RA, Dewel WC (1979) Studies on the tardigrades. IV. Fine structure of the hindgut of *Milnesium tardigradum* Doyère. J Morphol 161:79–109
Dewel RA, Roush BG, Dewel WC (1989) Fine structure of the midgut cells of the heterotardigrade *Echiniscus viridissimus*. Am Zool 28:149A
Dewel RA, Dewel WC, Roush BG (1992) Unusual cuticle-associated organs in the heterotardigrade, *Echiniscus viridissimus*. J Morphol 212:123–140
Dewel RA, Nelson DR, Dewel WC (1993) Tardigrada. In: Microscopic anatomy of invertebrates, Onychophora, Chilopoda, and lesser Protostomata, vol 12. Wiley, Hoboken, pp 143–183
Greven H (1979) Notes on the structure of vasa Malpighii in the eutardigrade *Isohypsibius augusti* (Murray 1907). Zesz Nauk Uniw Jagiellon Prace Zool 25:87–96
Greven H (1984) Tardigrada. In: Bereiter-Hahn J, Matoltsy AG, Richards KS (eds) Biology of the integument Invertebrates, vol 1. Springer, Berlin, pp 714–727
Greven H, Kuhlmann D (1972) Die Struktur des Nervengewebes von *Macrobiotus hufelandi* C.A.S. Schultze (Tardigrada). Z Zellforsch 132:131–146

Greven H, Peters W (1986) Localization of chitin in the cuticle of Tardigrada using wheat germ agglutinin-gold conjugate as a specific electron-dense marker. Tissue Cell 18:297–304

Guidetti R, Cesari M, Bertolani R, Altiero T, Rebecchi L (2019) High diversity in species, reproductive modes and distribution within the *Paramacrobiotus richtersi* complex (Eutardigrada, Macrobiotidae). Zool Lett 5:1. https://doi.org/10.1186/s40851-018-0113-z

Guidetti R, Rebecchi L, Bertolani R, Jönsson KI, Kristensen RM, Cesari M (2016) Morphological and molecular analyses on *Richtersius* (Eutardigrada) diversity reveal its new systematic position and lead to the establishment of a new genus and a new family within Macrobiotoidea. Zool J Linnean Soc 178:834–845

Halberg KA, Møbjerg N (2012) First evidence of epithelial transport in tardigrades: a comparative investigation of organic anion transport. J Exp Biol 215:497–507

Halberg KA, Persson D, Møbjerg N, Wanninger A, Kristensen RM (2009) Myoanatomy of the marine tardigrade *Halobiotus crispae* (Eutardigrada: Hypsibiidae). J Morphol 270:996–1013

Hashimoto T, Horikawa DD, Saito Y, Kuwahara H, Kozuka-Hata H, Shin-I T, Minakuchi Y, Ohishi K, Motoyama A, Aizu T, Enomoto A, Kondo K, Tanaka S, Hara Y, Koshikawa S, Sagara H, Miura T, Yokobori S, Miyagawa M, Suzuki Y, Kubo T, Oyama M, Kohara Y, Fujiyama A, Arakawa K, Katayama T, Toyoda A, Kunieda T (2016) Extremotolerant tardigrade genome and improved radiotolerance of human cultured cells by tardigrade-unique protein. Nat Commun 7. https://doi.org/10.1038/ncomms12808

Hyra M, Rost-Roszkowska MM, Student S, Włodarczyk A, Deperas M, Janelt K, Poprawa I (2016a) Body cavity cells of Parachela during their active life. Zool J Linnean Soc 178:878–887

Hyra M, Poprawa I, Włodarczyk A, Student S, Sonakowska L, Kszuk-Jendrysik M, Rost-Roszkowska MM (2016b) Ultrastructural changes in the midgut epithelium of *Hypsibius dujardini* (Doyère, 1840) (Tardigrada, Eutardigrada, Hypsibiidae) in relation to oogenesis. Zool J Linnean Soc 178:897–906

Jönsson KI, Rebecchi L (2002) Experimentally induced anhydrobiosis in the tardigrade *Richtersius coronifer*: phenotypic factors affecting survival. J Exp Zool 293:578–584

Koutsovoulos G, Kumar S, Laetsch DR, Stevens L, Daub J, Conlon C, Maroon H, Thomas F, Aboobaker AA, Blaxter B (2016) No evidence for extensive horizontal gene transfer in the genome of the tardigrade *Hypsibius dujardini*. Proc Natl Acad Sci USA 113:5053–5058

Kristensen RM (1976) On the fine structure of *Batillipes noerrevangi* Kristensen 1976. 1. Tegument and moulting cycle. Zool Anz 197:129–150

Kristensen RM (1978) On the structure of *Batillipes noerrevangi* Kristensen 1978 2. The muscle-attachments and the true cross-striated muscles. Zool Anz 200:173–184

Kristensen RM (1981) Sense organs of two marine arthrotardigrades (Heterotardigrada, Tardigrada). Acta Zool 62:27–41

Kristensen RM, Neuhaus B (1999) The ultrastructure of the tardigrade cuticle with special attention to marine species. Zool Anz 238:261–281

Marchioro T, Rebecchi L, Cesari M, Hansen JG, Viotti G, Guidetti R (2013) Somatic musculature of Tardigrada: phylogenetic signal and metameric patterns. Zool J Linnean Soc 169:580–603. https://doi.org/10.1111/zoj.12079

Marcus E (1929) Tardigrada. In: Bronn HG (ed) Klassen und Ordnungen des Tierreichs, vol V. Akademische Verlagsgesellschaft, Leipzig, pp 1–608

Møbjerg N, Dahl C (1996) Studies on the morphology and ultrastructure of the Malpighian tubules of *Halobiotus crispae* Kristensen, 1982 (Eutardigrada). Zool J Linnean Soc 116:85–99

Pelzer B, Dastych H, Greven H (2007) The osmoregulatory/excretory organs of the glacier dwelling eutardigrade *Hypsibius klebelsbergi* Mihelčič 1959 (Tardigrada). Mitt Hamb Zool Mus Inst 104:61–72

Persson DK, Halberg KA, Jørgensen A, Møbjerg N, Kristensen RM (2012) Neuroanatomy of *Halobiotus crispae* (Eutardigrada: Hypsibiidae): tardigrade brain structure supports the clade Panarthropoda. J Morphol 273:1227–1245

Persson DK, Halberg KA, Jørgensen A, Møbjerg N, Kristensen RM (2014) Brain anatomy of the marine tardigrade *Actinarctus doryphorus* (Arthrotardigrada). J Morphol 275:173–190

Rebecchi L, Altiero T, Bertolani R (2002) Banding techniques on tardigrade chromosomes: the karyotype of *Macrobiotus richtersi* (Eutardigrada, Macrobiotidae). Chromosom Res 10:437–443

Rebecchi L, Rossi V, Altiero T, Bertolani R, Menozzi P (2003) Reproductive modes and genetic polymorphism in different populations of the tardigrade *Richtersius coronifer* (Eutardigrada, Macrobiotidae). Invertebr Biol 122:19–27

Redi CA, Garagna S (1987) Cytochemical evaluation of the nuclear DNA content as a tool for taxonomical studies in eutardigrades. In: Bertolani R (ed) Biology of tardigrades, vol 1. Mucchi, Modena, pp 73–80

Rost-Roszkowska MM, Poprawa I, Wójtowicz M, Kaczmarek Ł (2011) Ultrastructural changes of the midgut epithelium in *Isohypsibius granulifer granulifer* Thulin, 1928 (Tardigrada: Eutardigrada) during oogenesis. Protoplasma 248:405–414

Shaw K (1974) The fine structure of muscle cells and their attachments in the tardigrade *Macrobiotus hufelandi*. Tissue Cell 6:431–445

Walz B (1973) Zur Feinstruktur der Muskelzellen des Pharynx-Bulbs von Tardigraden. Z Zellforsch 140:389–399

Walz B (1974) The fine structure of somatic muscles of Tardigrada. Cell Tissue Res 149:81–89

Walz B (1975) Ultrastructure of muscle cells in *Macrobiotus hufelandi*. Mem Ist Ital Idrobiol 32:425–444

Walz B (1978) Electron microscopic investigation of cephalic sense organs of the tardigrade *Macrobiotus hufelandi* C.A.S. Schultze. Zoomorphologie 89:1–19

Walz B (1979) Cephalic sense organs of Tardigrada. Current results and problems. Zesz Nauk Uniw Jagiellon Prace Zool 25:161–168

Węglarska B (1975) Studies on the morphology of *Macrobiotus richtersi* Murray, 1911. Mem Ist Ital Idrobiol 32:445–464

Węglarska B (1987a) Studies on the excretory system of *Isohypsibius granulifer* Thulin (Eutardigrada). In: Bertolani R (ed) Biology of tardigrades, vol 1. Mucchi, Modena, pp 15–24

Węglarska B (1987b) Morphology and ultrastructure of the excretory system in *Dactylobiotus dispar* (Murray) (Eutardigrada). In: Bertolani R (ed) Biology of tardigrades, vol 1. Mucchi, Modena, pp 25–33

Węglarska B (1987c) Yolk formation in *Isohypsibius* (Eutardigrada). Zoomorphology 107:287–292

Wiederhoft H, Greven H (1996) Notes on head sensory organs of *Milnesium tardigradum* Doyère, 1840 (Apochela, Eutardigrada). Zool Anz 238:338–346

Yoshida Y, Koutsovoulos G, Laetsch DR, Stevens L, Kumar S, Horikawa DD, Ishino K, Komine S, Kunieda T, Tomita M, Blaxter M, Arakawa K (2017) Comparative genomics of the tardigrades *Hypsibius dujardini* and *Ramazzottius varieornatus*. bioRxiv 15:e2002266. https://doi.org/10.1101/112664

Chapter 7
Tardigrade Ecology

Diane R. Nelson, Paul J. Bartels, and Noemi Guil

Abstract Understudied, but ubiquitous in every biome on Earth, tardigrades have been found in marine, freshwater, and terrestrial habitats, although ecological studies are still rather scarce, especially those that utilize quantitative replicate samples for statistical analysis. In this chapter, we discuss what is currently known about the ecology of tardigrades in the three major habitat categories: marine, freshwater, and terrestrial (including limnoterrestrial). Marine tardigrades have been described from all seas, occurring in intertidal and subtidal areas down to the abyss, while freshwater tardigrades inhabit a diversity of lotic, lentic, and subterranean ecosystems. Terrestrial tardigrade habitats include substrates such as mosses, lichens, and liverworts on rocks, soil, and tree trunks, as well as leaf litter and soil, whereas limnoterrestrial species are found in both freshwater and terrestrial environments. Tardigrades display various morphological adaptations for their associated environments. The high variability within and between samples (patchiness) limits data interpretation and reproducibility of previous studies of population dynamics and zonation. Long-term ecological studies are essential for future research. In addition, our knowledge of the biodiversity and biogeography of tardigrades is rapidly changing with the discovery of cryptic species and integrative taxonomy using combined morphological and molecular information, now considered essential for subsequent work with tardigrades.

D. R. Nelson (✉)
East Tennessee State University, Johnson City, TN, USA
e-mail: nelsond@etsu.edu

P. J. Bartels
Warren Wilson College, Asheville, NC, USA
e-mail: pbartels@warren-wilson.edu

N. Guil
Museo Nacional de Ciencias Naturales, Madrid, Spain

R. O. Schill (ed.), *Water Bears: The Biology of Tardigrades*, Zoological Monographs 2,
https://doi.org/10.1007/978-3-319-95702-9_7

7.1 Habitats

Ubiquitous in nature, tardigrades have been discovered in every biome on Earth, from the Arctic to the Antarctic, deserts to grasslands, forests to tundra, mountain tops to valleys, and freshwater to marine environments, including the deep sea. They have even been found "in hot, radioactive springs and ice cathedrals inside the Greenland ice cap" (Kristensen and Sørensen 2004). Although most of the species thus far have been described from terrestrial habitats, all water bears (as the name implies) must be surrounded by a film of water to be active, as discussed in previous chapters. As their primary habitat is associated with the substratum, tardigrades are considered as permanent and very important members of benthic, epibenthic, and epiphytic meiofauna (Ramazzotti and Maucci 1983; Nelson et al. 2010, 2015).

Tardigrades are notoriously understudied, and although the number of publications has increased in recent years, ecological studies are still rather scarce. The use of quantitative replicate samples within habitats is essential for ecological analysis, due to the high degree of spatial variability in tardigrade species richness and abundance between and within samples (e.g., Kathman and Cross 1991; Meyer 2006; Bartels and Nelson 2007). The statistical significance of environmental factors in determining tardigrade distributions must be tested by multivariate statistical analyses (e.g., Degma et al. 2005; Schuster and Greven 2007, 2013). Until recently, most of the previous research concentrated on presence/absence data to determine species richness in various habitats, with little or no statistical analysis.

It is also important to note that tardigrade sampling is still very sporadic and Eurocentric (Meyer 2013; Kaczmarek et al. 2014, 2015a). This is especially clear for marine tardigrades for which a new interactive map was published (Kaczmarek et al. 2015b). The distribution of the known species may not be an indication of their abundance worldwide, but rather that of the number and distribution of tardigradologists, their selection of collecting sites, and the difficulty of extracting tardigrades from certain habitats (Guil and Cabrero-Sañudo 2007; Nelson et al. 2015). Here we discuss what is currently known about the ecology of tardigrades in the three major habitat categories: marine, freshwater, and terrestrial (including limnoterrestrial).

7.1.1 Marine Habitats

According to Kaczmarek et al. (2015b) and the current checklist of tardigrades (Guidetti and Bertolani 2005; Degma and Guidetti 2007; Degma et al. 2009–2015), 197 species of marine tardigrades had been described, but Bartels et al. (2016) estimated that the total global number will likely be close to 1000. Generally smaller in size than freshwater and terrestrial species, almost all marine tardigrades, with few exceptions, belong to the class Heterotardigrada, including all members of the order Arthrotardigrada as well as the family Echiniscoididae in the order

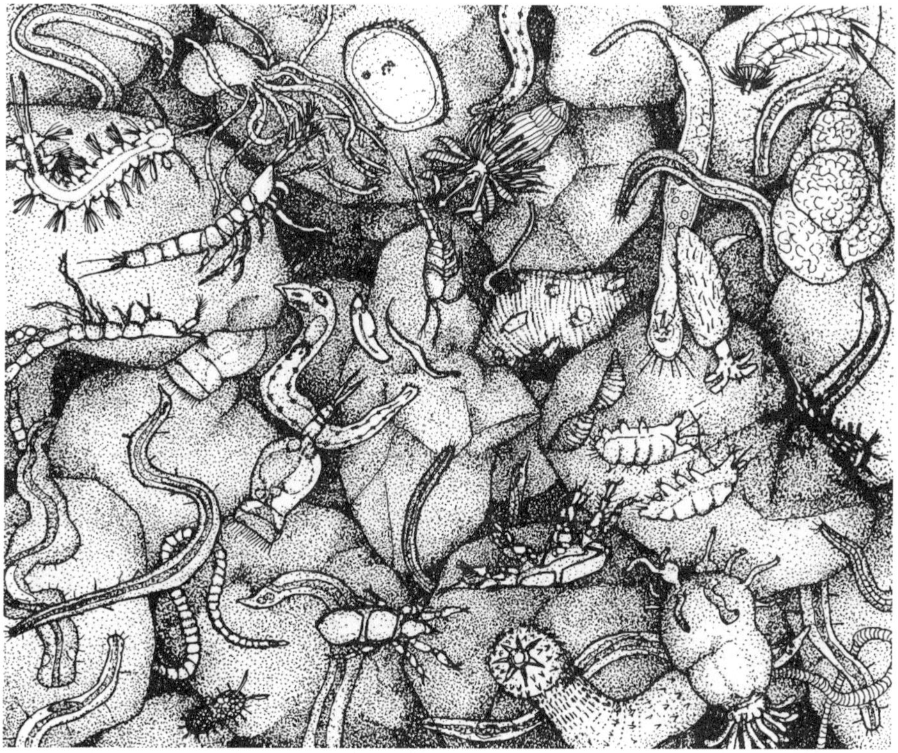

Fig. 7.1 Artistic illustration of a marine interstitial community. Two arthrotardigrades can be seen near the vertical midline toward the right of the image. From Giere (2009)

Echiniscoidea. Marine tardigrades are rare in the class Eutardigrada, with none in the Apochela and only four species in the order Parachela (*Thulinius itoi*, *Halobiotus arcturulius*, *Halobiotus crispae*, and *Halobiotus stenostomus*). In addition, *Ramajendas renaudi* is known to occur in brackish waters. The eutardigrades are secondarily adapted to the marine environment. Some freshwater eutardigrade species may occasionally be washed into brackish or marine environments, i.e., *Pseudobiotus kathmanae* has been found in a salt marsh in Louisiana (D. Nelson pers. comm.).

Marine tardigrades are interstitial in the psammon community or live on the surfaces of different substrates, detritus, or other organisms (Fig. 7.1). Generally, they are not holoplanktonic, but some can drift with the current or swim weakly just above the surface to settle in a new spot. Marine tardigrades have been described from all seas, occurring in intertidal and subtidal areas down to 5730 m in the abyss (Hansen et al. 2003; J. Hansen pers. comm.), in crevices on manganese nodules, abyssal mud, and deep-sea ooze. Research effort is severely limited for these animals compared to limnoterrestrial tardigrades (Vicente and Bertolani 2013), and there is noticeable bias in published literature toward intertidal versus subtidal habitats. This reflects the increased effort and cost needed to collect and extract tardigrades from depth.

Marine tardigrades occur interstitially in the intertidal zone (e.g., *Batillipes*) or more commonly in subtidal coarse coralline sand (Halechiniscidae). In finer sands and mud, they tend to be epibenthic as a result of decreased oxygen availability (Giere 2009). Some Halechiniscidae (e.g., *Tholarctus*, *Lepoarctus*, *Neostygarctus*, *Zioellea*, and *Tanarctus*) are considered semibenthic, sometimes drifting or weakly swimming above the substrate (Kristensen and Renaud-Mornant 1983). Other species are associated with organic slime growing on algae, including *Styraconyx sargassi* on *Sargassum* seaweed (Giere 2009), while other species are symbiotic with other invertebrates (see section on symbiosis in this chapter). In the abyss, *Moebjergarctus manganis* is a free-living species associated with manganese nodules at depths of 4140–4170 m in the eastern South Pacific, a very specialized microhabitat (Bussau 1992). Representative diversity of marine heterotardigrades is shown in Fig. 7.2.

Tardigrades have also been found in marine caves, which are presumably more protected than other marine habitats with less environmental fluctuation (Boesgaard and Kristensen 2001; Grimaldi de Zio and Gallo D'Addabbo 2001). A very small number of species may be cave specialists, e.g., *Trogloarctus trionyches* (Coronarctidae) in mud sediments with deep-sea-like conditions (Villora-Morena 1996). *Actinarctus neretinus* has been found in Australian and Mediterranean caves, suggesting that this and other cave species may be Tethys Sea relics (Boesgaard and Kristensen 2001; Grimaldi de Zio and Gallo D'Addabbo 2001). Very high species diversity and a number of new species were reported from 2 Australian caves; among just 67 specimens, there were 16 species, 12 of which were new to science (Jørgensen et al. 2014).

7.1.2 Freshwater Habitats

Inland aquatic (limnic) systems are generally categorized as either lentic or lotic habitats. Lotic systems (flowing waters) include rivers, streams, cold and hot springs, and other permanent or intermittent bodies of water. Lentic habitats are standing waters such as lakes, ponds, wetlands (including swamps, marshes), and temporary/ephemeral ponds. Most of these are freshwater environments, but varying salinities may occur such as in brackish waters and hypersaline lakes. Although the fewest number of species have been described as freshwater, hydrophilous and hygrophilous tardigrades inhabit a diversity of lotic, lentic, and subterranean ecosystems that are perennial, ephemeral, or temporary. "Hydrophilous" tardigrades are those that live only in permanent freshwater habitats. Most hydrophilous eutardigrade genera occur in both lentic and lotic habitats. Some tardigrades are classified as "hygrophilous" species, i.e., they can colonize both freshwater and terrestrial microhabitats.

The Heterotardigrada are very rare in freshwater. In the order Arthrotardigrada, *Styraconyx hallasi* is the only marine species that has also been found in freshwater,

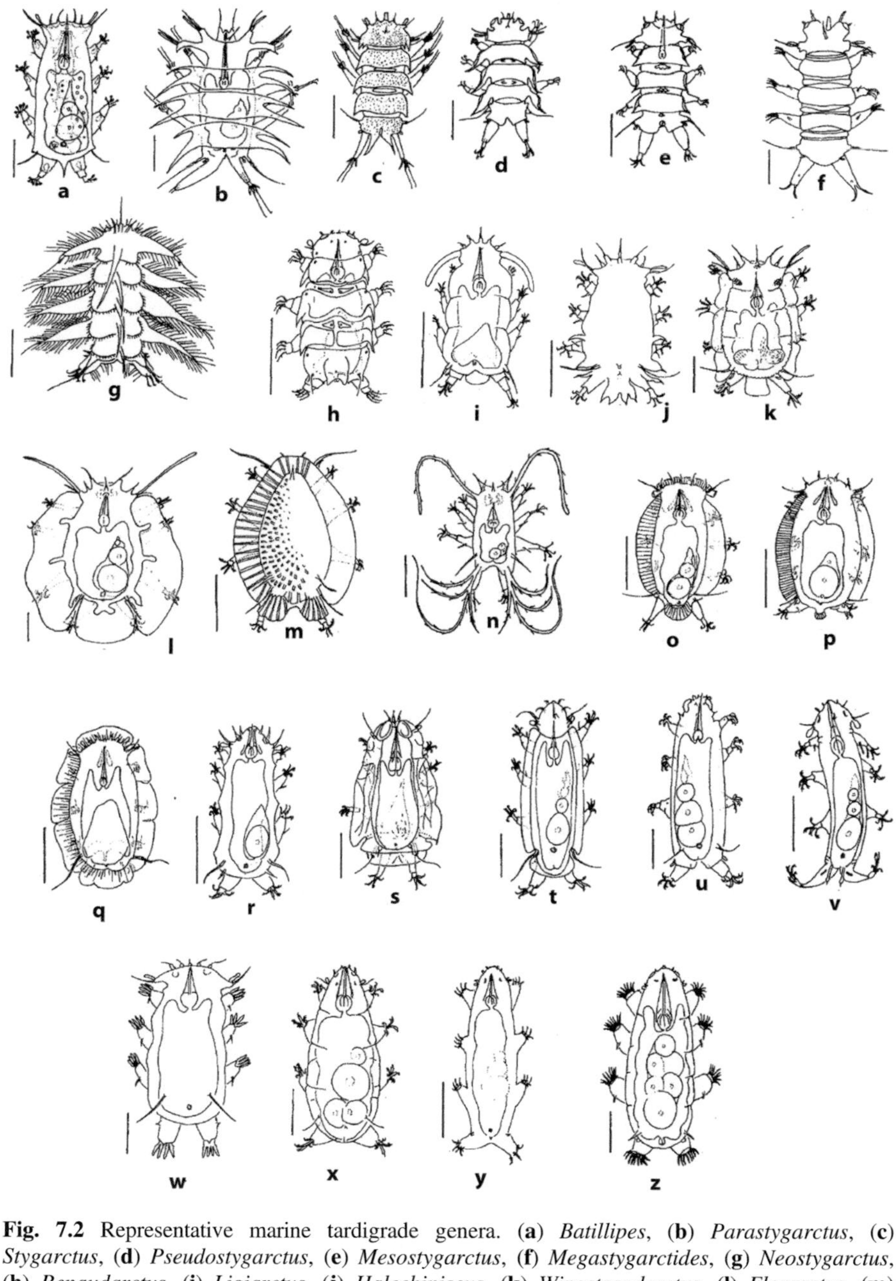

Fig. 7.2 Representative marine tardigrade genera. (**a**) *Batillipes*, (**b**) *Parastygarctus*, (**c**) *Stygarctus*, (**d**) *Pseudostygarctus*, (**e**) *Mesostygarctus*, (**f**) *Megastygarctides*, (**g**) *Neostygarctus*, (**h**) *Renaudarctus*, (**i**) *Ligiarctus*, (**j**) *Halechiniscus*, (**k**) *Wingstrandarctus*, (**l**) *Florarctus*, (**m**) *Actinarctus*, (**n**) *Tanarctus*, (**o**) *Raiarctus*, (**p**) *Rhomboarctus*, (**q**) *Parmursa*, (**r**) *Syraconyx*, (**s**) *Tholoarctus*, (**t**) *Lepoarctus*, (**u**) *Clavarctus*, (**v**) *Angursa*, (**w**) *Orzeliscus*, (**x**) *Archechiniscus*, (**y**) *Anisonyches*, (**z**) *Echiniscoides*. Scale bars = 50 μm. From Renaud-Mornant (1988)

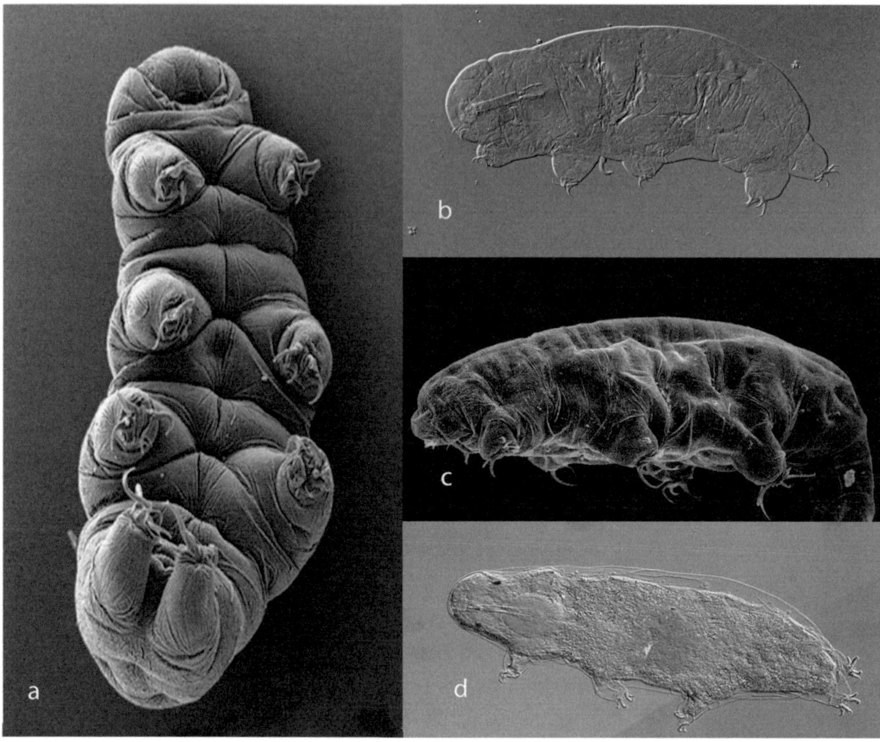

Fig. 7.3 Representative freshwater tardigrades. (**a**) Scanning electron micrograph (SEM) of *Hypsibius dujardini* adult, ventral view, ~150 μm long, an extremely common and widespread freshwater tardigrade. (**b**) Phase contrast (PhC) micrograph of *Thulinius romanoi* with unique sculptured cuticle from stream sediments in the Great Smoky Mountain National Park, Tennessee/North Carolina, USA. (**c**) Scanning electron micrograph (SEM) of *Pseudobiotus kathmanae* from sediments in the Ocoee River, Tennessee, USA. (**d**) Differential interference contrast (DIC) micrograph of aquatic *Murrayon pullari* from Disko Island, Greenland. (**a**) Photo by Vicky Madden and Bob Goldstein, University of North Carolina-Chapel Hill, North Carolina, USA, from Tenlen et al. (2013), (**b**) from Bertolani et al. (2014a), (**c**) from Nelson et al. (2015), (**d**) photo by N Guil

in a Greenland spring (Kristensen 1977). In the heterotardigrade order Echiniscoidea, *Carphania fluviatilis* (family Carphaniidae) is freshwater (Binda 1978), while only a few species of the family Echiniscidae occasionally occur in limnic habitats. The vast majority of the class Eutardigrada are limnoterrestrial (freshwater or terrestrial). (For descriptions, illustrations, and keys to aquatic eutardigrades, see Bertolani 1982; Nelson and McInnes 2002; Pilato and Binda 2010; Nelson et al. 2010, 2015, 2016.) Representative freshwater tardigrades are illustrated in Fig. 7.3.

7.1.2.1 Lotic Habitats

Regardless of their temporal status, all riverine systems are part of the watershed and transport not only water but also sediments (containing tardigrades and other meiofauna), nutrients, chemicals, and debris. Tardigrades in this habitat may be collected from aquatic plants as well as the surface sediments (see chapter on techniques).

In streams, invertebrate communities generally vary in composition longitudinally downstream, depending on a number of factors (e.g., current velocity, water temperature, substrate, stability of both aquatic and riparian vegetation, dissolved substances, competition, zoogeography, food, disturbance history, and human impact). The physical characteristics of stream flow (stream hydraulics) are apparently the most important environmental factor governing the zonation of stream benthos on a worldwide scale (Statzner and Higler 1986).

An important microhabitat, seldom examined for tardigrades, is the hyporheic zone, the region beneath and adjacent to streams and rivers where surface and groundwater mix, linking aquatic and terrestrial systems and serving as a transition area between surface water and groundwater systems. This zone, containing meiofauna common to both surface and subsurface systems, is critically important in stream nutrient cycling, moderating stream temperatures, trapping heavy metals and other contaminants, improving surface water quality, and forming unique habitats within streams. Bacteria, fungi, and protozoa that live on or in sediments sequester nutrients, making the hyporheic zone an area of high productivity for a diverse community of invertebrates. Core samples must be taken to find tardigrades in this zone, but the tardigrade community has been rarely examined as it is considered a minor component of the meiofauna. Shallow groundwaters of hyporheic zones and springs in southeastern New York yielded six taxa in five genera (*Dactylobiotus*, *Murrayon*, *Thulinius*, *Pseudobiotus*, *Isohypsibius*), but tardigrades were scarce at most sites ($<1\%$ of the invertebrate community except in springs and one site in the hyporheic zone of a creek). Additionally, seeps, shallow phreatic waters, and streams in caves in New York contained no tardigrades (Strayer et al. 1994), but this may have been a result of the techniques used. However, a few tardigrades have been found in hypogeal caves (Bertolani 2002; N. Guil pers. comm.).

For freshwater tardigrades, the two non-eutardigrade families are represented by only one species, and both are endemic: *Carphania fluviatilis* (Heterotardigrada, Carphaniidae) occurs only in sediments in a river in Sicily, while *Thermozodium esakii* (Mesotardigrada, Thermozodiidae) is reported only once from a hot sulfur spring near Nagasaki, Japan. The latter type locality was reportedly destroyed, either by an earthquake or during World War II, and subsequent efforts to find any mesotardigrades in Japan, Europe, or the USA have been unsuccessful. In addition, no type specimen exists. Thus, the class Mesotardigrada is considered dubious. Warm homothermic springs on Disko Island, Greenland, however, were very productive, with high species diversity and four new species with unusual characters (Kristensen 1982).

Hydrophilous eutardigrade genera frequently found in rivers and streams include *Pseudobiotus*, *Dactylobiotus*, and *Thulinius*, as well as the hygrophilous genera *Hypsibius*, *Isohypsibius*, *Bertolanius*, *Doryphoribius*, and *Murrayon*. *Eohypsibius nadjae* was originally found in a homothermic spring in Greenland and subsequently in other areas including the Faroe Islands, Alps, and Apennines (Italy). Three major ecological studies have been conducted in Tennessee and North Carolina (USA). In the Ocoee River (Tennessee), six species in the genera *Pseudobiotus*, *Murrayon*, *Dactylobiotus*, *Doryphoribius*, *Hypsibius*, and *Isohypsibius* were identified (Kathman and Nelson 1987). In a small stream on a university campus (Tennessee), 5 species in the genera *Pseudobiotus*, *Murrayon*, *Diphascon*, *Hypsibius*, and *Isohypsibius* were present in significant numbers with 10 other species being rare (Nelson et al. 1987). As part of an extensive survey of tardigrades in the Great Smoky Mountains National Park (Tennessee/North Carolina), 22 species (in the genera *Dactylobiotus*, *Diphascon*, *Doryphoribius*, *Hypsibius*, *Isohypsibius*, *Microhypsibius*, *Mixibius*, *Pseudobiotus*, *Thulinius*) were found predominantly or exclusively in 13 streams (Bartels and Nelson 2006; Bertolani et al. 2014a). Another multihabitat study, in a glacial valley in the Apennines, Italy, yielded 39 species, with 7 from stream mosses and 4 from stream sediments. Only *Pseudobiotus megalonyx* was found in a temporary pool, while eight species were present in permanent pool sediments. *Murrayon pullari* and *Hypsibius dujardini* were noted in all three aquatic habitats (Bertolani and Rebecchi 1996). Other random small surveys of lotic and lentic habitats in the Canadian and European High Arctic, as well as Mount Kenya and Mount Kilimanjaro in Africa, conducted in the late 1980s–early 1990s (e.g., Van Rompu and De Smet 1991) revealed low diversity and abundance of aquatic tardigrades in the areas sampled.

7.1.2.2 Lentic Habitats

Large lentic systems (lakes) can be formed artificially by dams or naturally by volcanic craters, earthquakes, glaciers, and sinkholes. Benthic sediments with suitable particle size and sufficient oxygen provide favorable habitat for aquatic eutardigrades, even at considerable depths in cold mountain lakes at high altitudes (150 m in Lake Geneva, Switzerland). For example, tardigrades in the genera *Dactylobiotus*, *Hypsibius*, *Isohypsibius*, *Murrayon*, *Pseudobiotus*, and *Thulinius* thrive in Lake Tahoe (California, USA), with cold, oxygenated water and a sandy-pebble substrate, whereas warm lakes with clay particles are devoid of tardigrades (Schuster et al. 1978). In Antarctica, tardigrades associated with cyanobacterial mats, benthic moss pillars, and epiphytes in lakes encountered low predation pressure, minimal competition, and perennial abundance of food, allowing large populations to develop. Species diversity in Signy Island shallow lakes was higher than that reported in temperate lakes (McInnes and Pugh 1999).

A rather unique habitat is found in activated sludge ponds in sewage treatment plants typically located along rivers. Microorganisms that reduce the dissolved organic matter (DOM) in wastewater include bacteria, ciliates, rotifers, and tardigrades, which are indicative of a well-functioning sewage system. In an abstract, Utsugi (2001) reported a single species, *Isohypsibius myrops*, living and reproducing in aeration tanks in several Japanese sewage treatment plants. In the few cases where activated sludge has been examined in the USA (Texas and Tennessee), the tardigrade populations were monospecific members of the genus *Thulinius* (D. Nelson pers. comm.). Recently Sobczyk et al. (2015) reported *Thulinius ruffoi* from activated sludge in a Polish wastewater system. Thus, activated sludge appears to be a seldom studied, though potentially cosmopolitan habitat, for eutardigrades.

Temporary pools vary in size, shape, and longevity, lasting from hours to months. Often overlooked, they may contain a diverse community of organisms, including tardigrades that undergo anhydrobiosis or encystment to survive long periods of desiccation (see previous chapter). Temporary ponds are often shallow, well-lit, predator-free, and rich in nutrients, making them favorable habitats. Many of these organisms show drastically decreased life cycle times. *Pseudobiotus* and *Dactylobiotus* species with large numbers of eggs in the exuvia have been found in temporary ponds in Tennessee (D. Nelson pers. comm.). Organisms that colonize such habitats may travel as resistant spores or eggs carried on the wind or by waterfowl (as demonstrated for rotifers); they may persist in resistant stages (cryptobiotic eggs) in the soil; or they may be carried in by floodwaters (see the section on dispersal in this chapter). These habitats are very poorly known and need further examination for tardigrades.

Cryoconite holes are microcaverns formed on glaciers when solar radiation is absorbed by the accumulation of dark particles on the surface (Fig. 7.4). As the ice melts, a specialized habitat is then available for micrometazoans, including tardigrades and rotifers that feed on the entrapped microflora. These melt holes contain abundant algal, cyanobacteria, heterotrophic bacteria, protozoans, nematodes, rotifers, and tardigrade communities. (For a review of the microorganisms in cryoconite holes, see Kaczmarek et al. 2015c.) Three darkly pigmented and extremely cold stenothermic tardigrade species (*Hypsibius klebelsbergi*, *H. janetscheki*, and

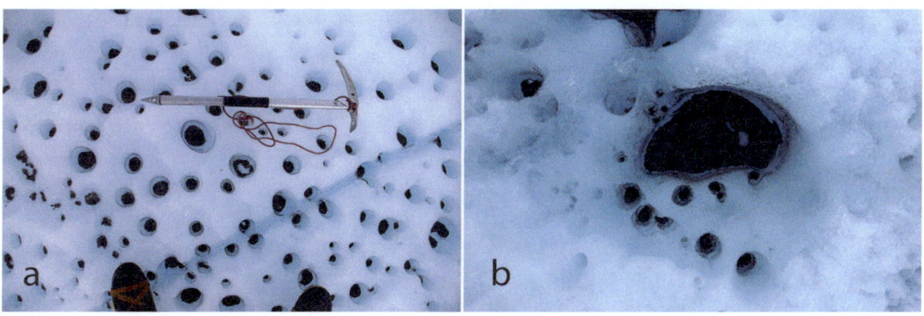

Fig. 7.4 Photographs of cryoconite holes in glaciers in Greenland. (**a**) Photo by Nozomu Takeuchi, Chiba University, Japan, (**b**) photo by Krzysztof Zawierucha, Adam Nickiewicz University, Poland

H. thaleri) have been exclusively reported from this habitat. The heavily pigmented tardigrade cuticle absorbs heat and serves as protection from UV radiation (for review, see Dastych et al. 2003; Dastych 2004a, b; Greven et al. 2005). In their review of the fauna of cryoconite fauna, Zawierucha et al. (2015) noted that other tardigrade species such as *Pilatobious recamieri* are frequent in cryoconite holes in the Holarctic but are primarily known from non-cryoconic habitats. Although studies on these extreme microcosms are limited, climate change and the melting of polar ice could threaten these sensitive communities in the future.

7.1.3 Terrestrial Habitats

Terrestrial habitats include a vast diversity of places considered suitable for tardigrades, including such disparate substrata as mosses, lichens, and liverworts on rocks, soil, tree trunks, and dead wood, as well as on walls, sidewalks, roofs, and other construction made by humans, ferns and flowering plants (such as those from the genera *Silene*, *Saxifraga*, and *Androsace*), leaf litter from a great variety of trees and bushes, and soil (from different landscapes, e.g., sand dunes, acid pine forest, tropical forest, or deserts). Terrestrial moss tufts can also be species rich, especially in Arctic and Antarctic areas. In general, terrestrial habitats must provide sufficient aeration (since tardigrades are sensitive to low oxygen levels), alternate wet and dry periods (exposure to water, sun and wind), and sufficient food (Ramazzotti and Maucci 1983). Cryptobiosis (mainly anhydrobiosis and cryobiosis) allows tardigrades to colonize a diversity of microhabitats through a combination of passive dispersal of cryptobionts, cryptobiotic eggs, and active adults with other biological characteristics (e.g., reproductive modes, survival rates, ecological requirements) (see previous chapter for detailed information on cryptobiosis).

In the past, ecological classifications of terrestrial tardigrades were based on assumptions about environmental parameters (Ramazzotti and Maucci 1983). Moisture requirements were the basis for four categories of tardigrade species: xerophilous (in dry habitats), eurytopic (with broad humidity tolerances), hygrophilous (in humid environments), and hydrophilous (in wet habitats, including lotic and lentic waters as explained above). Another classification of tardigrade species was based on altitudinal zones and/or vegetation: lowland (0–200 m), upland (201–500 m), foreland (501–1000 m), montane (above 1000 m), and tychoalpine or euryzonal (species present in all altitudinal zones) (Dastych 1987). However, these and other classifications had a short existence because they were based on very limited knowledge of tardigrade ecology.

Based on the 2015 species list (Degma et al. 2009–2015), about 80% of the total tardigrade species described thus far have been found in terrestrial environments. In the Heterotardigrada, only species in the family Echiniscidae occur in terrestrial environments, compared with all superfamilies, families, and the majority of genera of Eutardigrada. Few genera are associated primarily with specific terrestrial habitats, although some are ascribed only to environments closely related to soil (e.g.,

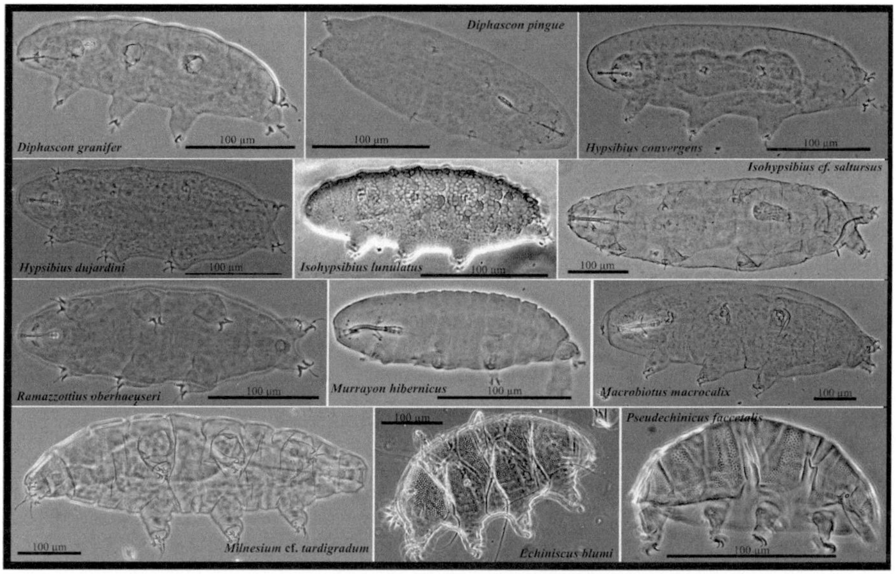

Fig. 7.5 Representative terrestrial tardigrade diversity collected in a Spanish survey. Species name indicated for each animal. Scale bar = 100 μm. Photos by N Guil

decomposed leaf litter) or accidentally found in other habitats (e.g., a *Hexapodibius* species in a moss located on a beach dune during a flood, N. Guil, pers. comm.). Soil-related genera occur in the families Isohypsibiidae (e.g., *Hexapodibius*, *Apodibius*, and certain species of *Doryphoribius*), Macrobiotidae (e.g., *Xerobiotus*), and Necopinatidae (*Necopinatum*). Representative terrestrial tardigrade diversity is shown in Fig. 7.5.

The apparent concentration of biodiversity in terrestrial habitats is probably due to differential sampling effort, primarily favoring terrestrial environments because of the ease of access, sample collection, and processing (Guil and Cabrero-Sañudo 2007). Within terrestrial habitats, the majority of sampling effort has been concentrated on mosses, lichens, and liverworts collected from rocks, tree trunks, and soil. The largest number of tardigrade species has been found in mosses (Ramazzotti and Maucci 1983), although there is no direct correlation between moss species and tardigrade species, except for coastal dune mosses. Moss and lichen growth form may be a more important factor than moss or lichen species, with thin spreading moss and foliose lichen usually being more productive than thick cushion moss or crustose lichen. In comparison, leaf litter has been studied primarily in beech forests (Hallas and Yeates 1972; Guidetti et al. 1999) but neglected under other tree species, shrubs, and bushes (Guil and Sánchez-Moreno 2013). In many cases, leaf litter, humus, and soil have been collected together, making differentiation of habitats for each species difficult to determine for each layer (Hallas and Yeates 1972; Nelson and Bartels 2013). In the few studies that have examined only the soil habitat, high

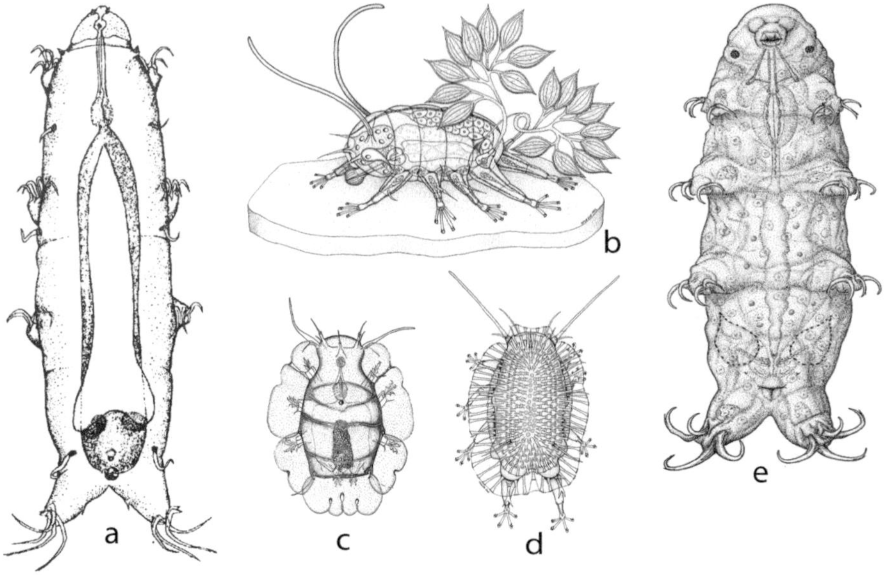

Fig. 7.6 Adaptations in various marine tardigrades. (**a**) vermiform body shape in the deep water, mud-dwelling *Coronarctus tenellus,* (**b**) buoyant bodies in *Tanarctus bubulubus*, (**c**, **d**) cuticular extensions or "alae" in (**c**) *Florarctus* and (**d**) *Actinarctus*, (**e**) enlarged Malpighian tubules (dotted lines) in *Halobiotus crispae*. (**a**) From Renaud-Mornant (1974), (**b**) from Jørgensen and Kristensen (2001), (**c**, **d**) from Renaud-Mornant (1988), (**e**) from Møbjerg and Dahl (1996)

diversity and abundance have been reported (e.g., Ito and Abe 2001; Harada and Ito 2006; Guil et al. 2015).

7.2 Adaptations

7.2.1 Marine Adaptations

Morphological adaptations are associated with the various marine habitats (e.g., Grimaldi de Zio et al. 1983) and are the basis of an updated key to the genera (Fontoura et al. 2017). Marine tardigrades have telescopic legs, with up to 13 claws, or 4 toes with complex claws, or 4–6 toes with round or rod-shaped adhesive discs inserted on toes. Interstitial species (e.g., *Batillipes*, *Orzeliscus*) have adhesive structures such as toe pads or paddles that allow them to tightly adhere to shifting sand grains, whereas those living in deep-sea muds have cylindrical, wormlike bodies with reduced legs (e.g., *Coronarctidae*, Fig. 7.6a). Semibenthic species have a variety of structures that facilitate swimming and drifting, including bizarre floats (e.g., *Tanarctus bubulubus*) with 18–20 balloon-like appendages attached to the fourth pair of legs (Jørgensen and Kristensen 2001, Fig. 7.6b) or cuticular

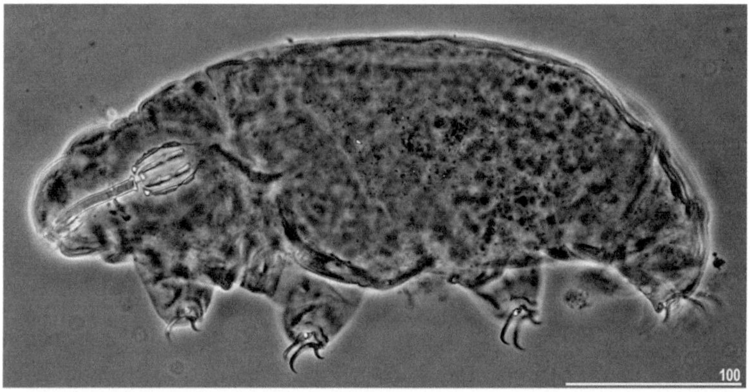

Fig. 7.7 Phase contrast (PhC) photomicrograph of lateral adult *Dactylobiotus luci* from *Sphagnum-Carex* fens extending into a permanent marsh pool at 4225 m a.s.l. in the Rwenzori Mountains, Uganda. From Kaczmarek et al. (2008)

extensions (e.g., Floractinae Fig. 7.6c, *Actinarctus* Fig. 7.6d). These projections increase surface area and may contribute to passive dispersal. Lipid droplets may also increase buoyancy (Grimaldi de Zio et al. 1983). Species living epibenthically or on algae often have elongated appendages and claws with multiple hooks. Adaptations of parasitic species are discussed under the Symbiosis section (this chapter). The intertidal tardigrade *Echiniscoides sigismundi groenlandicus* is hypothesized to undergo all four types of cryptobiosis: anhydrobiosis, cryobiosis, osmobiosis, and anoxybiosis (Møbjerg et al. 2011, N. Møbjerg pers. comm.). Anhydrobiotic capabilities were recently documented for one additional marine species, *Styraconyx haploceros*, which lives on intertidal lichens (Jørgensen and Møbjerg 2014). However, *Batillipes pennaki*, which lives on beaches, did not survive dessication experiments. *Halobiotus crispae*, a eutardigrade which has secondarily colonized marine habitats, possesses considerably enlarged Malpighian tubules attributed to a secondary shift to seawater, and these vary in size with the stage of cyclomorphosis (Crisp and Kristensen 1983; Møbjerg and Dahl 1996) (Fig. 7.6e).

7.2.2 Freshwater Adaptations

The true limnic taxa of eutardigrades usually have long legs and very long claws (Fig. 7.7). Four genera (*Dactylobiotus*, *Macroversum*, *Pseudobiotus*, and *Thulinius*) are exclusively freshwater and do not undergo cryptobiosis (except possibly in their eggs, but this has not been proven experimentally). *Dactylobiotus* is often epiphytic on filamentous algae or submerged plants in lentic waters; *Macroversum* has been found once in sandy, pebbly sediments of a stream; *Pseudobiotus* (mostly lotic) and *Thulinius* (both lotic and lentic) have been frequently found in the sediments in

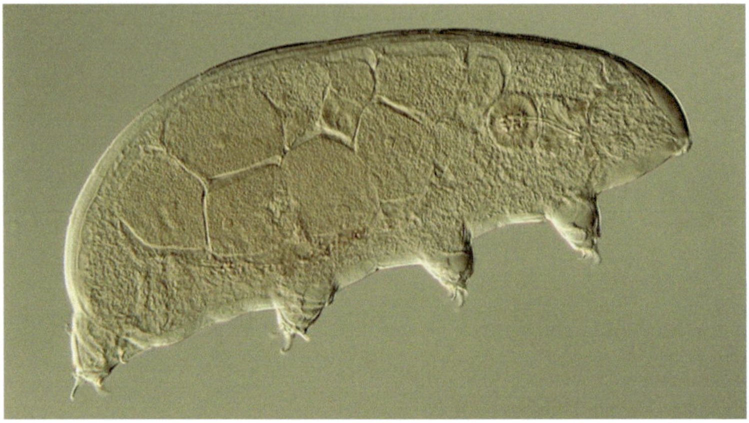

Fig. 7.8 The limnic boreo-alpine eutardigrade *Borealibius zetlandicus*, reported from various habitats such as marine samples, freshwater sediments, submerged mosses, moss on soil, soil, and cyanobacteria. Specimen collected from submerged moss in a spring in the Italian Alps at 2270 m a.s.l. From Rebecchi et al. (2009)

lakes, rivers, and streams. Other genera have stumpy legs, each bearing a pair of double claws (with primary and secondary branches), and have one or more species that are typically or often aquatic, i.e., *Bertolanius*, *Borealibius* (Fig. 7.8), *Murrayon*, *Doryphoribius*, *Microhypsibius*, *Mixibius*, *Ramajendas*, *Acutuncus*, *Hypsibius*, and especially *Isohypsibius*. Most occur in aquatic sediments. All genera with *Isohypsibius*-type claws are limnic or have limnic or limnoterrestrial species (except *Halobiotus*, which is marine). Several species of the genus *Hypsibius* (e.g., *H. dujardini* and *H. convergens*) are not obligate limnic organisms but are mainly or often found in freshwater habitats. *Milnesium tardigradum*, *Paramacrobiotus richtersi*, and *Macrobiotus hufelandi*, as well as some species of *Eohypsibius*, *Eremobiotus*, *Ramazzottius*, *Astatumen*, and *Diphascon*, are eurytopic terrestrial taxa that can also inhabit freshwater. All limnoterrestrial tardigrades are probably capable of cryptobiosis.

The eutardigrade *Borealibius zetlandicus* is an unusual eutardigrade that inhabits various microenvironments, but its adaptive mechanisms are unknown. It has been found in marine samples from the Barents Sea, freshwater sediments, submerged mosses and *Sphagnum* sp., terrestrial substrates, such as mosses, soil (animals and cysts), and on a colony of cyanobacteria (Pilato et al. 2006). The wide range of habitats available for this species represents an exception for tardigrade species, which are more usually restricted to only one (e.g., marine) or two (e.g., freshwater and terrestrial) major habitats.

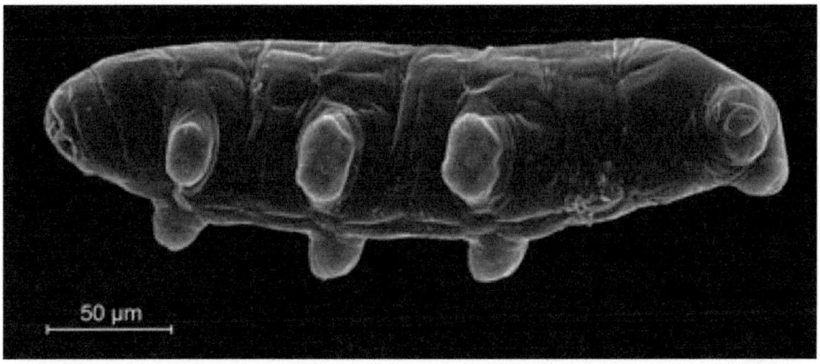

Fig. 7.9 The pioneer eutardigrade *Apodibius confusus* with modified legs for crawling in sandy soils. From Hohberg et al. (2011)

7.2.3 Terrestrial Adaptations

Since most of the terrestrial tardigrade genera and species occupy diverse habitats, specific morphological adaptations for those habitats are not evident. However, in certain leaf litter and soil genera, particular morphological adaptations evolved. Interstitial eutardigrades living in inorganic soils are characterized by very short legs with small claws and a reduction or absence of the hind claws. The soil genera *Apodibius* and *Hexapodibius* (Isohypsibiidae) and *Necopinatum* (Necopinatidae) exhibit different degrees of reduction in their claws and/or legs (Bertolani and Biserov 1996). *Apodibius* has very reduced legs with no claws (Fig. 7.9), whereas *Hexapodibius* and *Necopinatum* have short stubby legs with claws reduced or absent. Also, *Xerobiotus* (family Macrobiotidae) and some species of *Doryphoribius* (family Isohypsibiidae, e.g., *Doryphoribius pilatoi*) have reduced claws.

These adaptations, which have evolved several times (Dabert et al. 2013; Guil et al. 2013; Bertolani et al. 2014b), as indicated by their presence in different superfamilies (Isohypsibioidea, Hypsibioidea, and Macrobiotoidea), are probably related to the tiny interstitial spaces available in soil habitats. In addition, their vermiform shape is an adaptation for more effective wormlike movement between the soil particles. However, whether these genera occur exclusively in edaphic environments is unknown as a result of the low sampling effort in this environment. The factors driving exclusivity in a soil-based environment, and the evolutionary causes underlying the parallel evolution of leg and claw reduction, have not yet been determined. The Itaquasconinae genera (e.g., *Itaquascon*, *Mesocrista*) in the family Hypsibiidae (superfamily Hypsibioidea) are more common in leaf litter and soil environments and seldom found on mosses and lichens on rocks and trees. In this subfamily, the buccal-pharyngeal apparatus morphology is generically distinctive, which could be related to feeding habits. In other groups, different morphological variations in the buccal-pharyngeal apparatus are related to feeding habits (e.g., the carnivorous *Milnesium*) (Hallas and Yeates 1972; Guidetti et al. 2013; Guil and

Sánchez-Moreno 2013) (see chapter on taxonomy and functional anatomy for further information on the buccal-pharyngeal apparatus).

7.3 Population Dynamics

7.3.1 Marine Population Dynamics

Intertidal interstitial marine tardigrades are usually found within the top few centimeters of the surface but may occur as deep as 180 cm in the substrate (Renaud-Mornant 1988), often co-occurring with nematodes, harpacticoid copepods, turbellarians, and other meiofaunal organisms. They are usually much less numerous than nematodes and copepods, but occasionally the numbers can be quite high, e.g., 3500 specimens of *Batillipes pennaki* per 100 cm^3 of beach sand in one account (Giere 2009). (For a review of *B. pennaki*, see Zawierucha et al. 2013.) As many as 402 individuals of the barnacle-dwelling commensal *Echiniscoides sigismundi* have been found in a single barnacle, with an estimate of 40,000 individuals "per running coastal rock meter" (Kristensen and Hallas 1980).

Patchy distributions are frequently reported for meiofauna in general (Findlay 1981). Seasonal changes in abundance have been related to water drainage and temperature, with lowest abundance values in winter (Crisp and Hobart 1954; de Zio and Grimaldi 1966; Pollock 1970; Martinez 1975; Renaud-Mornant 1988; da Rocha et al. 2004), but these results need to be verified with appropriate statistical sampling and analysis. Short-term, sometimes abrupt, changes in abundance have been associated with opportunistic reproduction during favorable climatic conditions and/or disturbance of beach habitats (Pollock 1970).

7.3.2 Freshwater Population Dynamics

Little is known of the population dynamics of freshwater tardigrades in limnic (lotic and lentic) systems. Often the lack of replicate samples and the high variability (0–100+ per cm^3) within and between samples (due to patchiness) limit the data interpretation and reproducibility of previous studies of freshwater tardigrades (Kathman and Nelson 1987; Meyer 2006). Most of these studies involved only a single year of sampling. Only one (Nelson et al. 1987) resampled the same aquatic site after several years to determine the effect of habitat disturbance on tardigrade populations. Other environmental effects, such as drought, can impact tardigrade densities, necessitating sampling in multiple years. Some authors have indicated the highest density of tardigrades in the autumn or spring, with lowest numbers in the summer, but this may depend in part on the location (latitude, longitude, altitude). For example, *Dactylobiotus grandipes* was present in high numbers in the autumn in the high altitude, cold-water Lake Tahoe, California (Schuster et al. 1978), but populations peaked in the spring for *Murrayon pullari*, *Pseudobiotus kathmanae*, and *Hypsibius dujardini* in a stream in Tennessee (Nelson et al. 1987).

Egg-laying may be year-round (Kathman and Nelson 1987) or limited to the season of peak abundance (Schuster et al. 1978). In a study of shallow groundwaters in southeastern New York, USA, tardigrades were collected with different nonquantitative techniques. Six taxa of tardigrades were scarce at most sites (<1% of the invertebrate community), but they were abundant at two of the five sites, constituting 18% of invertebrates in a spring and 5% in the hyporheic zone of a creek. The density of all four abundant taxa in the creek peaked in early spring, accompanied by egg-laying, but the populations declined in the summer. The summer minimum may be due to a decline in population size, a seasonal movement of animals from the hyporheic zone to adjacent habitats, or inadequate replicate samples or collection techniques (Strayer et al. 1994). In sandy substrates, tardigrade abundance may increase with depth and dominate the meiofauna of deeper sediments, but often tardigrades have been lumped together and not identified even to genus, so little quantitative data are available. Obviously further research is essential to determine abundance, seasonality, and population dynamics of freshwater tardigrades.

7.3.3 Terrestrial Population Dynamics

More information is available on population dynamics of terrestrial tardigrades than in other environments. However, the density of individuals in tardigrade populations is highly variable, and the factors that control tardigrade densities are not yet understood. For a review of previous studies, see Schuster and Greven (2007, 2013).

Estimates of tardigrade densities in soil habitats in Japan varied greatly, from 8050 to 75,500 per m^2, similar to those of oribatid mites or collembolans (Ito 1999). Core samples to depths of 10 cm showed that tardigrade densities were highest at depths of 4–6 cm, but tardigrades were also found in deep soils at depths to 40 cm (Ito and Abe 2001), probably depending on phreatic water level. Comparable numbers were found for litter and soil samples in Denmark (Hallas and Yeates 1972). Other estimates of soil tardigrades in older literature are even higher, with up to 300,000 tardigrades per m^2. In young soils at a post-coal-mining site in Germany, densities of the carnivore *Paramacrobiotus richtersi* ranged up to 33,600 individuals per m^2 (Hohberg 2006). In an investigation of tardigrade colonization and succession at a similar site (Hohberg et al. 2011), the distribution of *Apodibius confusus*, a soil-algivore, initially was very patchy with small-scale densities of up to 2281 individuals per m^2 in the top 10 cm^2. Over the subsequent 2-year period, distribution became more homogeneous with densities up to 349,700 individuals per m^2, probably related to the lack of competitors and predators in young soils and the establishment of appropriate food resources such as nematodes and algae. In moss samples at the University of Swansea (UK), the variability in density ranged more widely, from 36,000 to 2,280,000 specimens per m^2, with higher densities attributed to the tremendous reproductive potential of tardigrades (Morgan 1977). Population densities of terrestrial Antarctic tardigrades, often higher than those reported from temperate zones, ranged from 11,000 per m^2 to 874,000 per m^2, with a maximum density of 14,130,000 per m^2 in a mat of foliose algae in seal wallows with high nutrient input (Jennings 1976a, b). In the maritime Antarctic South Sandwich Islands, species richness was low (6 taxa), but population densities reached 4,000,000 per m^2 (McInnes and Convey 2005).

A seasonal and altitudinal comparative study of leaf litter tardigrades in beech forests was conducted in the USA and Italy (Guidetti et al. 1999). In the USA (Tennessee, Roan Mountain) tardigrade densities ranged from a low of 20,000 per m^2 in summer to 98,000 per m^2 in winter, with the highest densities at the lowest altitudes. In contrast, in Italy (Modenese Apennine), densities ranged from 7000 per m^2 at the lowest altitude to 17,000 per m^2 at the intermediate altitude, with 13,000 per m^2 at the highest altitude. On both continents, the highest densities and species richness were in the upper layer of the leaf litter, at depths of 3–5 cm. The species composition in soil and leaf litter was similar on both continents but differed greatly from that in nearby moss and lichen communities. Similarly, in a Spanish survey, the leaf litter community and abundances were different from that on tree trunk or rock communities (Guil et al. 2009a).

Fluctuations in population densities have been attributed to environmental variables such as humidity, fungal population dynamics, precipitation, temperature, soil type, habitat and landscape type, and air pollution in various studies (e.g., Franceschi et al. 1962–1963; Morgan 1977; Dastych 1987, 1988; Steiner 1994; Jönsson 2003, 2007; Schuster and Greven 2007, 2013; Guil et al. 2009a; Kaczmarek et al. 2011). The long-term population dynamics of tardigrades in a moss carpet in Germany was studied over a period of 54 months (Schuster and Greven 2007, 2013). The total number of individuals was 19,909, belonging to 26 species, of which 3 species were the most abundant and occurred in all months. The tardigrade community was relatively robust over time, with species diversity highest in winter, although some species showed a decline in the winter and an increase in spring to fall. The mean number of individuals of *M. hufelandi* was negatively correlated with rainfall and positively correlated with temperature, and both *P. richtersi* and *D. pingue* were positively correlated with rainfall. However, variables in the microenvironment may be more important than mesoclimatic factors when abundances are analyzed, but mesoclimatic variables may determine overall tardigrade diversity values (Guil et al. 2009b).

Temporal variations with a unimodal pattern with maximum peaks in spring, or with a bimodal pattern with two peaks (spring, fall), have been also reported (Franceschi et al. 1962–1963; Morgan 1977; Steiner 1994). Morgan (1977) found not only a seasonal pattern for *Macrobiotus hufelandi* and *Echiniscus testudo* but also a second long-term cycle in population dynamics for *M. hufelandi*, superimposed on the seasonal cycle. Tardigrade populations are usually dominated by females, while males peak in winter or at the beginning of spring. Frequently, gonochoric species have a sex ratio of 1:1 in their populations; however, some populations exhibit a sex ratio of 1:4, probably because gonochoric and parthenogenetic populations can coexist within the same species (see previous chapter on reproduction). The data on temporal variations are questionable due to patchiness in tardigrade populations and sampling errors in collections. In a single moss cushion of *Hypnum cupressiforme*, tardigrade individuals and species displayed an aggregated horizontal distribution (within-substrate heterogeneity), with the number of species per sample being random (Degma et al. 2011) (see also Meyer 2006). In addition, the horizontal microdistribution of tardigrades was not correlated with the moisture distribution within the moss cushion. In one re-collection study (Nelson and McGlothlin 1996) of tardigrades on Roan Mountain (Tennessee, USA) over 15 years apart, 18 species of the 21 original species reported by Nelson (1975) were discovered in 2 of Nelson's 6 sites at 3 elevations. In this pristine environment at high elevation, the presence of the tardigrade species remained almost constant more than a decade. Long-term studies of tardigrade population dynamics, with appropriate statistical design and analysis, are necessary to determine the climatic factors and cyclical patterns that affect their distribution and abundance.

7.4 Zonation

7.4.1 Marine Zonation

Zonation has been reported in some marine tardigrades. For example, species composition shifts from littoral to deep sea to submarine caves (Grimaldi de Zio et al. 1984; Grimaldi de Zio and Gallo D'Addabbo 2001, Fig. 7.10). In the Mediterranean, only 4 species were reported to be exclusively intertidal, while 53 were exclusively subtidal, and 20 were common to both intertidal and subtidal (Grimaldi de Zio and Gallo D'Addabbo 2001). In a comprehensive review of marine tardigrade zoogeography, Kaczmarek et al. (2015b) recognized intertidal, shallow sublittoral, deep sublittoral, and abyssal zones and reported 62 exclusively intertidal species, 44 exclusively subtidal species, 57 that occurred in both intertidal and subtidal zones, and only 22 species exclusively from deep subtidal and abyssal zones, but they also noted that sampling is strongly biased toward the intertidal. On a smaller scale within one beach where multiple species co-occurred, zonation appeared to be related to depth and elevation (de Zio and Grimaldi 1966; Pollock 1970; Lindgren 1971; da Rocha et al. 2004) (Fig. 7.11). On a sandy beach on Itamaraca Island (Brazil), *Stygarctus bradypus* colonized the upper mid-littoral zone, with the population concentrated between 10 and 20 cm depth throughout the year; its distribution was not influenced by granulometry or interstitial water (Verçosa et al. 2009).

Intertidal tardigrades, like other meiofauna, migrate with tide level. Beaches are stratified with respect to water saturation and oxygen, and this stratification is dynamic due to tidal fluctuations. Meiofauna, including tardigrades, have been reported to move daily both horizontally and vertically (Giere 2009).

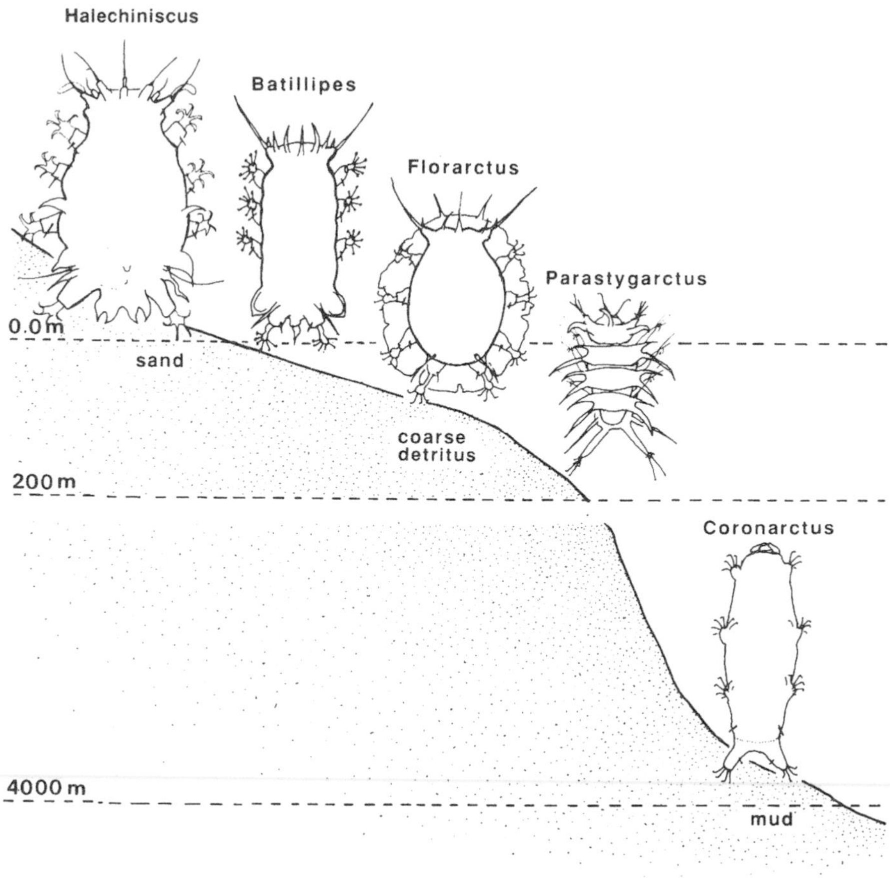

Fig. 7.10 Representative marine genera along a depth profile from intertidal to shallow subtidal to deep sea. From Giere (2009)

A complex interaction of abiotic and biotic factors is responsible for structuring meiobenthic communities (Fig. 2.16, Giere 2009). Sediment particle size, temperature, and salinity are the primary factors, but biotic factors may play a role as well (Coull 1999). Sediment size appears to be the preeminent factor affecting tardigrade species distributions (Hansen et al. 2001; Grimaldi de Zio and Gallo D'Addabbo 2001).

7.4.2 Freshwater Zonation

Tardigrade zonation in freshwater habitats is relatively unknown, with species inhabiting all of the various zones in lakes and streams (Strayer et al. 1994; Strayer

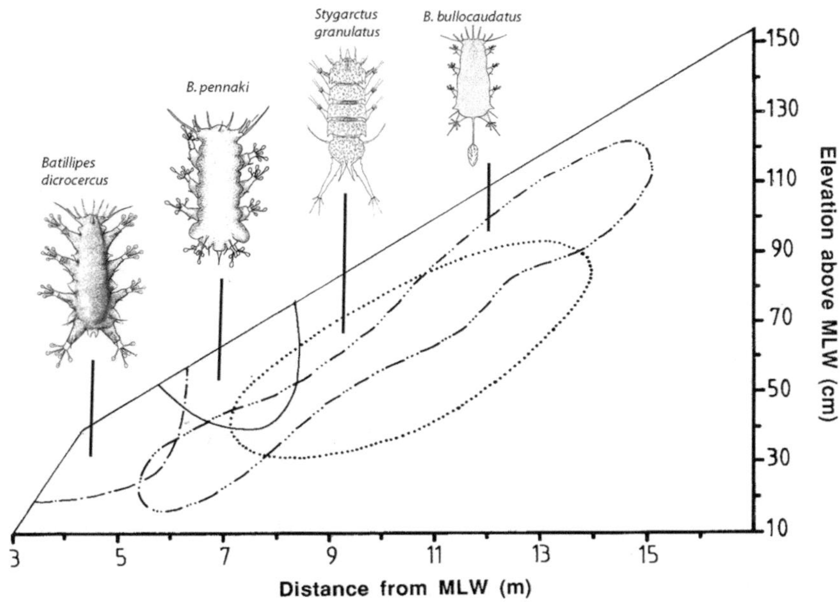

Fig. 7.11 Intertidal zonation in a beach at Woods Hole, Massachusetts, USA, based on distance and elevation from MLW. Seasonal changes in distributions were also investigated. From Pollock (1970)

and Findlay 2010). In lakes, tardigrades have been reported from littoral, limnetic, profundal, and benthic zones, but no species are restricted to a particular zone. On Signy Island (South Orkney Islands, maritime Antarctica), tardigrades were found at all depths in the shallow lakes, probably due to wind turbulence mixing, although some species showed specific depth profiles (McInnes and Pugh 1999). In streams, invertebrate communities in surface water, the hyporheic zone, and groundwater generally vary in composition longitudinally downstream, depending on a number of factors (e.g., current velocity, water temperature, substrate, stability of both aquatic and riparian vegetation, dissolved substances, competition, zoogeography, food, and disturbance history). The physical characteristics of stream flow ("stream hydraulics") are the most important environmental factors governing the zonation of stream benthos (Statzner and Higler 1986).

7.4.3 Terrestrial Zonation

On a microscale level, very few investigations have been conducted to determine the zonation of terrestrial tardigrades, such as the microdistribution of tardigrades within a moss cushion. In the moss *Grimmia alpicola*, the only tardigrade species with a stratified distribution was *Echiniscus viridissimus*, with a significantly higher frequency at the top layer of the moss, regardless of moisture conditions (Nelson and Adkins 2001). In contrast, in the moss *Grimmia pulvinata*, the eutardigrade *Paramacrobiotus richtersi* was reported to migrate within the moss cushion, but the heterotardigrade *Echiniscus testudo* did not migrate from the canopy layer when drying occurred (Wright 1991); however, the data set was extremely small, and the reported difference may not be significant (Nelson and Adkins 2001). Variables such as availability of food, hardness of leaves, or water retention capacity could influence tardigrade microdistribution within mosses (Greven and Schűttler 2001; Degma et al. 2011). In a study of soil microdistribution, species in the *Diphascon*, *Itaquascon*, *Mesocrista*, and *Platicrista* genera were located in the first 3 cm of the soil, whereas species in the *Macrobiotus* and *Minibiotus* genera occurred in higher abundances in deeper soil layers, below 4 cm (Ito and Abe 2001). In addition, cryptogram-dwelling species may show a zonation pattern from the terrestrial to the marine littoral zone that may be related to salt tolerance of cryptograms (Kinchin 1994) (Fig. 7.12).

7.5 Dispersal

7.5.1 Marine Dispersal

Many species of marine tardigrades appear to be cosmopolitan, suggesting substantial dispersal capabilities. The mechanisms of long-distance dispersal, however, remain very poorly understood. Dispersal in meiofauna may be passive (sediment displacement) or active (as in copepod emergence and reentry behavior) (Commito and Tita 2002). Yet, the lack of cryptobiotic states would seem to limit passive dispersal, and it is unclear if the weak swimming ability of semibenthic tardigrades allows much active dispersal. For interstitial burrowers, active dispersal is very unlikely. A comparison of geographic ranges for interstitial species and semibenthic species has not been reported.

The cuticular extensions and elaborate appendages of semibenthic tardigrades are thought to aid passive dispersal (Grimaldi de Zio et al. 1984; Jørgensen and Kristensen 2001). Active emergence from the sand may occur in *Florarctus salvati* using lateral cuticular wings and in *Batillipes bullacaudatus* using an inflatable tail (Giere 2009). Kristensen and Sørensen (2004) observed *Tanarctus bubulubus* with its spectacular balloons inflated, floating upside down in the water column.

The discovery of *Echiniscoides sigismundi* eggs in an exuvium of its host barnacle suggested that this species may spread through the shed exuvia of their barnacle host or during larval barnacle dispersal (Kristensen and Hallas 1980). Meiofaunal dispersal might also occur through the ballast water of marine ships as well as via the barnacles or algal lawn on the underwater surface of ships (Giere 2009).

Mats of microalgae and cyanobacteria, algae such as *Sargassum*, coconuts, and other vegetation have been discussed as possible meiofaunal rafts (Giere 2009;

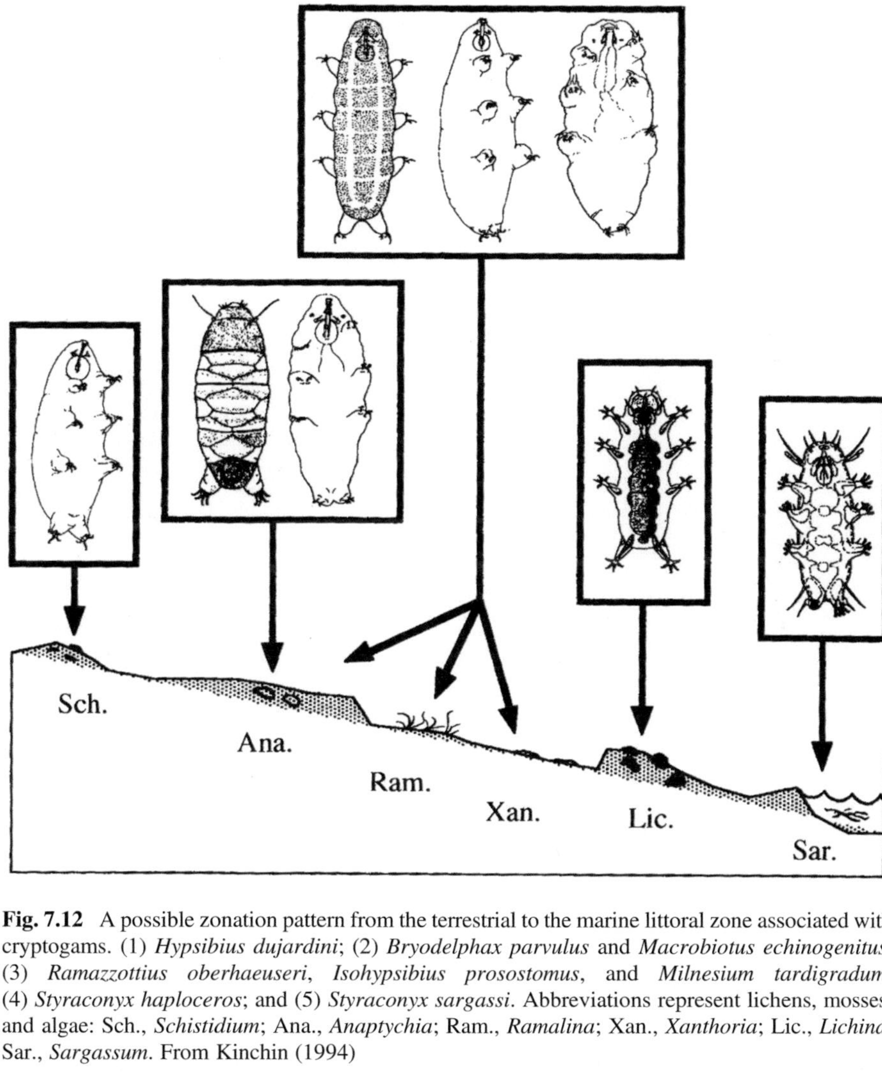

Fig. 7.12 A possible zonation pattern from the terrestrial to the marine littoral zone associated with cryptogams. (1) *Hypsibius dujardini*; (2) *Bryodelphax parvulus* and *Macrobiotus echinogenitus*; (3) *Ramazzottius oberhaeuseri*, *Isohypsibius prosostomus*, and *Milnesium tardigradum*; (4) *Styraconyx haploceros*; and (5) *Styraconyx sargassi*. Abbreviations represent lichens, mosses, and algae: Sch., *Schistidium*; Ana., *Anaptychia*; Ram., *Ramalina*; Xan., *Xanthoria*; Lic., *Lichina*; Sar., *Sargassum*. From Kinchin (1994)

Arroyo et al. 2006). *Styraconyx sargassi* has been found on *Sargassum* (Thulin 1942), and a new species, *Styraconyx turbinarium*, was found on rafting *Turbinaria* (Bartels et al. 2015); these brown algae are known to float for great distances. Floating plastic debris is an additional man-made mechanism of dispersal of barnacles and other macrofauna (Barnes 2002), and these could carry a large meiofauna complement.

Marine snow, a term used to describe a floating aggregate of detritus, phytoplankton, micrometazoans, bacteria, inorganic matter, and mucus, may have global significance in the production of organic matter and is perhaps the most significant mode of long-distance dispersal (Giere 2009). A rich assemblage of meiofauna has been found associated with these aggregates (Shanks and Edmonson 1990; Shanks and Walters 1997), although not tardigrades as yet.

7.5.2 Freshwater Dispersal

Distinctly freshwater species undergo limited or no cryptobiosis, and information on their dispersal is based on assumptions without experimental evidence. Rivers and streams are thought to actively transport adult and juvenile tardigrades, as well as cysts and eggs, downstream during heavy rainfall, flooding, and snow melt. Passive transport is assumed to occur easily, based on the wide geographic distributions of some freshwater species. The activities of larger aquatic and terrestrial animals may account for some of this transport, although no hard evidence is available. The mechanism for repopulation of aquatic habitats by tardigrades after their apparent disappearance in the summer or after habitat disturbances by nature or humans is unknown (Nelson et al. 1987).

7.5.3 Terrestrial Dispersal

Dispersal in terrestrial tardigrades is intrinsically related to the microscopic size of individuals and their eggs, their ability to enter into cryptobiosis (Fig. 7.13), and their ability to be easily transported passively by vectors such as wind, rain, or other animals or plants (Guil 2011). Active tardigrade dispersal is limited due to their small size; i.e., they do not crawl very far, and crawling only allows microdispersal capabilities. Cryptobiosis permits tardigrades to be transported through any climatic condition (e.g., across different atmospheric layers) and for long periods of time. Aerial transport of microbiota (microarthropods, moss spores, pollen, fungal spores, bacteria, algal and lichen propagules, nematodes, and tardigrade cysts) has been documented in the lower air levels of Antarctica (Marshall and Convey 1997; Hawes et al. 2007; Pearce et al. 2009) and in a Norway alpine glacier (Flø and Hågvar 2013). The influence of human activity and migration on tardigrade dispersal through transport of plants and soil has not been investigated but may also be a factor in long-distance movements of tardigrades across the globe. Pioneer success depends on both dispersibility and the ability to survive and reproduce upon arrival, as ably demonstrated by cryptobiotic tardigrades. Passive transport such as aerial transport has been used for decades to support the erroneous idea that most tardigrade species are cosmopolitan. Certainly the phylum at large is cosmopolitan (Pilato and Binda 2001), but individual species may be endemic, restricted, widespread, or possibility cosmopolitan.

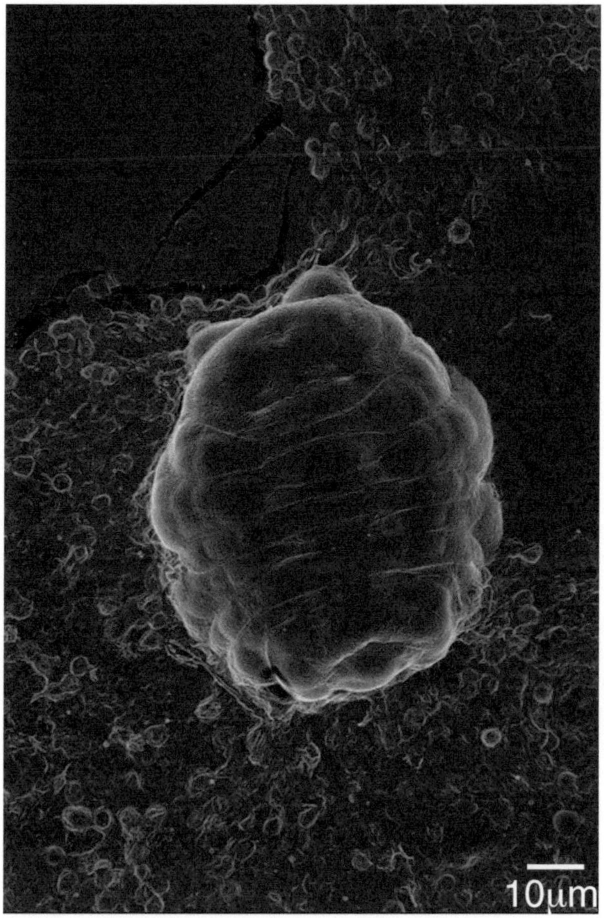

Fig. 7.13 Scanning electron micrograph (SEM) of the eutardigrade *Hypsibius exemplaris* in the "tun" state during anhydrobiosis. Photo by Thomas Boothby, University of North Carolina-Chapel Hill, North Carolina, USA

7.6 Biodiversity

Traditionally the term "species diversity" has included both species richness (the number of species) and species evenness (the abundance of each species). Diversity has been measured by calculations such as Shannon-Wiener, Simpson's, and Similarity Indices although sometimes the term "diversity" has been used to indicate only the number of species, especially in surveys of tardigrades in various regions. The popular current term "biodiversity" is most frequently quantified as the number of species. Less than 2 million animal species have been described worldwide, but the number of undescribed species is undoubtedly much higher, especially in small invertebrates such as tardigrades. Estimates of the total number of all extant species range from 5 million to 30 million. Currently the global total number of tardigrade species has been estimated at 2654 (Bartels et al. 2016), but the number of described species is just over 1200 (Guidetti and Bertolani 2005; Degma and Guidetti 2007; Degma et al. 2009–2015).

Biodiversity patterns have been analyzed only in a few tardigrade studies, but these assessments are largely inaccurate without valid taxonomic information (integrating morphological and molecular data) and without adequate sampling methods and appropriate statistical analysis. In general, previous studies have focused on variables such as altitude, species of vegetation substrate (e.g., moss, lichen), type of habitat (e.g., rocks, trees, leaf litter, soil), environmental factors (rainfall, temperature), or presence and abundance of other invertebrates (e.g., rotifers, nematodes) (e.g., Dastych 1987, 1988). Altitude is frequently cited as an important factor in species distributions (richness) (see Kaczmarek et al. 2011). In fact, altitude is not a single factor but comprises many other variables such as humidity, insolation, vegetation, temperature, exposure, and other factors. However, no statistically significant differences in patterns have been found in tardigrade species richness and altitude per se, although contradictory patterns have been reported in various studies. This was confirmed in a study of terrestrial tardigrades based on analysis of published worldwide data, although a comparison of latitudes of previous studies indicated that the northern higher latitudes (Japan and Newfoundland) exhibited lower tardigrade species richness than lower latitudes (Europe and USA) (Guil et al. 2009b). Based on the limited studies in Antarctica, terrestrial tardigrade richness is very low, and the ecosystems are relatively simple, due to the extreme environmental conditions (Tsujimoto et al. 2014). In comparison, freshwater tardigrades can be abundant in maritime and continental Antarctica. Obviously regional tardigrade diversity is largely unknown and under-researched in Antarctica as well as other regions of the world. In the Arctic (e.g., Svalbard) and in the mountains in temperate zones and South America, tardigrades appear to be larger and more diverse and have a greater density, but these data are not yet published (L. Kaczmarek pers. comm.). In general, the trend in most terrestrial organisms is toward higher species diversity in the tropics, whereas in most marine groups, the Indo-Pacific is the biodiversity hotspot. Whether or not tardigrades display a standard or nonstandard latitudinal diversity gradient is an important question that needs further examination.

Although no statistical correlation has been found between tardigrade species richness and moss species, microhabitats and substrates are related to both species diversity (richness and evenness) and community structure (Guil et al. 2009b; Nelson and Bartels 2013; Guil and Sánchez-Moreno 2013). Humidity and rainfall influenced both species richness and diversity indices in moss tardigrades (Schuster and Greven 2007), as well as the type of vegetation (Guil et al. 2009a), nematode frequency and soil porosity (Harada and Ito 2006) (Fig. 7.14), and pollution (Peluffo et al. 2007). The first study of the effect of fire on terrestrial tardigrade biodiversity showed a reduction in species richness due to habitat loss but rapid postfire recolonization (Vicente et al. 2013). In Spain local biodiversity patterns were studied to assess the relationships between habitat type and diversity of terrestrial tardigrade communities (Guil et al. 2009b). In that study, statistically supported trends indicated that leaf litter habitats showed high species richness and low abundances, rock habitats (mosses and lichens) showed intermediate species richness and high

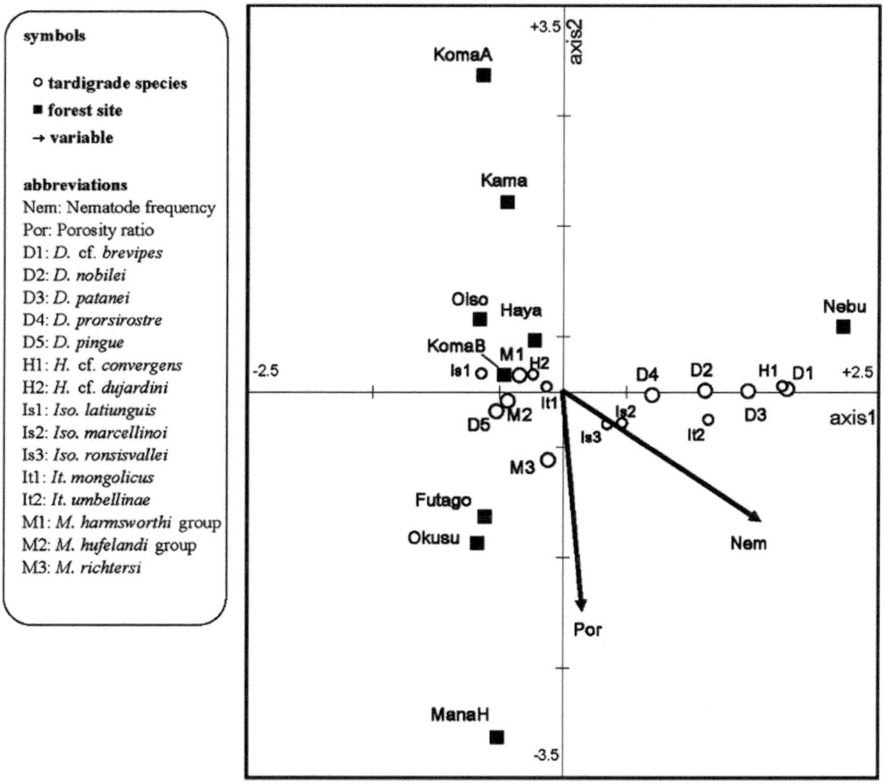

Fig. 7.14 Canonical correspondence analysis (CCA) displaying the correlations between tardigrade species and environmental factors, such as nematode frequency and soil porosity. From Harada and Ito (2006)

abundances, and trunk habitats (mosses and lichens) showed low numbers of both richness and abundance.

7.7 Biogeography (Distribution)

This topic is covered thoroughly in the chapter on Biogeography in this book, so only basic information is given here.

7.7.1 Marine Biogeography (Distribution)

Although detailed genetic studies to assess cryptic species have just begun, Kaczmarek et al. (2015b) reported that 68% of marine species are endemic and only 11% are cosmopolitan using a definition of endemism as a distribution covering only one or two adjacent FAO areas (the UN's designation of major fishing areas), although this is clearly strongly influenced by sampling bias. A molecular study of *Echiniscoides* (Faurby et al. 2012) found high species diversity in a group once thought to be cosmopolitan. In conjunction with the remarkably large number of new species recently reported from Australian marine caves (Jørgensen et al. 2014), it is clear that many species remain undiscovered, and much remains to be learned about marine tardigrade biogeography.

7.7.2 Freshwater Biogeography (Distribution)

With regard to true limnic taxa, the two non-eutardigrade families are represented by only one species, and both are endemic: the heterotardigrade *Carphania fluviatilis* and the mesotardigrade *Thermozodium esakii*.

Among the Eutardigrada, no families with limnic species are restricted to a biogeographic region. In fact, the true limnic genera *Dactylobiotus* and *Pseudobiotus* are considered cosmopolitan and composed of species reported to be cosmopolitan but may contain cryptic species (e.g., *D. dispar*, *D. ambiguus*, *P. megalonyx*, and *P. kathmanae*, the latter often erroneously cited as *P. augusti*), as well as Palearctic species (e.g., *D. ampullaceus*, *D. haplonyx*, *D. parthenogeneticus*, and *D. selenicus*), and endemic species, such as *D. caldarellai* (Chile), *D. dervizi* (Komandorskiye Islands), *D. grandipes* and *D. kansae* (North America), *D. luci* (Uganda), *D. lombardoi* (Chile), *D. octavi* (Greenland), *D. vulcanus* (Oceania) and *P. hirsutellus* (Israel), *P. matici* (Italy), *P. spinifer* (South Korea), and *P. vladimiri* (Japan). The true limnic genus *Thulinius* has a Holarctic distribution, with *T. saltursus* and *T. romanoi* being endemic to North America and *T. itoi* to Japan. The monospecific genus, *Macroversum*, is endemic to Sicily, whereas another monospecific genus, *Borealibius*, has a boreal-alpine distribution.

Several limnic species are found in the limnoterrestrial cosmopolitan genera *Isohypsibius* and *Hypsibius*. Some species of these genera have a very wide distribution, are cosmopolitan, or are species complexes (e.g., *I. granulifer*, *H. dujardini*, *H. convergens*), while other species are endemic to small areas, e.g., *H. klebelsbergi* (Austria) and *I. baldii* (a single small lake in Northern Italy). The aberrant eutardigrade *Eohypsibius nadjae*, described from a cold mud volcano in Greenland, was later found in cold springs in the Faroe Islands and Northern Italy and may be an Arctic relict. In contrast, *Acutuncus antarcticus* is restricted to Antarctica but is widespread in both maritime and continental Antarctica and subantarctic islands

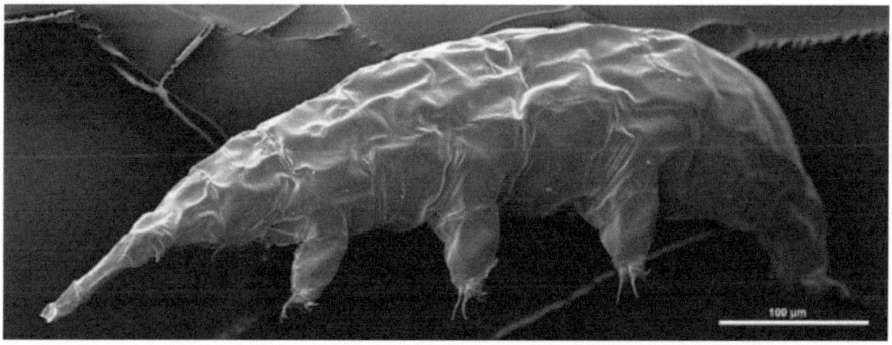

Fig. 7.15 Scanning electron micrograph (SEM) of simplex stage of *Bergtrollus dzimbowski* (Milnesiidae) endemic to the Lyngen Alps, ca. 1000 m a.s.l., Norway. Note the strikingly long, protrusible proboscis ("snout") that was observed moving from side to side as if searching for food. From Dastych (2011)

(McInnes 1995) in freshwater lakes as well as in terrestrial environments such as mosses (Tsujimoto et al. 2014). This is now believed to be a species complex.

7.7.3 Terrestrial Biogeography (Distribution)

Based on the current information available, some terrestrial tardigrade families, subfamilies, and genera are endemic to one biogeographic region. For example, in the family Milnesiidae, the genus *Milnesioides* has thus far only been reported from Australia, the genus *Limmenius* only from New Zealand, and the genus *Bergtrollus* only from Norway (Fig. 7.15). Other endemic genera have been described from Europe, Africa, Asia, North America, or South America.

The discovery of synonymies and cryptic species in certain species or group of species based on molecular data is changing our understanding of the global biogeography of limnoterrestrial tardigrades previously established from literature based on morphological information (Pilato and Binda 2001; Jørgensen et al. 2007; Faurby et al. 2008; Michalczyk et al. 2012a, b; Bertolani et al. 2014b). Based on cluster analysis, principal component analysis, and parsimony analysis of endemism (PAE), the Laurasian fauna is considered to be derived and the Gondwanan group basal (McInnes and Pugh 2007), although the reverse was originally proposed based on plate tectonics (McInnes and Pugh 1998).

In the past, many very similar species (cryptic species or species complexes) were identified as one taxon, which complicated identifications and biodiversity studies. Species originally considered as having widespread or cosmopolitan distributions, such as *Milnesium tardigradum, Hypsibius dujardini* and *Macrobiotus hufelandi*, have now been separated into various species based on a combination of molecular data and careful morphological analysis. This integrative approach uses morphological analyses of animals (with light microscopy), eggs (with light microscopy and scanning electron microscopy), and voucher specimens, which are also used for molecular analysis of *cox1* for DNA barcoding (Bertolani et al. 2011, 2014b). Cryptic speciation can result in morphologically similar but molecularly different species. In contrast, molecular homogeneity was found in the highly variable morphology of the *Echiniscus blumi-canadensis* group of species (Jørgensen et al. 2007; Guil and Giribet 2009). Morphological information suggests endemic and/or narrow ranges for many tardigrade distributions; however, some molecular data (although limited for most tardigrades) has shown transatlantic (Europe and USA) or worldwide distributed haplotypes in some genera (COI and ITS2 sequences of *Macrobiotus* and *Milnesium*) (N. Guil pers. comm.). Morphological and molecular information that reach incongruent results regarding species identifications stresses the necessity for integrative approaches for biogeographical analyses. Until then, the global distributions of tardigrade species remain largely undetermined, possibly ranging from endemic to cosmopolitan depending on the species (Guil 2011, N. Guil pers. comm.). This line of integrative research, especially with the development of molecular analyses, is increasing in frequency.

7.8 Trophic Webs

7.8.1 Marine Trophic Webs

Very little is known about trophic interactions in marine tardigrades. Most marine tardigrades probably feed on algal cells, including macroalgae and diatoms, using paired piercing stylets and a muscular, sucking pharynx. Others may be detritivores, bacteriovores, or ectoparasites (Kristensen and Sørensen 2004). The retractable mouth cone and stout stylets of *Dipodarctus borrori* were suggested as possible adaptations for a predatory life style (Pollock 1995), but if this is true, the thin buccal tube would greatly limit range of prey, and no further substantiation has been documented.

Since tardigrades usually make up a very small component of marine meiofauna, their specific role in these systems has not been studied. An extensive literature does exist for marine meiofauna considered collectively (Giere 2009). Polychaetes, bivalves, various crustaceans, fish, and birds rely heavily on meiofauna, sometimes depending solely on them at least during some phase of their life cycle (Coull 1990, 1999) (Fig. 7.16). One study found that about 75% of marine meiofauna was utilized by organisms at higher trophic levels (Danovaro et al. 2007).

Although no exclusion experiments have been conducted to document competition, competitive interactions have been suggested based on nonoverlapping distributions. Competition may impact zonation in beach tardigrades (Pollock 1970; Lindgren 1971; Martinez 1975) and may affect vertical distribution patterns in *Echiniscoides* species living in barnacles (Kristensen and Hallas 1980).

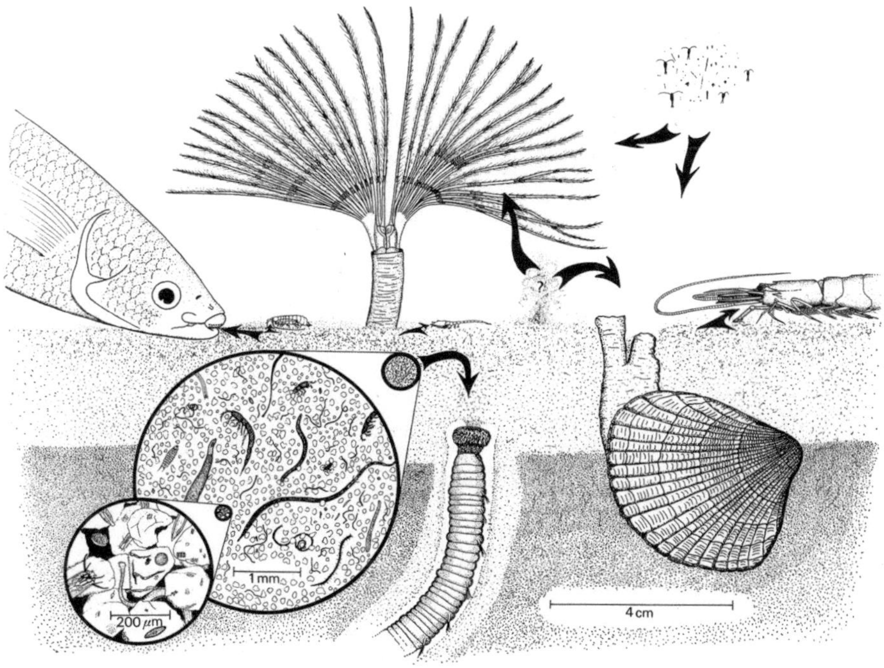

Fig. 7.16 Artistic illustration of a hypothetical marine benthic food web indicating possible connections between the meiofauna and macrofauna. Meiofauna may constitute a significant dietary component of organisms at higher trophic levels. From Platt (1981)

7.8.2 Freshwater Trophic Webs

Limnoterrestrial tardigrades can be predators, prey, or primary consumers in food webs (Nelson et al. 2015). Predators of aquatic tardigrades include oligochaetes, nematodes, rotifers, insect larvae (e.g., chironomids and plecopterans), as well as other tardigrades. Some species of tardigrades prey on other micrometazoans, mainly nematodes and rotifers, but also on other tardigrades of different species. Many species of tardigrades are herbivorous, feeding on plant material such as algae and mosses, while others may feed on bacteria and detritus.

Lecophagus antarcticus is an aquatic predatory hyphomycete fungus that attacks both bdelloid rotifers and tardigrades (*Acutuncus antarcticus*) in cyanobacterial mats and sediments at shallow lake margins in maritime Antarctic lakes (McInnes 2003). Rotifers and tardigrades were trapped on adhesive pegs from the fungal vegetative hyphae, which penetrated the prey's body cavity. This fungal species may also be terrestrial but secondarily adapted to high prey densities and stable environments of maritime Antarctic lake margins. Other species in the carnivorous genus *Lecophagus* are terrestrial, reported in mosses, mossy soils, and leaf litter.

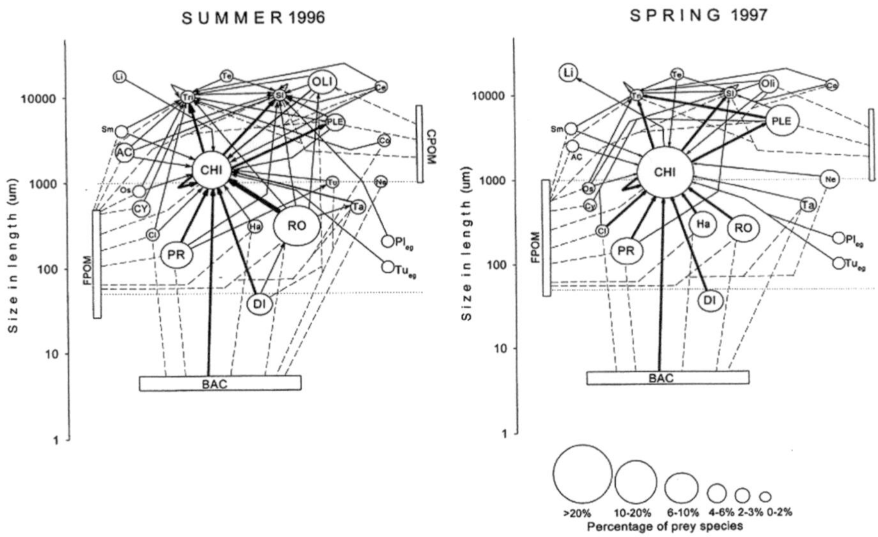

Fig. 7.17 Trophic web from an acidic stream in England. The *y* axis is body length of representative organisms. Circle size represents abundance. Ta = tardigrades; CHI = chironomid midges. From Schmid-Araya and Schmid (2000)

Trophic interactions within food webs are difficult to determine. Detailed studies of food webs conducted in an iron-rich, acidic stream system (Schmid-Araya and Schmid 2000; Schmid-Araya et al. 2002) (Fig. 7.17) showed that meiofauna, including tardigrades, increase web complexity since many stream invertebrates feed on meiofauna and organic matter. Tardigrades were noted as prey for rotifers and chironomid larvae. Marked differences in web complexity were found between the summer/autumn and winter/spring periods. Meiofauna accounted for most of the links in the web and thus may be crucial to the understanding of food web properties and ecosystem processes in streams.

7.8.3 Terrestrial Trophic Webs

Among the terrestrial tardigrades, some are carnivores, feeding on protists, nematodes, rotifers, and even other tardigrade species (Doncaster and Hooper 1961; Morgan 1977) (Fig. 7.18). Others are herbivores or detritivores. Tardigrades are also the prey of other predators in their habitats, e.g., nematodes, collembolans, mites, spiders, and insect larvae.

Trophic interactions among terrestrial tardigrades are largely unknown except for some predator-prey interactions both between tardigrade species and between tardigrades and other organisms (nematodes and rotifers). Within soil environments, a "super-predator" role was discovered for two tardigrade species, *Paramacrobiotus*

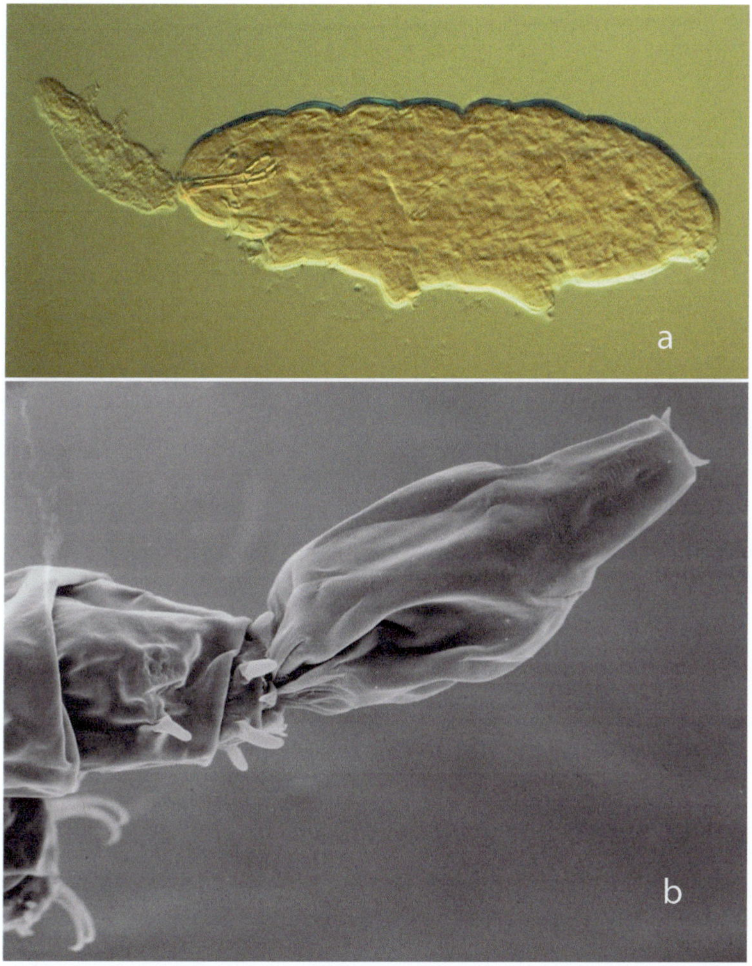

Fig. 7.18 Predation in tardigrades. (**a**) Differential interference contrast (DIC) micrograph of *Paramacrobiotus tonollii* feeding on *Minibiotus intermedius* from Great Smoky Mountains National Park, USA, (**b**) scanning electron micrograph (SEM) of *Milnesium* sp. ingesting a bdelloid rotifer from Roan Mountain, Tennessee/North Carolina, USA. (**a**) Photo by P Bartels; (**b**) from Nelson (1982)

richtersi and *Macrobiotus harmsworthi*, which suppressed nematode communities, despite lower abundance of tardigrades with respect to nematode populations (Sanchez-Moreno et al. 2008). This conclusion was congruent with the tardigrade feeding behavior and rates described previously for *P. richtersi* (Hohberg and Traunspurger 2005). Tardigrade buccal-pharyngeal morphologies are correlated with their feeding habits (Hallas and Yeates 1972; Guidetti et al. 2012, 2013; Guil and Sánchez-Moreno 2013). Three basic groups have been defined: carnivores, herbivores, and microbivores (Guil and Sánchez-Moreno 2013). Carnivorous and

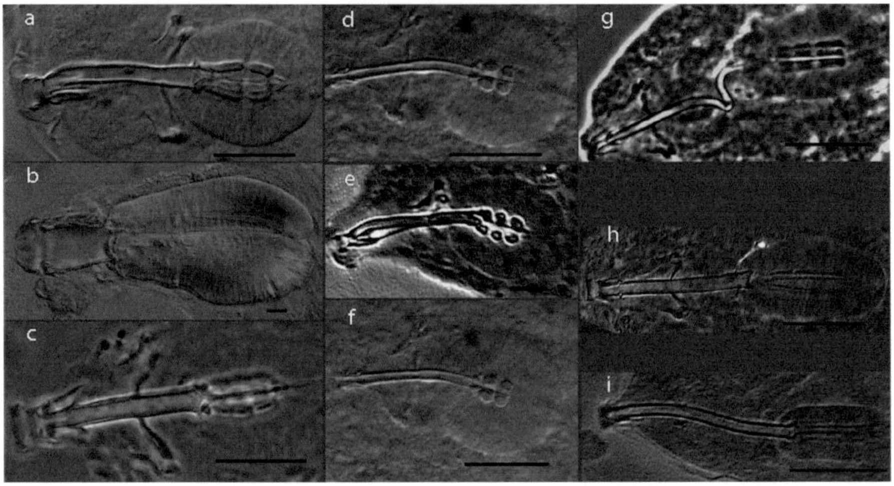

Fig. 7.19 Diversity in buccal-pharyngeal apparatus feeding types. Carnivorous [(**a**) *Macrobiotus*, (**b**) *Milnesium*, (**c**) *Isohypsibius*], herbivorous [(**d**) *Hypsibius*, (**e**) *Minibiotus*, (**f**) *Ramazzottius*], and microbivores [(**g**) *Diphascon*, (**h**) *Platicrista*, (**i**) *Astatumen*]. Scale bar = 50 μm. Photos by N Guil

omnivorous tardigrades (e.g., *Macrobiotus*, *Milnesium*) have short and wide buccal tubes with strong stylets and large pharynxes, while herbivores (e.g., *Echiniscus*, *Hypsibius*) have narrow buccal tubes, weak stylets, and large stylet furca and apophyses for the insertion of the stylet muscles. Tardigrades that feed on bacteria, fungi, and/or detritus, which have long, narrow, and partially flexible buccal tubes and small, weak stylets, are classified as microbivores (e.g., *Diphascon*, *Sarascon*) (Fig. 7.19). (See also the chapter on taxonomy and anatomy.) Based on these three groups, some diversity patterns were discerned in the leaf litter environments studied in Spain (Guil and Sánchez-Moreno 2013). Carnivorous tardigrade species inhabited coniferous needles, while herbivores were associated with xerophilous litter habitats and microbivores with mixed leaf litter. As expected, high abundance of carnivorous tardigrades was significantly related to low abundance of microbivore and herbivore groups, which was constant across all types of leaf litter. In comparison, in Argentina, two new species of *Milnesium* were found: one with a longer, thinner buccal tube that fed on rotifers and one with a shorter, wider buccal tube that fed on *Macrobiotus* tardigrades (Roszkowska et al. 2015). Other studies have compared the numbers of tardigrades, nematodes, and rotifers within a moss community in an attempt to determine the relationships between these organisms, which are frequently present together.

Polymerase chain reaction (PCR), a highly sensitive detection method, was used to identify the sequence of a chloroplast gene (rbcL) in the gut of active tardigrades in moss (Schill et al. 2011). The PCR method was also used to determine food preferences. In addition, digestion rate was followed over a 48-h period with autofluorescent emission of green light from the chloroplasts.

7.9 Symbiosis

7.9.1 Marine Symbiosis

Commensal relationships have been described for several marine tardigrades. Commensals appear to be facultative and are also found free living in interstitial or algal habitats. *Echiniscoides andamanensis* and the *E. sigismundi* complex are frequently found associated with barnacles, most often *Semibalanus balanoides*. There they live among the plates of the barnacles and probably feed on associated algae. In addition to physical protection, the crevices of the barnacle plates may provide shelter from temperature fluctuations especially at low tide (Faurby et al. 2012). Species of *Halechiniscus* has been found in the washings of oysters and *Actinarctus doryphorus* on the epidermis of an echinoid (*Echinocyamus pusillus*).

Ectoparasitism is known in four species: *Pleocola limnoriae*, *Tetrakentron synaptae*, *Echiniscoides hoepneri* (Kristensen and Hansen 2005), and *Styraconyx qivitoq* (Kristensen and Higgins 1984). *Pleocola limnoriae* is considered a facultative parasite on the pleopods of the isopod *Limnoria lignorum*, while the others are obligate parasites and show various adaptations for parasitism. *Echiniscoides hoepneri* feeds on embryos in the brood chamber of the barnacle *Semibalanus balanoides* with extremely long stylets that grow longer with each molt (Fig. 7.20a). *Tetrakentron synaptae*, an ectoparasite found solely on the tentacles of the holothurian *Leptosynapta galliennei*, is dorsoventrally flattened with reduced sensory appendages but with enlarged stylet glands and robust claws (with multiple hooks per claw) that assist with piercing host cells (van der Land 1975; Kristensen and Sørensen 2004) (Fig. 7.20b). Dwarf males appear at high population densities in both *T. synaptae* and *E. hoepneri*. *Styraconyx qivitoq* feeds on its host bryozoans with elongate stylets.

Most members of the genus *Florarctus* and *Wingstrandarctus* have epicuticular vesicles associated with the buccal apparatus that house symbiotic bacteria (Fig. 7.20c). These species occur in clean coral sand with very little algae, and nutrient exchange with these bacteria may be their major form of nutrition. Some evidence of regulating bacterial populations in these vesicles has been described (Kristensen 1984). A case of symbiotic bacteria was reported in a florarctid marine tardigrade, with bacteria population control by tardigrades and secretions from bacteria passing across the tardigrade cuticle (Kristensen 1984).

One final interspecific interaction has been seen in *Tanarctus bubulubus*. The entire dorsal surface of these animals is mucous covered and apparently this provides adhesion for coccoliths from the protistan coccolithophore, *Emiliania huxleyi*. These coccoliths are a large part of many marine sediments, and they may provide chemical or mechanical camouflage for these epibenthic animals (Jørgensen and Kristensen 2001).

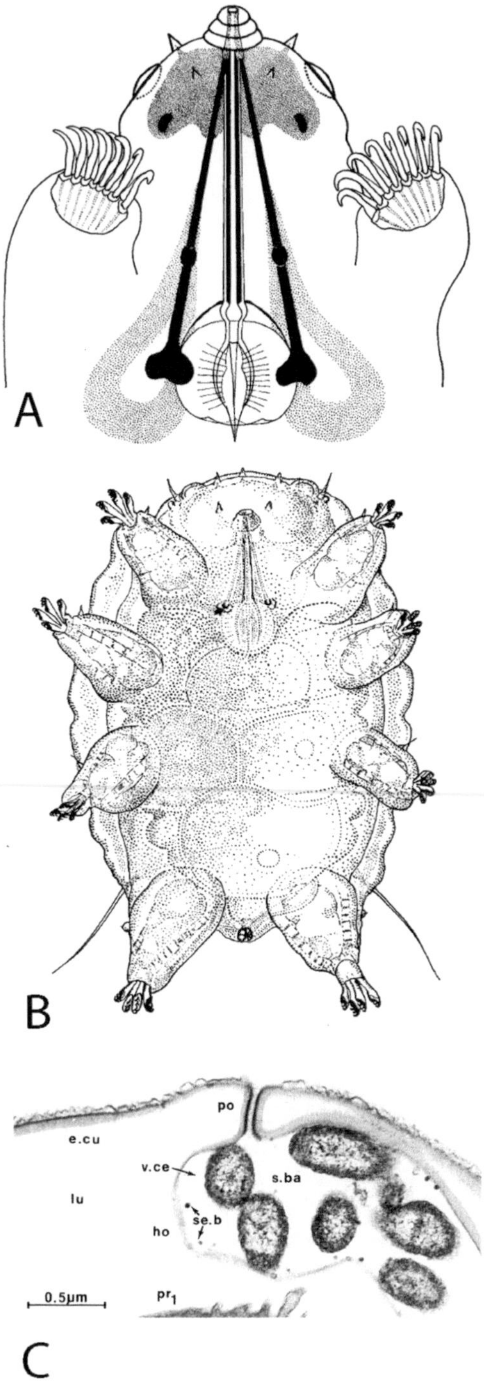

Fig. 7.20 Symbiosis in marine tardigrades. (**a**) enlarged stylets in *Echiniscoides hoepneri*, (**b**) flattened body shape in *Tetrakentron synaptae*, (**c**) bacteria in cephalic vesicles of *Florarctus antillensis*. Abbreviations: v.ce. = cephalic vesicle; po = pore; s.ba. = symbiotic bacteria; se.b = secreted granulae from bacteria. (**a**) From Kristensen and Hallas (1980), (**b**) from Kristensen and Sørensen (2004), (**c**) from Kristensen (1984)

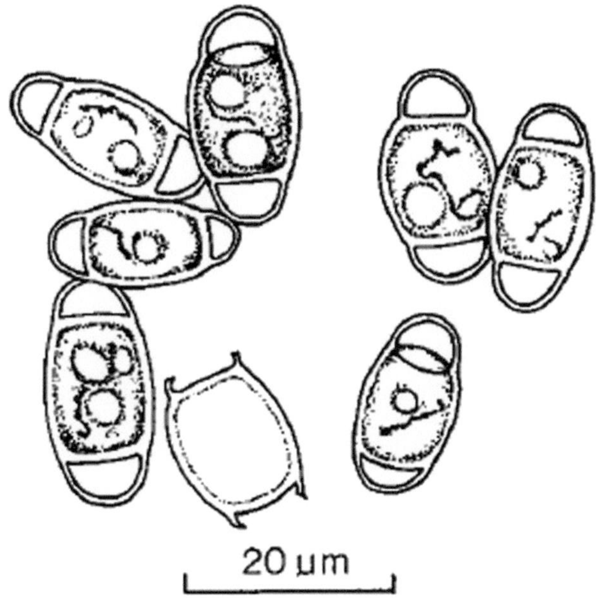

Fig. 7.21 X-bodies found in *Dactylobiotus dispar* and *Hypsibius dujardini* from Svalbard in the Arctic Ocean. From Van Rompu and De Smet (1991)

7.9.2 Freshwater Symbiosis

Parasitic protozoa and fungi infect tardigrades, but little is known about their role in the ecology of water bears. An unidentified fungus was found on many dead specimens of *Thulinius saltursus* in Lake Tahoe, California (USA), but it is unknown if the fungus was associated with the mid-October depletion in the population (Schuster et al. 1978).

Fungal or protozoan inclusions called "X-bodies" were present in *Dactylobiotus dispar* and *Hypsibius dujardini* from freshwater samples on an island in the archipelago of Svalbard in the Arctic Ocean (Van Rompu and De Smet 1991) (Fig. 7.21). These structures have a characteristic barrel-shaped midsection filled with cytoplasm and empty hemispherical ends and seem to be restricted to eutardigrades in arctic or subarctic habitats (Hallas 1977). They are similar to those found in *Mesocrista spitzbergensis* from Spitsbergen reported as a parasitic protozoan (Richters 1904) and similar to "microsporidians" of the genus *Pleistophora* in Greenland (Petersen 1951). X-bodies were also found in an individual of *Bertolanius smreczynskii* from Axel Heiberg Island (NWT Canada) (Węglarska 1970) and in *H. dujardini* and *Adropion scoticum* from Finland (Hallas 1977).

Microsporidian infections have also been found in midgut epithelial cells of *Isohypsibius granulifer granulifer* (Rost-Roszkowska et al. 2013). Membranes called phagophores develop around microsporidians or organelles containing microsporidians. The phagophores together with lysosomes are responsible for autophagy, and if the parasites are not completely digested, they may be eliminated into the gut lumen for further breakdown.

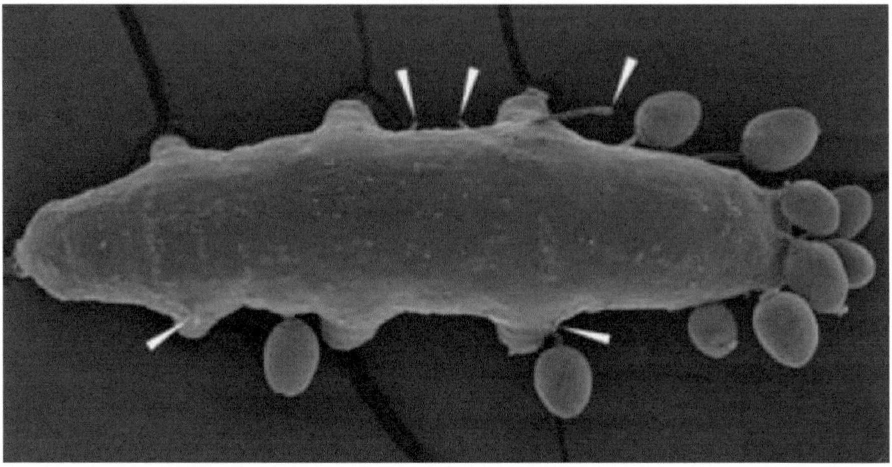

Fig. 7.22 Scanning electron micrograph (SEM) of *Ramazzottius* cf. *oberhaueseri* with symbiotic ciliate *Pyxidium tardigradum*. From Vicente et al. (2008)

7.9.3 Terrestrial Symbiosis

Terrestrial tardigrades have numerous symbiotic relationships with fungi, protozoans, and bacteria (Vecchi et al. 2016) but are mostly frequently parasitized by fungi, especially in moist habitats and in the laboratory. The life cycle of the chytridiomycete *Sorochytrium milnesiophthora* was described from infected populations of *Milnesium* cf. *tardigradum* (Dewel et al. 1985). As an endoparasite, *S. milnesiophthora* invades the cuticle of the tardigrade and grows in the body cavity until it kills its host and consumes virtually all cellular contents. Then it produces zoospores to initiate an infection in a new host (Dewel and Dewel 1987). Species of the fungus *Ballocephala* infect other tardigrades during a similar life cycle in which the conidia attach to the cuticle and the infection penetrates into the body cavity (Saikawa and Oyama 1992). Other fungal parasites include the genera *Macrobiotophthora*, *Harposporium*, and *Haptoglossa* (Baumann 1961; Saikawa et al. 1991). Unidentified fungi have been reported on eutardigrades in other studies (D. Nelson pers. comm.). Reaching a peak during the wettest months (Morgan 1977), fungal infections seem to be controlled by dry-wet periods in the natural environment, characteristic of most terrestrial environments where tardigrades occur.

In the case of protistans, the ciliate symphoriont *Pyxidium tardigradum* has been found on a number of eutardigrade species (Ciobanu et al. 2015) (Fig. 7.22). A detailed study of this ciliate on *Macrobiotus* cf. *hufelandi* and *Ramazzottius* cf. *oberhaeuseri* indicated that the ciliate did not penetrate the host but fed on bacteria and other organisms from the surrounding water (Vicente et al. 2008). On average, however, the ciliates added 14% to the volume of the host, which may inflict an energetic cost during locomotion and may interfere with reproduction. Thus, the relationship is more likely parasitic than commensal.

Living tardigrades (*Echiniscus molluscorum*) have also been collected from the feces of the land snail *Bulimulus exilis* in Puerto Rico (Fox and Garcia-Moll 1962). All of the tardigrade developmental stages were present in the feces, and infested snails harbored 1–30 tardigrades. However later the species was also found free-living in mold growing on a cement wall where the snails were resting (Fox 1966). Thus, it is unclear whether *E. molluscorum* is a commensal, a parasite, or a free-living species possibly transported by snails.

Bacteria are one of the food items utilized by tardigrades, but they can also be found over the surface of tardigrades. It is unknown if the bacteria are feeding on the tardigrade secretions. Some bacteria species have been considered also as part of tardigrade natural intestinal flora. Plant pathogens *Xanthomonas* and *Pseudomonas* were found in the gut of *Macrobiotus hufelandi* (Krantz et al. 1999), and later *Xanthomonas campestris* pv. *raphani* were reported to spread to plants by *M. hufelandi* and to cause infection in radishes, a potential economic impact of tardigrades on plants (Benoit et al. 2000). In addition, viruses have been also observed in tardigrade tissue, but the function and effect are unknown (Kinchin 1994).

7.10 Future Research in Tardigrade Ecology

Our knowledge of tardigrade ecology is still in its infancy. We still know little about patterns of biogeography, dispersal, trophic relationships, competition, and infectious agents. Although we have no good estimates of total tardigrade biodiversity, we know that many species are rare and that populations are patchily distributed. Long-term and large-sample studies are therefore required to further our knowledge of tardigrade ecology.

Biodiversity estimates have been assisted in recent years with the use of species richness estimators such as Chao, ACE, and ICE (Hansen et al. 2001; Bartels and Nelson 2007; Colwell et al. 2012; Nelson and Bartels 2013) (Fig. 7.23). As these become more widely used, comparisons of biodiversity patterns will be possible.

Factors determining species distribution patterns have begun to be elucidated with multivariate statistics such as principal component analysis, discriminant analysis, and general linear models (e.g., Guil et al. 2009a, b), and this is beginning to move us beyond anecdotal evidence. Newer tools such as MaxEnt and Partition Tree Analysis integrating with geographic information systems offer additional promise (Phillips and Dudík 2008; Elith et al. 2011; Rubal et al. 2016) for niche modeling.

Molecular information is providing new insights into the systematics and ecology of tardigrades. Until recently, morphological and morphometric data based on a limited number of characters have been used to determine species and other taxonomic units in all biological studies on tardigrades, including ecological investigations. Integrative taxonomy, using combined morphological and molecular

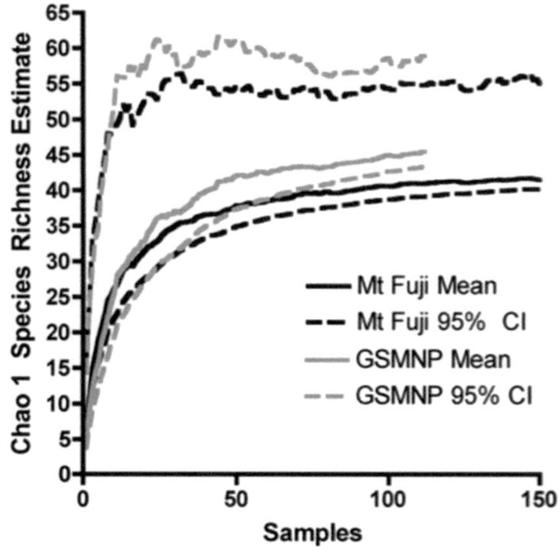

Fig. 7.23 Chao 1 species richness estimates from soil-dwelling tardigrades of the Great Smoky Mountains National Park, USA, and Mt. Fuji, Japan, indicating higher diversity in the Smokies. From Nelson and Bartels (2013)

information, is essential for future work with tardigrades. In addition to genetic analysis, information on reproductive strategies, feeding behavior, microhabitats, and other independent data can be incorporated into a more complete understanding of the taxa and the patterns and processes involved in tardigrade evolution. New data from molecular studies are indicating the existence of previously unknown cryptic species, which are indistinguishable by morphology alone. This can lead to underestimating biodiversity. In other cases, however, molecular data has been used to confirm that a group of previously named species could be synonymized into one species displaying morphological variation. Without complete knowledge of which species are present, ecological studies will remain questionable.

References

Arroyo NL, Aarnio K, Bonsdorff E (2006) Drifting algae as a means of re-colonizing defaunated sediments in the Baltic Sea, a short-term microcosm study. Hydrobiologia 554:83–95

Barnes DKA (2002) Invasions by marine life on plastic debris. Nature 416:808–809

Bartels PJ, Nelson DR (2006) A large-scale, multihabitat inventory of the Phylum Tardigrada in the Great Smoky Mountains National Park, USA: a preliminary report. Hydrobiologia 558:111–118

Bartels PJ, Nelson DR (2007) An evaluation of species richness estimators for tardigrades of the Great Smoky Mountains National Park, Tennessee and North Carolina, USA. Zool J Limnol 66 (Suppl 1):104–110

Bartels PJ, Fontoura P, Nelson DR (2015) New records of marine tardigrades from Moorea, French Polynesia with the description of *Styraconyx turbinarium* sp. nov. (Arthrotardigrada, Halechiniscidae). Zootaxa 3955:389–402

Bartels PJ, Apodaca JJ, Mora C, Nelson DR (2016) A global biodiversity estimate of a poorly known taxon: Phylum Tardigrada. Zool J Linnean Soc 178(4):730–736

Baumann H (1961) Der Lebensablauf von *Hypsibius* (*H.*) *convergens* Urbanowicz (Tardigrada). Zool Anz 167:362–381

Benoit TG, Locke J, Marks JR, Beasley CW (2000) Laboratory transmission of *Xanthomonas campestris* pv. *raphani* by a tardigrade (Parachela: Macrobiotidae). Fla Entomol 83:197–199

Bertolani R (1982) Tardigradi. Guide per il riconoscimento delle specie animali delle acque interne Italiane. Consiglio Nazionale Delle Ricerche, Verona, Italy, 104 p [In Italian]

Bertolani R (2002) First report of tardigrades within hypogeal caves. Boll Accad Gioenia Sci Nat 35:649–654

Bertolani R, Biserov VI (1996) Leg and claw adaptations in soil tardigrades, with erection of two new genera of Eutardigrada, Macrobiotidae: *Pseudohexapodibius* and *Xerobiotus*. Invertebr Biol 115(4):299–304

Bertolani R, Rebecchi L (1996) The tardigrades of Emilia (Italy). II. Monte Rondinaio. A multihabitat study on a high altitude valley of the northern Apennines. Zool J Linnean Soc 116:3–12

Bertolani R, Rebecchi L, Giovannini I, Cesari M (2011) DNA barcoding and integrative taxonomy of *Macrobiotus hufelandi* C.A.S. Schultze 1834, the first tardigrade species to be described, and some related species. Zootaxa 2997:19–36

Bertolani R, Bartels PJ, Guidetti R, Cesari M, Nelson DR (2014a) Aquatic tardigrades in the Great Smoky Mountains National Park, North Carolina and Tennessee, U.S.A., with the description of a new species of *Thulinius* (Tardigrada, Isohypsibiidae). Zootaxa 3754(5):524–536

Bertolani R, Guidetti R, Marchioro T, Altiero T, Rebecchi L, Cesari M (2014b) Phylogeny of Eutardigrada: new molecular data and their morphological support lead to the identification of new evolutionary lineages. Mol Phylogenet Evol 76:110–126

Binda MG (1978) Risistemazione di alcuna tardigradi con l'istituzione di un nuovo genere di Oreellidae e della nuova famiglia Archechiniscidae. Animal 5:307–314

Boesgaard TM, Kristensen RM (2001) Tardigrades from Australian marine caves with a redescription of *Actinarctus neretinus* (Arthrotardigrada). Zool Anz 240:253–264

Bussau C (1992) New deep-sea Tardigrada (Arthrotardigrada, Halechiniscidae) from a manganese nodule area of the eastern South Pacific. Zool Scr 21:79–92

Ciobanu D, Roszkowska M, Moglan I, Kaczmarek Ł (2015) The first record of *Pyxidium tardigradum* Van der Land, 1964 (Ciliophona) in Romania. Zootaxa 3941:595–599

Colwell RK, Chao A, Gotell NJ, Lin S-Y, Mao CX, Chazdon RL, Longino JT (2012) Models and estimators linking individual-based and sample-based rarefaction, extrapolation, and comparison of assemblages. J Plant Ecol-UK 5:3–21

Commito JA, Tita G (2002) Differential dispersal rates in an intertidal meiofauna assemblage. J Exp Mar Biol Ecol 268:237–256

Coull BC (1990) Are members of the meiofauna food for higher trophic levels? T Am Microsc Soc 109:233–246

Coull BC (1999) Role of meiofauna in estuarine soft-bottom habitats. Aust J Ecol 24:327–343

Crisp DJ, Hobart J (1954) A note on the habitat of the marine tardigrade *Echiniscoides sigismundi* (Schultze). Ann Mag Nat Hist 7:554–560

Crisp M, Kristensen RM (1983) A new marine interstitial eutardigrade from east Greenland, with comments on habitat and biology. Vidensk Meddr Dansk Naturh Foren 144:99–114

da Rocha CM, Fonsêca-Genevois V, de Castro JV, Bezerra TNC, Venekey V, Bothelo AP (2004) Environmental interactions of *Batillipes pennaki* (Tardigrada, Heterotardigrada) in a tropical sandy beach (Itamaracá, Pernambuco, Brazil). Meiofauna Mar 13:79–86

Dabert M, Dastych H, Hohberg K, Dabert J (2013) Phylogenetic position of the enigmatic clawless eutardigrade genus *Apodibius* Dastych, 1983 (Tardigrada), based on 18S and 28S rRNA sequence data from its type species *A. confusus*. Mol Phylogenet Evol 70:70–75

Danovaro R, Scopa M, Gambi C, Frascheti S (2007) Trophic importance of subtidal metazoan meiofauna: evidence from in situ exclusion experiments on soft and rocky substrates. Mar Biol 152:339–350

Dastych H (1987) Altitudinal distribution of Tardigrada in Poland. In: Bertolani R (ed) Biology of tardigrades. Selected symposia and monographs, U.Z.I., Mucchi, Modena, Italy, pp 169–176

Dastych H (1988) The Tardigrada of Poland. Monografie Fauny Polski 16:1–255

Dastych H (2004a) *Hypsibius thaleri* sp. nov., a new species of a glacier-dwelling tardigrade from the Himalayas, Nepal (Tardigrada). Mitt Hamb Zool Mus Inst 101:169–183

Dastych H (2004b) Redescription of the glacier tardigrade *Hypsibius janetscheki* Ramazzotti, 1968 (Tardigrada) from the Nepal Himalayas. Entomol Mitt Zool Mus Hamburg 14:181–194

Dastych H (2011) *Bergtrollus dzimbowski* gen. n., sp. n., a remarkable new tardigrade genus and species from the nival zone of the Lyngen Alps, Norway (Tardigrada: Milnesiidae). Entomol Mitt Zool Mus Hamburg 15:335–359

Dastych H, Kraus HJ, Thaler K (2003) The description and notes on the biology of the glacier tardigrade *Hypsibius klebelsbergi* Michlčič, 1959 (Tardigrada) based on material from Ötztal Alps, Austria. Mitt Hamb Zool Mus Inst 100:73–100

de Zio S, Grimaldi P (1966) Ecological aspects of Tardigrada distribution in South Adriatic beaches. Veröff Inst Meeresforsch Bremerhav Sonderb 2:87–94

Degma P, Guidetti R (2007) Notes to the current checklist of Tardigrada. Zootaxa 1579:41–53

Degma P, Šimurka M, Gulánová S (2005) Community structure and ecological macrodistribution of moss-dwelling water bears (Tardigrada) in central European oak-hornbeam forests (sw Slovakia). Ekológia (Bratislava) 24(Suppl 2):59–75

Degma P, Bertolani R, Guidetti R (2009–2015) Actual checklist of Tardigrada species (2009–2015, Ver. 28: 31–03–2015). http://www.tardigrada.modena.unimo.it/miscellanea/Actual%20check list%20of%20Tardigrada.pdf. Accessed 20 Oct 2015

Degma P, Katina S, Sabatovičová L (2011) Horizontal distribution of moisture and Tardigrada in a single moss cushion. J Zool Syst Evol Res 49(Suppl 1):71–77

Dewel WC, Dewel RA (1987) Study of a moss community containing *Milnesium tardigradum* parasitized by a chytridiomycetous fungus. In: Bertolani R (ed) Biology of tardigrades. Selected symposia and monographs U.Z.I., 1. Mucchi, Modena, Italy, pp 45–56

Dewel RA, Joines J, Bond J (1985) A new chytridiomycete parasitizing the tardigrade *Milnesium tardigradum*. Can J Bot 63:1525–1534

Doncaster CC, Hooper DJ (1961) Nematodes attacked by protozoa and tardigrades. Nematologica 6:333–335

Elith J, Phillips SJ, Hastie T, Dud ík M, Chee YE, Yates CJ (2011) A statistical explanation of MaxEnt for ecologists. Divers Distrib 17:43–57

Faurby S, Jonsson KI, Rebecchi L, Funch P (2008) Variation in anhydrobiotic survival of two eutardigrade morphospecies: a story of cryptic species and their dispersal. J Zool 275:139–145

Faurby S, Jørgensen A, Kristensen RM, Funch P (2012) Distribution and speciation in marine intertidal tardigrades: testing the roles of climatic and geographical isolation. J Biogeogr 39:1596–1607

Findlay SEG (1981) Small-scale spatial distribution of meiofauna on a mud and sandflat. Estuar Coast Shelf Sci 12:471–484

Flø & Hågvar (2013) Aerial dispersal of invertebrates and mosses close to a receding alpine glacier in southern Norway. Arct Antarct Alp Res 45:481–490

Fontoura P, Bartels PJ, Jorgensen A, Kristensen RM, Hansen JG (2017) A dichotomous key to the genera of marine heterotardigrades. Zootaxa 4292:1–45

Fox I (1966) On the significance of tardigrades in the digestive tract of snails. Proceedings of the first international congress of parasitology, Rome, Italy, 21–26 Sept, 1964, vol 2, p 1064

Fox I, Garcia-Moll I (1962) *Echiniscus molluscorum*, new tardigrade from the feces of the land snail, *Bulimulus exilis* (Gmelin) in Puerto Rico (Tardigrada: Scutechiniscidae). J Parasitol 48:177–181

Franceschi T, Loi ML, Pierantoni R (1962–1963) Risultati di una prima indagine ecologica condotta su popolazioni di Tardigradi. Boll Mus Ist Biol Genova 32:69–93

Giere O (2009) Meiobenthology: the microscopic motile fauna of aquatic sediments, 2nd edn. Springer, London

Greven H, Schüttler L (2001) How to crawl and dehydrate on moss. Zool Anz 240:341–344

Greven H, Dastych H, Kraus HJ (2005) Notes on the integument of the glacier-swelling tardigrade *Hypsibius klebelsbergi* Mihelčič, 1959 (Tardigrada). Mitt Hamb Zool Mus Inst 102:11–20

Grimaldi de Zio S, Gallo D'Addabbo M (2001) Further data on the Mediterranean Sea tardigrade fauna. Zool Anz 240:345–360

Grimaldi de Zio S, De Lucia RM, Gallo MD, Grimaldi P (1983) Marine tardigrades ecology. Oebalia 9:15–31

Grimaldi de Zio S, Morone de Lucia MR, D'Addabbo Gallo M (1984) Relazione tra morfologia ed ecologia nei tardigradi marini (Heterotardigrada-Arthrotardigrada). Cah Biol Mar 25:67–73

Guidetti R, Bertolani R (2005) Tardigrade taxonomy: an updated check list of the taxa and a list of characters for their identification. Zootaxa 845:1–46

Guidetti R, Bertolani R, Nelson DR (1999) Ecological and faunistic studies on tardigrades in leaf litter of beech forests. Zool Anz 238:215–223

Guidetti R, Altiero T, Marchioro T, Amadè LS, Avdonina AM, Bertolani R, Rebecchi L (2012) Form and function of the feeding apparatus in Eutardigrada (Tardigrada). Zoomorphology 131:127–148

Guidetti R, Bertolani R, Rebecchi L (2013) Comparative analysis of the tardigrade feeding apparatus: adaptive convergence and evolutionary pattern of the piercing stylet system. J Limnol 72(s1):24–35. https://doi.org/10.4081/jlimnol.2013.e4

Guil N (2011) Molecular approach to micrometazoans. Are they here, there and everywhere? In: Fontaneto D (ed) Biogeography of microorganisms: Is everything small everywhere? Cambridge University Press, Cambridge, pp 284–306

Guil N, Cabrero-Sañudo F (2007) Analysis of the species description process of a little known invertebrate group: the limnoterrestrial tardigrades (Bilateria, Tardigrada). Biodivers Conserv 16:1063–1086

Guil N, Giribet G (2009) Fine scale population structure in the *Echiniscus blumi-canadensis* series (Heterotardigrada, Tardigrada) in an Iberian mountain range—When morphology fails to explain genetic structure. Mol Phylogenet Evol 1:606–613

Guil N, Sánchez-Moreno S (2013) Fine-scale patterns in micrometazoans: tardigrade diversity, community composition and trophic dynamics in leaf litter. Syst Biodivers 11:181–193

Guil N, Hortal J, Sánchez-Moreno S, Machordom A (2009a) Effects of macro and micro-environmental factors on the species richness of terrestrial tardigrade assemblages in an Iberian mountain environment. Landsc Ecol 24:375–390

Guil N, Sánchez-Moreno S, Machordom A (2009b) Local biodiversity patterns in micrometazoans: Are tardigrades everywhere? Syst Biodivers 7:259–268

Guil N, Machordom A, Guidetti R (2013) High level of phenotypic homoplasy amongst eutardigrades (Tardigrada) based on morphological and total evidence phylogenetic analyses. Zool J Linnean Soc 169:1–26

Guil N, Rodrigo E, Machordom A (2015) Soil tardigrade biodiversity with the description of a new eutardigrade genus and it phylogenetic position. Syst Biodivers 13:234–256. https://doi.org/10.1080/14772000.2014.986554

Hallas T (1977) Survey of the Tardigrada of Finland. Ann Zool Finnici 14:173–183

Hallas T, Yeates G (1972) Tardigrada of the soil and litter of a Danish beech forest. Pedobiologia 12:287–304

Hansen JG, Jørgensen A, Kristensen RM (2001) Preliminary studies of the tardigrade fauna of the Faroe Bank. Zool Anz 240:385–393

Hansen JG, Kristensen RM, Arbizu PM (2003) Deep-sea tardigrades from the Angola Basin. Abstract in: First international DIVA workshop, Wihlemshaven, June 5–6, 2003, p 23

Harada H, Ito M (2006) Soil-inhabiting tardigrade communities in forests of central Japan. Hydrobiologia 558:119–127

Hawes TC, Worland MR, Convey P, Bale JS (2007) Aerial dispersal of springtails on the Antarctic Peninsula: implications for local distribution and demography. Antarct Sci 19:3–10

Hohberg K (2006) Tardigrade species composition in young soils and some aspects on life history of *Macrobiotus richtersi* J. Murray, 1911. Pedobiologia 50:267–274

Hohberg K, Traunspurger W (2005) Predator-prey interaction in soil food web: functional response, size-dependent foraging efficiency, and the influence of soil texture. Biol Fertil Soils 41:419–427

Hohberg K, Russell DJ, Elmer M (2011) Mass occurrence of algal-feeding tardigrade *Apodibius confusus*, in the young soils of a post-mining site. J Zool Syst Evol Res 49(Suppl 1):62–65

Ito M (1999) Ecological distribution, abundance and habitat preference of terrestrial tardigrades in various forests on the northern slope of Mt. Fuji, Central Japan. Zool Anz 238:225–234

Ito M, Abe W (2001) Micro-distribution of soil inhabiting tardigrades (Tardigrada) in a sub-alpine coniferous forest of Japan. Zool Anz 240:403–407

Jennings PG (1976a) The Tardigrada of Signy Island, South Orkney Islands, with a note on the Rotifera. Br Antarct Surv Bull 44:1–25

Jennings PG (1976b) Tardigrada from the Antarctic Peninsula and Scotia Ridge Region. Br Antarct Surv Bull 44:75–95

Jönsson KI (2003) Population density and species composition of moss-living tardigrades in a boreo-nemoral forest. Ecography 26:356–364

Jönsson KI (2007) Long-term experimental manipulation of moisture conditions and its impact on moss-living tardigrades. J Limnol 66(Suppl 1):119–125

Jørgensen A, Kristensen RM (2001) A new tanarctid arthrotardigrade with buoyant bodies. Zool Anz 240:425–439

Jørgensen A, Møbjerg N (2014) Notes on the cryptobiotic capability of the marine arthrotardigrades *Styraconyx haploceros* (Halechiniscidae) and *Batillipes pennaki* (Batillipedidae) from the tidal zone in Roscoff, France. Mar Biol Res 11(2):214–217. https://doi.org/10.1080/17451000.2014.904883

Jørgensen A, Møbjerg N, Kristensen RM (2007) A molecular study of the tardigrade *Echiniscus testudo* (Echiniscidae) reveals low DNA sequence diversity over a large geographical area. J Limnol 66(Suppl 1):77–83

Jørgensen A, Boesgaard TM, Møbjerg N, Kristensen RM (2014) The tardigrade fauna of Australian marine caves with descriptions of nine new species of Arthrotardigrada. Zootaxa 3802(4):401–443

Kaczmarek Ł, Michalczyk Ł, Eggermont H (2008) *Dactylobiotus luci*, a new freshwater tardigrade (Eutardigrada: Macrobiotidae) from the Rwenzori Mountains, Uganda. Afr Zool 43:150–155

Kaczmarek Ł, Gołdyn B, Wełnicz W, Michalczyk Ł (2011) Ecological factors determining Tardigrada distribution in Costa Rica. J Zool Syst Evol Res 49(Suppl 1):78–83

Kaczmarek Ł, Michalczyk Ł, McInnes SJ (2014) Annotated zoogeography of non-marine Tardigrada. Part I: Central America. Zootaxa 3763(1):1–62

Kaczmarek Ł, Michalczyk Ł, McInnes SJ (2015a) Annotated zoogeography of non-marine Tardigrada. Part II: South America. Zootaxa 3923(1):1–107

Kaczmarek Ł, Bartels PJ, Roszkowska M, Nelson DR (2015b) The zoogeography of marine Tardigrada. Zootaxa 4037(1):1–189

Kaczmarek Ł, Jakubowska N, Celewicz-Gołdyn S, Zawierucha K (2015c) The microorganisms of cryoconite holes (algae, Archaea, bacteria, cyanobacteria, fungi, and Protista: a review. Polar Record. Cambridge University Press, 28 p. doi:https://doi.org/10.1017/S0032247415000637

Kathman RD, Cross SF (1991) Ecological distribution of moss-dwelling tardigrades on Vancouver Island, British Columbia, Canada. Can J Zool 69:122–129

Kathman RD, Nelson DR (1987) Population trends in the aquatic tardigrade *Pseudobiotus augusti* (Murray). In: Bertolani R (ed) Biology of tardigrades. Selected symposia and monographs U.Z.I., 1. Mucchi, Modena, Italy, pp 155–168

Kinchin IM (1994) The biology of tardigrades. Portland Press, London, 186 p
Krantz S, Benoit TG, Beasley CW (1999) Phytopathogenic bacteria associated with Tardigrada. Zool Anz 138:259–260
Kristensen RM (1977) On the marine genus *Styraconyx* (Tardigrada, Heterotardigrada, Halechiniscidae) with description of a new species from a warm spring on Disko Island, West Greenland. Astarte 10:87–91
Kristensen RM (1982) New aberrant eutardigrades from homothermic springs on Disko Island, West Greenland. In: Nelson DR (ed) Proceedings of the third international symposium on the Tardigrada. East Tennessee State University Press, Johnson City, TN, pp 203–220
Kristensen RM (1984) On the biology of *Wingstrandarctus corallinus* nov. gen. et spec., with notes on symbiontic [*sic*] bacteria in the subfamily Florarctinae (Arthrotardigrada). Vidensk Meddr Dansk Naturh Foren 145:201–218
Kristensen RM, Hallas TE (1980) The tidal genus *Echiniscoides* and its variability, with erection of Echiniscoididae fam. n. Zool Scripta 9:113–127
Kristensen RN, Hansen J (2005) Tardigrada (Water Bears). In: Rhode K (ed) Marine parasitology, Ch 5, Minor groups and fossils. CSIRO Publishing, Australia, pp 230–235
Kristensen RM, Higgins RP (1984) Revision of *Styraconyx* (Tardigrada: Halechiniscidae) with description of two new species for Disko Bay, West Greenland. Smithsonian Contrib Zool 391:1–40
Kristensen RM, Renaud-Mornant J (1983) Existence d'arthrotardigrades semi-benthiques de genres nouveaux de la sous-famille des Styraconyxinae subfam. nov Cah. Biol Mar 24:337–353
Kristensen RM, Sørensen MV (2004) Phylum: Tardigrada (Water Bears). In: Grzimek's animal life encyclopedia, vol 2: protostomes. Thompson Gale, Detroit, pp 115–123
Lindgren EW (1971) Psammolittoral marine tardigrades from North Carolina and their conformity to worldwide zonation patterns. Cah Biol Mar 12:481–496
Marshall WA, Convey P (1997) Dispersal of moss propagules on Signy Island, maritime Antarctic. Polar Biol 18:373–383
Martinez EA (1975) Marine meiofauna of a New York City beach with particular reference to Tardigrada. Estuar Coast Mar Sci 3:337–348
McInnes S (1995) Tardigrades from Signy Island, South Orkney Islands, with particular reference to freshwater species. J Nat Hist 29:1419–1445
McInnes S (2003) A predatory fungus (Hyphomycetes: *Lecophagus*) attacking Rotifera and Tardigrada in maritime Antarctic lakes. Polar Biol 26:79–82
McInnes S, Convey P (2005) Tardigrade fauna of the South Sandwich Islands, maritime Antarctic. Zootaxa 1058:43–49
McInnes S, Pugh P (1998) Biogeography of limnoterrestrial Tardigrada, with particular reference to the Antarctic fauna. J Biogeogr 25:31–36
McInnes S, Pugh P (1999) Zonation in Antarctic lake-dwelling benthic meiofauna, with emphasis on the Tardigrada. Zool Anz 238:283–288
McInnes S, Pugh P (2007) An attempt to revisit the global biogeography of limnoterrestrial Tardigrada. J Limnol 66(Suppl 1):90–96
Meyer H (2006) Small-scale spatial distribution variability in terrestrial tardigrade populations. Hydrobiologia 558:133–139
Meyer H (2013) Terrestrial and freshwater Tardigrada of the Americas. Zootaxa 3747(1):1–71
Michalczyk Ł, Welnicz W, Frohme M, Kaczmarek Ł (2012a) Redescriptions of three *Milnesium* Doyère, 1840 taxa (Tardigrada: Eutardigrada: Milnesiidae), including the nominal species for the genus. Zootaxa 3154:1–20
Michalczyk Ł, Wełnicz W, Frohme M, Kaczmarek Ł (2012b) Corrigenda of Zootaxa, 3154: 1–20 Redescriptions of three *Milnesium* Doyère, 1840 taxa (Tardigrada: Eutardigrada: Milnesiidae), including the nominal species for the genus. Zootaxa 3393:66–68
Møbjerg N, Dahl C (1996) Studies on the morphology and ultrastructure of the Malpighian tubules of *Halobiotus crispae* Kristensen, 1982 (Eutardigrada). Zool J Linnean Soc 116:85–99

Møbjerg N, Halberg KA, Jørgensen A, Persson D, Bjørn M, Ramløv H, Kristensen RM (2011) Survival in extreme environments – on the current knowledge of adaptations in tardigrades. Acta Physiol 202:409–420. https://doi.org/10.1111/j.1748-1716.2011.02252.x

Morgan C (1977) Population dynamics of two species of Tardigrada, *Macrobiotus hufelandi* (Schultze) and *Echiniscus* (*Echiniscus*) *testudo* (Doyère), in roof moss from Swansea. J Anim Ecol 46:263–279

Nelson DR (1975) Ecological distribution of Tardigrada on Roan Mountain, Tennessee–North Carolina. Mem Ist Ital Idrobiol 32(Suppl):225–276

Nelson DR (1982) Developmental biology of the Tardigrada. In: Harrison RW, Cowden R (eds) Developmental biology of freshwater invertebrates. Alan R. Liss, Inc, New York, NY, pp 363–398

Nelson DR, Adkins RG (2001) Distribution of tardigrades within a moss cushion: Do tardigrades migrate in response to changing moisture conditions? Zool Anz 240:493–500

Nelson DR, Bartels PJ (2013) Species richness of soil and leaf litter tardigrades in the Great Smoky Mountains National Park (North Carolina/Tennessee, USA). J Limnol 72(s1):144–151

Nelson DR, McGlothlin KM (1996) A new species of *Calohypsibius* (Phylum Tardigrada, Eutardigrada) from Roan Mountain, Tennessee–North Carolina, U.S.A. Zool J Linnean Soc 116:167–174

Nelson DR, McInnes SJ (2002) Tardigrada. In: Rundle SD, Robertson AL, Schmid-Araya JM (eds) Freshwater meiofauna: biology and ecology. Backhuys Publishers, Leiden, The Netherlands, pp 177–215

Nelson DR, Kincer CJ, Williams TC (1987) Effects of habitat disturbances on aquatic tardigrade populations. In: Bertolani R (ed) Biology of tardigrades. Selected symposia and monographs U. Z.I., 1. Mucchi, Modena, Italy, pp 141–153

Nelson DR, Guidetti R, Rebecchi L (2010) Chapter 14: Tardigrada. In: Thorp JH, Covich AP (eds) Ecology and classification of North American freshwater invertebrates. Elsevier, London, pp 455–484

Nelson DR, Guidetti R, Rebecchi L (2015) Chapter 17: Phylum Tardigrada. In: Thorp JH, Rogers DC (eds) Thorp & Covich's freshwater invertebrates, Vol 1: Ecology and general biology. Elsevier, London, pp 347–380

Nelson DR, Guidetti R, Rebecchi L (2016) Chapter 15: Tardigrada. Keys to nearctica fauna. In: Thorp JH, Rogers DC (eds) Thorp and Covich's freshwater invertebrates, vol II. Elsevier, London

Pearce DA, Bridge PD, Hughes KA, Sattler B, Psenner R, Russel NJ (2009) Microorganisms in the atmosphere over Antarctica. FEMS Microbiol Ecol 69:143–157

Peluffo JR, Rocha AM, Peluffo MCM (2007) Species diversity and morphometrics of tardigrades in a medium-sized city in the Neotropical Region: Santa Rosa (La Pampa, Argentina). Anim Biodivers Conserv 30:43–51

Petersen B (1951) The tardigrade fauna of Greenland. A faunistic study with some few ecological remarks. Meddr Grønland, København 150(5):1–94

Phillips SJ, Dudík M (2008) Modeling of species distributions with MaxEnt: new extensions and a comprehensive evaluation. Ecography 31:161–175

Pilato G, Binda MG (2001) Biogeography and limno-terrestrial tardigrades: are they truly incompatible binomials? Zool Anz 240:511–516

Pilato G, Binda MC (2010) Definition of families, subfamilies, genera and subgenera of the Eutardigrada, and keys to their identification. Zootaxa 2404:1–54

Pilato G, Guidetti R, Rebecchi L, Lisi O, Hansen J, Bertolani R (2006) Geonemy, ecology, reproductive biology and morphology of the tardigrade *Hypsibius zetlandicus* (Eutardigrada: Hypsibiidae) with erection of *Borealibius* gen. n. Polar Biol 29:595–603

Platt HM (1981) Chapter 16: Meiofaunal dynamics and the origin of the metazoa. In: Foreu PL (ed) The evolving biosphere. British Museum (Natural History) and Cambridge University Press, Cambridge, pp 207–216

Pollock LW (1970) Distribution and dynamics of interstitial Tardigrada at Woods Hole, Massachusetts, U SA. Ophelia 7:145–165

Pollock LW (1995) New marine tardigrades from Hawaiian beach sand and phylogeny of the family Halechiniscidae. Invertebr Biol 114:220–235

Ramazzotti G, Maucci W (1983) Il Phylum Tardigrada. III edizione riveduta e aggiornata. Mem Ist Ital Idrobiol 4l:1–1012 [An English translation, edited by Dr. Clark Beasley (deceased), may be obtained from Dr. Diane R. Nelson, East Tennessee State University, TN, USA]

Rebecchi L, Boschini D, Cesari M, Lencioni V, Bertolani R, Guidetti R (2009) Stress response of a boreo-alpine species of tardigrade, *Borealibius zetlandicus* (Eutardigrada, Hypsibiidae). J Limnol 68(1):64–70

Renaud-Mornant J (1974) Un nouvelle famille de Tardigrades marins abyssaux: les Cornarctidae fam. nov. (Heterotardigrada). C R Acad Sci Paris 278:3087–3090

Renaud-Mornant J (1988) Chapter 33: Tardigrada. In: Higgins RP, Thiel H (eds) Introduction to the study of meiofauna. Smithsonian Institution Press, Washington, DC, pp 357–364

Richters R (1904) Arktische Tardigraden. In: Romer F, Schaudinn F (eds) Arktische Tardigraden. Fauna Arctica 3(3):495–508

Rost-Roszkowska MM, Poprawa I, Kaczmarek Ł (2013) Autophagy as the cell survival in response to a microsporidian infection of the midgut epithelium of *Isohypsibius granulifer granulifer* (Eutardigrada: Hypsibiidae). Acta Zool-Stockholm 94(3):273–279

Roszkowska M, Ostrawska M, Kaczmarek Ł (2015) The genus *Milnesium* Doyère 1840 (Tardigrada) in South America with the descriptions of two new species from Argentina and discussion of the feeding behaviour in the family Milnesiidae. Zool Stud 54:12. https://doi.org/10.1186/s40555-014-0082-7

Rubal M, Veiga P, Fontoura P, Santos E, Sousa-Pinto I (2016) Biodiversity of marine tardigrades from the northern coast of Portugal (Iberian Peninsula). Zool J Linnean Soc 178(4):747–754

Saikawa M, Oyama M (1992) Electron microscopy on infection of tardigrades by *Ballocephala sphaerospora* and *B. verrucospora* conidia. T Mycol Soc Japan 33:305–311

Saikawa M, Oyama M, Yamaguchi Y (1991) *Harposporium anguillulae* and *Haptoglossa intermedia* parasitizing tardigrades. T Mycol Soc Japan 32:501–508

Sanchez-Moreno S, Ferris H, Guil N (2008) Role of tardigrades in the suppressive service of a soil food web. Agric Ecosyst Environ 124:187–192

Schill RO, Jönsson KI, Pfannkuchen M, Brümmer F (2011) Food of tardigrades: a case study to understand food choice, intake and digestion. J Zool Syst Evol Res 49(Suppl 1):66–70

Schmid-Araya JM, Schmid PE (2000) Trophic relationships: integrating meiofauna into a realistic benthic food web. Freshw Biol 44:149–163

Schmid-Araya JM, Hildrew AG, Robertson A, Schmid PE, Winterbottom J (2002) The importance of meiofauna in food webs: evidence from an acid stream. Ecology 83(5):1271–1285

Schuster R, Greven H (2007) A long-term study of population dynamics of tardigrades in the moss *Rhytidiadelphus squarrosus* (Hedw.) Warnst. J Limnol 66(Suppl 1):141–151

Schuster R, Greven H (2013) Reproductive traits of *Macrobiotus hufelandi* during a long-term field study with notes on *Paramacrobiotus richtersi* and *Diphascon pingue* (Eutardigrada). J Limnol 72:166–174

Schuster RO, Toftner EC, Grigarick AA (1978) Tardigrada of Pope Beach, Lake Tahoe, California. Wasmann J Biol 35:115–136

Shanks AL, Edmonson EW (1990) The vertical flux of metazoans (holoplankton, meiofauna, and larval invertebrates) due to their association with marine snow. Limnol Oceanogr 35:455–463

Shanks AL, Walters K (1997) Holoplankton, meroplankton, and meiofauna associated with marine snow. Mar Ecol Prog Ser 156:75–86

Sobczyk M, Michno K, Kosztyła P, Stec D, Michalczyk Ł (2015) Tolerance to ammonia of *Thulinius ruffoi* (Bertolani 1981), a tardigrade isolated from a sewage treatment plant. Bull Environ Contam Toxicol 95:721–727. https://doi.org/10.1007/s00128-015-1593-6

Statzner B, Higler B (1986) Stream hydraulics as a major determinant of benthic invertebrate zonation patterns. Freshw Biol 16:127–139

Steiner WA (1994) The influence of air pollution on moss-dwelling animals. 4. Seasonal and long-term fluctuations on rotifer, nematode and tardigrade populations. Rev Suisse Zool 101 (4):1017–1031

Strayer D, Findlay S (2010) Ecology of freshwater shore zones. Aquat Sci 72:127–163

Strayer D, Nelson DR, O'Donnell EB (1994) Tardigrades from shallow groundwaters in south-eastern New York, with the first record of *Thulinia* from North America. T Am Microsc Soc 113:325–332

Tenlen JR, McCaskill S, Goldstein B (2013) RNA interference can be used to disrupt gene function in tardigrades. Dev Genes Evol 223:171–181

Thulin G (1942) Ein neur mariner Tardigrad. Medd Göteborgs Mus Zool Afd 99:1–10

Tsujimoto M, McInnes S, Convey P, Imura S (2014) Preliminary description of tardigrade species diversity and distribution pattern around coastal Syowa Station and inland Sør Rondane Mountains, Dronning Maud Land, East Antarctica. Polar Biol 37:1361–1367

Utsugi K (2001) Tardigrades living in the sewage treatment plants. Zool Sci 18(Suppl):46 (abstract only)

van der Land J (1975) The parasitic marine tardigrade *Tetrakentron synaptae*. Mem Ist Ital Idrobiola 32(Suppl):413–423

Van Rompu EA, De Smet WH (1991) Contribution to the fresh water Tardigrada from Barentsøya, Svalbard (78 degrees 30 minutes North). Fauna Norvegica, Series A 12:29–39

Vecchi M, Vicente F, Guidetti R, Bertolani R, Rebecchi L, Cesari M (2016) Interspecific relationships of tardigrades with bacteria, fungi and protozoans, with a focus on the phylogenetic position of (Ciliophora). Zool J Linnean Soc 178(4):846–855

Verçosa MM, Dos Santos ECL, De Souza JRB, da Rocha CMC (2009) Distribuição de *Stygarctus bradypus* Schulz, 1951 (Tardigrada, Heterotardigrada) na margem sul da Ilha de Itamaracá (PE, Brasil). Atlantica (Rio Grande) 31(1):25–33

Vicente F, Bertolani R (2013) Considerations on the taxonomy of the Phylum Tardigrada. Zootaxa 3626:245–248

Vicente F, Michalczyk Ł, Kaczmarek Ł, Boavida MJ (2008) Observations on *Pyxidium tardigradum* (Ciliophora), a protozoan living on Eutardigrada: infestation, morphology and feeding behavior. Parasitol Res 103:1323–1331

Vicente F, Cesari M, Serrano A, Bertolani R (2013) The impact of fire on terrestrial tardigrade biodiversity: a first case-study from Portugal. J Limnol 72(s1):152–159

Villora-Morena S (1996) A new genus and species of the deep-sea family Coronarctidae (Tardigrada) from a submarine cave with a deep-sea like condition. Sarsia 81:275–283

Węglarska B (1970) *Hypsibius (Isohypsibius) smreczynskii* spec. nov., a new species of freshwater tardigrade. Zeszyty Nauk Uniw Jagiellonsk. Prace Zool 16:107–114

Wright J (1991) The significance of four xeric parameters in the ecology of terrestrial Tardigrada. J Zool 224:59–77

Zawierucha K, Grzelak K, Kotwicki L, Michalczyk Ł, Kaczmarek Ł (2013) *Batillipes pennaki* Marcus, 1946, a new addition to the Thai tardigrade fauna, with an overview of literature on the species. Pakistan J Zool 45(3):801–808

Zawierucha K, Kolicka M, Takeuchi N, Kaczmarek Ł (2015) What animals can live on cryoconite holes? A faunal review. J Zool 295:119–169

Chapter 8
Reproduction, Development and Life Cycles

Tiziana Altiero, Atsushi C. Suzuki, and Lorena Rebecchi

Abstract In tardigrades reproduction occurs only through eggs, fertilized or unfertilized, and therefore only through gametes. Tardigrades exploit several reproductive modes, amphimixis, self-fertilization and thelytokous parthenogenesis (both apomixis and automixis). These modes are often in close relationship with the colonized environment. As regards sexuality, tardigrades can be gonochoristic (bisexual or unisexual) or hermaphroditic. The anatomy of the reproductive apparatus of males, females and hermaphrodites and the maturative patterns of male and female germinal elements are presented and discussed, as well as the ultrastructure of spermatozoa and eggs, including their phylogenetic implications. In addition, mating and fertilization patterns, embryonic and post-embryonic development, sexual dimorphism and parental care are considered and discussed. Finally, vegetative reproduction does not occur in tardigrades, and their capability to regenerate is limited to a physiological tissue restoration of a few cells.

8.1 Introduction

8.1.1 Sexuality and Reproductive Modes

Tardigrades are unable to propagate via vegetative reproduction. Their reproduction occurs only by means of gametes, in particular through fertilized or unfertilized eggs. Heterochromosomes have never been identified in tardigrades. Both amphimixis due to cross- or self-fertilization and parthenogenesis take place in tardigrades, often in

T. Altiero
Department of Education and Humanities, University of Modena and Reggio Emilia, Reggio Emilia, Italy

A. C. Suzuki
Department of Biology, University of Keio, Yokohama, Japan

L. Rebecchi (✉)
Department of Life Sciences, University of Modena and Reggio Emilia, Modena, Italy
e-mail: lorena.rebecchi@unimore.it

R. O. Schill (ed.), *Water Bears: The Biology of Tardigrades*, Zoological Monographs 2,
https://doi.org/10.1007/978-3-319-95702-9_8

close relationship with the environment colonized (Bertolani et al. 1990; Bertolani 2001).

Tardigrades probably originated in the sea and marine species (in practice all of them heterotardigrades) are gonochoristic (bisexual), with a sex ratio close to 1:1. Only a hermaphroditic species is known in marine environments. Therefore, in marine habitats there is only amphimixis, and parthenogenesis does not occur. The successive colonization of freshwater and terrestrial habitats leaded tardigrades to change their reproductive modes in order to adapt to less stable environments, with respect to the marine one. Limnic and terrestrial species (both eutardigrades and heterotardigrades) are mainly unisexual (i.e. composed of only females) and reproduce via parthenogenesis that can be automictic or apomictic and often associated with polyploidy. Limnic and terrestrial species can be also bisexual with a sex ratio close to 1:1 and characterized by amphimixis. In addition, even hermaphroditism occurs. It is the less frequent sexual condition in tardigrades and has been found in several families both in limnic and terrestrial eutardigrades (Bertolani 2001).

8.2 Anatomy of the Reproductive Apparatus

An unpaired gonad (ovary, testis or ovotestis), one or two gonoducts (oviduct or vasa deferents, respectively), a gonopore or a cloaca and, eventually, spermatheca/seminal receptacle and seminal vesicles form the reproductive apparatus of tardigrades (Fig. 8.1). The gonad is always a dorso-caudal sack, dorsally overlying to the midgut and anteriorly suspended at the cuticle by one or two ligaments. In heterotardigrades a basal membrane limits the gonad, whereas in eutardigrades a discontinuous epithelium with bundles of contractile-like filaments has been described. Other evident anatomical differences in the reproductive apparatus between the two classes of tardigrades will be described in the next paragraphs.

8.2.1 Female Reproductive Apparatus

The single ovary present in tardigrade females varies in size with animal age and reproductive stage. For instance, the ovary (Figs. 8.1 and 8.2) size increases considerably when oocytes increase in their volume as a consequence of vitellogenesis. In some species the ovary can be extended from the pharynx up to the posterior end of the animal. Anteriorly, it is attached to the dorsal body wall through a single and median ligament in heterotardigrades and two ligaments in eutardigrades (Fig. 8.1a). The oviduct originates from the posterior part of the ovary, extends beyond it and leads directly outside through the gonopore (heterotardigrades) or leads into the rectum and successively in the cloaca (eutardigrades) (Bertolani 1983a). Six cuticular plates surround the female gonopore of heterotardigrades, making it like a

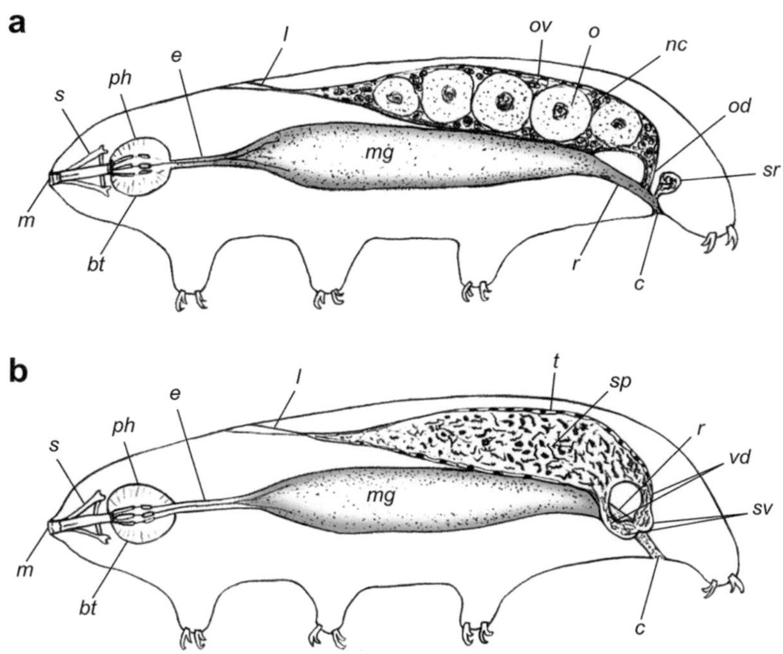

Fig. 8.1 Reproductive and digestive apparatuses in female (**a**) and male (**b**) eutardigrades. Abbreviations: bt = Buccal tube; m = mouth; s = stylet; ph = pharynx; e = esophagus; mg = midgut; r = rectum; c = cloaca; l = dorsal ligaments of the gonad; ov = ovary; o = oocyte; nc = nurse cell; ov = oviduct; sr = spermatheca; t = testis; sp = sperms; vd = vas deferents; sv = seminal vesicles

rosette with six petals (Fig. 8.3a). Gonopore is located anteriorly to the anus between the third and the hind legs.

In heterotardigrades, females of the most part of marine species and those of the terrestrial family Oreellidae have two external cuticular seminal receptacles located ventrolaterally. Each receptacle has a long, slightly curved genital duct that opens at a distance from the gonopore (Kristensen 1984; Jørgensen et al. 1999). These organs, with their content of spermatozoa, are abandoned with the exuvia during the moulting process. The new cuticular receptacles, without spermatozoa, appear as thin cuticular organs in newly moulted females of the marine species *Actinarctus doryphorus* and again filled of spermatozoa when the ovary contains one mature oocyte (Jørgensen et al. 1999). Cuticular seminal receptacles have never been observed in the intertidal family Batillipedidae; an internal epithelial spermatheca has been found only in *Batillipes pennaki* (see Grimaldi de Zio and D'Addabbo Gallo 1975). An internal epithelial spermatheca (Fig. 8.2c) that opens into the rectum was observed also in a few terrestrial eutardigrade species of the genus *Macrobiotus*, *Ramazzottius* and *Xerobiotus* (see Bertolani 1983a; Rebecchi and Bertolani 1988; Rebecchi 1997).

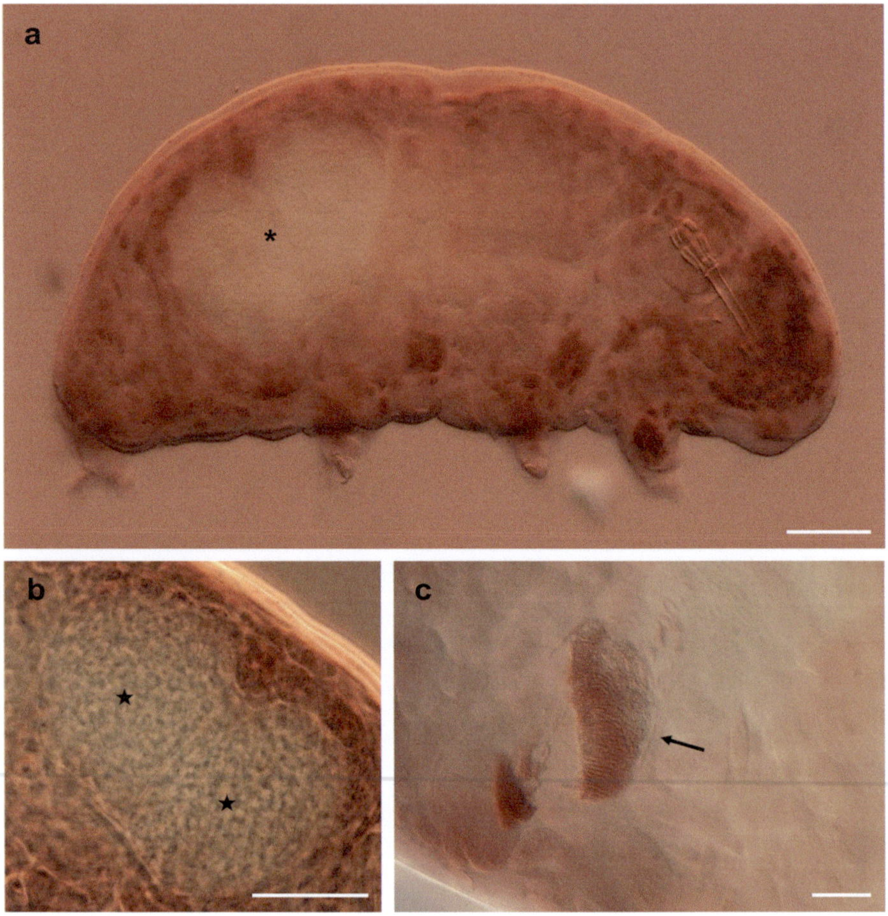

Fig. 8.2 Female reproductive apparatus of the eutardigrade *Macrobiotus* sp. (**a**) *In toto* female with two oocytes into the ovary (asterisk); (**b**) ovary with two oocytes (stars); (**c**) spermatheca (arrow). (**a**, **c**) Orcein staining and differential interference contrast (DIC), (**b**) orcein and phase contrast. Scale bars = 10 μm

8.2.2 Male Reproductive Apparatus

The testis of marine heterotardigrades is often small and triangle-shaped with an anteriorly oriented apex (e.g. *Opydorscus*, *Wingstrandarctus*, *Tanarctus*, *Florarctus*; see Rebecchi et al. 2000a). In parachelan eutardigrades, the sac-shaped testis can change in size, similarly to the ovary, depending to the developmental stage of the male germinal cells (Figs. 8.1 and 8.4). The only exception is represented by apochelan eutardigrades (*Milnesium* species), which always have a small testis composed by two fusiform portions united medially to form an inverted V (Rebecchi and Nelson 1988; Suzuki 2006). A basal membrane surrounds the male

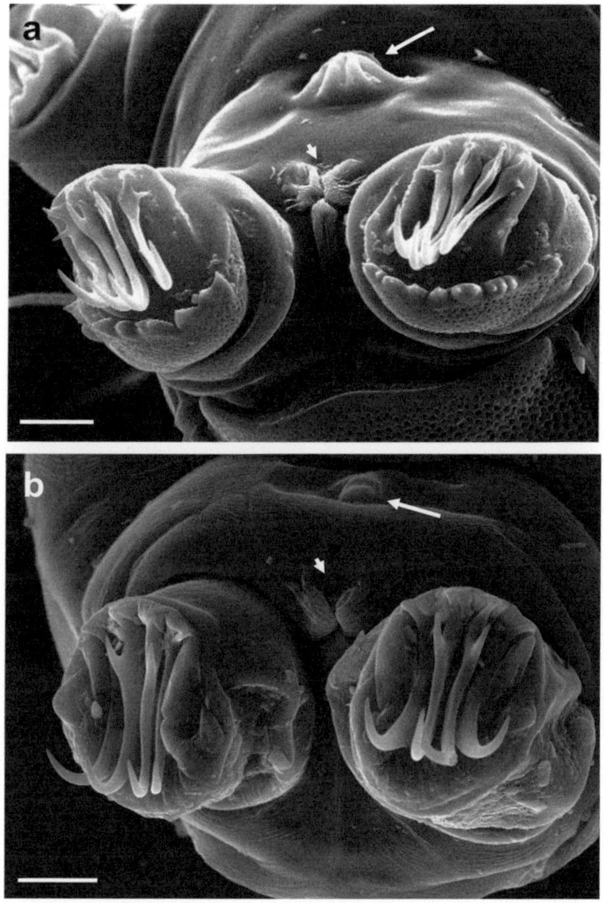

Fig. 8.3 Sexual dimorphism in the heterotardigrade *Echiniscus* sp. (**a**) Female with anus (arrow head) and rosette-shape gonopore (arrow); (**b**) male with anus (arrow head) and tubular-shaped gonopore (arrow). (**a**, **b**) Scanning electron microscopy. Scale bars = 10 μm

reproductive apparatus of the heterotardigrades (e.g. *Batillipes*, *Tetrakentron*, *Halechiniscus*; Bertolani 1983b; Rebecchi et al. 2000a), whereas in the terrestrial eutardigrade *Milnesium*, a discontinuous squamous epithelium showing cells with bundles of contractile-like filaments forms the wall of the testis (Dewel et al. 1993).

The two very thin deferents turn ventrally around the midgut, join medially and open into the rectum and successively in the cloaca in eutardigrades. In heterotardigrades the two deferents join to form a short common duct that opens into the gonopore. Differently from the female gonopore, the male gonopore of the heterotardigrades is closer to the anus and generally tubular (Fig. 8.3b). It can show some species-specific peculiar morphologies. For instance, a round or oval-shaped opening, sometimes covered anteriorly by a cuticular fold, is present in *Zioella*, while in other genera, the male gonopore is located on a small ovoid papilla (Bertolani 1983b; Rebecchi et al. 2000a). In some species of both heterotardigrades and eutardigrades, males have two small internal seminal vesicles. In eutardigrades, the seminal vesicles are due to a small swelling of the distal part of each deferent

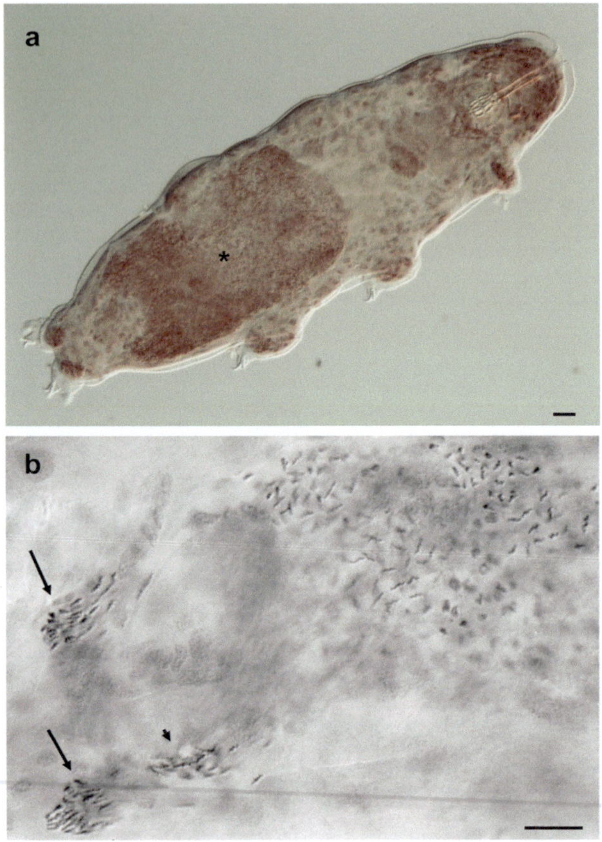

Fig. 8.4 Male reproductive apparatus of eutardigrades. (**a**) In toto male of *Macrobiotus* gr. *hufelandi* with the large sac-shape testis (asterisk) containing germinal cells; (**b**) testis of *Hypsibius* cf. *dujardini* with one of the two deferents (arrow head) and seminal vesicles full of spermatozoa (arrows). (**a**, **b**) Orcein staining and DIC. Scale bars = 10 μm

such as in *Diphascon* and *Platicrista*, whereas in the heterotardigrades of the genera *Wingstrandarctus*, *Styraconyx* and *Florarctus*, they look as two caudal lateral bulges of the testis. In male eutardigrades, spermatozoa within the spermiducts and seminal receptacles are motile and clumped in bundle (Rebecchi et al. 2000a).

8.2.3 Hermaphroditic Reproductive Apparatus

In hermaphrodite eutardigrades, a single sac-like ovotestis develops dorsally and posteriorly, overlying the midgut. The ovotestis is followed by a short gonoduct, and it is built of a single layer of cells. Within the ovotestis no septum or other kinds of morphological barriers separating clusters of female and male germinal cells have been observed (Fig. 8.5). In *Macrobiotus joannae* and *Bertolanius weglarkae*, the ovotestis varies in size in relationship to the body size, the degree of maturation of the germinal elements and the predominant type (female or male) of gamete. The anterior part of the ovotestis ends with two blind appendices connected with two

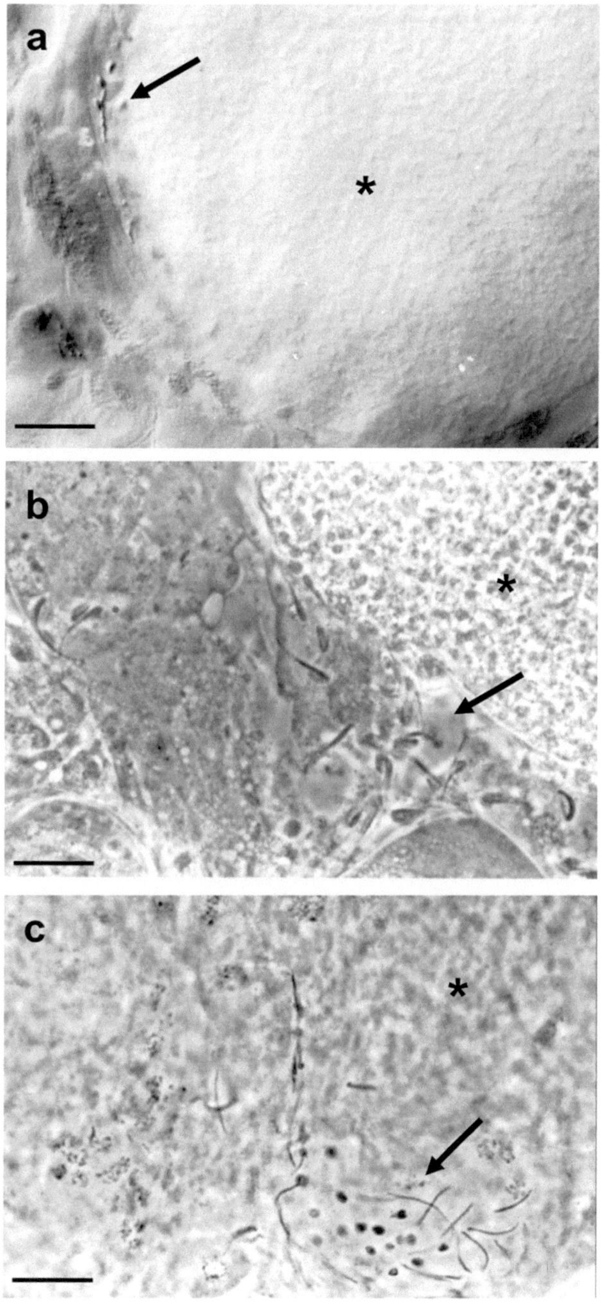

Fig. 8.5 Ovotestis of hermaphroditic eutardigrades with an oocyte (asterisk) at the end of the vitellogenesis and male germinal cells (arrow) at different stage of development around it; (**a, b**) *Macrobiotus joannae*; (**c**) *Isohypsibius monoicus.* (a) Orcein and DIC; (**b**) In vivo and phase contrast; (**c**) orcein and phase contrast. Scale bars = 10 μm

ligaments (protrusions of the gonad wall cells), through which the ovotestis is anchored to the dorsal cuticle (Węglarska 1987; Rebecchi et al. 2000b; Poprawa et al. 2015a). In *M. joannae*, a caudal sack-like diverticulum of the ovotestis containing a large number of spermatozoa and spermatids was evidenced (Rebecchi et al. 2000b). In *Isohypsibius granulifer granulifer*, the apical part of the gonad, germarium, is filled with oogonia; the central part, vitellarium, contains developing oocytes and nurse cells; the last part of the gonad is filled with spermatogonia (Węglarska 1987).

The ovotestis of *Orzeliscus* cf. *belopus* (the only hermaphroditic marine heterotardigrade) is a sac-like structure without septum to separate male and female germ cells, but fully formed and mature male cells are located only at the periphery in both anterior and posterior regions of the gonad. In addition, a single gonoduct and a pair of seminal receptacles are present laterally and posteriorly to the gonopore. The gonopore has the shape of a rosette-like structure (Suzuki and Kristensen 2014).

8.2.4 Sexual Dimorphism

Secondary sex characters have been observed in many heterotardigrade species and a few eutardigrade ones. In both classes, males are generally smaller than females, even though the body size cannot represent a useful criterion to distinguish sex in tardigrades, as mature males may be larger than immature females (Bertolani 1992; Rebecchi and Nelson 1988). Other than the already mentioned sexual dimorphism in the shape of the heterotardigrade gonopore in heterotardigrades (Fig. 8.3), two kinds of dwarf males have been evidenced in the marine heterotardigrade *Tetrakentron synaptae*: stationary pygmy males having a flattened body and vagile males with longer secondary clavae and unflattened body (Kristensen 1980). Small and large males have been described in another marine heterotardigrade *Tholoarctus natans*; the small ones have no secondary or accessory spurs on the external digits, and the large ones have spurs on the external claws (Kristensen and Renaud-Mornant 1983). Additional sexual dimorphism may be exhibited by some heterotardigrades in the size of the cephalic appendages, especially the clavae, which are often much longer and larger in males than in females, and may play a role in mating (Bertolani 1992).

In a few eutardigrade species, mature males have a basal hook-shaped modified claw with a small spur on the front legs; the appearance of modified claws occurs only with the moulting preceding the mating since they are probably used in attaching the females during copulation (Bertolani 1992; Rebecchi and Nelson 1988; Suzuki 2008). Classical example are represented by the secondary branch of the inner claw of the limnic *Pseudobiotus* species and primary and secondary branch of the front legs of *Milnesium* and *Milnesioides* species (Fig. 8.6). Adult females of the genus *Pseudobiotus* moult producing modified claws on the hind legs before laying eggs; these claws are shorter than normal ones in order do not to destroy or damage eggs laid within the exuvia (Bertolani 1992). A lateral conical papilla is

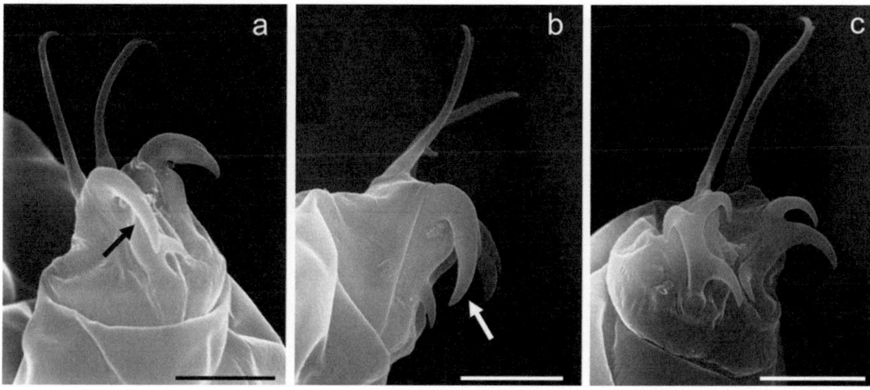

Fig. 8.6 Sexual dimorphism in the claw of *Milnesium* sp. from Yokohama, Japan. (**a**, **b**) First claw pair of a mature male with hook (arrow); (**c**) first claw pair of a female. Scanning electron microscopy. Scale bars = 10 μm

present on the hind legs of males of *Ramazzottius oberhaeuseri* and, rarely, in *Macrobiotus* species (Bertolani 1992). Since some of these sex characters in eutardigrades are related to reproductive cycle, they cannot be used to distinguish sex features and to evaluate sex ratio.

8.3 Maturative Patterns of the Gonad

8.3.1 Maturative Pattern of the Ovary

In tardigrades, oocyte maturation usually begins after the second moulting, rarely even after the first one. Asynchronous maturation of oocytes occurs in most marine heterotardigrades, which lay a single egg at a time with the most mature oocyte located at the posterior part of the ovary (Rebecchi et al. 2000a). Only females of *Batillipes* spp. had showed a different kind of oocyte maturation. In practice, a synchronous maturation of oocytes (up to 8) occurs in spring-summer, while in autumn-winter oocytes mature asynchronously (Pollock 1970). In non-marine heterotardigrades and in limnic and terrestrial eutardigrades, oocytes development is synchronous, and groups of 2 up to more than 40 eggs are laid, as evidenced in the terrestrial eutardigrade *Paramacrobiotus richtersi* (currently attributable to *Paramacrobiotus fairbanksi*; Altiero et al. 2006, 2015; Guidetti et al. 2019).

Maturation of oocytes and their oviposition occur synchronously with moulting and several times throughout adult female life of both non-marine heterotardigrades (echiniscids) and eutardigrades, which are iteroparous. In particular, four maturation stages of the ovary, corresponding to the oogenesis phases, have been identified in eutardigrades: pre-vitellogenesis, early vitellogenesis, late vitellogenesis and mature oocyte (Węglarska 1975, 1979; Rebecchi and Bertolani 1994; Poprawa 2005a; Hyra et al. 2016; Poprawa et al. 2015a, b, c). Both in young and adult females right after the oviposition, the small ovary contains undifferentiated cells in pre-vitellogenesis, very similar to each other. Within an ovary at the early vitellogenesis, two kinds of cells are present: oocytes at the beginning of the auxocytosis and nurse cells. The first ones are distinguishable by cytoplasm rich in vacuoles and nucleus with uncoiled chromatin and a large nucleolus. Females in *simplex* stage (i.e. without the sclerified parts of the buccal-pharyngeal apparatus) have gonads at the middle-late vitellogenesis, which have grown in size as a consequence of the synthesis/accumulation of yolk in oocytes. At this stage, within oocytes a large nucleolus is visible, and chromosomes are poorly distinguishable. Moreover, nurse cells with a large nucleus are recognizable among the oocytes. Finally, at the fourth maturative stage, the ovary extends for nearly all the animal's length leaving little space to the midgut and contains mature oocytes, filled with spheres of yolk. At this stage, first metaphase chromosomes can be visible in oocytes.

8.3.2 Maturative Pattern of the Testis

The maturation pattern of the testis was poorly analysed, especially that of marine species. The testis of newborns is undifferentiated, and its maturation may occur already after the first moulting. The only exception known has been described in the marine ectoparasite heterotardigrade *Tetrakentron synaptae*, in which dwarf males showed mature germinal cells within the gonad since birth (Kristensen 1980). The arrangement of male germinal elements in *Batillipes* and *Wingstrandarctus* is zonal and uniform, respectively (Kristensen 1979, 1984), while in *Actinarctus* spermatids surround a nurse cell, and males are probably iteroparous (Jørgensen et al. 1999).

In males, conversely to the egg maturation in females, the spermatozoan maturation does not seem correlated with moulting. Males can be iteroparous or semelparous, and their testis displays two pattern of spatial arrangement of the developing germinal elements, i.e. zonal and uniform, and three temporal maturation patterns have been described (Rebecchi and Bertolani 1994).

Firstly, a continuous pattern in which spermatogenesis occurs continuously in time and mature spermatozoa, spermatids and spermatogonia are always present in the testis (e.g. *Paramacrobiotus*, *Minibiotus*, *Xerobiotus*, *Ramazzottius*, *Diphascon*, *Platicrista*).

Secondly, a progressive zonal maturation pattern with gonad completely filled with only spermatozoa, as it takes place in semelparous males of the eutardigrade *Pseudobiotus* and in marine heterotardigrades.

Thirdly, a cyclical maturation pattern in which the testis contains either exclusively spermatozoa or immature germinal cells as observed in the eutardigrade *Bertolanius*; these males are therefore iteroparous as females.

8.3.3 Maturative Patterns of the Ovotestis

Hermaphroditic tardigrades are iteroparous and often protandrous. In the eutardigrades *Macrobiotus joannae* and *Bertolanius weglarskae*, the maturative pattern of the ovotestis can be subdivided into four stages. The first stage includes three phases (undifferentiated, male and pre-vitellogenic phases), whereas stages 2–4 correspond to the steps of the vitellogenesis during which groups of 3–8 oocytes mature in strict relationship to the moulting cycle. Apart from the first two phases of stage 1, all other stages simultaneously possess male and female germinal elements. The gonad looks as an exclusively "male gonad" in relatively small specimens, indicating that the male phase is present only in the first reproductive cycle (Rebecchi et al. 2000b). Only in the limnic species *Isohypsibius monoicus*, the ovotestis shows an alternation of predominantly male and predominantly female germ cell maturation indicating a cyclic maturation (Węglarska 1987). The contemporary presence of mature oocytes and spermatozoa in close contact with each other evidences that tardigrades are simultaneous hermaphrodites and that self-fertilization can occur even though it may lead to loss of genetic variability. Although the marine *O.* cf. *belopus* is considered a simultaneous hermaphrodite with both mature eggs and a large number of mature male gametes in the ovotestis, this species should have cross-fertilization elucidated by the presence of spermatozoa in the seminal receptacles, with the modified sperm shape implying the additional maturation process after ejaculation (Suzuki and Kristensen 2014).

In general, hermaphroditic individuals have a higher energy cost than gonochoristic species, being energy allocated for both male and female gametes. However, self-fertilizing hermaphroditism has the advantage to allow the colonization of a new area by a single specimen (Bertolani 2001).

8.4 Gametogenesis and Gamete Morphology

8.4.1 Oogenesis

The ultrastructural process of the oogenesis in tardigrades is still understudied. After Węglarska's studies (1979, 1982, 1987, 1989) on the details of oogenesis in the eutardigrades *Paramacrobiotus* cf. *richtersi* and *Isohypsibius granulifer*, we have waited for about 25 years before to know more about oogenesis of other eutardigrades species (Suzuki 2006; Poprawa 2005a; Poprawa et al. 2015a, b, c; Hyra et al. 2016). A meroistic oogenesis, with nurse cells (trophocytes) and follicular cells, has been evidenced in tardigrades. Stable cytoplasmic bridges connect cystocytes (both trophocytes and the oocyte) each other forming a cluster. During pre-vitellogenesis, all cystocytes of a cluster synthesize and accumulate macromolecules and organelles, which will eventually transfer by cytoplasmic bridges to the only cell that will differentiate into an oocyte. A mixed vitellogenesis takes place in

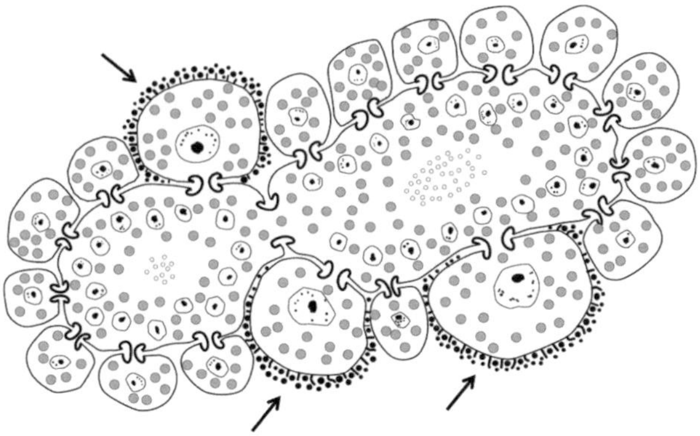

Fig. 8.7 Representation of *Milnesium* sp. ovary showing the relationship of the germ cell cluster. Yolk granules (grey circle) are formed in every cell, but only large oocytes (arrows) are growing to be eggs with chorion

tardigrades: a part of the yolk is produced by the different kind of cells within the gonad (both ovary and ovotestis), and another part is produced by cells/organs located outside of the gonad. In particular, within the gonad yolk material is produced by oocytes (autosynthesis) and nurse cells. Outside of the gonad, the yolk is synthesized by storage cells and cells of the midgut epithelium; this yolk is absorbed by oocytes via micropinocytosis (Węglarska 1979; Poprawa 2005a, Hyra et al. 2016; Rost-Roszkowska et al. 2011, Poprawa et al. 2015a, b, c). A different pattern was observed in the apochelan *Milnesium* sp. (formerly shown as *M. tardigradum*) in which four large multinuclear cells, connected each other through cytoplasmic bridges, are surrounded by many mononuclear cells (oocytes; Fig. 8.7). Multinuclear cells and most of mononuclear ones play a role in vitellogenesis (nurse cells), whereas only some of the mononuclear cells develop in eggs (Suzuki 2006).

8.4.2 Eggs Shell Morphology

Laid eggs of tardigrades are often spherical or oval in shape, usually between 50 and 100 μm in diameter and sometimes even up to 235 μm when the size of shell processes are considered (Fig. 8.8). Those of the marine heterotardigrades are surrounded by a sticky smooth shell and are laid freely. The eggs of non-marine heterotardigrades and those of eutardigrades are surrounded by a rigid and sclerified shell that can be smooth when laid in the exuvia (both in eutardigrades and heterotardigrades except for *Oreella*) or ornamented when laid freely (eutardigrades). In particular, in some genera of limnic and terrestrial eutardigrades (e.g. *Macrobiotus*, *Paramacrobiotus*, *Murrayon*, *Minibiotus*, *Dactylobiotus*, *Acutuncus*, *Bertolanius*, *Eohypsibius*, *Ramazzottius*, *Hebesuncus* and *Fractonotus*), as well as in the heterotardigrade *Oreella*, the egg shell shows species-specific ornamentation with a variety of shape processes, reticulation and pits (Fig. 8.8a–e). These morphological traits have an essential taxonomical significance, especially in the genera of eutardigrades in which animals are morphologically indistinguishable

Fig. 8.8 Morphology of the egg shell of eutardigrade species. (**a**, **b**) *Macrobiotus macrocalix*; (**c**, **d**) *Paramacrobiotus* cf. *richtersi*; (**e**) *Ramazzottius oberhaeuseri*; (**f**) exuvium of *Hypsibius dujardini* containing eggs with smooth shell. (**a–e**) Scanning electron micrographs; (**f**) orcein staining and DIC. Scale bars: (**b**, **d**, **e**) = 5 μm; (**a**, **c**) = 10 μm; (**f**) = 50 μm

to each other (Bertolani and Rebecchi 1993; Bertolani et al. 1996). So far, the egg shell of tardigrades does not exhibit a mycropilar opening. In some eutardigrade species, egg envelope formation starts during middle or late vitellogenesis. The choriogenesis occurs with synthesis and secretion activity of the oocyte and somatic cells of the ovary wall. Afterwards, the oocyte produces precursors of the vitelline envelope which are released in the space between oolemma and chorion. At the end of the oogenesis, the egg capsule, with a protective function, is completely formed, and it is composed of two envelopes: a very thin vitelline envelope, which overlaps the oolemma, and a thick egg shell (chorion). The vitelline envelope consists of polysaccharides, proteins and lipids. The chorion is three-layered: a thin inner layer, the middle labyrinthine layer and an outer layer (Węglarska 1975, 1982; Poprawa 2005b, 2011; Rost-Roszkowska et al. 2011).

8.4.3 Spermatogenesis

Spermatogenesis usually begins after the first or second moulting of the animal. Spermatogonia and spermatocytes can be distinguished each other by the larger size or by the typical star-shaped metaphases of the latter (Bertolani 1971; Altiero and Rebecchi 2003). Data on spermiogenesis at ultrastructural level in tardigrades are still very limited and often too anecdotal to allow reasonable conclusion (Wolburg-Buchholz and Greven 1979; Kristensen 1979; Rebecchi 1997; Jørgensen et al. 1999; Greven and Kristensen 2001). During the spermatogenesis, a spermatogonium undergoes to mitosis leading to two primary spermatocytes, which after two subsequent meiotic divisions give rise to a multinucleate cell. Thereafter, this multinucleate cell turns into a cluster of eight spermatids joined together by cytoplasmic bridges. These cytoplasmic bridges have been evidenced both in eutardigrades and heterotardigrade and both in males and in hermaphrodites. In the eutardigrade *Xerobiotus pseudohufelandi*, spermatids are sphere-shaped (Fig. 8.10a) and have a large nucleus, and their chromatin appears condensed. Afterwards, spermatids elongate, the acrosome originates from the Golgi apparatus, the amount of cytoplasm decreases, and the nucleus shape changes from round/comma to helicoidal. Lastly, spermatids differentiate into spermatozoa: the nucleus moves from the centre of the spermatid cytoplasm outwards and coils; the flagellum develops from the unique centriole present; the number of mitochondria decreases until only two (Rebecchi 1997). In marine species of the genera *Echiniscoides* and *Batillipes*, early spermatids contain two unmodified mitochondria closely attached and an evident acrosome located close to the nucleus. Instead, in *Actinarctus*, five spermatid stages are recognized, and the spermatid stage 5 is only found in the posterior part of the testis (i.e. in the ventral seminal vesicles) and in the gonoducts (Jørgensen et al. 1999).

8.4.4 Ultrastructure of the Spermatozoa

The spermatozoa of tardigrades are always flagellate with a classical axoneme ultrastructure (9 + 2; Figs. 8.9, 8.10, 8.11 and 8.12). A noteworthy morphological heterogeneity in the spermatozoa of eutardigrades has been evidenced with several different structural features that may be associated with colonized habitat or with phylogeny (Figs. 8.9 and 8.10c–e). Nevertheless, the spermatozoa of eutardigrades are thin and filiform with a length that comprises between 25 and 100 μm, according

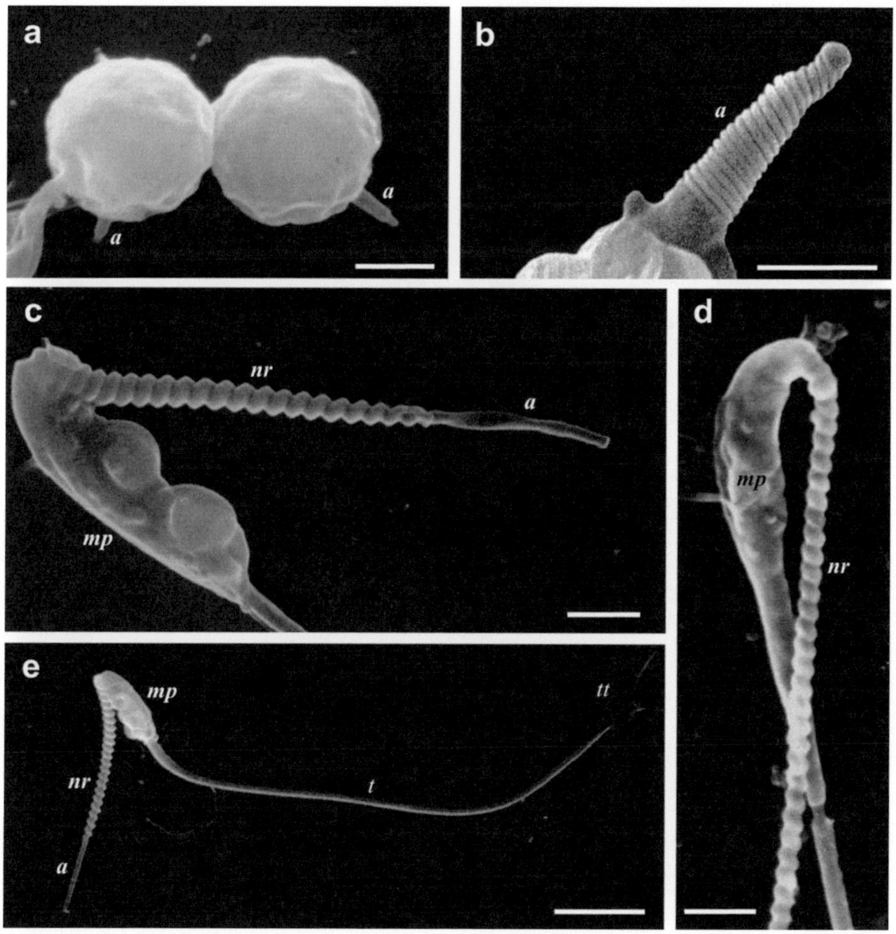

Fig. 8.9 Male germinal cells of eutardigrade species. (**a**) Roundish spermatids of *Diphascon nobilei*; (**b**) acrosome of the spermatozoon of *Bertolanius volubilis*; (**c**) head (acrosome and nuclear region) and middle piece of the male gamete within the testis of *Xerobiotus pseudohufelandi*; (**d**) nuclear region and midpiece of the male gamete of *Mesobiotus harmsworthi*; (**e**) in toto spermatozoon of *Macrobiotus macrocalix*. (**a–e**) Scanning electron microscopy. Scale bars: (**a–d**) = 1 μm; (**e**) = 5 μm. Abbreviations: a = Acrosome; nr = nuclear region; mp = middle piece; t = tail; tt = terminal tuft

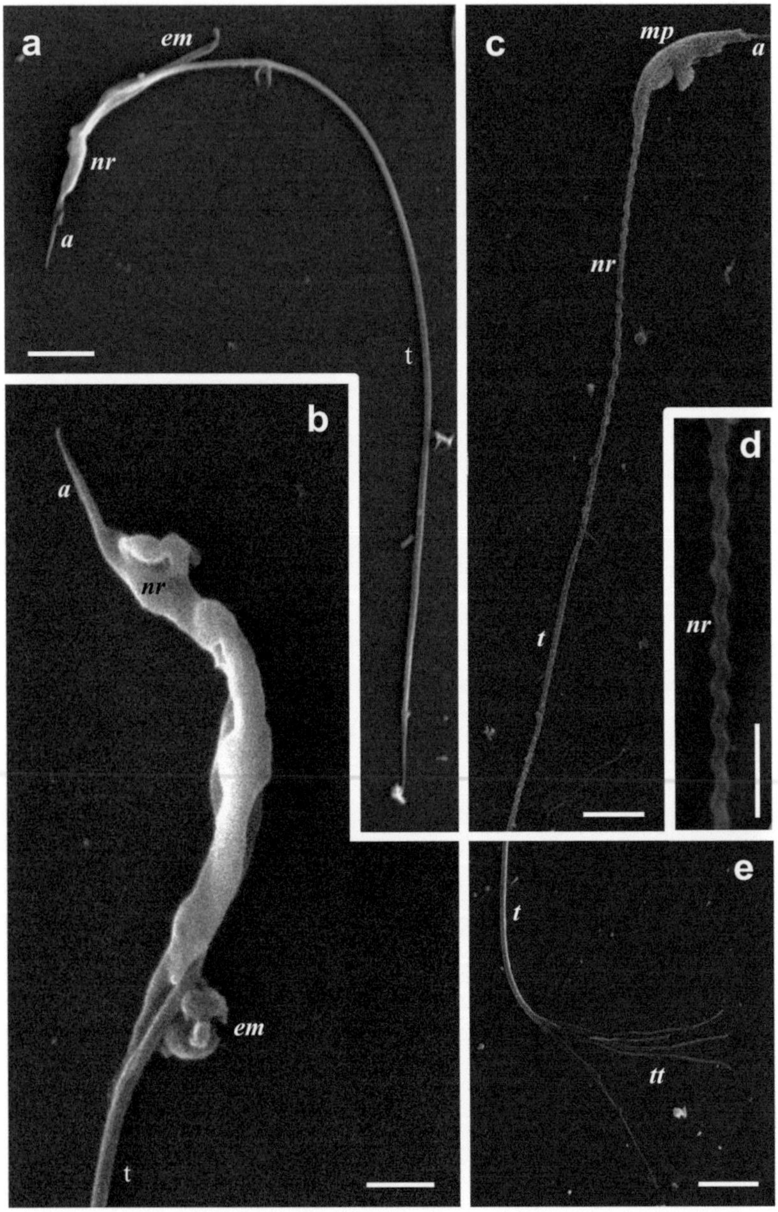

Fig. 8.10 Ultrastructure of tardigrade spermatozoa. (**a**) *In toto* sperm cell of the moss-dwelling heterotardigrade *Echiniscus duboisi*. (**b**) Head of the sperm cell of *Echiniscus duboisi*; (**c**) middle piece and nuclear region of the male gamete of the moss-dwelling eutardigrade *Minibiotus furcatus*; (**d**) detail of the nuclear region of the male gamete of *Minibiotus furcatus;* (**e**) terminal tuft of the tail of the male gamete of *Minibiotus furcatus.* (**a–e**) Scanning electron microscopy. Scale bars: (**a**, **c**, **e**) = 2 μm (**b**, **d**) = 1 μm. Abbreviations: a = Acrosome; em = external mitochondria; nr = nuclear region; mp = middle piece; t = tail; tt = terminal tuft

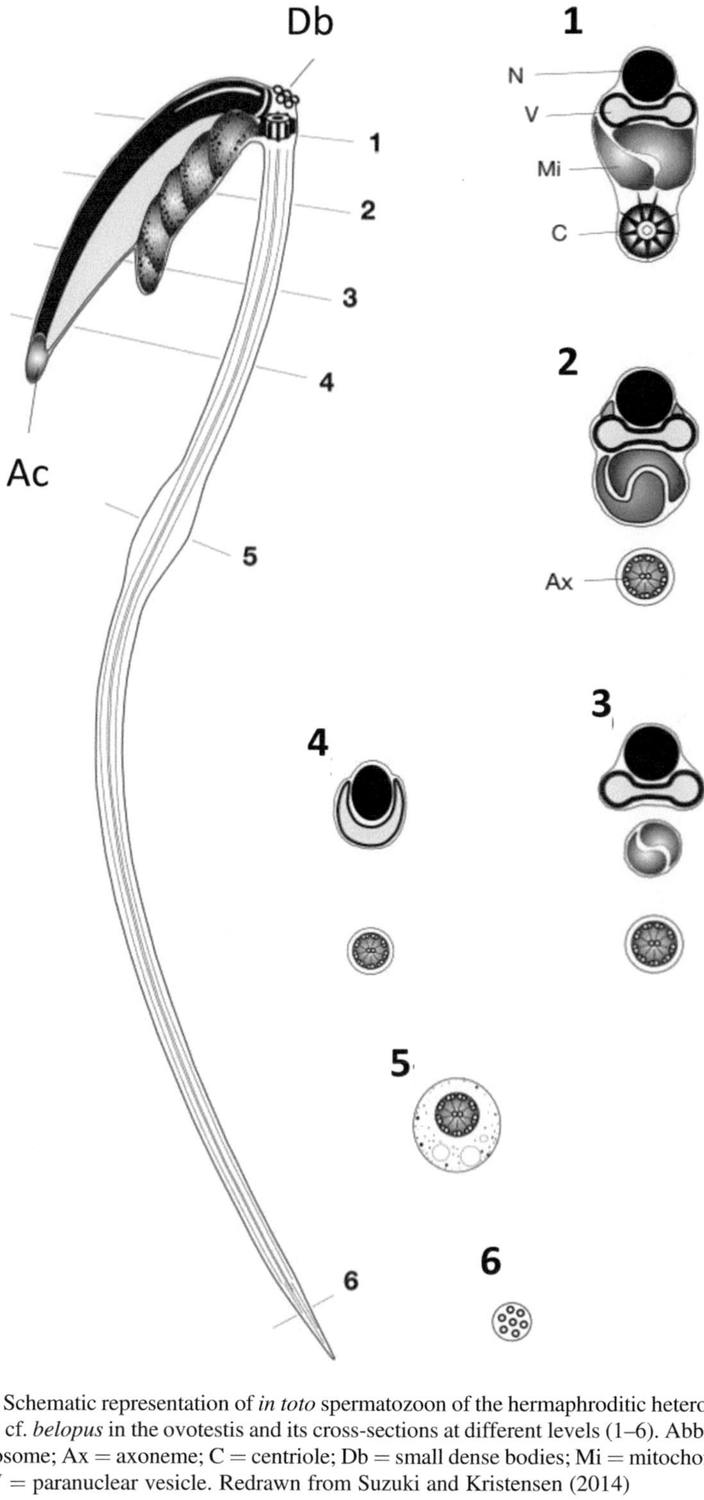

Fig. 8.11 Schematic representation of *in toto* spermatozoon of the hermaphroditic heterotardigrade *Orzeliscus* cf. *belopus* in the ovotestis and its cross-sections at different levels (1–6). Abbreviations: Ac = Acrosome; Ax = axoneme; C = centriole; Db = small dense bodies; Mi = mitochondria; N = nucleus; V = paranuclear vesicle. Redrawn from Suzuki and Kristensen (2014)

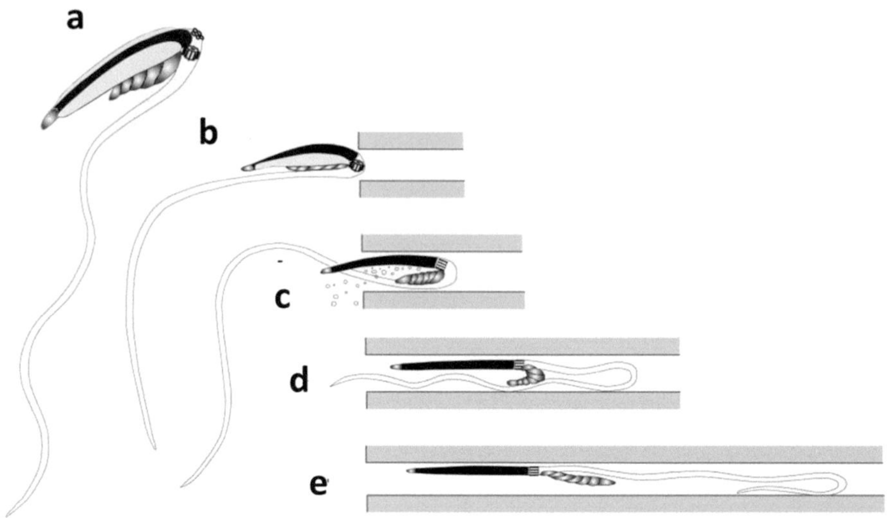

Fig. 8.12 (**a**) Schematic representation of the spermatozoon of *Orzeliscus* cf. *belopus* in front of the gonopore; (**b–e**) scenario of the morphological modification of the male gamete in the duct of the female receptacle. Redrawn from Suzuki and Kristensen (2014)

to the species. They are made up by the following regions: an elongated coiled head in which acrosome and nucleus can be easily recognized, a neck or an elongated midpiece (sometimes absent) and a flagellum. The acrosomes can be (1) conical and corkscrew-shaped, with tightly coiled spires (e.g. *Bertolanius*, *Adropion*); (2) -rod-shaped, with smooth surface, and relatively short (e.g. *Macrobiotus*, *Xerobiotus*, *Ramazzottius*); (3) cylindrical and very thin, with smooth surface and longer than the nucleus (e.g. *Paramacrobiotus*, *Mesobiotus*); and (4) very short and comma-shaped (e.g. *Pseudobiotus*, *Isohypsibius*, *Doryphoribius*, *Platicrista*, *Dipascon*, *Isohypsibius*). The acrosome is always bilayered and fits on the anterior part of the nucleus like a cup; in *Bertolanius* species the acrosome shows a narrow longitudinal canal and caudal small vesicles (Rebecchi and Guidi 1995; Rebecchi et al. 2000a).

In all eutardigrade species, the spermatozoon nucleus is always made up by condensed and electron-dense chromatin. In *Bertolanius*, the nucleus is cylindrical and surrounded by cytoplasm organized into ovoidal elements delimited by own cytomembrane. In all other eutardigrade species, the nucleus is always coiled even though with differences in the coil morphology. For instance, in *Platicrista* and *Adropion*, the corkscrew-shaped nucleus is made up by 3–4 coils (Rebecchi et al. 2000a). In *Macrobiotus*, *Xerobiotus* and *Ramazzottius* species, the nucleus exhibit tightly coils increasing in diameter caudally, and it is surrounded by a thin layer of cytoplasm, while in *Paramacrobiotus* the nucleus is longer than that of the previous species, but it is weakly coiled and contains fibrils running parallel to the nucleus (Rebecchi et al. 2011). These fibres could have an important role in facilitating the movement of the spermatozoa. In *Doryphoribius* the drill-shaped long nucleus has a constant diameter pitch (Baccetti 1987; Rebecchi 2001), while in the filiform and thin nucleus of *Pseudobiotus*, the helical shape is due to a thin helical ridge wrapped around a central axial rod (Rebecchi and Bertolani 1999). In *Isohypsibius*, a "condensed body" with unknown function surrounds on one side the whole length of the helicoidal drill-shaped nucleus from the end of the acrosome to the beginning of the tail (Wolburg-Buchholz and Greven 1979; Rebecchi et al. 2000a).

A short- and cup-shaped neck with large mitochondria around the only centriole, as in *Bertolanius*, *Platicrista* and *Adropion*, is located between nuclear region and tail (Rebecchi and Guidi 1995; Guidi and Rebecchi 1996; Rebecchi et al. 2000a). The male gametes of *Macrobiotus*, *Xerobiotus*, *Ramazzottius* and *Paramacrobiotus* have a rod-shaped elongated midpiece in which ovoidal elements surround a mitochondrial sleeve which envelope the only centriole and the first part of the axoneme (Rebecchi 1997, 2001; Rebecchi and Guidi 1995; Rebecchi and Bertolani 1999; Rebecchi et al. 2000a, 2011). The midpiece (and mitochondria) is absent in limnic species with filiform sperm cells of the genera *Pseudobiotus*, *Doriphoribius* and *Isohypsibius* (Rebecchi and Bertolani 1999; Rebecchi et al. 2000a, 2011; Rebecchi 2001).

The spermatozoan tail has a constant diameter and splits terminally into a tuft of 9–11 fine elements (Rebecchi et al. 2000a). Two exceptions are known. In *Paramacrobiotus* the proximal part of the axoneme is surrounded by nine outer accessory fibres (Rebecchi et al. 2011), while in *Doriphoribius* the tail is flanked for the most part of its length by an undulating membrane that is internal supported by a palisade of microtubules (Baccetti 1987; Rebecchi et al. 2000a). In eutardigrades, the spermatozoa within the gonoducts and seminal vesicles are always motile, clumped in bundles and oriented with the head towards the cloaca. In particular, the testicular spermatozoa of *Macrobiotus*, *Paramacrobiotus*, *Xerobiotus* and *Mesobiotus* are folded back between head and midpiece resembling a nutcracker.

Marine and non-marine heterotardigrade spermatozoa, considered of primitive type, are shorter (14–30 μm in length) than those of eutardigrades. They have a globose or slightly elongated head without coils, "free" mitochondria and a tapering tail without tuft and do not have a clear midpiece (Figs. 8.10a, b, 8.11, 8.12). The spermatozoa of terrestrial Echiniscidae (e.g. *Pseudechiniscus*, *Novechiniscus*, *Echiniscus*, *Antechiniscus*, *Proechiniscus*) are made up by (1) a cylindrical acrosome with smooth surface containing an electron-dense core surrounded by cytomembranes, (2) a cylindrical nucleus with electron-dense chromatin surrounded by a thin layer of cytoplasm, (3) an electron-dense sickle-shaped band which represents the beginning of the centriole located within a tapering tail and (4) a voluminous vesicle, located laterally to the centriole and representing the beginning of a long free mitochondria (Rebecchi 2001; Rebecchi et al. 2003a). The spermatozoa of the marine heterotardigrade *Echiniscoides* have an elongated acrosome, an osmiophilic cylindrical nucleus and two mitochondria shifted parallel to the flagellum and then considered "free" (Greven and Kristensen 2001). A similar pattern was evidenced in the marine *Batillipes* species, even though the acrosome is considered aberrant and the nucleus shows a "snout" of condensed chromatin in contact with three lamellated rings (Kristensen 1979; Rebecchi et al. 2000a). Within the ovotestis, the spermatozoa (Fig. 8.11) of the hermaphrodite species *O.* cf. *belopus* show acrosome, head, centriole and flagellum; the head, incorporating a cylindrical nucleus, a paranuclear vesicle and two mitochondria, is attached at the angle to the centriole forming a half-headed arrow shape (Suzuki and Kristensen 2014).

8.4.4.1 Storage of Spermatozoa Within the Female Reproductive Apparatus

The spermatozoa stored in spermatheca of the female eutardigrades or in the seminal receptacles of female heterotardigrades are immobile and appear to have a different shape with respect to those within the testis. The spermatozoa of *Xerobiotus pseudohufelandi* within the spermatheca are straight, have a midpiece thin and without hemispherical swelling, and the tail is reduced to a short stub without terminal tuft (Rebecchi 1997).

In marine heterotardigrades, cuticular and external seminal receptacles and their content in spermatozoa are renewed at each moulting (Kristensen 1984; Jørgensen et al. 1999). In *Actinarctus*, male gametes undergo strong postcopulatory modification as in the seminal receptacle, the acrosome is oriented straight forward, the nucleus becomes rod-shaped and the vesicle in the head region disappear; in addition spermatozoa are immotile and imbedded in an osmiophilous secretion (Jørgensen et al. 1999). In *Orzeliscus* cf. *belopus*, the spermatozoa are located outside the epicuticle of the seminal receptacle and are modified after their discharge since within the duct the cell was no longer bent backwards and the paranuclear vesicle was not visible (Fig. 8.12, Suzuki and Kristensen 2014).

8.5 Mating and Fertilization

Information about mating and fertilization in tardigrades are limited. In the marine and gonochoristic heterotardigrade *Batillipes noerrevangi*, the male stimulates the female with his sense organs, and the female lays her eggs on a sand grain (Kristensen 1979). This species has external fertilization which occurs at the time of oviposition as in *Actinarctus* (Jørgensen et al. 1999). In the limnic eutardigrade *Pseudobiotus megalonyx*, one or more males copulate a single female. In particular, up to nine males of *P. megalonyx* clinch a female with their modified internal claws on the front legs and near-mature oocytes in the ovary not yet completely released by the female in the old cuticle (von Erlanger 1895; von Wenck 1914). According to von Erlanger (1895), the male gametes are inserted in the cloaca of old cuticle, while according to Henneke (1911), the males penetrated the old cuticle with their buccal stylets and then introduce sperms into the old cuticle, and from there spermatozoa rapidly reach the female cloaca. In any case, insemination in *P. megalonyx* takes place in a protected environment and fertilization is internal (Rebecchi et al. 2000a). In another bisexual eutardigrade, *Isohypsibius dastychi*, mating includes a mutual stimulation between male and female which precedes sperm ejaculation and egg oviposition (Bingemer et al. 2016). For instance, the male holds the female with his first pair of legs, and the female stimulates the male by moving the stylets and contracting the sucking pharynx. Spermatozoa are ejaculated from the male seminal vesicles through the cloaca. During the third moulting, the female develops eggs, and after the cuticle and stylets have been rebuilt, the gravid female remains inside the exuvia, and she is ready for mating; if mating does not occur, egg absorption is observed (Bingemer et al. 2016). Lastly, copulation has not yet been observed in *Milnesium*, even though a sexual behaviour was observed, in which a rare male in thelytokous population approached a female, who had already oviposited in the old cuticle before ecdysis, and the male walked around the female for several minutes, pushing her with his mouth (Suzuki 2008).

As regards the hermaphroditic tardigrades, self-fertilization was demonstrated in the limnic eutardigrade *Isohypsibius monoicus* in which individuals kept isolated from birth to death had been able to reproduce (Altiero and Rebecchi 2001). Marine hermaphrodite *Orzeliscus* cf. *belopus* should have copulation and cross-fertilization as described above, even though it is unclear how spermatozoa could have been transferred to the seminal receptacles (Suzuki and Kristensen 2014).

External fertilization can be presumed for all marine and terrestrial heterotardigrade species on the basis of sperm morphology. In addition, the presence of two external seminal receptacles on the ventral cuticle of the females and far from the female gonopore represents another evidence of external fertilization in heterotardigrades. The more specialized morphology of eutardigrade spermatozoa and the presence of an internal epithelial spermatheca connected to the rectum may represent adaptations to internal fertilization. In line with internal fertilization, the contemporary presence in eutardigrade spermatozoa of a helicoidal head and a tufted tail may help stabilize and direct motion in non-canalized tracts (Rebecchi et al. 2000a). In *Paramacrobiotus* species, the fibrils running parallel to the nucleus could have an important role in facilitating the movement of the spermatozoa. Moreover, other structures, as the accessory fibres, may represent additional motor elements, allowing the acquisition of the internal fertilization (Rebecchi et al. 2011).

8.6 Parthenogenesis

Parthenogenesis is unknown in tardigrades colonizing marine habitats, whereas it is very frequent in tardigrades colonizing limno-terrestrial habitats. Among terrestrial heterotardigrades, many species of the genus *Echiniscus* reproduce via parthenogenesis and among eutardigrades the limnic species of the family Murrayidae, some limnic species of *Pseudobiotus*, *Thulinius* and *Hypsibius*, and several limno-terrestrial or terrestrial species of Macrobiotidae. Their parthenogenesis can be automictic or apomictic and in the latter case is often associated with polyploidy (triploidy, the most frequent, and rarely tetraploidy; Bertolani 1982; Rebecchi et al. 2003b). The family Murrayidae represents an example of evolution and persistence of asexual lineages; males have never been found to date indicating that the entire line made up of three genera and more than 20 described species differentiated without sexual reproduction (Guidetti et al. 2000).

Only thelytokous and continuous parthenogenesis has been found in tardigrades, and heterogony has never been found. Nevertheless, the appearance of males with modified claw in the first pair of legs in a thelytokous strain of *Milnesium* sp. was detected (Suzuki 2008). It is unknown whether these males can actually function in sexual reproduction; however, they might allow some possibility of genetic exchange among clonal populations. No environmental factors that generate males were determined (Suzuki 2008). When animals reproduce by automixis for several generations, the loss of genetic variability, unless of mutations, leads to a complete homozygosity. On the other hand, the heterozygosity level of animals reproduce by apomixis may be kept or increased in time, as a consequence of mutation events (Bertolani 2001).

The success of the parthenogenesis in tardigrades from non-marine habitats may be due to the advantage that this kind of reproduction offers to the colonization of new habitats by means of passive dispersal. A relationship among the evolution of parthenogenesis and cryptobiosis and the colonization of terrestrial environment may be recognized in tardigrades (Bertolani 2001). In limnic and terrestrial habitats, many species of tardigrades can survive drying and freezing, entering anhydrobiosis and cryobiosis, respectively. These adaptive strategies allow tardigrades to be subjected both to a lower environmental selective pressure and to passive dispersal. By means of passive dispersal, even a single tardigrade can colonize a new territory if it is parthenogenetic or has been previously fertilized (Bertolani 2001). Since parthenogenesis is more widespread than cross- and self-fertilization, it is favourably selected, and its permanence across many generations may have been aided by anhydrobiosis. Moreover, parthenogenesis and hermaphroditism have never been noted simultaneously in populations attributable to the same tardigrade species; therefore the widespread occurrence of parthenogenesis in limno-terrestrial habitats has been selected for over hermaphroditism, which only evolves when parthenogenesis is absent (Bertolani 2001).

8.7 Development

8.7.1 Embryonic Development

Literature data on tardigrade development has mainly referred on Marcus' work (1929a, b) for a long time. We waited for nearly 70 years for a reinvestigation on tardigrade embryology by Eibye-Jacobsen (1996/97), then followed by other more recent studies (Hejnol and Schnabel 2005; Gabriel and Goldstein 2007; Gabriel et al. 2007; Gross et al. 2015; Smith et al. 2016).

Embryonic development time in tardigrades lasts from 5 to more than 100 days according to the species and obviously to the rearing/environmental temperature. Tardigrades produce homolecithal eggs. The embryo undergoes an irregular indeterminate cleavage pattern without early fate determination. In particular, Eibye-Jacobsen (1996/97) found a total and equal first cleavage that becomes asynchronous from the second cleavage onwards in the eutardigrade *Halobiotus crispae* and in the heterotardigrade *Echiniscoides sigismundi*. This is in line with successive studies carried out on other two eutardigrades, namely, *Thulinius stephaniae* and *Hypsibius dujardini* (see Hejnol and Schnabel 2005; Gabriel and Goldstein 2007; Gabriel et al. 2007; Gross et al. 2015). In the latter species, the authors describe a radial and stereotyped and highly regulative cleavage pattern, including nuclear migrations that can predict the orientation of the embryonic axis, and a stereotyped pattern of stem cell-like asymmetric divisions. The blastula lacks the blastocoel, even though a small blastocoel seems to be present in *H. dujardini* (see Gabriel et al. 2007). In the embryos of *T. stephaniae*, the gastrulation starts after the 64-cell stage with the immigration of single blastomeres at two spatially distinct regions on the ventral surface of the embryo (Hejnol and Schnabel 2005). Mesodermal precursor cells proliferate and form bands along the right and left side of the prospective pharynx and midgut. Later, the mesodermal bands develop into four pairs of somites without a cavity and disconnected with the archenteron, but more recently endo-mesodermal pouches have been described as the earliest morphological evidence of segmentation in *H. dujardini* embryos (Gabriel and Goldstein 2007; Gabriel et al. 2007; Gross et al. 2015). Primordial of all major organs were present from about 7 days after egg laying in *Halobiotus crispae*, when the embryo orientation is determined, and maturation continues after hatching (Eibye-Jacobsen 1996/97). Expression patterns of key developmental genes, including axis-determining segmentation and Hox genes, are recently evaluated (Gabriel and Goldstein 2007; Smith et al. 2016).

8.7.2 Post-Embryonic Development

As soon as embryonic development is complete, the tardigrade emerges from the egg with the action of the stylets and the pharyngeal pump. By sucking water through the corion into the gut, the embryo increases the inner hydrostatic pressure and breaks the egg shell also with the help of the stylets and the hind legs. Post-embryonic development takes place by growing of the single cells and animal moulting without any evidence of cell number increase. Tardigrades are not considered true eutelic animals, having been observed mitosis in some of their tissues.

In eutardigrades, newborns are very similar to the adults, apart from the smaller size, the immature gonad, slight differences in the claws and buccal apparatus anatomy (Bertolani 1990) and sometimes for the still immature eyespots and Malpighian tubules, as in *Halobiotus crispae* (see Eibye-Jacobsen 1996/97). For instance, in *Richtersius* cuticular pores are present in the newborns and are lost with first moulting, a novelty in the life cycle of eutardigrades (Guidetti et al. 2016). Another example from *Milnesium* shows increasing number of secondary branch with juvenile growth into adult (Morek et al. 2016).

In the heterotardigrade echiniscids, newborns show some anatomical differences from the adults. Anus, gonopore and the two external claws (out of four in the adults) per leg are absent in the juveniles. The juveniles need two moultings before they become adults with anus, gonopore and four claws per leg (Bertolani 1990). Sometimes, the number of cuticular filaments and spines of heterotardigrade specimens can increase with the age, apart from *Mopsechiniscus* in which it decreases. A similar pattern was detected in marine heterotardigrades in which two or four post-embryonic developmental stages (juvenile stages) have been identified (Bertolani 1990). For instance, in a new species of *Florarctus*, two distinct juvenile stages during which the anus is formed and the gonads mature prior to the development of the gonopore have been recognized. These juvenile stages are followed by one adult stage in female and two adult stages in males (Hansen et al. 2016). The morphology of the cuticular structures changes during the post-embryonic development showing an adult and juvenile appearance. Nevertheless, a pedomorphosis phenomenon could be evidenced in males having a mature gonad and juvenile characters of the cuticular appendages (Hansen et al. 2016).

8.8 Parental Care

Limnic and terrestrial tardigrades produce either smooth shelled eggs laid in the moulted cuticle (exuvia) or ornamented shelled eggs normally laid freely (i.e. not within the exuvia); instead marine tardigrades laid smooth shelled eggs freely. When the eggs are laid freely, parental care are not observed. On the contrary, when the eggs are laid within the exuvia, some forms of parental care can occur, even though they are very limited and mainly observed in limnic species of eutardigrades. *Pseudobiotus* females carry their exuvia filled with their eggs, keeping it attached to the caudal part of the body until the eggs hatch. Sometimes two subsequent exuviae filled with eggs of different ages can be found (for review see Bertolani 1990). Specimens of the hermaphrodite *Borealibius zetlandicus* can carry their exuvia filled with eggs, and adults were observed within the old cuticle alongside embryonated eggs and/or newborns (Pilato et al. 2006). Moreover, females of *Isohypsibius annulatus* attach the exuvia containing the eggs to the anterior-dorsal part of the body, and females of *Murrayon* genus lay their eggs within the exuvia of cladocerans (Bertolani et al. 2009).

8.9 Life Cycles

8.9.1 Life History Traits

Although certain useful information might be estimated from field sampling, most of the life history traits from birth to death of a tardigrade would be only available by laboratory culture. The oldest record of tardigrade rearing (Von Wenck 1914) still provides us important knowledge about mating behaviour and embryology. The necessity of culture for further study of tardigrades has already been emphasized in the earlier monograph "Handbuch der Zoologie" (Richters and Krumbach 1926). However, investigations on life history traits of tardigrades had not increased until around the 1960s, and the second phase of lab-culture works has begun in the new century (Table 8.1). All of these studies have been done on eutardigrades and in particular on parthenogenetic tardigrades, even though a few studies on gonochoristic animals [e.g. *Paramacrobiotus tonollii*, *Macrobiotus sapiens* (see Lemloh et al. 2011) and *Isohypsibius dastychi* (see Bingemer et al. 2016)] are recently available (Table 8.2). On the contrary, in spite of the mention about rearing of *Batillipes* and *Echiniscoides* by Marcus (1929b), there is practically no detailed record of heterotardigrades culture so far.

The reproductive life cycle of a eutardigrade female includes the following steps: embryonic development, hatching, juvenile growth with successive moults, achievement of sexual maturity, egg production, egg laying and adult moulting (Fig. 8.13). From studies under controlled conditions, life history traits such as life span, age at first oviposition, clutch size (number of eggs per oviposition), total number of eggs per life span and hatching phenology can be extracted (Table 8.2). Considerable variations in life history traits may occur within populations of the same species or among different species, as evidenced in Table 8.2. For instance, Dougherty (1964) recorded detailed life history traits of an Antarctic tardigrade, *Hypsibius arcticus*, which is possibly regarded as *Acutuncus antarcticus* (see Dastych 1991), although his culture might not maintained well because of a progressive deterioration, that differs from those recorded by Tsujimoto et al. (2015, 2016a, b) and Altiero et al. (2015) using different populations of the same species, *A. antarcticus*. On the contrary, two populations of *Paramacrobiotus kenianus* did not show any significant differences with respect to their longevity, oviposition and hatching phenology (Schill 2013). A carnivorous species, *Milnesium tardigradum*, was at first reared by Baumann (1964), but their life history was elucidated in detail by a cultured Japanese strain (Suzuki 2003, 2008). However, the latter might not be actually *M. tardigradum* but *Milnesium* sp. with [3-3]-[3-3] claw system, which is not distinguished by DNA analysis from Tübingen strain of *Milnesium* sp. (see Hengherr et al. 2008). Although Tübingen strain was suggested to have longer life span (up to 107 days) than Hiyoshi strain (up to 58 days), the author afterword observed some individuals of Hiyoshi strain lived up to 4 months (at 23 °C), which is similar data with Tübingen's one. Therefore, more statistical data are necessary for precise comparison between these populations.

The active life span of tardigrades has been estimated from 18 to 30 months (Ramazzotti and Maucci 1983) even though from lab cultures the longest recorded longevity was 518 days (ca. 17 months), from a clonal strain (CDMr01) of the terrestrial and carnivorous *Paramacrobiotus fairbanksi* (see Altiero et al. 2006) and 18 months for the *pseudosimplex 1* stage of the marine eutardigrade *Halobiotus crispae* (see Kristensen 1982). Hengherr et al. (2008) also carried out an experiment on tardigrade longevity concluding that repeated desiccation-rehydration events has no effect on the total active life.

Table 8.1 Records on tardigrade rearing

Author/s (Year)	Species	Habitat	Food	Ref.[a]
Von Wenck (1914)	*Macrobiotus lacustris*[b]	Freshwater	Green algae	–
Węglarska (1957)	*Dactylobiotus dispar*	Freshwater	Algae (*Chlorosphaera*, Diatomes)	–
Baumann (1961)	*Hypsibius convergens*	Green algae (*Blidingia*)	Green algae (*Chlorella pyrenoidosa*)	1
Ammermann (1962, 1967)	*Hypsibius dujardini*	Freshwater	Green algae (*Chlorella* sp.?)	2
Baumann (1964)	*Milnesium tardigradum*	Moss	Bacteria, protozoa, rotifers	3
Dougherty (1964)	*Hypsibius arcticus*[c]	Freshwater	Green algae	4
Baumann (1966)	*Ramazzottius oberhaeuseri*	Moss	Green algae (*Pseudochlorella* sp.)	5
Sayre (1969)	*Isohypsibius myrops*[d]	Freshwater (moist sphagnum)	Nematode (*Panagrellus redivivus*)	6
Baumann (1970)	*Macrobiotus hufelandii*	Moss	Green algae (*Pseudochlorella* sp.)	7
Bertolani and Buonagurelli (1975)	*Dactylobiotus parthenogeneticus*	Freshwater	Pond water, algae and Erd-Schreiber solution	8
Altiero and Rebecchi (2001)	*Paramacrobiotus fairbanksi*	Leaf litter	Nematodes[g]	
	Macrobiotus joannae	Leaf litter	Nematodes[g]	9
	Diphascon cf. *scoticum*	Freshwater	Green algae (*Scenedesmus acutus*)	9
	Isohypsibius monoicus	Freshwater	Natural sediment with algae	9
Suzuki (2003)	*Milnesium* sp.[e]	Moss	Rotifer (*Lecane inermis*)	10
Hohberg (2006)	*Paramacrobiotus* cf. *richtersi*	Soil	Nematodes[h]	11
Gabriel et al. (2007)	*Hypsibius dujardini*	Freshwater	Green algae (*Chlorococcum* sp.)	12
Horikawa et al. (2008)	*Ramazzottius varieornatus*	Moss	Green algae (*Chlorella vulgaris*)	13
Altiero et al. (2006, 2010)	*Paramacrobiotus fairbanksi*	Leaf litter	Nematodes[g]	14
Hengherr et al. (2008)	*Milnesium* sp.[f]	Moss	Rotifer (*Philodina citrina*)	15
Lemloh et al. (2011)	*Paramacrobiotus tonolli*	Moss	Green algae (*Chlorogonium elongatum*) + rotifer (*Philodina citrina*)	16
	Macrobiotus sapiens	Moss	Green algae (*Chlorogonium elongatum*)	16
Kagoshima et al. (2013)	*Acutuncus antarcticus*	Moss	Green algae and/or cyanobacteria	–

(continued)

Table 8.1 (continued)

Author/s (Year)	Species	Habitat	Food	Ref.[a]
Schill (2013)	*Paramacrobiotus kenianus*	Moss	Rotifer (*Philodina citrina*)	17
	Paramacrobiotus palaui	Moss	Rotifer (*Philodina citrina*)	17
Altiero et al. (2015)	*Acutuncus antarcticus*	Freshwater sediment	Green alga (*Chlorococcum* sp.)	18
Poprawa et al. (2015b)	*Dactylobiotus parthenogeneticus*	Freshwater pond	Green algae (*Chlorella* sp., *Chlorococcum* sp.)	–
Stec et al. (2015)	*Macrobiotus paulinae*	Moss	Green algae and rotifers[i]	–
Poprawa et al. (2015c)	*Macrobiotus polonicus*	Moss	Alga (*Chlorella* sp.), nematode (*C. elegans*)	–
Tsujimoto et al. (2015)	*Acutuncus antarcticus*	Moss/ freshwater	Green algae (*Chlorella vulgaris*)	19
Bingemer et al. (2016)	*Isohypsibius dastychi*	Post-mining site	Green algae (*Chlorella vulgaris*)	20
Ito et al. (2016)	*Isohypsibius myrops*	Sewage sludge	Green algae (*Parachlorella beyerinckii*), rotifers (*Lecane inermis*)	21
Kosztyła et al. (2016)	*Diphascon higginsi*	Leaves in pond	Green algae and rotifers[i]	–
	Hypsibius cf. *dujardini*	Leaves in pond	Green algae and rotifers[i]	–
	Isohypsibius pushkini	Leaves in pond	Green algae and rotifers[i]	–
	Milnesium cf. *alpigenum*	Moss	Rotifers and green algae[i]	–
	Paramacrobiotus cf. *richtersi*	Moss	Rotifers and green algae[i]	–
	Thulinius ruffoi	Sewage sludge	Green algae and rotifers[i]	–

[a]Reference number correlate with Table 8.2
[b]Originally described as *Macrobiotus lacustris*; today this name is not valid. On the basis of the original drawing, specimens are attributable to *Pseudobiotus megalonyx*
[c]*Acutuncus antarcticus*
[d]Tentatively identified. Sayre's animal tolerated against dehydration
[e]Hiyoshi-1 strain, originally described as *M. tardigradum*
[f]Tubingen strain, originally described as *M. tardigradum*
[g]*Pristionchus iheritieri*, *Panagrolaimus rigidus*, *Caenorhabditis elegans*
[h]*Pelodera teres*, *Acrobeloides nanus*
[i]Algae: *Chloroccoum* sp., *Chlorella* sp. Rotifer: *Lecane inermis*

Sexual maturity is reached with the second or third moulting, rarely even with the first one. Reared females lay from 1 to 44 eggs at each oviposition, depending on the species, age, and nutritional state of the specimens. Ramazzotti and Maucci (1983)

Table 8.2 Life history traits of cultured tardigrades

Species	Life span (days)	Max. longevity (days)	Age at first oviposition (days)	No. of eggs/ clutch	No. of ovipositions/ animal	No. of eggs/ animal	Hatching success (%)	Hatching time (days)	Temp. (°C)	Ref.[a]	Reproductive modes
Acutuncus antarcticus	69.2 ± 36.4	162	9.3 ± 1.1	1–10	7.5	34.4 ± 22.6	97.6	8.5 ± 0.7	15	19	P
Acutuncus antarcticus	88.8 ± 20.0	130	17.1 ± 3.6	1.8 ± 0.8	6.4 ± 1.0	13.0 ± 3.7	67.1 ± 22.7	8.8 ± 2.6	14	18	P
Dactylobiotus dispar		–	–	411	–	–	–	–	19	8	P
Diphascon (Adropion) scoticum	–	263	–	1–11	15	–	–	5–14	14	9	P
Hypsibius arcticus	–	160	21–45	2–13	11	84	69.0	15–39	4–7	4	P
Hypsibius convergens	–	159	–	2–12	–	–	–	5–10	–	1	–
Hypsibius dujardini	–	–	14	3–10	–	–	–	5	21–23	2	P
Hypsibius dujardini	–	--	13.6 ± 0.8	3.4 ± 1.9	-	–	–	4–4.5	10–18	12	P
Isohypsibius dastychi	–	–	44.9 ± 6.5	–	–	–	80.0	13.1 ± 1.5	12	20	A/G
Isohypsibius dastychi	–	–	28.2 ± 4.6	–	–	–	66.0	7.6 ± 1.1	20	20	A/G
Isohypsibius myrops (strain Im1)	18.8 ± 7.0	~22	10.3 ± 0.7	19.0 ± 7.6	3.5 ± 2.8	67.3 ± 61.3	80.0 ± 29.4	3.6 ± 0.6	22/23	21	P
Isohypsibius monoicus	–	197	–	1–3	3	–	–	12–20	14	9	A/H
Macrobiotus joannae	–	266	–	1–7	–	–	–	14–39	20	9	A/H
Macrobiotus sapiens	83.0 ± 33.5	145	16.5 ± 3.8	5.1 ± 3.1	–	–	78	11.9 ± 2.7	20	16	A/G
Macrobiotus hufelandii	–	84	31	1	3	3	–	26–31	20	7	P
Milnesium tardigradum	–	–	41	1–6	–	–	–	4–12	–	3	P
Milnesium sp. (Hiyoshi)	42.7 ± 11.8	58	15.3 ± 1.6	7.0 ± 2.2	3.3 ± 1.6	22.9 ± 12.9	77.2	11.2 ± 2.4	25	10	P
Milnesium sp. (Tübingen)	82.7 ± 2.7	107	–	–	–	–	–	–	20	15	P
Paramacrobiotus kenianus (population 1)	125 ± 35	204	10	–	–	138 ± 71	33	7	20	17	P
Paramacrobiotus kenianus (population 2)	141 ± 54	212	10	–	–	124 ± 78	51	8	20	17	P
Paramacrobiotus palaui	97 ± 31	187	10	–	–	42 ± 54	54	9	20	17	P
Paramacrobiotus richtersi (Berzdorf)	–	–	64.2 ± 1.7	4.0 ± 0.8	–	–	100	28.1 ± 1.2	16	11	

Paramacrobiotus fairbanksi (CGMr01)	194.9 ± 164.4	518	76.9 ± 16.4	11.0 ± 7.1	3.4 ± 2.1	37.8 ± 29.3	83.1 ± 12.7	41.0 ± 9.5	14	14	P
Paramacrobiotus fairbanksi (CGMr02)	137.3 ± 136.4	457	70.7 ± 19.4	8.0 ± 6.2	2.0 ± 1.2	17.8 ± 16.0	51.0 ± 36.0	60.0 ± 16.7	14	14	P
Paramacrobiotus tonollii	69.0 ± 45.1	237	24.4 ± 4.4	6.5 ± 4.3	–	–	82.2	16.7 ± 6.7	20	16	A/G
Ramazzottius oberhaeuseri	–	70	–	1–11	7	–	–	8–20	–	5	P
Ramazzottius varieornatus	35.0 ± 16.4	87	9	–	–	7.85	82.5	5.7 ± 1.1	25	13	P

P = parthenogenesis; A/G = amphimixis, gonochoric; A/H = amphimixis, hermaphroditic
Data are shown as either average ± SD, the range or simple available data
Minus sign indicates no data
[a]Correlate with reference numbers in Table 8.1

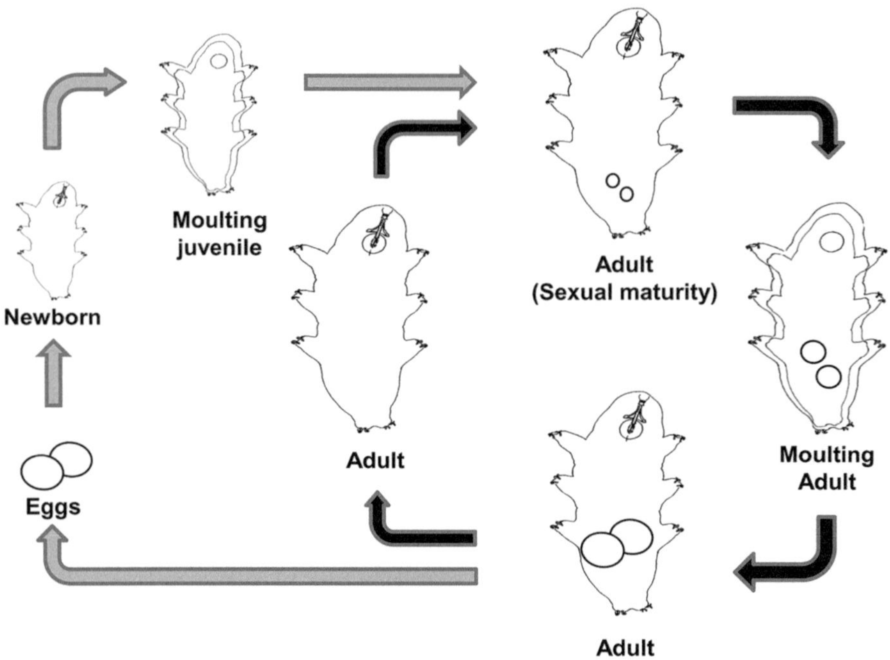

Fig. 8.13 Schematic representation of the steps of the life cycle of a eutardigrade laying smooth eggs within the old cuticle (exuvia)

reported a maximum of 60 eggs at each oviposition, but this data have not been confirmed by laboratory cultured species. Egg production continues throughout adult life, although moulting can occur without egg production, up to 190 eggs (Table 8.2). The development time of eggs varies from 4 to 87 days, depending on rearing conditions (e.g. temperature, oxygen amount) and species/populations, but strong variations have been evidenced even in a same clonal lineage (Table 8.2).

8.9.2 Moulting Process

Tardigrades grow up even after sexual maturity, through periodical moultings, occurring throughout their life. Body length increases at every moulting until maximum size is reached, although a decrease in size may be caused by the lack of food. The moulting process lasts 5–10 days, and it involves the expulsion through the buccal opening of the whole cuticular lining of the foregut together with the buccal tube, placoids, stylets and stylet supports (Guidetti et al. 2012). After that, animals are in the so-called "simplex" stage; they cannot feed because of the closure of the mouth opening and the absence of every sclerified buccal-pharyngeal apparatus. In the meanwhile, the epithelial tissue of the oesophagus starts the reconstruction of the posterior part of the buccal apparatus, the underlying epidermis of body cuticle, and hindgut lining begins to synthesize these new cuticular structures, and the claw (pedal) glands in each leg regenerate new claws. Finally, animal can release its old cuticle (exuvia, including the lining of the hindgut), which many species of tardigrades use for laying their eggs. For that species, a synchronization between moulting and oviposition is present (Nelson et al. 2015).

8.9.3 Resting Eggs

Experimental cultures of *Paramacrobiotus fairbanksi*, a leaf litter-dwelling species, have allowed to document the presence of resting eggs in tardigrades (Altiero et al. 2010). Resting eggs, needing an environmental cue to hatch (dehydration followed by rehydration), represent only a small portion (<10%) of all laid eggs. The remaining 90% of laid eggs, morphological indistinguishable from resting eggs, involved subitaneous eggs (>58%) that hatch in 30–40 days and delayed-hatching eggs (ca. 30%), which completed their development in 41–62 days, together with a very small fraction of abortive eggs (Altiero et al. 2010). Winter and summer eggs with differently shaped egg processes have been evidenced in *Bertolanius nebulosus* from Greenland: the former type may be interpreted as resting eggs (Hansen and Katholm 2002). Resting eggs, together with the high variability in hatching time of tardigrade eggs, might be considered as a bet-hedging adaptive strategy to cope with unpredictable environments, such as those colonized by tardigrades.

8.9.4 Rearing Tardigrades for Life History Trait Analysis

Cultured tardigrades, mainly belonging to eutardigrades, came from freshwater sediments, leaf litter or moss (Table 8.1). Reared herbivorous tardigrades fed on green algae, e.g. *Chlorella* spp., *Scenedesmus acutus*, *Chlorococcum* sp. and *Chlorogonium elongatum* (see Table 8.1). The actual food preference of four herbivorous tardigrade species (*Macrobiotus sapiens*, *M. persimilis*, *Richtersius coronifer* and *Echiniscus granulatus*) was examined by sequence analysis of a chloroplast gene from mosses and algae, revealing that algae seem to be a more likely food for these moss-dwelling species (Schill et al. 2011). Reared carnivorous tardigrades fed on rotifers or nematodes (see Table 8.1).

Pool cultures of tardigrade species can be kept in a flask with food source, but to collect data on life history traits, eggs must be collected immediately after their deposition and kept individually till hatching, as well as each newborn must be reared singularly from hatch to death. The recent culture system generally consists of appropriate containers (such as plastic Petri dishes) with 1.2–2% agar coat or with the bottom scratched with fine sand to aid tardigrade locomotion, water layer, food according to the rearing species and incubator for temperature and light/dark conditioning. The medium has to be changed periodically (1–2 weeks), and pool cultures have to be divided in order to prevent overcrowding.

Examples from recent culture protocols are shown as follows.

Paramacrobiotus fairbanksi (from Altiero and Rebecchi 2001, with a slight modification). Animals are reared in small plastic dishes (15 mm diameter and 7 mm height) with bacto agar (Difco, 1.2% w/v in water) and some drops of natural mineral water (San Benedetto). They feed on three species of nematodes, *Pristionchus iheritieri*, *Panagrolaimus rigidus* and *Caenorhabditis elegans*, which are cultured utilizing the bacterium *Escherichia coli.* The diploid strain seemed to prefer 20 °C, whereas the triploid one seemed to prefer 14 °C. Newborns were fed with algae.

Milnesium tardigradum (from Suzuki 2003, with a slight modification). Animals are reared in plastic dishes (35 mm diameter). The bottom of each dish is coated with 2.0% agar (Agar Noble, Difco) in KCM solution (7 mg KCl, 8 mg $CaCl_2$ and 8 mg $MgSO_4x7H_2O$ in 1000 ml of water), and a layer of water (Milli-Q, Millipore) is added. To simplify, agar solution can be prepared in water, then a layer of natural mineral water (e.g. Volvic) is added. Prey species of rotifer *Lecane inermis* is maintained with rice grain immersed in KCM solution. Bdelloid rotifers can also be used as food. This method is applicable to other milnesiid species.

Ramazzottius varieornatus (from Horikawa et al. 2008, with personal communication from Kunieda). Animals are reared in plastic dishes (35 or 90 mm diameter) with bacto agar (1.5% in either Milli-Q or Volvic water). Solution with *Chlorella vulgaris*, purchased from Chlorella Industry Co., Ltd. (Tokyo, Japan), is diluted ca. 1:100 in Volvic and added to the dish (1–2 mm height). Temperature should be kept at 20–25 °C. Recently, *Acutuncus antarcticus* also proved to be cultured well by this method (Tsujimoto et al. 2015) at 5–15 °C.

References

Altiero T, Rebecchi L (2001) Rearing tardigrades: results and problems. Zool Anz 240:217–221

Altiero T, Rebecchi L (2003) First evidence of achiasmatic male meiosis in the water bears *Richtersius coronifer* and *Macrobiotus richtersi* (Eutardigrada, Macrobiotidae). Hereditas 139:116–120

Altiero T, Rebecchi L, Bertolani R (2006) Phenotypic variations in the life history of two clones of *Macrobiotus richtersi* (Eutardigrada, Macrobiotidae). Hydrobiologia 558:33–40

Altiero T, Bertolani R, Rebecchi L (2010) Hatching phenology and resting eggs in *Paramacrobiotus richtersi* (Eutardigrada, Macrobiotidae). J Zool 280:290–296

Altiero T, Giovannini I, Guidetti R, Rebecchi L (2015) Life history traits and reproductive mode of the tardigrade *Acutuncus antarcticus* under laboratory conditions: strategies to colonize the Antarctic environment. Hydrobiologia 761:277–291

Ammermann D (1962) Parthenogenese bei dem Tardigraden *Hypsibius dujardini* (Doy.). Naturwissenschaften 49:115

Ammermann D (1967) Die Cytologie der Parthenogenese bei dem Tardigraden *Hypsibius dujardini*. Chromosoma 23:203–213

Baccetti B (1987) The evolution of the sperm cell in the phylum Tardigrada (Electron microscopy of Tardigrades. 5). In: Bertolani R (ed) Biology of Tardigrades, selected symposia and monographs, UZI, vol 2. Mucchi, Modena, pp 87–91

Baumann H (1961) Der Lebensablauf von *Hypsibius (H.) convergens* Urbanowicz (Tardigrada). Zool Anz 167:362–381

Baumann H (1964) Über den Lebenslauf und die Lebensweise von *Milnesium tardigradum* Doyère (Tardigrada). Veröff Überseemus Bremen 3:161–171

Baumann H (1966) Lebenslauf und Lebensweise von *Hypsibius (H.) oberhaeuseri* Doyère (Tardigrada). Veröff Überseemus Bremen 3:245–258

Baumann H (1970) Lebenslauf und Lebensweise von *Macrobiotus hufelandii* Schultze (Tardigrada). Veröff Überseemus Bremen 4:29–43

Bertolani R (1971) Contributo alla cariologia dei Tardigradi. Osservazioni su *Macrobiotus hufelandi*. Atti Accad Naz Lincei Rend Ser 8^{a}(50):772–775

Bertolani R (1982) Cytology and reproductive mechanisms in tardigrades. In: Nelson DR (ed) Proc. Third International Symposium on the Tardigrada. East Tennessee State Univ. Press, Johnson City, Tennessee, pp 93–114

Bertolani R (1983a) Tardigrada. In: Adiyodi KG, Adiyodi RG (eds) Reproductive biology of invertebrates. Oogenesis, oviposition, and oosorption, vol 1. Wiley, Chichester, pp 431–441

Bertolani R (1983b) Tardigrada. In: Adiyodi KG, Adiyodi RG (eds) Reproductive biology of invertebrates. Spermatogenesis and sperm function, vol 2. Wiley, Chichester, pp 387–396

Bertolani R (1990) Tardigrada. In: Adiyodi KG, Adiyodi RG (eds) Reproductive biology of invertebrates. Fertilization, development and parental care, vol 4, Part B. Wiley, Chichester, pp 49–60

Bertolani R (1992) Tardigrada. In: Adiyodi KG, Adiyodi RG (eds) Reproductive biology of invertebrates. Sexual differentiation and behaviour, vol 5. Wiley, Chichester, pp 255–266

Bertolani R (2001) Evolution of the reproductive mechanisms in tardigrades. A review. Zool Anz 240:247–252

Bertolani R, Buonagurelli GP (1975) Osservazioni cariologiche sulla partenogenesi meiotica di *Macrobiotus dispar* (Tardigrada). Atti Accad Naz Lincei Rend Ser 8^{a}(58):782–786

Bertolani R, Rebecchi L (1993) A revision of the *Macrobiotus hufelandi* group (Tardigrada, Macrobiotidae), with some observations on the taxonomic characters of eutardigrades. Zool Scr 22:127–152

Bertolani R, Rebecchi L, Beccaccioli G (1990) Dispersal of *Ramazzottius* and other tardigrades in relation to type of reproduction. Invertebr Reprod Dev 18:153–157

Bertolani R, Rebecchi L, Claxton SC (1996) Phylogenetic significance of egg shell variation in tardigrades. Zool J Linnean Soc 116:139–148

Bertolani R, Altiero T, Nelson DR (2009) Tardigrada (Water Bears). In: Likens GE (ed) Encyclopedia of Inland Waters, vol 2. Elsevier, Oxford, pp 443–455

Bingemer J, Hihberg K, Schill RO (2016) First detailed observations on tardigrade mating behaviour and some aspects of the life history of *Isohypsibius dastychi* Pilato, Bertolani & Binda 1982 (Tardigrada, Isohypsibiidae). Zool J Linnean Soc 1787:856–862

Dastych H (1991) Redescription of *Hypsibius antarcticus* (Richters, 1904), with some notes on *Hypsibius arcticus* (Murray, 1907) (Tardigrada). Mitt Hamb Zool Mus Inst 88:141–159

Dewel RA, Nelson DR, Dewel WC (1993) Tardigrada. In: Harrison FW, Rice ME (eds) Microscopic anatomy of invertebrates, vol 12. Wiley, New York, pp 143–183

Dougherty EC (1964) Cultivation and nutrition of micrometazoa. II. An Antarctic strain of the tardigrade *Hypsibius arcticus* (Murray, 1907) Marcus, 1928. Trans Am Microsc Soc 83:7–11

Eibye-Jacobsen J (1996/97) New observation on the embryology of the Tardigrada. Zool Anz 235:201–216

Gabriel WN, Goldstein B (2007) Segmental expression of Pax3/7 and Engrailed homologs in tardigrade development. Dev Genes Evol 217:421–433

Gabriel WN, McNuff R, Patel SK et al (2007) The tardigrade *Hypsibius dujardini*, a new model for studying the evolution of development. Dev Biol 312:545–559

Greven H, Kristensen RM (2001) Some aspects of spermiogenesis and spermatozoa ultrastructure in *Echiniscoides sigismundi* (Heterotardigrada: Echiniscidae). Zool Anz 240:331–339

Grimaldi de Zio S, D'Addabbo Gallo M (1975) Reproductive cycle of *Batillipes pennaki* Marcus (Heterotardigrada) and observation on the morphology of the female genital apparatus. Stn Zool Napoli 39:212–225

Gross V, Treffkorn S, Mayer G (2015) Tardigrada. In: Wanninger A (ed) Evolutionary biology of invertebrates Ecdysozoa 1: Non Tetraconata, vol 1. Springer, Wien, pp 36–52

Guidetti R, Rebecchi L, Bertolani R (2000) Cuticle structure and systematics of the Macrobiotidae (Tardigrada, Eutardigrada). Acta Zool 81:27–36

Guidetti R, Altiero T, Marchioro T, Sarzi-Amadè L, Avdonina AM, Rebecchi L (2012) Form and function of the feeding apparatus in Eutardigrada (Tardigrada). Zoomorphology 131:127–148

Guidetti R, Rebecchi L, Bertolani R, Jönsson KI, Kristensen RM, Cesari M (2016) Morphological and molecular analyses on *Richtersius* (Eutardigrada) diversity reveal its new systematic position and lead to the establishment of a new genus and a new family within Macrobiotoidea. Zool J Linnean Soc 178:834–845

Guidetti R, Cesari M, Bertolani R, Altiero T, Rebecchi L (2019) High diversity in species, reproductive modes and distribution within the *Paramacrobiotus richtersi* complex (Eutardigrada, Macrobiotidae). Zool Lett 5:1. https://doi.org/10.1186/s40851-018-0113-z

Guidi A, Rebecchi L (1996) Spermatozoan morphology as a character for tardigrade systematics: comparison with sclerified parts of animals and eggs in eutardigrades. Zool J Linnean Soc 116:101–113

Hansen JG, Katholm AK (2002) A study of the genus *Amphibolus* from Disko Island with special attention on the life cycle of *Amphibolus nebulosus* (Eutardigrada: Eohypsibiidae). In: Hansen JG (ed) Arctic biology field course, Godhaven, Zoological Museum, University of Copenhagen. H.C.Ø. TRYK, Copenhagen, pp 129–163

Hansen GJ, Kristensen RM, Jörgensen A, Accogli G, D'Addabbo R, Gallo M (2016) Postembryonic development, paedomorphosis, secondary sexual dimorphism and population structure of a new *Florarctus* species (Tardigrada, Heterotardigrada). Zool J Linnean Soc 1787:871–877

Hejnol A, Schnabel R (2005) The eutardigrade *Thulinia stephaniae* has an indeterminate development and the potential to regulate early blastomere ablations. Development 132:1349–1361

Hengherr S, Brümmer F, Schill RO (2008) Anhydrobiosis in tardigrades and its effects on longevity traits. J Zool 275:216–220

Henneke J (1911) Beiträge zur Kenntnis der Biologie und Anatomie der Tardigraden (*Macrobiotus macronyx* Duj.). Z Wiss Zool 97:721–752

Hohberg K (2006) Tardigrade species composition in young soils and some aspects on life history of *Macrobiotus richtersi* J. Murray, 2011. Pedobiologia 50:267–274

Horikawa DD, Kunieda T, Abe W, Watanabe M, Nakahara Y, Yukuhiro F, Sakashita T, Hamada N, Wada S, Funayama T, Katagiri C, Kobayashi Y, Higashi S, Okuda T (2008) Establishment of a rearing system of the extremotolerant tardigrade *Ramazzottius varieornatus*: a new model animal for astrobiology. Astrobiology 8:549–556

Hyra M, Poprawa I, Włodarczyk A, Student S, Sonakowska L, Kszuk-Jendrysik M, Rost-Roszkowska MM (2016) Ultrastructural changes in the midgut epithelium of *Hypsibius dujardini* (Doyère, 1840) (Tardigrada, Eutardigrada, Hypsibiidae) in relation to oogenensis. Zool J Linnean Soc 1787:897–906

Ito M, Saigo T, Abe W, Kubo T, Kunieda T (2016) Estabishment of an isogenic strain of the desiccation sensitive tardigrade *Isohypsibius myrops* (Parachela, Eutardigrada) and its life history. Zool J Linnean Soc 1787:863–870

Jørgensen A, Møbjerg N, Kristensen RM (1999) Ultrastructural studies on spermiogenesis and postcopulatory modifications of spermatozoa of *Actinarctus doryphorus* Sculz, 1935 (Arthrotardigrada: Halechiniscidae). Zool Anz 238:235–257

Kagoshima H, Imura S, Suzuki AC (2013) Molecular and morphological analysis of Antarctic tardigrade, *Acutuncus antarcticus*. J Limnol 72(s1):15–23

Kosztyła P, Stec D, Morek W, Gąsiorek P, Zawierucha K, Michno K, Ufir K, Małek D, Hlebowicz K, Laska A, Dudziak M, Frohme M, Prokop ZM, Kaczmarek Ł, Michalczyk Ł (2016) Experimental taxonomy confirms the environmental stability of morphometric traits in a taxonomically challenging group of microinvertebrates. Zool J Linnean Soc 178:765–775

Kristensen RM (1979) On the fine structure of *Batillipes noerrevangi* Kristensen, 1978 (Heterotardigrada). 3. Spermiogenesis. Zesz Nauk Uniw Jagiellon Pr Zool 25:97–105

Kristensen RM (1980) Zur Biologie des marinen Heterotardigraden *Tetrakentron synaptae*. Helgol Meer 34:165–177

Kristensen RM (1982) The first record of cyclomorphosis in Tardigrada based on a new genus and species from Arctic meiobenthos. Z Zool Syst Evol – forsh 20:249–270

Kristensen RM (1984) On the biology of *Wingstrandarctus corallinus* nov. gen. et spec., with notes on the symbiontic bacteria in the subfamily Florarctinae (Arthrotardigrada). Vidensk Meddr Dansk Naturhist Foren 145:201–218

Kristensen RM, Renaud-Mornant J (1983) Existence d'Arthrotardigrades semi-benthiques de genres nouveaux de la sous-famille des Styraconyxinae, subfam. nov. Cah Biol Mar 24:337–354

Lemloh M-L, Brümmer F, Schill RO (2011) Life-history traits of the bisexual tardigrades *Paramacrobiotus tonollii* and *Macrobiotus sapiens*. J Zool Syst Evol Res 49(s1):58–61

Marcus E (1929a) Zur embryologie der tardigrade. Zool Jahrb 50:333–384

Marcus E (1929b) Tardigrada. In: Bronn HG (ed) Klassen und Ordnungen des Tier-Reichs. Akademische Verlagsgesellschaft, Bd 5, IV-3, Leipzig, pp 1–608

Morek W, Stec D, Gąsiorek P, Schill RO, Kaczmarek Ł, Michalczyk Ł (2016) An experimental text of eutardigrade preparative methods for light microscopy. Zool J Linnean Soc 178:785–793

Nelson DR, Guidetti R, Rebecchi L (2015) Phylum Tardigrada. In: Thorp J, Rogers DC (eds) Ecology and general biology, IV edition: Thorp and Covich's freshwater invertebrates, Chapter 17. Academic Press, New York, pp 347–380

Pilato G, Guidetti R, Rebecchi L, Lisi O, Hansen JG, Bertolani R (2006) Geonemy, ecology, reproductive biology and morphology of the tardigrade *Hypsibius zetlandicus* (Eutardigrada: Hypsibiidae) with erection of *Borealibius* gen. n. Polar Biol 29:595–603

Pollock LW (1970) Reproductive anatomy of some marine Heterotardigrada. Trans Am Microsc Soc 89:308–316

Poprawa I (2005a) The ovary structure, previtellogenic and vitellogenic stages in parthenogenetic species *Dactylobiotus dispar* (Murray, 1907) (Tardigrada: Eutardigrada). Tissue Cell 37:385–392

Poprawa I (2005b) The structure and the formation of egg shells in the parthenogenetic species *Dactylobiotus dispar* Murray, 1907 (Tardigrada: Eutardigrada). Folia Biol 53:173–177

Poprawa I (2011) Ultrastructural studies of the formation of the egg capsule in the hermaphroditic species, *Isohypsibius granulifer granulifer* Thulin, 1928 (Eutardigrada: Hypsibiidae). Zool Sci 28(1):37–40

Poprawa I, Hyra M, Kszuk-Jendrysik M, Rost-Roszkowska MM (2015a) Ultrastructural changes and programmed cell death of trophocytes in the gonad of *Isohypsibius granulifer granulifer* Thulin, 1928 (Tardigrada, Eutardigrada, Isohypsibiidae). Micron 70:26–33

Poprawa I, Hyra M, Rost-Roszkowska MM (2015b) Germ cell cluster organization and oogenesis in the tardigrade *Dactylobiotus parthenogeneticus* Bertolani, 1982 (Eutardigrada, Murrayidae). Protoplasma 252:1019–1029

Poprawa I, Schlechte-Welnicz W, Hyra M (2015c) Ovary organization and oogenesis in the tardigrade *Macrobiotus polonicus* Pilato, Kaczmarek, Michalczyk & Lisi, 2003 (Eutardigrada, Mcrobiotidae): ultrastructural and histochemical analysis. Protoplasma 252:857–865

Ramazzotti G, Maucci W (1983) Il Phylum Tardigrada. III Edizione riveduta e aggiornata. Mem Ist Ital Idrobiol 41:1–1012

Rebecchi L (1997) Ultrastructural study of spermiogenesis and the testicular and spermathecal spermatozoon of the gonochoristic tardigrade *Xerobiotus pseudohufelandi* (Eutardigrada, Macrobiotidae). J Morphol 234:11–24

Rebecchi L (2001) The spermatozoon in tardigrades: evolution and relationships with the environment. Zool Anz 240:525–533

Rebecchi L, Bertolani R (1988) New cases of parthenogenesis and polyploidy in the genus *Ramazzottius* (Tardigrada, Hypsibiidae) and a hypothesis concerning their origin. Invertebr Reprod Dev 14:187–196

Rebecchi L, Bertolani R (1994) Maturative pattern of ovary and testis in eutardigrades of freshwater and terrestrial habitats. Invertebr Reprod Dev 26:107–118

Rebecchi L, Bertolani R (1999) Spermatozoon morphology of three species of Hypsibiidae (Tardigrada, Eutardigrada) and phylogenetic evaluation. Zool Anz 238:319–328

Rebecchi L, Guidi A (1995) Spermatozoon ultrastructure in two species of *Amphibolus* (Eutardigrada, Eohypsibiidae). Acta Zool 76:171–176

Rebecchi L, Nelson DR (1988) Evaluation of a secondary sex character in eutardigrades. Invertebr Biol 117:194–198

Rebecchi L, Guidi A, Bertolani R (2000a) Tardigrada. In: Jamieson BGM (ed), Progress in male gamete ultrastructure and phylogeny. Reproductive biology of invertebrates, Adiyodi KG, Adiyodi RG series, vol IX, Part B, Chapter 6, pp 267–291

Rebecchi L, Guidi A, Bertolani R (2000b) Maturative pattern of the ovotestis in two hermaphrodite species of eutardigrades. Invertebr Reprod Dev 37:25–34

Rebecchi L, Guidi A, Bertolani R (2003a) The spermatozoon of the Echiniscidae (Tardigrada, Heterotardigrada) and its phylogenetic significance. Zoomorphology 122:3–9

Rebecchi L, Rossi V, Altiero T, Bertolani R, Menozzi P (2003b) Reproductive modes and genetic polymorphism in the *Richtersius coronifer* (Eutardigrada, Macrobiotidae). Invertebr Biol 122:19–27

Rebecchi L, Altiero T, Guidi A (2011) The ultrastructure of the tardigrade spermatozoon: a comparison between *Paramacrobiotus* and *Macrobiotus* species (Eutardigrada). Invertebr Zool 8:63–77

Richters F, Krumbach T (1926) Tardigrada. In: Kükenthal W, Krumbach T (eds) Handbuch der Zoologie, vol 3. Walter de Gruyter & Co, Berlin, pp 1–68

Rost-Roszkowska MM, Poprawa I, Wójtowicz M, Kaczmarek Ł (2011) Ultrastructural changes of the midgut epithelium in *Isohypsibius granulifer granulifer* Thulin, 1928 (Tardigrada: Eutardigrada) during oogenesis. Protoplasma 248:405–414

Sayre RM (1969) A method for culturing a predaceous tardigrade on the nematode *Panagrellus redivivus*. Trans Am Microsc Soc 88:266–274

Schill RO (2013) Life-history traits in the tardigrade species *Paramacrobiotus kenianus* and *Paramacrobiotus palaui*. J Limnol 72(s1):160–165

Schill RO, Jönsson KI, Pfannkuchen M, Brümmer F (2011) Food of tardigrades: a case study to understand food choice, intake and digestion. J Zool Syst Evol Res 49(s1):66–70

Smith FW, Boothby TC, Giovannini I, Rebecchi L, Jockusch E, Goldstein B (2016) The compact body plan of tardigrades evolved by the loss of a large body region. Curr Biol 26:224–229

Stec D, Smolak R, Kaczmarek Ł, Michalczy Ł (2015) An integrative description of *Macrobiotus paulinae* sp. nov. (Tardigrada: Eutardigrada: Macrobiotidae: *hufelandi* group) from Kenya. Zootaxa 4052(2):501–526

Suzuki AC (2003) Life history of *Milnesium tardigradum* Doyère (Tardigrada) under a rearing environment. Zool Sci 20:49–57

Suzuki AC (2006) Ovarian structure in *Milnesium tardigradum* (Tardigrada, Milnesiidae) during early vitellogenesis. Hydrobiologia 558:61–66

Suzuki AC (2008) Appearance of males in a thelytokous strain of *Milnesium* cf. *tardigradum* (Tardigrada). Zool Sci 25:849–853

Suzuki AC, Kristensen RM (2014) Spermatozoa in the reproductive system of a hermaphroditic marine tardigrade, *Orzeliscus belopus* (Tardigrada: Arthrotardigrada). Zool Anz 253:497–511

Tsujimoto M, Suzuki AC, Imura S (2015) Life history of the Antarctic tardigrade, *Acutuncus antarcticus*, under a constant laboratory environment. Polar Biol 38:1575–1581

Tsujimoto M, Komori O, Imura S (2016a) Effect of lifespan and age on reproductive performance of the tardigrade *Acutuncus antarcticus*: minimal reproductive senescence. Hydrobiologia 772 (1):93–102

Tsujimoto M, Imura S, Kanda H (2016b) Recovery and reproduction of an Antarctic tardigrade retrieved from a moss sample frozen for over 30 years. Cryobiology 72(1):78–81

von Erlanger R (1895) Beiträge zur Morphologie der Tardigraden. I. Zur Embryologie eines Tardigrade *Macrobiotus macronyx* Dujardin. Morphol Jahrb 22:491–513

Von Wenck W (1914) Entwicklungsgeschichtliche Untersuchungen an Tardigraden (*Macrobiotus lacustris* Duj.). Zool Jahrb Abt Anat Ontog Tiere 37:465–514

Węglarska B (1957) On the encystation in Tardigrada. Zool Pol 8:315–323

Węglarska B (1975) Studies on the morphology of *Macrobiotus richtersi* Murray, 1911. Mem Ist Ital Idrobiol 32:445–464

Węglarska B (1979) Electron microscope study on previtellogenesis and vitellogenesis in *Macrobiotus richtersi* J. Murr. (Eutardigrada). Zesz Nauk Uniw Jagiellon Prace Zool 25:169–190

Węglarska B (1982) Ultrastructural study of the formation of egg envelops in *Macrobiotus richtersi* (Eutardigrada). In: Nelson DR (ed), Proceedings third international symposium Tardigrada, East Tennessee State Univ Press, Johnson City, Tennessee, pp 115–128

Węglarska B (1987) Yolk formation in *Isohypsibius* (Eutardigrada). Zoomorphology 107:287–292

Wolburg-Buchholz K, Greven H (1979) On the fine structure of the spermatozoon of *Isohypsibius granulifer* Thulin 1928 (Eutardigrada) with reference to its differentiation. Zesz Nauk Uniw Jagiellon Prace Zool 25:191–197

Chapter 9
Environmental Adaptations: Encystment and Cyclomorphosis

Roberto Guidetti and Nadja Møbjerg

Abstract Stressful environmental conditions generally limit animal survival, growth, and reproduction and may induce dormancy in the form of various resting stages. Tardigrades represent one of a few animal phyla in which different forms of dormancy are frequently encountered. One of these forms, cryptobiosis, a quick response to sudden changes in the environment, has gained a great deal of attention, whereas much less is known of the slower emerging form of dormancy, diapause. In this review we present the current knowledge of diapause in tardigrades.

Diapause in tardigrades, represented by encystement and cyclomorphosis, is likely controlled by exogenous stimuli, such as temperature and oxygen tension, and perhaps also by endogenous stimuli. These stimuli initiate and direct successive phases of deep morphological transformations within the individual. Encystment is characterized by tardigrades that lie dormant—in diapause—within retained cuticular coats (exuvia). The ability to form cysts is likely widespread but presently only confirmed for a limited number of species.

In tardigrades, cyclomorphosis was first reported as a characteristic of the marine eutardigrade genus *Halobiotus*. This phenomenon is characterized by pronounced seasonal morphological changes and in *Halobiotus* involves stages with an extra protecting cuticle. Cyst formation in moss-dwelling limnic species may also occur as part of a seasonal cyclic event and can thus be viewed as part of a cyclomorphosis. Therefore, whereas diapause generally seems to be an optional response to environmental changes, it may also be an obligate part of the life cycle.

The evolution of encystment and cyclomorphosis finds its starting point in the molting process. Both phenomena represent an adaptation to environmental constraints. Notably, the evolution of diapause is not necessarily an alternative to cryptobiosis, and some tardigrades may enter both forms of dormancy. The simultaneous occurrence of

R. Guidetti
Department of Life Sciences, University of Modena and Reggio Emilia, Modena, Italy
e-mail: roberto.guidetti@unimore.it

N. Møbjerg (✉)
Department of Biology, University of Copenhagen, Copenhagen, Denmark
e-mail: nmobjerg@bio.ku.dk

R. O. Schill (ed.), *Water Bears: The Biology of Tardigrades*, Zoological Monographs 2,
https://doi.org/10.1007/978-3-319-95702-9_9

several adaptive strategies within tardigrades has largely increased the resistance of these enigmatic animals toward extreme environmental stress.

9.1 Introduction

Tardigrades are renowned for their ability to handle environmental extremes deleterious to most living organisms. They may do so by entering various forms of dormancy or in some cases by staying active and enduring large fluctuations in abiotic factors (reviewed in, e.g., Guidetti et al. 2011; Møbjerg et al. 2011; Wełnicz et al. 2011). Numerous recent studies have been devoted to the latent life phenomenon, cryptobiosis (see Chaps. 10 and 11), whereas less attention has been put on other adaptive strategies. Notably, their ecdysozoan heritage offers tardigrades a unique possibility to use molts in other remarkable adaptations. In this chapter, we put focus on encystment and cyclomorphosis. Both phenomena involve profound morphological changes and are associated with unconventional molts, not directly associated with growth of the animal. The morphological changes are so profound that they easily lead to taxonomic misinterpretations. For example, *Echinursellus longiuguis*, described by Iharos (1968) and considered a missing link between the two heterotardigrade orders, Arthrotardigrada and Echiniscoidea, was later found to be a cyst or cyclomorphic stage of the eutardigrade genus *Pseudobiotus* (Kristensen 1987).

We consider encystment and certain stages of cyclomorphosis within tardigrades to be stages of diapause. The term "diapause" is, however, not always univocal, and definitions tend to depend on the animal group considered. Here, we consider the phenomenon to be defined by a suspension of growth and development accompanied by a reduction in metabolic activity in response to adverse environmental conditions.

Encystment is likely widespread among tardigrades but only confirmed for a limited number of species (Table 9.1), and we are presently far from a complete understanding of the phenomenon. Encysted tardigrades have a characteristic oval contracted appearance (Figs. 9.1, 9.2, and 9.3). They are surrounded by retained, often opaque, exuvia or cuticular involucra that may have orange, purple-red, or brown colors. These retained involucra are often referred to as "extra cuticles" or "cuticle layers." Generally, as holds for the normal molting process, each time a new cuticle is synthesized, a new modified buccal-pharyngeal apparatus is also formed. Claw morphology of the cyst cuticles may be modified with respect to that of the active animal (Fig. 9.1). Noticeably, claws, as well as legs, may be completely missing from cyst cuticles (Figs. 9.4 and 9.5).

The other phenomenon that we highlight in this chapter, cyclomorphosis, refers to cyclic changes in morphology within the life cycle of an animal, a phenomenon well-known from other animals, such as cladoceran crustaceans and rotifers. In tardigrades, this phenomenon was first reported by Kristensen (1982) as a characteristic of the marine eutardigrade genus *Halobiotus*. Cyclomorphosis in *Halobiotus* includes the production of a moving cyst, characterized by retention of the old cuticle, and alterations of the buccal-pharyngeal apparatus, claws, as well as maturational state of the gonad.

Table 9.1 Tardigrade species in which encystment has been reported

Eutardigrada
Eohypsibioidea
Bertolanius nebulosus (Dastych, 1983) (in Westh and Kristensen 1992)
Bertolanius volubilis (Durante Pasa and Maucci, 1975) (in Guidetti et al. 2006)
Bertolanius smreczynskii (Węglarska, 1970) (in Węglarska 1970)
Bertolanius weglarskae (Dastych, 1972) (in Rebecchi and Bertolani 1984; Marley and Wright 1996; Hansen and Katholm 2002)
Macrobiotoidea
Macrobiotus echinogenitus Richters, 1903 (in Murray 1907a, d)
Macrobiotus hufelandi C.A.S. Schultze, 1834 (in Murray 1907d)
Murrayon pullari (Murray, 1907) (in Marcus 1929)
Dactylobiotus dispar (Murray, 1907) (in Murray 1907b; Węglarska 1957; Szymańska 1995)
Dactylobiotus macronyx (Dujardin, 1851) (in Lauterborn 1906)
Dactylobiotus parthenogeneticus Bertolani, 1982 (in Guidetti et al. 2006)
Richtersius coronifer (Richters, 1903) (in Heinis 1910)
Hypsibioidea
Borealibius zetlandicus (Murray, 1907) (in Stark and Kristensen 1999)
Diphascon species (in Murray 1907c)
Diphascon bullatum Murray, 1905 (in Ramazzotti 1959)
Hypsibius convergens (Urbanowicz, 1925) (in Marcus 1929)
Hypsibius dujardini (Doyère, 1840) (in Marcus 1929)
Ramazzottius species (in Murray 1907d)
Isohypsibioidea
Doryphoribius macrodon Binda, Pilato & Dastych, 1980 (in Stark and Kristensen 1999)
Eremobiotus ovezovae Biserov, 1992 (in Biserov 1992)
Isohypsibius chiarae Maucci, 1987 (in Manicardi 1989)
Isohypsius laevis McInnes, 1995 (in McInnes and Pugh, 1999)
Isohypsibius prosostomus Thulin, 1928 (in Marcus 1929)
Isohypsibius tuberculatus (Plate, 1888) (in Ramazzotti 1959)
Pseudobiotus megalonyx (Thulin, 1928) (in von Wenck 1914)
Thulinius ruffoi (Bertolani, 1982) (in Schill et al. 2007)
Heterotardigrada
Echiniscoidea
Echiniscus species (in Murray 1907d)
Echiniscus perarmatus Murray, 1907 (in Murray 1907d)
Echiniscoides sigismundi (Schultze, 1865) (in Clausen et al. 2014)

9.2 Encystment

Cysts have been found in species belonging to both Heterotardigrada and Eutardigrada. In spite of great interest toward this phenomenon at the beginning of the twentieth century (Lauterborn 1906; Murray 1907a, b; Richters 1909; Heinis 1910; Thulin 1911, 1928; von Reukauf 1912; von Wenck 1914; Nederström 1919; Rahm 1925, 1926, 1927; Richters and Krumbach 1926; Marcus 1929), only a

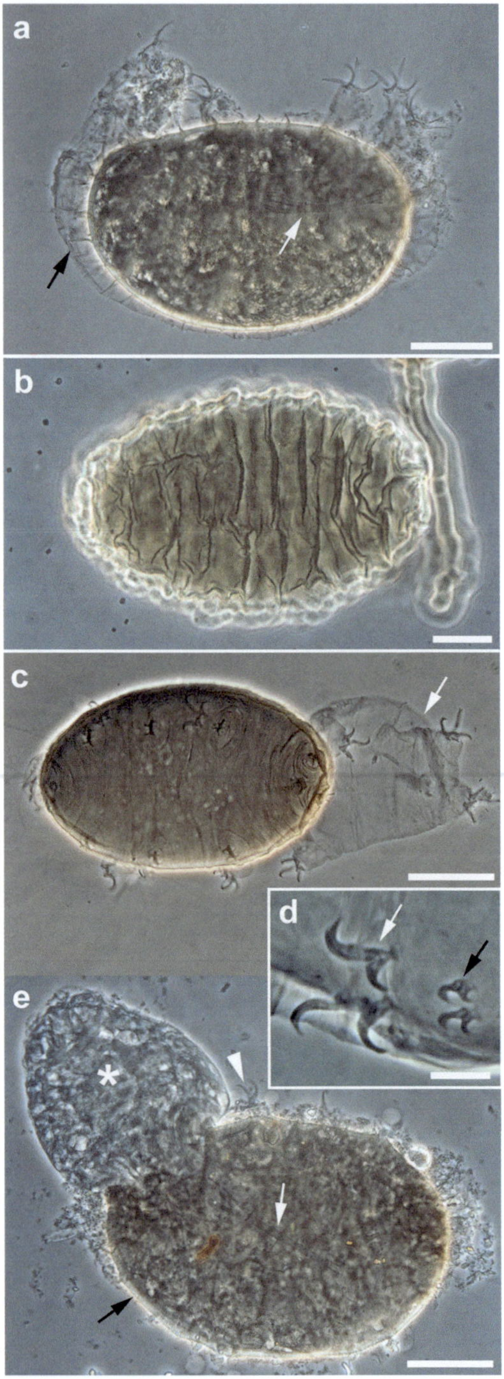

Fig. 9.1 Tardigrade cysts (light microscopy, phase contrast). (**a**) *Thulinius augusti* (black arrow = old cuticle; white arrow = buccal tube). (**b**) *Hypsibius convergens*. (**c**) *Hypsibius convergens* (white arrow = old cuticle). (**d**) Enlargement of c (white arrow = claws of old cuticle; black arrow = claws of cyst cuticle). (**e**) *Isohypsibius* sp. (arrow head = claws of old cuticle; asterisk = animal inside the cyst. Note that the posterior part of the cyst has been breached; white arrow = buccal tube; black arrow = cyst cuticle). Scale bars: **a**, **c**, **e** = 50 μm; **b** = 25 μm; **d** = 10 μm

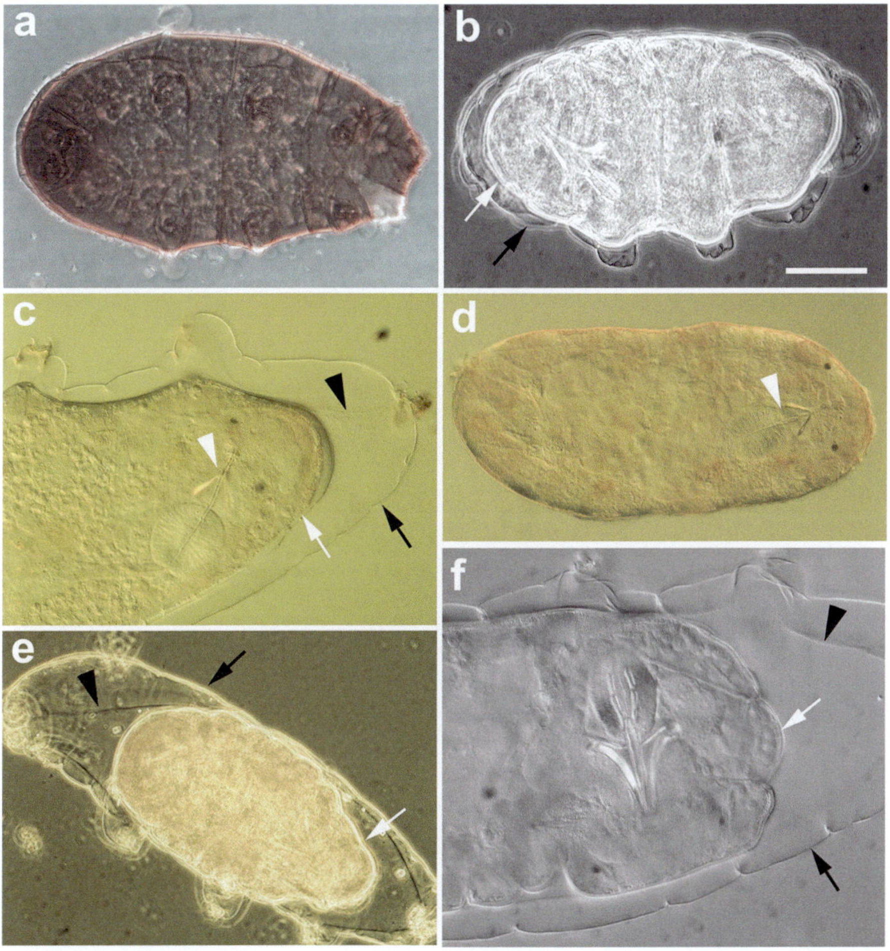

Fig. 9.2 *Bertolanius volubilis*. (**a**) Type 2 cyst (light microscopy, DIC). (**b**) Type 2 cyst without the old cuticle (black arrow = sarcophagus cuticle; white arrow = animal cuticle; light microscopy, phase contrast). (**c**) Type 2 cyst without the old cuticle (black arrow = old cuticle; black arrow head = sarcophagus cuticle; white arrow = mummy cuticle; white arrow head = modified buccal-pharyngeal apparatus; light microscopy, DIC). (**d**) Type 2 cyst without the old and sarcophagus cuticles (white arrow head = modified buccal-pharyngeal apparatus; light microscopy, DIC). (**e**) Type 1 cyst (black arrow = old cuticle; black arrow head = sarcophagus cuticle; white arrow = animal cuticle; light microscopy, phase contrast). (**f**) Type 1 cyst (black arrow = old cuticle; black arrow head = sarcophagus cuticle; white arrow = animal cuticle; DIC). Scale bars = 40 μm

limited number of more recent studies have been devoted to this area (Węglarska 1957, 1970; Westh and Kristensen 1992; Stark and Kristensen 1999; Guidetti et al. 2006, 2008, 2011; Clausen et al. 2014). Notably, more studies are needed in order to clarify the presence of encystment among tardigrade species, the evolutionary significance of the phenomenon as well as its underlying physiology.

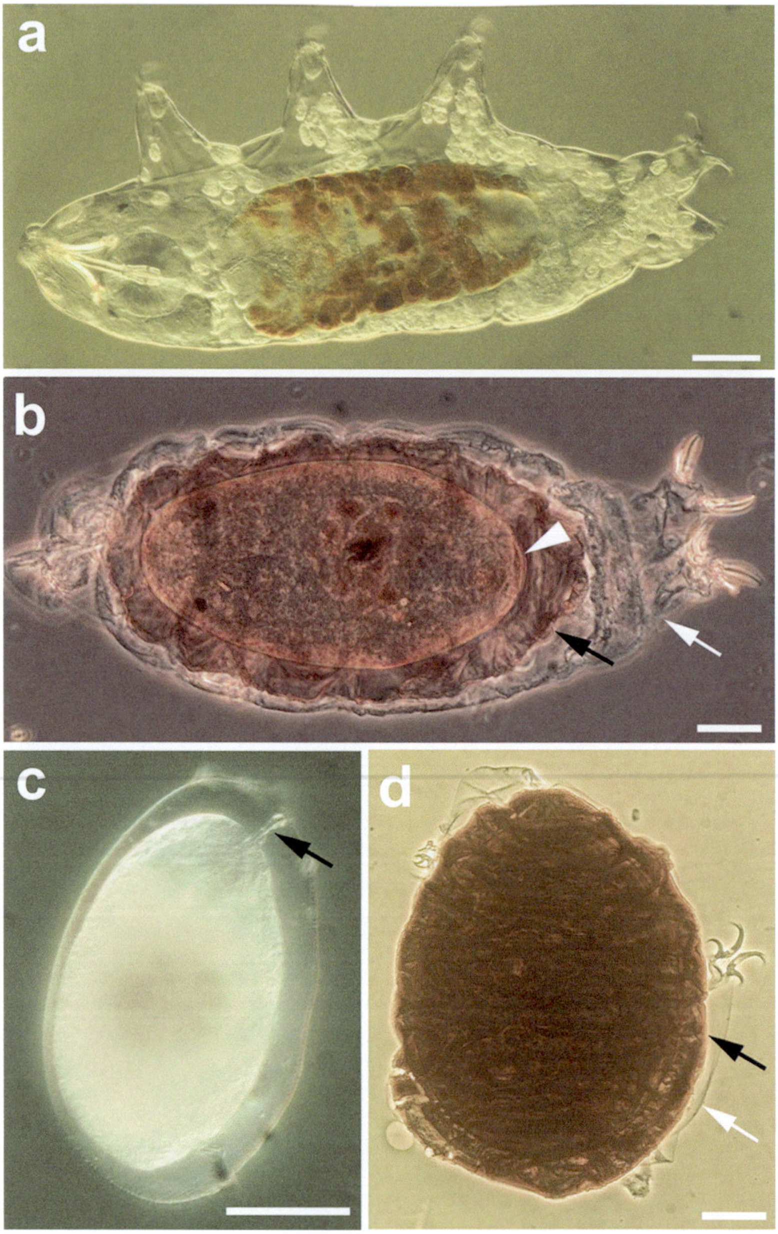

Fig. 9.3 *Dactylobiotus* specimens. (**a–c**) *Dactylobiotus parthenogeneticus*. (**a**) Active animal in vivo (light microscopy, DIC). (**b**) Cyst with all involucra (white arrow = old cuticle; black arrow = sarcophagus cuticle; arrow head = mummy cuticle; light microscopy, phase contrast). (**c**) Cyst with mummy cuticle (black arrow = buccal tube; light microscopy, DIC). (**d**) Cyst of *Dactylobiotus dispar* (white arrow = old cuticle; black arrow = sarcophagus cuticle; light microscopy, phase contrast). Scale bars = 35 μm

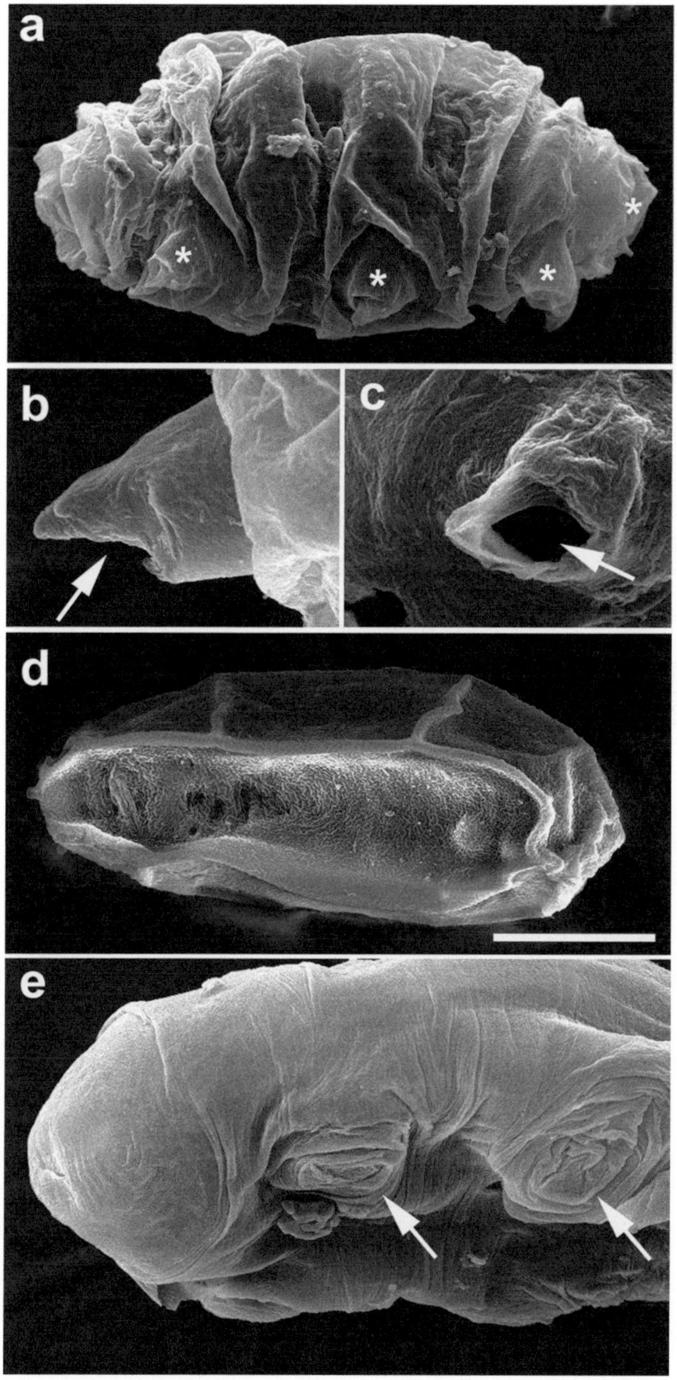

Fig. 9.4 Tardigrade cysts (scanning electron microscopy). (**a–d**) *Dactylobiotus parthenogeneticus*. (**a**) Sarcophagus cuticle (asterisk = legs without claws). (**b**) Leg of the sarcophagus cuticle, lateral view (arrow = usual position of claws when present). (**c**) Leg of the sarcophagus cuticle, ventral

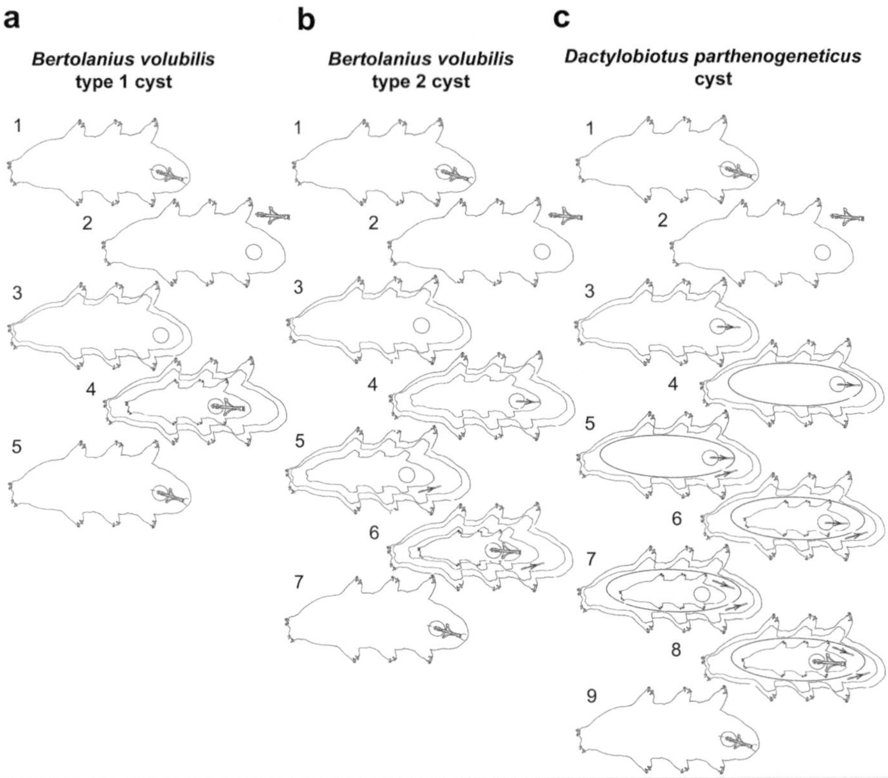

Fig. 9.5 Schematic representation of steps in the encystment processes of *Bertolanius* and *Dactylobiotus* species. (**a**) Type 1 cyst of *Bertolanius volubilis*. (**b**) Type 2 cyst of *Bertolanius volubilis*. (**c**) Cyst of *Dactylobiotus parthenogeneticus*

The first authors to report on encystment in tardigrades were Lauterborn (1906) and Murray (1907a), although Doyère (1840) probably described an encysted tardigrade without recognizing it, and Richters suggested that Lazzaro Spallanzani already in the eighteenth century had found tardigrade cysts (in Murray 1907b, d). More recently, encystment and cyst morphology have been studied in detail in two eutardigrade genera, i.e., the limnic *Dactylobiotus* (Murrayidae) and the moss-dwelling/freshwater *Bertolanius* (Eohypsibiidae) (Węglarska 1957; Westh and Kristensen 1992; Szymańska 1995; Guidetti et al. 2006), as well as in the marine tidal heterotardigrade *Echiniscoides sigismundi* (M. Schultze, 1865) (Echiniscoididae) (Clausen et al. 2014).

Fig. 9.4 (continued) view (arrow = usual position of claws when present). (**d**) Mummy cuticle, the folds are an artifact due to specimen preparation (**e**) Sarcophagus cuticle of type 2 cyst of *Bertolanius volubilis* (arrows = legs without claws). Scale bars = 10 μm

An encysted tardigrade with its series of surrounding cuticles resembles an onion or a Russian matryoshka doll (Guidetti et al. 2006) (Figs. 9.1 and 9.5). The number of cuticular layers enclosing the animal may vary, but some events and features are common to all tardigrade cysts. Encystment begins, as the normal molting process, with discharging of sclerified parts of the buccal-pharyngeal apparatus (the animal is in the *simplex stage*; Fig. 9.5). Subsequently, between one and three new cuticles are serially synthesized. The animal remains within these cuticles and does not undergo a normal ecdysis. Tardigrade cysts thus contain up to four cuticular layers, when including the animal cuticle (Fig. 9.5). These cuticles have been referred to as external/old cuticle, sarcophagus cuticle, mummy cuticle, and animal cuticle (Westh and Kristensen 1992; Hansen and Katholm 2002; Guidetti et al. 2006). One of the outer cyst cuticles, generally, becomes harder and colored with time, with pigmentation varying both among and within species (Figs. 9.1, 9.2, and 9.3).

During encystment tardigrades contract their body longitudinally and progressively decrease body movements. Internal fluids and organs may become opaque and glassy, and histolysis involving partial regression of organs has been suggested by several authors (Murray 1907a; Richters 1909; Heinis 1910; Westh and Kristensen 1992), whereas others disagree or have been unable to confirm this hypothesis (Marcus 1929; Węglarska 1957; Guidetti et al. 2006). Newly synthesized cuticular structures of the buccal-pharyngeal apparatus, claws as well as body cuticles often appear simpler, with an absence or reduction of selected features, as compared to active animals (Figs. 9.1c, d, 9.2 and 9.3). The mouth, as well as cloaca/anus, is obstructed by folds of the cyst cuticles. Thus the modified buccal-pharyngeal apparatus of an encysted tardigrade is, obviously, not used for feeding. Encystment is terminated as the tardigrade resynthesizes a normal cuticle and a normal buccal-pharyngeal apparatus, breaks through the cuticular involucra, and leaves the cyst (Fig. 9.5). A more detailed description of the morphological changes occurring during encystment in selected species follows below.

9.2.1 Encystment in the Moss Dwelling Limnic Eutardigrade Genus **Bertolanius**

More than one cyst type has been reported for species within the genus *Bertolanius* (Westh and Kristensen 1992; Rebecchi and Bertolani 1994; Guidetti et al. 2006, 2008). As holds for all tardigrades, encystment in *Bertolanius* begins with the discharge of buccal-pharyngeal sclerified parts and entrance into the *simplex stage* (Fig. 9.5). The animal subsequently produces a second cuticle while remaining within the old cuticle. In *B. nebulosus*, the tardigrade may contract and remain protected by this single-layered cyst cuticle, forming what has been termed a *white cyst* (Westh and Kristensen 1992). The formation of a single-layered cyst is a characteristic of cyclomorphosis in *Halobiotus* (see below). The tardigrade inside the white *B. nebulosus* cyst is normally immobile, but, if disturbed, it may move about in the cyst, and it eventually develops a new buccal-pharyngeal apparatus, perforates the cyst wall, and leaves the cyst. Notably, *B. nebulosus* also forms a more complicated cyst type, termed a *red cyst*, due to hardening and coloring of the external/old cuticle (Westh and Kristensen 1992). In red cysts, the second cuticle (termed the *mummy cuticle*) has a smooth surface and is without protuberances for legs and claws (Westh and Kristensen 1992, with elaborations in the student report by Hansen and Katholm 2002). A new, highly modified buccal-pharyngeal apparatus is subsequently synthesized. A third cuticle with normal legs and claws, as well as a functional buccal-pharyngeal apparatus is synthesis prior to excystation. Cysts similar to these red cysts were reported from *B. volubilis* and *B. weglarskae* (Rebecchi and Bertolani 1994). Interestingly, *B. nebulosus* leaving white cysts have been reported to lay *summer eggs*, whereas individuals leaving the red cysts seem to lay morphological different *winter eggs* (Hansen and Katholm 2002).

Detailed investigations on encystment in *B. volubilis* have shown that two types of cysts are also present in this species (Guidetti et al. 2006, 2008), i.e., *type 1* (corresponding to the white cyst) and *type 2* (corresponding to the red cyst) (Fig. 9.2). In type 1 cysts, a third cuticle with legs and claws, and a new buccal-pharyngeal apparatus, identical to that of a non-encysted animal, is synthesized. A type 1 cyst thus contains three cuticle layers, i.e., the external/old cuticle, the sarcophagus cuticle, and the cuticle lining the tardigrade itself (Fig. 9.2). These type 1 cysts are elongated, ovoid-shaped, and have a glassy, white-opaque, or yellow color. Type 2 cysts are characterized by the presence of a third, mummy cuticle, with smooth surface and legs but without claws. The latter differs from *B. nebulosus*, as this cuticle has leg protuberances (Figs. 9.2 and 9.4) (Guidetti et al. 2006, 2008). In *B. volubilis*, the second cuticle was termed the *sarcophagus cuticle* by Guidetti et al. (2006). In this stage, a simple, nonfunctional buccal-pharyngeal apparatus is formed, characterized by a buccal opening, without peribuccal lamellae, followed by a long, narrow, and flexible buccal tube (Fig. 9.2). Stylets are short, without furca; pharyngeal apophyses and placoids are also absent (Guidetti et al. 2006). These modified buccal-pharyngeal structures are subsequently discharged. The tardigrade may stay inactive for a very long time in this stage, before it synthesizes a new, fourth cuticle with legs and claws as well as a complete, functional buccal-pharyngeal apparatus. Type 2 cysts, with tardigrades that are ready to leave their cyst, thus comprise the animal with its own cuticle, a mummy, and a sarcophagus cuticle all surrounded by the old cuticle (Fig. 9.5). At the ultrastructural level, these four cuticles appear similar (Guidetti et al. 2006). In the type 2 cyst, the old cuticle becomes hard and colored.

In summary, species of *Bertolanius* seem to form two cyst types. The *white cysts* of *B. nebulosus* correspond to *type 1 cysts* of *B. volubilis*—white cysts, however, lack a sarcophagus cuticle. The *red cysts* of *B. nebulosus* correspond to *type 2 cysts* of *B. volubilis*—the former, however, lack a sarcophagus cuticle, and the mummy cuticle is without legs. Noticeably, the sarcophagus cuticle of *B. volubilis* has mostly been found in cysts induced under laboratory conditions, and it only seems present in a relatively small number of cysts collected in nature (Guidetti et al. 2008).

9.2.2 Encystment in the Limnic Eutardigrade Genus Dactylobiotus

Cysts have been described from three species of *Dactylobiotus*, i.e., *Dactylobiotus parthenogeneticus* Bertolani, 1982, *Dactylobiotus dispar* (Murray, 1907), and *Dactylobiotus macronyx* (Dujardin, 1851) (Lauterborn 1906; Murray 1907b; Węglarska 1957; Szymańska 1995; Guidetti et al. 2006).

A detailed account on encystment in *D. parthenogeneticus* was given by Guidetti et al. (2006). As holds for all tardigrades, the first step of encystment, in this species, is represented by a discharge of buccal-pharyngeal sclerified structures (Fig. 9.5). The animal enters a simplex stage and produces a second (sarcophagus) cuticle, which, due to contraction of the tardigrade, is characterized by a folded surface (Fig. 9.4). The sarcophagus cuticle has legs without claws—wide holes mark the distal end of the cuticle's legs (Fig. 9.4). In light microscopy, the surface of the sarcophagus cuticle appears reticulated with irregular meshes, whereas it, in scanning electron microscopy, appears smooth and uniform (Fig. 9.4). At the ultrastructural level, this cuticle is composed of an external electron-lucent heterogeneous layer, a trilaminar layer, and an internal wide layer containing large lacunae. During encystment the sarcophagus cuticle becomes harder and darker while losing its connection with the external/old cuticle. The animal further synthesizes new, highly modified cuticular structures for the buccal-pharyngeal apparatus (Fig. 9.5). A narrow and flexible buccal tube connects to the pharynx, which has an internal cuticular lining, but lacks apophyses and placoids (Fig. 9.3). Stylets and associated structures, such as sheaths and supports, are also absent. Successively, the animal synthesizes yet another wrinkled cuticle (the mummy cuticle) without both legs and claws (Figs. 9.4 and 9.5). At the ultrastructural level, the mummy cuticle is characterized by two trilaminar layers, separated by a nonhomogeneous layer, which in turn is constituted by electron-lucent areas crossed by dense, randomly oriented, convolute cordons. Sometimes these cordons assume a more regular organization and the layer appears striated (see also Szymańska 1995). It is worth noting that Szymańska (1995) reported on cuticle ultrastructure in *D. dispar* cysts and that the above description is in agreement with her results. In Szymańska (1995) one can also find valuable information on the ultrastructure of the epidermis during encystment.

Following the production of the mummy cuticle, the animal discharges sclerified structures of the buccal-pharyngeal apparatus and begins synthesizing new structures, similar to the previous ones (Fig. 9.5). At this point, the tardigrade may stay inactive for an extended period, before continuing the encystment process. Subsequent steps of the encystment process include yet another discharge of buccal-pharyngeal sclerified structures (Fig. 9.5), followed by synthesis of a new body cuticle with legs and claws, as well as a new functional buccal-pharyngeal apparatus identical to that of non-encysted animals. The tardigrade is now ready to break through the cuticular coats and leave the cyst (Fig. 9.5). At the end of the encystment process, the cyst has become ovoid in shape and has attained a brown-red color (Fig. 9.3). Cysts collected in nature often lack the external/old cuticle, which presumably has been lost.

9.2.3 Encystment in the Tidal Heterotardigrade Echiniscoides sigismundi

The marine heterotardigrade *Echiniscoides sigismundi* constitutes a cryptic species complex present worldwide in tidal zones (see Chap. 3). Encystment, involving synthesis of two to three new cuticles, was recently reported in this species (Clausen et al. 2014). In *E. sigismundi*, the first stage of the encystment process resembles a normal molt (Fig. 9.6a, b). Noticeably, the newly synthesized, second cuticle is structurally similar to the old cuticle and contains legs as well as claws (Fig. 9.6b). If applying the terminology used above for *B. volubilis*, this second cuticle (c2) would correspond to the sarcophagus cuticle. A new, fully functional buccal-pharyngeal apparatus is present at this stage, but instead of leaving the old cuticle (exuvium), the tardigrade produces yet another cuticle and retracts into a tun-like state within the now two-layered cyst wall (Fig. 9.6c). In the most developed cyst, found by Clausen et al. (2014), the tardigrade seemed to have produced yet another protective cuticle lacking claws. This cuticle would correspond to the mummy cuticle (c3). The animal inside this cyst, now lined by a fourth cuticle, had discharged sclerified parts of the buccal-pharyngeal apparatus into the second (sarcophagus) cuticle (Fig. 9.6d, e). When compared to the cysts of eutardigrades, *E. sigismundi* cysts seem to lack pigmentation and the second (sarcophagus) cuticle has claws. Furthermore, a new buccal-pharyngeal apparatus (functional or not) may not always be produced together with new cuticle layers. The finding of cysts in *E. sigismundi* confirmed earlier observations on *Echiniscus*, reporting that encystment also occurs within heterotardigrades (Murray 1907b; Ramazzotti and Maucci 1983).

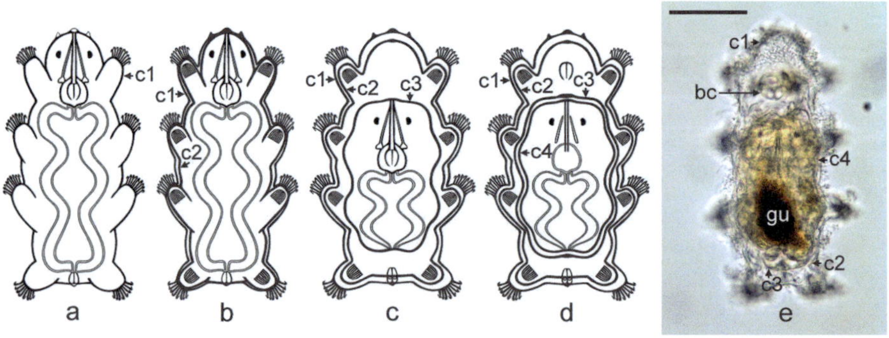

Fig. 9.6 Cyst formation in *Echiniscoides sigismundi*. (**a-d**) Schematic representation from Clausen et al. (2014). (**a**) Active animal. (**b**) Molting animal with closed mouth and anus. (**c**) The tardigrade has synthesized yet another cuticle (c3) and has retracted within the two cuticular coats (c1–c2), which both contain claws. It retains a fully developed buccal-pharyngeal apparatus. (**d**) Cyst with three cuticular coats (c1–c3). The animal has discharged sclerified parts of the buccal-pharyngeal apparatus into c2. c4 is the cuticle lining the tardigrade itself. (**e**) Light microscopy of *Echiniscoides sigismundi* cyst from Clausen et al. (2014) corresponding to the schematic representation in (**d**). Abbreviations: bc = bulb cuticle; gu = gut; scale bar 50 μm

9.2.4 Tardigrade Cyst Ecology

Diapause in tardigrades, in the form of encystment, is likely controlled by exogenous and perhaps also endogenous stimuli that initiate and direct the successive phases of the deep morphological transformations outlined above. Little is still known of these stimuli in tardigrades. The most common stimuli inducing diapause in insects and nematodes are temperature, dehydration, photoperiod, and food quantity and quality (Denlinger and Tanaka 1999; Sommerville and Davey 2002; Koštál 2006). It is commonly believed that encystment in tardigrades is induced by a general deterioration of environmental conditions. Reserve depletion, low oxygen tension, alterations in pH, osmotic stress, and temperature variation have all been highlighted as factors involved in tardigrade encystment (Murray 1907b; Von Wenck 1914; Marcus 1929, 1936; Węglarska 1957; Westh and Kristensen 1992; Szymańska 1995; McInnes and Pugh 1999; Guidetti et al. 2008; Clausen et al. 2014).

Encystment was induced in *D. dispar* under laboratory conditions by Węglarska (1957) and Szymańska (1995). Both authors argued that encystment is not an obligatory part of the life cycle in this species. Węglarska's observations suggested that encystment is caused by a general deterioration of environmental conditions, whereas Szymańska concluded that a decrease in pH is an important factor leading to encystment. Noticeable, the two conclusions are not necessarily in disagreement. Węglarska (1957) further reported that environmental changes had to occur gradually in order to induce encystment, as rapid habitat changes lead to death rather than encystment. It would take *D. dispar* approximately 12 days to produce a cyst, but once formed the animal could stay for more than 9 months in this state (Węglarska 1957). Węglarska (1957), moreover, observed that when she placed encysting *D. dispar* in clean water, the animals would reassume their active normal state rather than completing encystment, whereas animals left in water with decaying leaves completed encystment. Excystation appeared 14–81 days after placing cysts in tap water (at 20–25 °C). Cysts placed in water containing a tangle of algae did not excyst—when some of these cysts were subsequently placed in tap water, the animals left the cyst within 6–48 hours, while those kept in water with algae eventually died (Węglarska 1957). Szymańska (1995) cultured *D. dispar* at neutral pH (6.8–7.2 pH). Under these conditions the animals did not produce cysts, while lowering pH to 6.6 induced cysts. Szymańska hypothesized that encystment would protect the animals against acidified water, which could damage calcareous structures of the buccal-pharyngeal apparatus.

Starvation has been suggested as a factor, which could either induce or terminate encystment in tardigrades. It is not clear whether a depletion of energy supply normally occurs during encystment. Encysted tardigrades do not take in food through the digestive system, but they clearly have a metabolism as judged by measurements of a depressed metabolic rate (Pigòn and Węglarska 1953). Prior to entry into the cyst stage, tardigrades accumulate food reserves (lipids and polysaccharides, as glycogen) in their storage cells (Von Wenck 1914; Węglarska 1957; Szymańska 1995). Węglarska (1957) discussed von Wenck's (1914) hypothesis that exhaustion of reserve material was an encystment terminating factor in *Pseudobiotus megalonyx* (Thulin, 1928) (cited as *Macrobiotus lacustris* but later attributed to *P. megalonyx* by Thulin (1928); see also Bertolani (1976)), but Węglarska's work on *D. dispar* did not support von Wenck's hypothesis. Węglarska (1957) found excysting animals with storage cells having more reserve material than that of active animals. During encystation in *D. dispar*, numerous lipid droplets, vesicles, and multivesicular bodies accumulate in the cytoplasm of midgut epithelial cells (Rost-Roszkowska and Poprawa 2008). Exocytosis, into the gut lumen, by these cells could account for removal of wastes during encystation. Moreover, the numerous lipid droplets as well as the presence of a Golgi apparatus and cisterns of endoplasmic reticulum suggest intense lipid synthesis (Rost-Roszkowska and Poprawa 2008). Accordingly, storage as well as midgut cells likely play a role in accumulating and supplying nutrients to other tissues and organs during encystment in tardigrades.

The encysted state can last for several months, and tardigrades may, e.g., overwinter or endure dry summer periods in this state. Temperature has often been highlighted as a stimulus inducing and directing cyst formation. Murray (1907d) reported that in Scotland encystment of *D. dispar* took place at the beginning of winter, just as the shallow ponds with the tardigrades began to have an ice cover. During winter, the ponds froze completely, and in this situation, there were no active animals, only cysts. If substrate with cysts was placed in a warm room for some hours, active animals began to appear (Murray 1907d). A similar case was recorded in Greenland, where *B. nebulosus* was found within moss in the outlet stream of a small pond that during winter was frozen for 4–5 months (Westh and Kristensen 1992). Between 80% and 88% of the *B. nebulosus* population overwintered as white cysts. Yet, in a similar population present in a homothermic spring, which in summer would nearly dry out, most of the specimens died or formed the thick-walled red cysts (Westh and Kristensen 1992). Schill and co-workers (2007) reported that moss samples collected during a winter expedition to the Sinai Mountains (Egypt, 1750 m a.s.l) exclusively contained *Thulinius ruffoi* (Bertolani, 1982) in the cyst state (cysts were found in 46 samples out of 104 samples). During winter at this altitude, the temperature may drop below 0 °C at night, which would provide a possible explanation for the occurrence of cysts as a survival strategy.

Experimental and field observations on *B. volubilis* have shown that temperature represents a token stimulus for cyst induction (Guidetti et al. 2008). In a study by Guidetti and co-workers (2008), animals collected in nature were exposed to various temperatures (6 °C, 14 °C, and 20 °C) under laboratory conditions. Whereas *B. volubilis* collected in spring (warm season) mainly formed type 2 cysts in response to these temperatures, animals collected in the fall (cold season) also formed type 1 cysts (Guidetti et al. 2008). This would indicate that the animals were already sensitive (programmed) according to the approaching season. Photoperiod did not seem to affect the type of cysts formed by *B. volubilis*. Tardigrades kept at 14 °C entered the simplex stage after approximately 10 days, and it took them about 15–20 days to produce either type 1 or type 2 cysts (Guidetti et al. 2008). Tardigrades that had entered the type 1 cyst stayed encysted for about 20 days (at 14 °C) and 40 days (at 6 °C), respectively. Termination of the type 1 cyst state took place spontaneously. No tardigrades were observed to leave type 2 cysts within the 60 days of observations. However, if the most external thick involucra were broken, the animals would emerge from these cysts after a few days (Guidetti et al. 2008). Analyses of cyst occurrence in *B. volubilis* during a 2-year sampling period in the field confirmed that temperature was likely involved in cyst induction, maintenance, as well as excystation (Guidetti et al. 2008). Further evidence, supporting the hypothesis that temperature is an important stimulus directing cyst production in tardigrades, comes from the abovementioned study of the Greenlandic populations of *B. nebulosus*, in which warm periods seem to induce red cysts, whereas cold periods induce white cysts (Westh and Kristensen 1992).

Lastly, osmotic stress has recently been suggested to induce encystment in the marine *Echiniscoides sigismundi*, which inhabits barnacles, algae, and lichens in tidal zones worldwide (Clausen et al. 2014). In the study by Clausen and co-workers (2014), *E. sigismundi* cysts were found following laboratory treatments with hypo- and hypertonic solutions, suggesting that osmotic stress represents an adverse condition that ultimately may stimulate cyst formation. A continuous osmotic imbalance thus seems to be yet another diapause-inducing stimulus in tardigrades.

9.3 Cyclomorphosis

Kristensen (1982) used the term cyclomorphosis, during his erection of the marine eutardigrade genus *Halobiotus*, to describe cyclic changes in morphology occurring within single species (Fig. 9.7). In his paper, Kristensen (1982) presented a detailed historical account of the challenges with species descriptions and taxonomy within this genus that eventually led to recognition of the phenomenon.

Halobiotus crispae Kristensen, 1982 remains the only tardigrade for which cyclomorphosis has been clearly established (Kristensen 1982; Møbjerg et al. 2007; Halberg et al. 2013a). Cyclomorphosis, in this species, is characterized by pronounced seasonal changes within individual animals including the presence of extra cuticles, closing of the mouth/cloaca, differences in the buccal-pharyngeal apparatus and in claw size, presence of gut contents, as well as body color and speed of movements.

Animals in the so-called *active stage* are characterized by a fully developed gonad, large claws, six peribuccal sensory organs, and a buccal-pharyngeal apparatus with micro- and macro-placoids. The mouth and cloaca are open in this stage, the gut is normally filled, and the animals appear translucent, move relatively rapidly around, and reproduce.

The so-called *pseudosimplex 1 (P1) stage* is characterized by a double cuticle with closed mouth and cloaca (Fig. 9.7). As holds for tardigrades that are in the process of encystment, this stage is formed by an incomplete molt of an active animal, in which buccal-pharyngeal sclerified parts are discharged, but the old cuticle is retained (*pre-pseudosimplex*; Fig. 9.7). Subsequently, a new set of morphologically similar, but smaller, claws emerge beneath the existing one, and a new

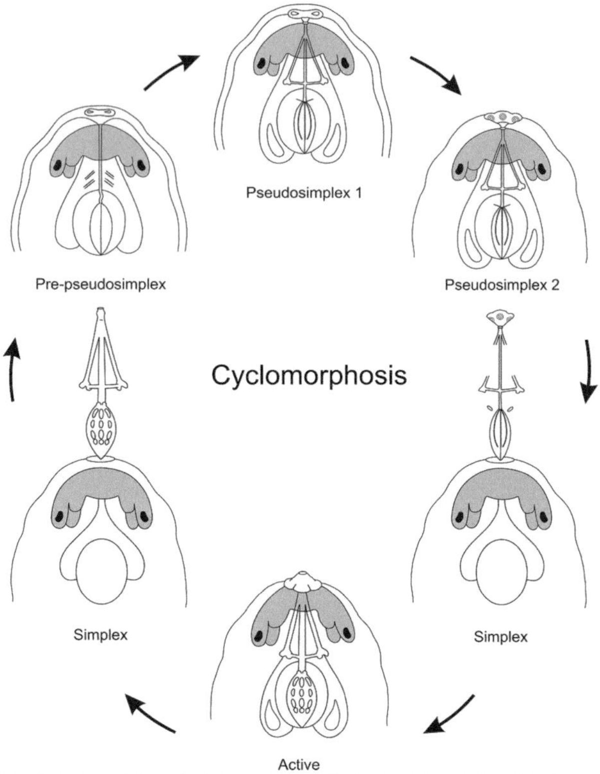

Fig. 9.7 Cyclomorphosis in *Halobiotus crispae* modified from Kristensen (1982)

modified buccal-pharyngeal apparatus is also formed: stylets and their supports are small, and, noticeably, the pharyngeal bulb is lined by straight cuticle-/bar-shaped placoids (Eibye-Jacobsen 1997, 2001; Kristensen 1982). Gut contents are usually absent or reduced, and the animals appear white, likely due to a high number of coelomocytes (storage cells). Gonads are reduced and sex determination is not possible in this stage. Contrary to the cysts described above, animals in the P1 stage are able to move (although with reduced speed) despite the presence of a double cuticle, indicating that muscles are still attached to attachment sites on the old cuticle. Along this line, Kristensen (1982) reported that he observed such connections in ultrathin sections. Animals in P1 are gregarious forming clusters often containing several hundred tardigrades. Similar to a semiterrestrial eutardigrade cyst wall, the outer cuticle of P1 may eventually turn dark and become sclerotized. Interestingly, animals in the P1 stage can, under laboratory conditions, be forced to retract into the outer cuticle and thus form a proper cyst (Fig. 9.8).

The animal in the pseudosimplex 1 stage eventually breaks through the outer cuticle as a *pseudosimplex 2 (P2) stage* with open mouth and cloaca and small claws (Fig. 9.7). P2 is a sexually ripening stage with an aberrant, but functional buccal-pharyngeal apparatus. The aberrant claws and buccal-pharyngeal apparatus of P2 would make species determination very hard if the transition between

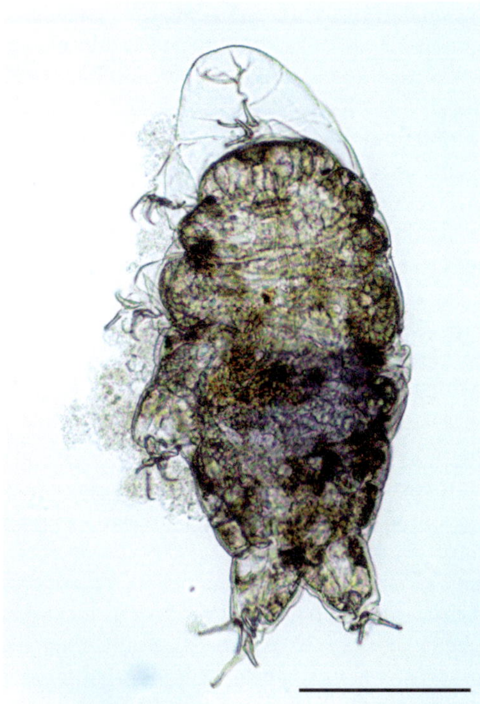

Fig. 9.8 *Halobiotus crispae* contracted within its P1 stage cuticle following exposure to $CuCl_2$ (Hygum et al. 2017). Scale bar = 100 μm

pseudosimplex and active stages was not directly observed (Kristensen 1982). A normal simplex or molting stage lacking stylets and placoids marks the transition from P2 to the active stage. This simplex stage is characterized by a closed mouth that lacks peribuccal sensory organs and a developing cuticle with new claws that correspond in shape and size to that of the emerging active stage.

9.3.1 Physiological and Ecological Significance of Cyclomorphosis in Halobiotus crispae

Halobiotus crispae is a holarctic species, and investigations into the seasonal appearance of cyclomorphic stages have been conducted at two localities: the type locality at Nipisat Bay, Greenland, and at the southernmost locality at Vellerup Vig, Denmark (Kristensen 1982; Møbjerg et al. 2007). Interestingly, marked differences are found in the time of year at which the different stages appear at these localities. The P1 stage is a hibernating stage occurring during the long winter in Greenland, while it in Denmark is the dominating stage during the summer period. The active stage, on the other hand, is present during summer in Greenland, while it is found during late winter and spring in Denmark. The P2 stage only appears for a very short period during spring at Nipisat, but this stage is the dominating stage during winter at Vellerup Vig. It was hypothesized that while P1 likely is an adaptation to withstand low temperatures during winter in Greenland, this stage possibly enables the tardigrades to tolerate heat stress and oxygen depletion during the Danish summer (Møbjerg et al. 2007).

Experimental investigations into the thermal tolerance of *Halobiotus crispae* from Vellerup Vig have subsequently revealed that the active stage is more sensitive to high temperatures than the pseudosimplex stages. Specifically, tardigrades in the active stage begin to die at a temperature of around 14 °C, whereas P1 and P2 start dying at around 25 °C (Halberg et al. 2013a). These experiments support the hypotheses that temperature is an important factor directing cyclomorphosis and at the same time a limiting factor for the geographic distribution of *H. crispae*. Entry into P1 during the Danish summer thus likely represents an adaptive strategy that ensures survival of the population during periods of heat stress (Halberg et al. 2013a).

Investigations into the physiological and ecological significance of cyclomorphosis have further revealed that P1 is freeze tolerant, whereas the active stage is not (Halberg et al. 2009). Both stages, however, avoid freezing and supercool to around −20 °C; thus, the active stage, nevertheless, tolerates quite low temperatures (Halberg et al. 2009). Experiments on osmotic stress tolerance have further shown that especially the active stage tolerates a large range of salinities, with specimens from the Greenlandic population at Nipisat remaining active in solutions ranging from distilled water to saltwater of up to 2000 mOsmkg^{-1} (Halberg et al. 2009). Interestingly, *H. crispae* and all other tardigrades investigated so far seem to be hyperosmotic regulators, keeping their body fluids hyperosmotic with respect to the external medium, ensuring a high water turnover (Møbjerg et al. 2011; Halberg et al. 2013b).

9.4 Concluding Remarks

Animals in diapause minimize exchange with the environment by producing, e.g., cocoons (e.g., Nemertea and Annelida; Càceres 1997; Diaz Cosin et al. 2006) and thick cuticles (e.g., dauer larva of nematodes; Càceres 1997), or by remaining within a cuticle of a previous instar (e.g., the pharate state of insects; Gullan and Cranston 2005). In tardigrades, diapause is associated with production of extra cuticles that isolate and protect the animal from the environment (Węglarska 1957; Kristensen 1982; Guidetti et al. 2006; Møbjerg et al. 2007; Halberg et al. 2009). Whereas diapause in many tardigrades seems to be an optional response to environmental change, it may also be an obligate and integrated part of a seasonal change, as holds for the P1 stage of *H. crispae*.

Rebecchi and Bertolani (1994) reported that cyst formation in *Bertolanius* is a cyclic event and as such they suggested that this genus also has cyclomorphosis. They argued that a stage similar to the P2 stage sometimes could be found in

B. volubilis and *B. weglarskae*. Specifically, they found specimens that had left their thick cyst cuticle and thus appeared with a modified buccal-pharyngeal apparatus, but also without claws on their legs. It was later suggested that such animals only would appear if cyst formation was interrupted prematurely and that these animals would not survive (Hansen and Katholm 2002). Hansen and Katholm (2002), nevertheless, considered the occurrence of the white/type 1 cyst in *B. nebulosus* as part of a cyclomorphosis. Along this line, encystment may occur as part of a cyclomorphosis, with individual animals undergoing cyclic and reversible morphological modifications. Accordingly, cyclomorphosis in *H. crispae* and encystment in other tardigrade species share many common features (Table 9.2).

Formation of cysts and pseudosimplex stages is a costly process, as the tardigrade produces new cuticular structures. Moreover, while the animals are encysted or in a pseudosimplex stage, the population cannot increase. Consequently, the production of diapause stages must be considered an expensive process at both individual and population levels. Nevertheless, diapause has evolved among tardigrades as a widespread adaptation strategy to survive environmental stress. The evolution of encystment and cyclomorphosis finds its starting point in the molting process (Figs. 9.5 and 9.7). The first step in the process of producing advanced diapause stages was likely an incomplete molt, characterized by delayed ecdysis, which

Table 9.2 Cyclomorphosis (P1 stage) vs encystment in tardigrades

	P1 stage of *H. crispae*[a]	Cyst[b]
Incomplete molt	Yes	Yes
Multiple cuticle layers	Yes	Yes
Buccal-pharyngeal apparatus	Modified and simpler	Modified and simpler
Morphological changes	Yes	High
Mobility	Reduced	Absent
Metabolic activity	Presumably lower	Low
Duration	Months	Months
Seasonality (predictable)	Yes	Yes
Possible inducing factors	Temperature	Temperature
Function (increased resistance to)[c]	Freezing/low O_2/high temperature	Low O_2/high temperature
Termination	Endogenous and environmental factors	Endogenous and environmental (low temperature) factors
Anhydrobiotic	No	Yes
Cryobiotic	No[d]	Yes
Cost for individual/ population	High	Very high

[a]Data referring to the cyclomorphic pseudosimplex 1 stage of *Halobiotus crispae* (see Kristensen 1982; Møbjerg et al. 2007; Halberg et al. 2009)
[b]Data referring to the type 2 cyst of *Bertolanius volubilis* (see Guidetti et al. 2006, 2008)
[c]Current hypotheses. Data not yet experimentally verified
[d]Freeze tolerant, but with a lower lethal temperature

allowed utilization of the old cuticle (exuvium) as a protecting coat. Interestingly, delayed ecdysis was observed in the eutardigrade *Paramacrobiotus areolatus* (Murray, 1907), when active animals were maintained under unnatural laboratory conditions (Crowe et al. 1971). A possible scenario in the evolution of encystment is the subsequent production of more cuticular involucra with protecting and isolating functions. The costly production of a nonfunctional buccal-pharyngeal apparatus with each new cuticle could be linked to developmental constraints, as production of these structures always parallels synthesis of the cuticle during a molt. The energetic costs seem to be reduced by the production of simplified cuticular structures.

Cyst formation and cyclomorphosis in tardigrades are examples of unique epigenetic phenotypic plasticities that evolved in order to withstand unfavorable environmental conditions. The morphological changes occurring during entrance to diapause in tardigrades are reversible, but they may be very drastic, inferring taxonomically problems as described for *H. crispae*. Even more drastically, encysted *Dactylobiotus* specimens can turn into an ovoid mass without any external or internal signs of being a tardigrade! (Fig. 9.3).

In many invertebrates, diapause is induced in advance of an advent of environmental adversities (Košťál 2006). Possible diapause-inducing environmental adversities for tardigrades likely include low oxygen tension and high temperatures (the two factors are strictly connected). *Bertolanius volubilis*, living in moss on rocks at 1200 m a.s.l., produces cysts in response to an increase in air temperature (Guidetti et al. 2008). *Isohypsibius laevis* McInnes, 1995, living in Antarctic lakes, passes the winter encysted as a possible response to the anoxic condition of the water (McInnes and Pugh 1999). *Halobiotus crispae* likely produces its resting state (P1) during warm periods in order to withstand oxygen depletion and heat stress (Møbjerg et al. 2007; Halberg et al. 2013a).

A common abiotic stress factor for semiterrestrial tardigrades is the periodic lack of water-associated terrestrial environments. Semiterrestrial species withstand periods of dehydration by entering anhydrobiosis—a form of cryptobiosis induced by dehydration. Notably, encystment is best studied in freshwater species, which generally have reduced or absent cryptobiotic abilities. However, most of the species that produce cysts (Table 9.1) can also enter anhydrobiosis. Thus, lack of free water cannot be considered the "environmental adversity" that tardigrades respond to when producing cysts, and it cannot be regarded as a driving force for the evolution of encystment. Experimental data demonstrate that specimens of *Bertolanius* are able to enter cryptobiosis, when already in a state of diapause, confirming that both dormancy phenomena (diapause and cryptobiosis) can be present within a single species (Westh and Kristensen 1992; Guidetti et al. 2008). Therefore, the evolution of diapause is not necessarily an alternative to cryptobiosis. On the contrary, the simultaneous occurrence of both adaptive strategies has largely increased tardigrades' resistance toward extreme environmental stress.

References

Bertolani R (1976) Osservazioni cariologiche su *Isohypsibius augusti* (Murray, 1907) e *I. megalonyx* Thulin, 1928 (Tardigrada) e ridescrizione delle due specie. Boll Zool 43:221–234

Bertolani R (1982) 15. Tardigradi (Tardigrada). Guide per il riconoscimento delle specie animali delle acque interne Italiane. Consiglio Nazionale Delle Ricerche, Verona, Italy, p 104

Biserov VI (1992) A new genus and three new species of tardigrades (Tardigrada: Eutardigrada) from the USSR. Boll Zool 59:95–103

Càceres CE (1997) Dormancy in invertebrates. Invertebr Biol 116:371–383

Clausen LKB, Andersen KN, Hygum TL, Jørgensen A, Møbjerg N (2014) First record of cysts in the tidal tardigrade *Echiniscoides sigismundi*. Helgol Mar Res 68(4):531–537. https://doi.org/10.1007/s10152-014-0409-0

Crowe JH, Newell IM, Thomson WW (1971) Cuticle formation in the tardigrade, *Macrobiotus areolatus* Murray. J Microsc 11:121–132

Denlinger DL, Tanaka S (1999) Diapause. In: Knobil E, Neill JD (eds) Encyclopedia of Reproduction, vol 1. Academic Press, San Diego, pp 863–872

Diaz Cosin DJ, Riuz MP, Ramajo M, Gutiérrez M (2006) Is the aestivation of the earthworm Hormogaster elisae a paradiapause? Invertebr Biol 125:250–255

Doyère M (1840) Mémoiresur les Tardigrades. Ann Sci Naturelles, Paris, Ser 2, Zool 14:269–361

Eibye-Jacobsen J (1997) Development, ultrastructure and function of the pharynx of *Halobiotus crispae* Kristensen, 1982 (Eutardigrada). Acta Zool 78:329–347

Eibye-Jacobsen J (2001) Are the supportive structures of the tardigrade pharynx homologous throughout the entire group? J Zool Syst Evol Research 39:1–11

Guidetti R, Boschini D, Rebecchi L, Bertolani R (2006) Encystment processes and "Matrioshka-like stage" cyst in a moss-dwelling and in a limnic species of eutardigrades (Tardigrada). Hydrobiologia 558:9–21

Guidetti R, Boschini D, Altiero T, Bertolani R, Rebecchi L (2008) Diapause in tardigrades: a study of factors involved in encystment. J Exp Biol 211:2296–2302

Gullan PJ, Cranston PS (2005) The Insects: an outline of entomology, 3rd edn. Blackwell, Oxford, p 505

Guidetti R, Altiero T, Rebecchi L (2011) On dormancy strategies in tardigrades. J Insect Physiol 57 (5):567–576

Halberg KA, Persson D, Ramløv H, Westh P, Kristensen RM, Møbjerg N (2009) Cyclomorphosis in Tardigrada: adaptation to environmental constraints. J Exp Biol 212(17):2803–2811

Halberg KA, Persson DK, Jørgensen A, Kristensen RM, Møbjerg N (2013a) Ecology and thermal tolerance of the marine tardigrade *Halobiotus crispae* (Eutardigrada: Isohypsibiidae). Mar Biol Res 9(7):716–724

Halberg KA, Larsen KW, Jørgensen A, Ramløv H, Møbjerg N (2013b) Inorganic ion composition in Tardigrada: cryptobionts contain large fraction of unidentified organic solutes. J Exp Biol 216:1235–1243

Hansen JG, Katholm AK (2002) A study of the genus *Amphibolus* from Disko Island with special attention on the life cycle of *Amphibolus nebulosus* (Eutardigrada: Eohypsibiidae). In: Hansen JG (ed) Arctic Biology Field Course Quqertarsuaq 2002. Zoological Museum University of Copenhagen, Copenhagen, pp 129–163

Heinis F (1910) Systematik und Biologie der Moosebewohnen den Rhizopoden, Rotatorien und Tardigraden usw. Archiv für Hydrobiologie und Plankton kund: 51–115

Hygum TL, Fobian D, Kamilari M, Jørgensen A, Schiøtt M, Grosell M, Møbjerg N (2017) Comparative investigation of copper tolerance and identification of putative tolerance related genes in tardigrades. Front Physiol 8:95. https://doi.org/10.3389/fphys.2017.00095

Iharos G (1968) The scientific results of the Hungarian soil zoological expeditions to South America: 6. Ein neue Tardigraden-Gattung von mariner Verwandtschaft aus dem chilenischen Altiplano. Opusc Zool 7:357–361

Koštál V (2006) Eco-physiological phases of insect diapause. J Insect Physiol 52:113–127

Kristensen RM (1982) The first record of cyclomorphosis in Tardigrada based on a new genus and species from Arctic meiobenthos. J Zool Syst Evol 20:249–270
Kristensen RM (1987) Generic revision of the Echiniscidae (Heterotardigrada) with a discussion of the origin of the family. In: Biology of tardigrades, pp 261–335
Lauterborn R (1906) Demostrationen aus der Fauna ders Oberreheins und seiner Umgebung. Verh Dtsch Zool Ges:265–268
Manicardi GC (1989) Two new species of soil moss eutardigrades (Tardigrada) from Canada. Can J Zool 67(9):2282–2285
Marcus E (1929) Tardigrada. In: Bronns HG (ed) Klassen und Ordnungen des Tierreichs, vol 5. Akademische Verlagsgesellschaft, Leipzig, Germany, p 608
Marcus E (1936) Tardigrada. In: Schulze FE, Kükenthal W, Heider K (eds) Das Tierreich, vol 66. Walter de Gruyter Berlin und Leipzig, Germany, pp 1–340
Marley NJ, Wright DE (1996) *Amphibolus weglarskae* (Dastych), a new addition to the Tardigrada of Iceland with an updated checklist of Icelandic species (Eohypsibiidae, Eutardigrada). Quekett J Microsc 37:541–545
Maucci W (1987) A contribution to the knowledge of the North American Tardigrada with emphasis on the fauna of Yellowstone National Park (Wyoming). In: Biology of Tardigrades. Selected Symposia and Monographs UZI, vol 1, pp 187–210
McInnes SJ, Pugh PJA (1999) Zonation in Antarctic lake-dwelling benthic meiofauna, with emphasis on the Tardigrada. Zool Anz 238:283–288
Møbjerg N, Jørgensen A, Eibye-Jacobsen J, Halberg KA, Persson D, Kristensen RM (2007) New records on cyclomorphosis in the marine eutardigrade *Halobiotus crispae* (Eutardigrada: Hypsibiidae). J Limnol 66(Suppl 1):132–140
Møbjerg N, Halberg KA, Jørgensen A, Persson D, Bjørn M, Ramløv H, Kristensen RM (2011) Survival in extreme environments – on the current knowledge of adaptations in tardigrades. Acta Physiol 202(3):409–420
Murray J (1907a) The encystment of *Macrobiotus*. Fortschr Zool 11:4–11
Murray J (1907b) Encystment of Tardigrada. Trans R Soc Edinb 45:837–854
Murray J (1907c) Some Tardigrada of the Sikkim Himalaya. J R Microsc Soc 3:269–273
Murray J (1907d) XXIV – Scottish Tardigrada, collected by the Lake Survey. Trans R Soc Edinb 45:641–668
Nederström P (1919) Die bisjetzt aus Finnland bekannten Tardigraden. Act Soc Fraun Flor Fenn 46 (8):1–15
Pigòn A, Węglarska B (1953) The respiration of Tardigrada: a study in animal anabiosis. Bull Acad Pol Sci Biol 1:69–72
Rahm G (1925) Die Cysten bildung bei den WasserbewohnendenTardigraden. Int Ver Theor Angew Limnol Verh 3:364–371
Rahm G (1926) Die Trockenstarre (Anabiose) der Moostierwelt. Biol Cent Bl 46:452–477
Rahm G (1927) Tardigrada Biolog d Tiere Deutschland 22:1–56
Ramazzotti G (1959) Tardigradi in terreni prativi. Atti Soc Ital Nat, Milano 98:199–210
Ramazzotti G, Maucci W (1983) Il Phylum Tardigrada. Terza edizione riveduta e corretta. Mem Isit Ital Idrobiol Dott Marco De Marchi 41:1–1012
Rebecchi L, Bertolani R (1994) Maturative pattern of ovary and testis in eutardigrades of freshwater and terrestrial habitats. Invertebr Reprod Dev 26:107–117
Richters F (1909) Tardigraden-Studien. Ber Senckenb Naturforsch Ges 40:28–45
Richters F, Krumbach T (1926) Tardigrada. Handbuch der Zoologie 3:1–68
Rost-Roszkowska M, Poprawa I (2008) Ultrastructure of the midgut epithelium in *Dactylobiotus dispar* (Tardigrada: Eutardigrada) during encystation. Zool Polon 53(1-4):19–25
Schill RO, Huhn F, Köhler HR (2007) The first record of tardigrades (Tardigrada) from the Sinai Peninsula, Egypt. Zool Middle East 42(1):83–88
Schultze M (1865) *Echiniscus sigismundi*, ein Arctiscoide der Nordsee. Arch Mikrosk Anat 1 (1):428–436

Sommerville RI, Davey KG (2002) Diapause in parasitic nematodes: a review. Can J Zool 80:1817–1840
Stark S, Kristensen RM (1999) Tardigrades in the soil of Greenland. Ber Polarforsch 330:44–63
Szymańska B (1995) Encystment in the tardigrade *Dactylobiotus dispar* (Murray, 1907) (Tardigrada: Eutardigrada). Part 1: Observation of leaving animals and structure of cyst. Zool Pol 40:91–102
Thulin G (1911) Beiträge zur Kenntnis der Tardigraden fauna Schwedens. Ark Zool 7(16):1–60
Thulin G (1928) Über die Phylogenie und das System der Tardigraden. Hereditas 11:207–266
Urbanowicz C (1925) Sur la variabilité de Macrobiotus oberhaeuseri. Bull biol Fr Bel 59:124–142
Von Reukauf EW (1912) Ein neuer Wasserbär, *Macrobiotus ferdinandi* (Reukauf). Zool Anz 39:352–353
von Wenck V (1914) Entwicklungs geschichtliche Untersuchungen an Tardigraden (*Macrobiotus lacustris* Duj.). Zool Jahrb Anat 37:465–514
Węglarska B (1957) On the encystation in Tardigrada. Zool Pol 8:315–325
Węglarska B (1970) Hypsibius (Isohypsibius) smreczynskii spec. nov., a new species of fresh-water tardigrade. Zesz Nauk Uniw Jagiellonsk Pr Zool 16:107–114
Wełnicz W, Grohme MA, Kaczmarek L, Schill R, Frohme M (2011) Anhydrobiosis in tardigrades – the last decade. J Insect Physiol 57:577–583
Westh P, Kristensen RM (1992) Ice formation in the freeze-tolerant eutardigrades *Adorybiotus coronifer* and *Amphibolos nebulosus* studied by differential scanning calorimetry. Polar Biol 12:693–699

Chapter 10
Environmental Adaptations: Desiccation Tolerance

Ralph O. Schill and Steffen Hengherr

Abstract Survival in microhabitats that experience extreme fluctuations in water availability and temperature requires extreme adaptations. Antonie van Leeuwenhoek was the first who describe the phenomenon of the resurrection of a desiccated rotifer in 1702. As with some rotifers and other small organisms, tardigrades enter a desiccated state known as anhydrobiosis to withstand such environmental conditions. This allows them to cope with the temporal variation of available water and to extend their lifespan in an anhydrobiotic state by up to 20 years without biological aging, according to the Sleeping Beauty hypothesis. Heat shock proteins serve as molecular chaperones to preserve or restore protein integrity, and tardigrade-specific intrinsically disordered proteins (TDPs) as well as metabolite help prevent the formation of damaging cellular compartments aggregates during water stress.

10.1 Life Without Water

The first to describe the phenomenon of the resurrection of a desiccated animal was the Dutch naturalist Antonie van Leeuwenhoek. He discovered that when dry and apparently lifeless dust from a roof gutter was rehydrated with clean water in a small glass, many "animalcules" (small organisms) became active within an hour (van Leeuwenhoek 1702). He noticed: "I confess, I never thought that there could be any living creature in a substance so dried as this was." These animals were probably of the bdelloid rotifer species *Philodina roseola* (Tunnacliffe and Lapinski 2003). In the past 300 years, various names were used for this phenomenon until David Keilin, who is most known for his research and rediscovery of cytochrome in the 1920s,

R. O. Schill (✉)
Institute of Biological Materials and Biomolecular Systems, University of Stuttgart, Stuttgart, Germany
e-mail: ralph.schill@bio.uni-stuttgart.de

S. Hengherr
Jakob-Friedrich-Schöllkopf School, Kirchheim u. T., Germany
e-mail: hen@jfs.de

R. O. Schill (ed.), *Water Bears: The Biology of Tardigrades*, Zoological Monographs 2,
https://doi.org/10.1007/978-3-319-95702-9_10

published a benchmark review and defined the term cryptobiosis. He called it "the state of an organism when it shows no visible signs of life and when its metabolic activity becomes hardly measurable, or comes reversibly to a standstill" (Keilin 1959). Of course, the difference between a "hardly measurable" metabolism and one that is at a "reversible standstill" is of considerable significance. The latter is difficult to interpret, because it ultimately means neither alive nor dead. Due to the fact that, for example, anhydrobiotic organism contains no or only a little water means that also no functional enzymes are available, and therefore no metabolism is possible. Despite this circumstance, the anhydrobiotes are able to survive desiccation and to continue to live normally after rehydration. Cryptobiosis results from desiccation (anhydrobiosis), low temperature (cryobiosis), lack of oxygen (anoxybiosis), high or low salt concentration (osmobiosis), or combinations of these (Keilin 1959). However, there are always similarities between these kinds of cryptobiotic states, for example, between the state of anhydrobiosis and cryobiosis. In both states, the amount of free water in the cells is reduced by desiccation or ice formation. Nevertheless, various studies suggest that the two conditions are not equivalent and that different mechanisms of survival have been developed (Crowe et al. 1992). The ability to enter cryptobiosis is quite common in nature. It occurs in many invertebrate taxa like sponges, crustaceans, rotifers, nematodes, and tardigrades (Crowe and Clegg 1973, 1978; Lapinski and Tunnacliffe 2003; Womersley 1987; Wright 2001). But also many prokaryotes such as bacteria and blue-green algae (Potts 2001), plant seeds (Alpert 2000; Chandler and Bartels 1999; Ingram and Bartels 1996; Priestley 1986; Vertucci and Farrant 1995), and also tissues of some higher plants have developed this ability. Particularly noteworthy here is the resurrection plant *Selaginella lepidophylla*, also known as the "Rose of Jericho" (Scott 2000; Tomos 1992).

10.2 Tardigrades and Their Longevity

10.2.1 Longevity in Anhydrobiosis

To date, few comparative studies have been carried out to obtain information on how long anhydrobiotes can remain in anhydrobiosis without losing their vitality after rehydration. Most studies were performed in the first half of the twentieth century (Baumann 1927; Fielding 1951; Franceschi 1948; Goodey 1923; Lee 1961; Rahm 1923; Steiner and Albin 1946). More information is available about parasitic nematodes of plants. For example, Steiner and Albin (1946) reported on two nematode species that successfully survived 28 and 30 years, respectively, in anhydrobiosis. Fielding (1951) also reported a nematode species that survived dehydrated for 20–28 years. Goodey (1923) and Lee (1961) in turn found a maximum time span of 9 and 10 years, respectively. There is much less information about tardigrades (Fig. 10.1a, b) and rotifers. One of the first who did extensive experiments was Baumann (1927). One species of the genus *Macrobiotus* was successfully

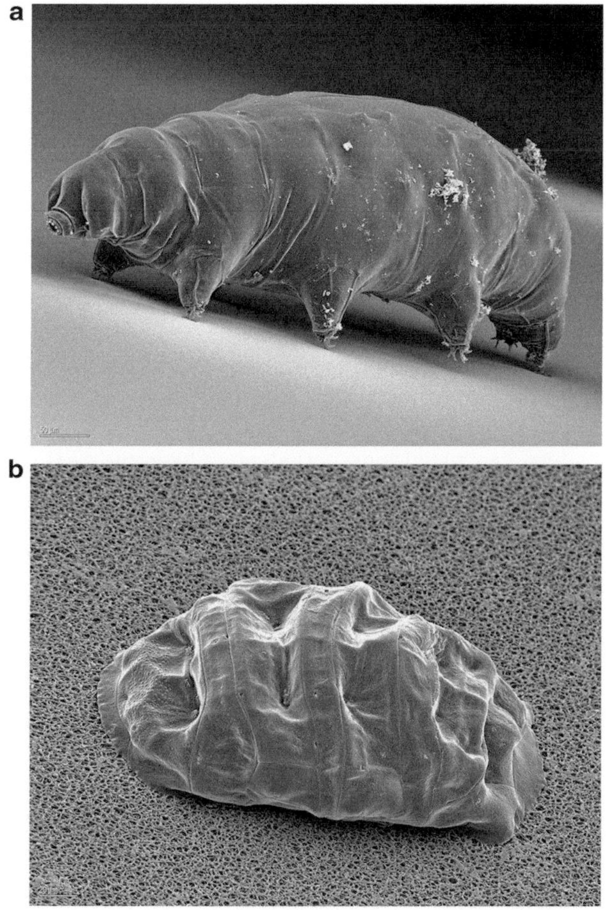

Fig. 10.1 (**a**) SEM picture of an active tardigrade of the species *Paramacrobiotus richtersi* and (**b**) anhydrobiotic tun state

rehydrated after 7 years (Baumann 1927). *Richtersius oberhaeuseri* survived up to 1604 days and *Echiniscus* sp. survived up to 1085 days in anhydrobiosis (Franceschi 1948). Franceschi (1948) even reported a tardigrade that came from a 120-year-old moss and showed briefly movement after rehydration before it died. Meanwhile, it is doubted whether the observed animal was alive. Presumably, passive rehydration caused the animal's supposed movements (Jönsson and Bertolani 2001). The longest reliable documented survival time in anhydrobiosis is known from the heterotardigrade *Echiniscus testudo* with 20 years (Jørgensen et al. 2007).

10.2.2 Sleeping Beauty Hypothesis

Very little is known about the effects of anhydrobiosis on life cycle and longevity achieved thereby. So far, little information is available, mainly about nematodes and rotifers (Ricci 2001; Ricci and Caprioli 2005; Ricci and Covino 2005; Ricci and Pagani 1997; Ricci et al. 1987; Wharton 2003; Wharton and Aalders 1999). Ricci and Pagani (1997) postulated three hypotheses of what effects anhydrobiosis could have on the life of organisms. The first hypothesis assumes that the time an animal spends in anhydrobiosis is ignored and biological aging is halted. This hypothesis is also referred to as the "Sleeping Beauty" model. The second hypothesis is that the internal clock and thus biological aging is slowed down. The third model describes unrestricted biological aging, even during anhydrobiosis. Studies on the rotifer species *Macrotrachela quadricornifera* and *Adineta ricciae* clearly demonstrated that the "Sleeping Beauty" model is correct and that the life cycle is not affected by occasional anhydrobiosis (Ricci and Caprioli 1998; Ricci et al. 1987). Even with nematodes, the internal clock seems to be stopped because, despite a short lifetime, they can be rehydrated again after months or years and can continue to live normally (Ricci and Covino 2005; Ricci et al. 1987). Since tardigrades, like nematodes and rotifers, can spend long periods in anhydrobiosis (Baumann 1927; Bertolani et al. 2004; Guidetti and Jönsson 2002), the "Sleeping Beauty" model has been investigated with the species *Milnesium tardigradum* (Hengherr et al. 2008). They were exposed to alternate dry periods of 7 days each compared to an active control group. The animals of the control group reached an age of 82.7 $\pm$ 2.7 days. The age of the animals exposed to the periodic drought periods, minus the times spent in anhydrobiosis, was 79.7 $\pm$ 5.7 days. The total time period including the dry periods was 133.2 $\pm$ 11.7 days. The result shows that the "Sleeping Beauty" model is valid for the tardigrades, and the time spent in anhydrobiosis is not counted as biological age (Hengherr et al. 2008). Nevertheless, there seems to be an upper limit to persist in anhydrobiosis. One possible explanation is that the anhydrobionts die during anhydrobiosis due to cellular damage from chemical aging (Schöneich 1999), since no repair mechanisms can become active in the desiccated state (Clegg 1967; Crowe and Madin 1975; Örstan 1998; Wright 2001).

10.2.3 Desiccation Tolerance in Embryonic Stages

In recent decades, a variety of experiments have shown that adult tardigrades can spend long periods, from months to several years, in the anhydrobiotic state. Bertolani et al. reported the successful hatching of four juveniles of the species *Richtersius oberhaeuseri* from 13 eggs, which were rehydrated after 9 years (Bertolani et al. 2004). This was the first study to show that not only adults have the ability to enter anhydrobiosis but also embryos. In the species *Ramazzottius oberhaeuseri* and *Milnesium tardigradum*, successful hatching after 1604 days of anhydrobiosis has been demonstrated (Rebecchi et al. 2006). The anhydrobiosis of embryos is considered to be an adaptation strategy, since *Paramacrobiotus richtersi* is also able to undergo a time-shifted hatching depending on the environmental conditions (Altiero et al. 2006). For rotifers of the species *Adineta vaga* (Örstan 1995) and other species (Ricci 1998), desiccation tolerance, depending on the stage

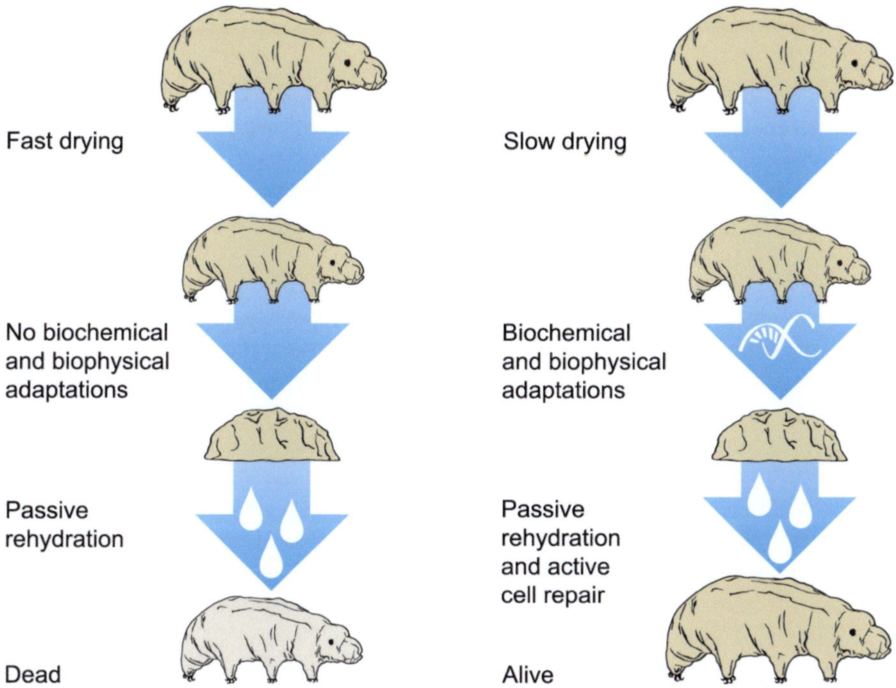

Fig. 10.2 A slow desiccation of a tardigrade results in biochemical and biophysical adaptations and a high survival rate after rehydration

of development, has already been demonstrated. The first comparative study in tardigrades has been done with the parthenogenetic tardigrade species *Milnesium tardigradum* which lives in microhabitats that dry out regularly (Ramazzotti and Maucci 1983; Wright 1991). It was shown that both factors, the humidity and the state of development (several stages from the blastula stage to the hatching juveniles), are decisive for desiccation tolerance (Schill and Fritz 2008). The less the embryos were developed, the lower was the survival rate for all humidities between 10 and 81%. When the developing embryos were dehydrated during the first 3 days of their development at low humidities, the survival rates were particularly low. The older they are, the better is the survival rate, especially for those which are close to hatching and which were dried slowly at high humidities (Schill and Fritz 2008). Tardigrades are thus successfully adapted to microhabitats in all stages of life, which are subject to regular dehydration and rehydration processes (Fig. 10.2).

10.3 Tardigrades and Stress Proteins

Heat shock reactions were first observed in 1962 as a puffing pattern in *Drosophila* larvae correlated with a temperature increase and later shown to produce heat shock proteins (Hsp) (Tissières et al. 1974). Although called heat shock proteins, their induction is not only due to temperature effects but to a whole series of other stressors. They play a critical role in intracellular protection against protein-denaturing factors, acting as molecular chaperones, helping to fold newly synthesized proteins, and preventing stress-induced denaturation or aggregation (proteotoxicity). Furthermore, they are involved in the renaturation and transmembrane transport of proteins. Their molecular and cellular physiological functions have been studied extensively in various fields of biology and have been the subject of numerous review articles (e.g., Feder and Hofmann 1999; Gething and Sambrook 1992; Morimoto 1993; Parsell and Lindquist 1993). Their classification is based on their respective molecular weights. The best known are the very conserved families Hp60, Hsp70, and Hsp90. Among them Hsp70 is one of the best studied families; their induction mechanism is by means of HSF (heat shock factor) and HSE (heat shock element). There is a whole range of heterogeneous low molecular weight Hsps with a molecular weight below 30 kDa (low molecular weight (LMW), which are also referred to as small Hsps).

10.3.1 Small Stress Proteins

Large amounts of two small stress proteins, artemin (Arrigo and Müller 2002; De Graaf et al. 1990) and p26 (Clegg et al. 1994, 1995), were found in stress-resistant, encysted embryos of the brine shrimp species *Artemia franciscana*. Both proteins have been extensively studied in the context of anhydrobiosis and osmobiosis (Chen et al. 2003, 2007; Clegg et al. 1995, 1999; Liang et al. 1997a; Liang and MacRae 1999; Warner et al. 2004; Willsie and Clegg 2002). p26 belongs to the small heat shock/α-crystallin protein family, which has molecular chaperone activity in vitro (Liang et al. 1997a, b) and probably also in vivo (Liang and MacRae 1999). During a long-term study over several years, *Artemia franciscana* embryos showed no evidence of misfolded proteins or protein aggregation (Clegg 1997; Clegg et al. 1999). The amount of artemin and p26 was between 10 and 15% of the non-yolk protein in these embryos. However, the proteins have been found only in the early stages during development, not in the adult animals (Clegg et al. 1999). There is strong evidence that p26 in particular plays an important role as a molecular chaperone and is responsible for the high tolerance of embryos (Clegg 2007). Other studies indicate that artemin might even play a role as a molecular chaperone for RNA (Warner et al. 2004). Several cDNA libraries and thus expressed sequence tags (EST) of *Milnesium tardigradum* were created, and thus a whole range of different stress protein families were identified (Reuner et al. 2009). The in silico analysis of two existing sequences showed that *Milnesium tardigradum* possesses proteins that contain a small Hsp/α-crystallin domain. Based on their amino acid sequence, the molecular weights were determined and the proteins designated as MtHsp19.5 and MtHsp17.2. Small Hsps form large complexes of several hundred kilodaltons during heat stress, which are able to stabilize the structure of other proteins. Expression analyses in *Milnesium tardigradum* showed that Mthsp17.2 is significantly upregulated by heat stress. In contrast to p26 in *Artemia*, however, no increased expression of Mthsp19.5 and Mthp17.2 was detectable in anhydrobiotic tardigrades. A cDNA library of *Milnesium tardigradum* also contained the complete, coding sequence for Hsp10 (chaperonin). Hsp10, like Hsp60, is a chaperone found in the mitochondria and cytoplasm. Hsp60 shows chaperone activity, while Hsp10 serves as a functional regulator (Reuner et al. 2009).

10.3.2 Hsp60 and Hsp70

The first tardigrade partial heat shock protein gene sequences with a molecular weight of 70 kDa were found in the species *Milnesium tardigradum* (Schill et al. 2004). All sequences showed a different expression pattern but were inducible by heat shock. Additional Hsp70 isoforms and one Hsp90 sequence were found later in a cDNA library (Reuner et al. 2009). Only one of three Hsp70 isoforms was significantly expressed during dehydration and is still found in the anhydrobiotic state. Maybe the RNA will be translated only after rehydration, so that the chaperones can fold new proteins or renature others. Accumulation of RNAs associated with anhydrobiosis is known from prokaryotic and eukaryotic cells (Albertson et al. 1990). For example, a high amount of mRNAs was detected in the cryptobiotic stages of the ciliates *Colpoda inflata* (Benítez and Gutiérrez 1997) and *Sterkiella histriomuscorum* (Tourancheau et al. 1999). Likewise, various fungi that form surviving spores are known to store mRNAs (Camonis et al. 1982). Byers et al. (Byers et al. 1991; Martinez-Guitarte et al. 2007) and Gutierrez et al. (Martinez-Guitarte et al. 2007) could also detect Hsp70 mRNA in anhydrobiotic stages of *Colpoda inflata* and *Colpoda nova*. However, two yeast studies with *Saccharomyces cerevisiae* and *Debaryomyces hansenii* showed that Hsp70 does not protect the yeast cells from dehydration stress during desiccation (Guzhova et al. 2008). This may also be the case for tardigrades in anhydrobiosis.

Unlike Hsp70, the cytoplasmic Hsp90 is not generally involved in the folding of new proteins (Nathan et al. 1997). Hsp90 differs from Hsp70, that most of the known substrates are signal transduction proteins (e.g., steroid receptors and signal kinases) (Picard et al. 1990; Xu and Lindquist 1993). Therefore, it plays an important role in the network of cellular signaling pathways and is part of a broader chaperone mechanism (Bose et al. 1996; Freeman and Morimoto 1996; Schumacher et al. 1996). The complexity of Hsp90 also makes it much harder to investigate its capabilities compared to Hsp70. In tardigrades, it is only significantly expressed in anhydrobiosis. Which function it has shortly before complete dehydration, or whether it is translated only after renewed rehydration, requires further studies.

10.4 Metabolites for Cell Protection

Several mechanisms probably play a role in the protection of living cells in the anhydrobiotic state. These include the maintenance of the biologically relevant structures of macromolecules through the accumulation of disaccharides and other metabolites (Buitink and Leprince 2004; Crowe et al. 1987; Crowe 2002). Drying and freezing of cells usually lead to a massive damage of cell proteins and membranes, which normally ends in cell death. However, there are a number of organisms that have the extraordinary ability to survive complete dehydration like tardigrades (Crowe et al. 1992). The removal of intracellular water causes drastic changes in inter- and intramolecular interactions. During desiccation, intracellular proteins and membranes compensate for the loss of hydrogen bonds with water through hydrogen bonding with other molecules. This leads to forced interactions between molecules that would normally not react in the presence of water. Protein–protein interactions induced by water loss can result in irreversible conformational changes and result in the loss of enzyme activity (Carpenter et al. 1987; Hanafusa 1969; Prestrelski et al. 1993). In membranes, water loss can lead to a phase transition from the biologically active liquid crystalline to the gel phase (Crowe et al. 1992, 1997). In addition, water loss can result in the fusion of cell organelles. In cells that exhibit dehydration tolerance, molecular interactions during drying are controlled by replacing lost hydrogen bonds with reversible molecular interactions. This means that biomolecules and cellular structures can be protected from degradation by dehydration and rehydrate to their previous natural conformation. In dehydration-sensitive cells, on the other hand, the lack of such control means that the biomolecules and cellular structures within these cells cannot regain their original functional conformation after the stressor has disappeared.

Many anhydrobiotic organisms, like some nematodes and yeast cells, accumulate trehalose (Crowe et al. 1997; Womersley 1987), while sucrose is accumulated in the tissues of higher plants, especially in seeds and pollen (Hoekstra 1986; Koster and Leopold 1988). However, there are also exceptions such as the plant *Myrothamnus flabellifolia* which concurrently enriches sucrose and trehalose (Bianchi et al. 1993; Drennan et al. 1993). As a reason for the stabilizing effect of dissacharides, the "water replacement" (or "water substitute") and "vitrification" (or "glassy state") hypotheses are discussed (Crowe 1971; Webb et al. 1965). These are not alternative, but rather complementary explanations. Both are necessary but do not seem to be sufficient to fully protect the membranes during drying and rehydration.

10.4.1 Water Replacement Hypothesis

In the early 1960s, S. J. Webb postulated the water replacement hypothesis (Webb et al. 1965). It explains the stabilizing effect of free hydroxyl groups of sugar molecules that are available to proteins during and after removal of water to form hydrogen bonds. This lowers the free enthalpy of the system, which stabilizes native proteins. When unprotected membranes dehydrate, the water molecules that help maintain the distance between the phospholipid headgroups are removed. This allows the lipid molecules to come closer together and increase the membrane phase transition temperature (Tm), resulting in a phase change. In rehydration, a renewed phase change occurs in membranes, resulting in a brief release of soluble substances in cells through the membranes (Crowe et al. 1997). By contrast, when membranes are dried in the presence of trehalose, the water replacement hypothesis postulates that sugar molecules prevent the closest approach of lipids through interaction with phospholipid head groups during dehydration (Crowe et al. 1988). When water is returned to such a system, there is no phase change and the membranes retain their barrier function (Crowe et al. 1992, 1994; Harrigan et al. 1990).

10.4.1.1 Trehalose in Anhydrobionts

Trehalose was detected in anhydrobiotic states of the freshwater sponge *Trochospongilla* sp. (gemmules), the bryozoan *Cristatella mucedo* (statoblasts), and embryos of the crustaceans *Daphnia magna, Daphnia pulex, Triops longicaudatus*, and *Triops cancriformis* (Hengherr et al. 2011; Hengherr and Schill 2011). In particular, anhydrobiotic embryos of the brine shrimp *Artemia franciscana* have been the subject of intense research over the past three decades. They increase to 15–18% trehalose on a dry weight basis as they enter the anhydrobiotic stage (Clegg 1986; Clegg and Conte 1980; Clegg and Drost-Hansen 1990). The chironomid larva *Polypedilum vanderplanki* also stores around 18% trehalose (Watanabe et al. 2002) and nematodes, depending on the species, 4 to 15% (Madin and Crowe 1975). The first quantitative detection of trehalose in tardigrades was performed in the species *Paramacrobiotus areolatus* (Crowe 1975). Later, trehalose was also detected in *Amphibolus nebulosus* and *Richtersius coronifer* (Westh and Ramlov 1988). Quantitative studies in *Richtersius coronifer* showed that the highest amount of trehalose is found in the anhydrobiotic state but with a rather low content of up to 2.3% trehalose based on dry weight (Westh and Ramlov 1991). In a comparative quantitative amperometric chromatography study with several tardigrade species, it was shown that in species of the genera *Paramacrobiotus* and *Macrobiotus*, trehalose is accumulated during dehydration, whereas in *Milnesium tardigradum* trehalose could not be detected either in the active or in the anhydrobiotic or the respective transitional stages (Hengherr et al. 2008). The highest trehalose amount was measured with $0.472 \pm 0.037\%$ dry weight in a *Paramacrobiotus* species (Hengherr et al. 2008). This is much less than measured by Westh and Ramlov (1991), possibly due to different measuring methods. Although small amounts of trehalose could be detected in the heterotardigrades *Echiniscus testudo* and *Echiniscus granulatus*, accumulation of trehalose did not occur during dehydration (Hengherr et al. 2008). The rotifer species *Philodina roseola* and *Adineta vaga*, which also have the ability to completely dehydrate, are known to have no trehalose and no corresponding gene for trehalose-6-phosphate synthase (Lapinski and Tunnacliffe 2003). Anhydrobiosis without incorporation of trehalose for cell stabilization seems to be possible, at least for the tardigrade species *Milnesium tardigradum*, as well as these rotifer species. Nevertheless, sugars appear to play an important role in the tolerance to desiccation stress in many microorganisms, invertebrates, and plants and to impart stability to dried biomolecules and membranes in vitro (Tunnacliffe et al. 2001).

10.4.2 Vitrification Hypothesis

The second important protective function of dissacharides, as well as other carbohydrates and proteins, is vitrification at low water content (Crowe et al. 1998; Crowe 2002; Sun and Leopold 1997). The "glassy state" hypothesis postulates the formation of glassy states to immobilize proteins, which greatly reduces the reaction kinetics. The glass of carbohydrates is a solid-like amorphous material which prevents diffusion-limiting destruction processes, such as membrane fusions. In order to convert proteins into a "solid" state, the glass transition temperature of the amorphous system must be lowered. In the temperature range just above the glass transition temperature, but below the eutectic point of the constituents contained, the product is in a highly viscous state. In this range, a gradual transformation of an unstable amorphous liquid into a metastable amorphous solid state occurs. Not only sugars but also proteins may be involved in cell stabilization. In model systems, proteins have had a considerable effect on the molecular properties of carbohydrate glasses (Bell and Hageman 1996; Kalichevsky et al. 1992; Wolkers et al. 1998). For example, the addition of proteins to a glass caused a much higher glass transition temperature (T_g) (Slade and Levine 1991) and a mean hydrogen bonding strength of the mixture compared to a pure sugar matrix (Wolkers et al. 1998). A mixture of protein and sugar is more dense than a pure carbohydrate or protein glass. In addition, proteins increase the temperature difference between the glass transition temperature and the critical temperature at which the dynamics of the glass transition from a solid-like to a liquid state. Proteins and carbohydrates probably interact through hydrogen bonds during the dry state in the cytoplasm of anhydrobiotes. By differential scanning calorimetry (DSC) measurements and Fourier transform infrared (FTIR) analyses, the "water replacement" and "vitrification" hypothesis could be examined for the first time in the drought-tolerant African chironomid larva *Polypedilum vanderplanki* (Sakurai et al. 2008). It was found that the anhydrobiotic larvae were in a glassy state; at a temperature above 65 ° C, the stabilizing glass melted. The previously mentioned high concentrations of trehalose have been detected throughout the organism, and the authors believe that trehalose

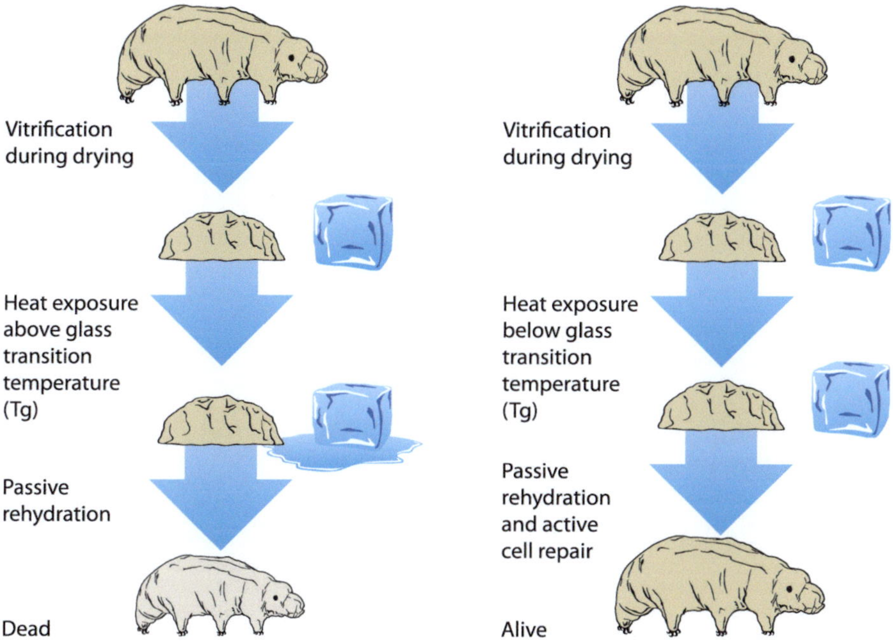

Fig. 10.3 As long as the heat exposure of a vitrified tardigrade is below the glass transition temperature (T_g), the protective function of the glassy state is available and the tardigrade becomes alive after rehydration

plays an important role in water replacement and intracellular glass formation. In order to detect possible vitrification in tardigrades, DSC measurements were used in combination with experiments on heat tolerance in several species of the genera *Paramacrobiotus, Macrobiotus, Echiniscus*, and *Milnesium* (Fig. 10.3). In the species of the genera *Paramacrobiotus* and *Macrobiotus*, a glass transition and thus a vitrified state could be detected (Hengherr et al. 2009). When temperature tolerance of these species is considered, in parallel, it can be seen that when the glass transition temperature (T_g) of about 80 °C is exceeded, the survival rates of the tardigrades drop sharply. This could indicate the loss of the protective function of the glassy state or the importance of vitrification in cellular protection. *Macrobiotus hufelandi* survived a short exposure to temperatures between 120 and 125 °C (Doyère 1842). Later, Baumann (1927) reported that anhydrobiotic tardigrades could tolerate 100 °C for 6 hours, and *Richtersius coronifer* survived temperatures up to approximately 70 °C for 60 minutes without any decrease in survival (Ramløv and Westh 2001). In the species *Milnesium tardigradum*, *Echiniscus granulatus*, and *Echiniscus testudo*, in which no or little trehalose was detected, no vitrification has been observed (Hengherr et al. 2009). It was found that exactly these species have a much better tolerance to high temperatures. At temperatures up to 90 °C, *Milnesium tardigradum*, *Echiniscus granulatus*, and *Echiniscus testudo* still showed survival rates of 90.0 ± 5.4% and 54.2 ± 13.7%, respectively, and 30.0% ± 11.2% after subsequent rehydration. The all-time record is held by *Milnesium tardigradum* at 100 °C with a survival rate of 91.7 ± 6.9%. Only higher temperatures led to a marked decline in survival rates, at 110 °C 1 ± 2% still survived (Hengherr et al. 2009). The temperature at which glass devitrifies (T_g) depends on several factors (Slade and Levine 1991). Interestingly, recent evidence suggests that various stress or late embryogenesis abundant (LEA) proteins that accumulate during dehydration may also be involved in glass formation and improve overall stability (Wolkers et al. 1999, 2001). This could at least explain why tardigrades have a glass transition and thus a vitrified state, although no or only small amounts of trehalose could be detected.

10.5 Tardigrade-Specific Intrinsically Disordered Proteins (TDPs)

10.5.1 Late Embryogenesis Abundant (LEA) Proteins

In plants, and more recently in several animals, induction of high levels of hydrophilic proteins, in particular the late embryogenesis abundant (LEA) proteins, has been associated with water stress. LEA proteins were first identified 30 years ago in plants, where they are produced during seed development (Galau et al. 1986; Grzelezak et al. 1982). However, their precise function is poorly understood. To be classified as molecular chaperones, LEA proteins must not only prevent aggregation, they must additionally form transient, non-covalent complexes (Ellis 2004). Unfavorable protein–protein interactions, however, can lead to irreversible conformational changes and, in enzymes, a loss of catalytic activity (Crowe et al. 1987). LEA proteins might simply function as "molecular shields," forming a physical barrier between partially unfolded neighboring proteins and preventing contact between them (Tunnacliffe and Wise 2007). LEA protein expression has more recently been demonstrated in other organisms and is also linked to desiccation stress and the acquisition of desiccation tolerance. So far these organisms include microorganisms (Battista et al. 2001), nematodes (Browne et al. 2002, 2004; Goyal et al. 2003, 2005), rotifers (Tunnacliffe et al. 2005), chironomid larvae (Kikawada et al. 2006), and Collembola (Bahrndorff et al. 2008). Research on LEA proteins has recently focused on tardigrades because of the evidence that these proteins play an important role in protecting cellular proteins. For the first time, putative LEA proteins have been detected in the anhydrobiotic state of the species *Macrobiotus hufelandi* (McGee et al. 2004). With high-throughput, high-accuracy proteomics in combination with a newly developed tardigrade-specific protein database (Schokraie et al. 2012), more than 3000 proteins were identified in *Milnesium tardigradum*. This comprehensive proteome resource includes protein families such as chaperones, antioxidants, ribosomal proteins, cytoskeletal proteins, transporters, protein channels, nutrient reservoirs, and developmental proteins. Within these proteins, also those belonging to the LEA family (group 3) were identified.

10.5.2 CAHS, SAHS, and MAHS Proteins

Tanaka et al. (2015) found in *Ramazzottius varieornatus* two novel mitochondrial heat-soluble proteins, RvLEAM (a group 3 LEA) and MAHS (mitochondrial abundant heat soluble), as potent mitochondrial protectants. The MAHS protein is also a heat-soluble protein that might have protective roles in tardigrades. In the tardigrade species *Ramazzottius varieornatus*, a further five abundant heat-soluble proteins have been found, but all of them showed no sequence similarity with LEA proteins. They formed two novel protein families, the cytoplasmic abundant heat soluble (CAHS) and secretory abundant heat soluble (SAHS) protein families, according to their localization (Yamaguchi et al. 2012). Both protein families were also found in *Hypsibius dujardini* and *Milnesium tardigradum*, but not in other phyla. Therefore LEA, CAHS, SAHS, and MAHS families have become referred to as the tardigrade-specific intrinsically disordered proteins (TDPs). Boothby et al. (2017) showed that the TDP genes are constitutively expressed at high levels or induced during desiccation in the tardigrade species *Hypsibius dujardini* and *Paramacrobiotus richtersi*. All TDPs form noncrystalline amorphous solids upon desiccation, and this vitrified state mirrors their protective capabilities.

10.6 DNA Damage During Anhydrobiosis

One reason for the decline in survival with increasing time in anhydrobiosis could be oxidative damage caused by reactive oxygen species (ROS) (Womersley 1987). Since enzymes do not work without water and therefore no metabolism can be detected, even energy-dependent repair systems cannot be activated. This causes an accumulation of DNA damage until it reaches a lethal limit, eventually leading to the death of the whole organism (Lindahl 1993). Deoxyribonucleic acids are a preferred biological target of ROS (Gros et al. 2002). For example, they may spontaneously result from the chemical degradation of various substances in the metabolism and have a deleterious effect on proteins, membranes, and DNA (Mattimore and Battista 1996). The phenomenon of induced double-strand breaks by dehydration is already known in bacteria (Billi 2009). Blasius et al. (2008), as well as Mattimore and Battista (1996), believe in effective repair of DNA as one of the most important survival mechanisms (Blasius et al. 2008; Mattimore and Battista 1996). To visualize and detect DNA damage, e.g., single- and double-strand breaks, incomplete excision repair, alkali-labile sites, and cross-linking, storage cells of anhydrobiotic tardigrades of the species *Milnesium tardigradum* were examined with single-cell gel electrophoresis (comet assay) (Neumann et al. 2009). The animals which spent

2 days in anhydrobiosis showed little DNA damage (2.09 ± 1.98% DNA) in the comet tail, compared with the DNA damage in the comet tail of active animals (0.44 ± 0.74%). With increasing time in anhydrobiosis, DNA damage increased. After 6 weeks of anhydrobiosis, 13.63 ± 6.41% DNA was found in the comet tail and 23.66 ± 7.56% DNA after 10 months (Neumann et al. 2009). The amount of DNA damage correlated with the duration of anhydrobiosis, since storage cells from animals that spend only 2 days in anhydrobiosis have relatively little damage and cells that are longer in anhydrobiosis show significantly greater damage. However, it is assumed that the DNA damage does not occur during the dehydration process but rather during anhydrobiosis. They can accumulate DNA damage over time. Of course, a repair process can only become activated after rehydration. The exact causes of the damage are not yet sufficiently known, but a direct or indirect involvement of reactive oxygen species and/or free radicals is likely.

In the last decade, many data from the tardigrade species *Echiniscus testudo*, *Milnesium tardigradum*, *Hypsibius dujardini*, *Paramacrobiotus richtersi*, and *Ramazzottius varieornatus* have been generated with state-of-the-art methods in genomics, transcriptomics, proteomics, and bioinformatics (e.g., Arakawa 2018; Arakawa et al. 2016; Beisser et al. 2012; Bemm et al. 2016; Boothby et al. 2015, 2017; Borner et al. 2014; Förster et al. 2009, 2011a, b; Kondo et al. 2015; Mali et al. 2008; Schokraie et al. 2010, 2012; Tanaka et al. 2015; Yamaguchi et al. 2012; Yoshida et al. 2017). This is an excellent basis for further studies to understand the mechanisms of desiccation tolerance in tardigrades.

References

Albertson NH, Nyström T, Kjelleberg S (1990) Functional mRNA half-lives in the marine Vibrio sp. S14 during starvation and recovery. J Gen Microbiol 136:2195–2199

Alpert P (2000) The discovery, scope, and puzzle of desiccation tolerance in plants. Plant Ecol 151:5–17

Altiero T, Rebecchi L, Bertolani R (2006) Phenotypic variations in the life history of two clones of *Macrobiotus richtersi* (Eutardigrada, Macrobiotidae). Hydrobiologia 558:33–40

Arakawa K (2018) The complete mitochondrial genome of *Echiniscus testudo* (Heterotardigrada: Echiniscidae). Mitochondrial DNA Part B 3:810–811

Arakawa K, Yoshida Y, Tomita M (2016) Genome sequencing of a single tardigrade *Hypsibius dujardini* individual. Sci Data 3:16006

Arrigo A-P, Müller WEG (2002) Small stress proteins. Springer, Berlin, I–XV, 1–270 pp

Bahrndorff S, Tunnacliffe A, Wise MJ, McGee B, Holmstrup M, Loeschcke V (2008) Bioinformatics and protein expression analyses implicate LEA proteins in the drought response of Collembola. J Insect Physiol 55:210–217

Battista JR, Park MJ, McLemore AE (2001) Inactivation of two homologues of proteins presumed to be involved in the desiccation tolerance of plants sensitizes *Deinococcus radiodurans* R1 to desiccation. Cryobiology 43:133–139

Baumann H (1927) Anabiosis of tardigrades. Zool Anz 72:175–179

Beisser D, Grohme M, Kopka J, Frohme M, Schill RO, Hengherr S, Dandekar T, Klau GW, Dittrich M, Müller T (2012) Integrated pathway modules using time-course metabolic profiles and EST data from *Milnesium tardigradum*. BMC Syst Biol 6:72

Bell LN, Hageman MJ (1996) Glass transition explanation for the effect of polyhydroxy compounds on protein denaturation in dehydrated solids. J Food Sci Technol 61:372–378

Bemm F, Weiss CL, Schultz J, Förster F (2016) Genome of a tardigrade: horizontal gene transfer or bacterial contamination? Proc Natl Acad Sci U S A 113:E3054–E3056

Benítez L, Gutiérrez JC (1997) Encystment – specific mRNA is accumulated in the resting cysts of the ciliate *Colpoda inflata*. Biochem Mol Biol Int 41:1137–1141

Bertolani R, Guidetti R, Jönsson KI, Altiero T, Boschini D, Rebecchi L (2004) Experiences with dormancy in tardigrades. J Limnol 63:16–25

Bianchi G, Gamba A, Limiroli R, Pzzi N, Ester R, Salamini F, Bartels D (1993) The unusual sugar composition in leaves of the resurrection plant *Myrothamnus flabellifolia*. Physiol Plant 87:223–226

Billi D (2009) Subcellular integrities in *Chroococcidiopsis* sp. CCMEE 029 survivors after prolonged desiccation revealed by molecular probes and genome stability assays. Extremophiles 13:49–57

Blasius M, Hubscher U, Sommer S (2008) *Deinococcus radiodurans*: what belongs to the survival kit? Crit Rev Biochem Mol Biol 43:221–238

Boothby TC, Tapia H, Brozena AH, Piszkiewicz S, Smith AE, Giovannini I, Rebecchi L, Pielak GJ, Koshland D, Goldstein B (2017) Tardigrades use intrinsically disordered proteins to survive desiccation. Mol Cell 65(975–984):e975

Boothby TC, Tenlen JR, Smith FW, Wang JR, Patanella KA, Nishimura EO, Tintori SC, Li Q, Jones CD, Yandell M, Messina DN, Glasscock J, Goldstein B (2015) Evidence for extensive horizontal gene transfer from the draft genome of a tardigrade. Proc Natl Acad Sci U S A 112:15976–15981

Borner J, Rehm P, Schill RO, Ebersberger I, Burmester T (2014) A transcriptome approach to ecdysozoan phylogeny. Mol Phylogenet Evol 80:79–87

Bose S, Weikl T, Bugl H, Buchner J (1996) Chaperone function of Hsp90-associated proteins. Science 274:1715–1717

Browne J, Tunnacliffe A, Burnell A (2002) Anhydrobiosis – plant desiccation gene found in a nematode. Nature 416:38

Browne JA, Dolan KM, Tyson T, Goyal K, Tunnacliffe A, Burnell AM (2004) Dehydration-specific induction of hydrophilic protein genes in the anhydrobiotic nematode *Aphelenchus avenae*. Eukaryot Cell 3:966–975

Buitink J, Leprince O (2004) Glass formation in plant anhydrobiotes: survival in the dry state. Cryobiology 48:215–228

Byers TJ, Kim BG, King LE, Hugo ER (1991) Molecular aspects of the cell cycle and encystment of Acanthamoeba. Rev Infect Dis 5:373–384

Camonis J, Julien J, Ayala J, Jaequet M (1982) Polyadenylated RNA population present in dormant spores of *Dictyostelium discoideum*. Cell Differ (1):55–61

Carpenter JF, Crowe LM, Crowe JH (1987) Stabilization of phosphofructokinase with sugars during freeze-drying characterization of enhanced protection in the presence of divalent cations. Biochim Biophys Acta 923:109–115

Chandler J, Bartels D (1999) Plant desiccation. In: Lerner HR (ed) Plant responses to environmental stresses: from phytohormones to genome reorganization. Marcel Dekker, New York, pp 575–590

Chen T, Amons R, Clegg JS, Warner AH, MacRae TH (2003) Molecular characterization of artemin and ferritin from *Artemia franciscana*. Eur J Biochem 270:137–145

Chen T, Villeneuve TS, Garant KA, Amons R, MacRae TH (2007) Functional characterization of artemin, a ferritin homolog synthesized in *Artemia* embryos during encystment and diapause. FEBS J 274:1093–1101

Clegg JS (1967) Metabolic studies of crytobiosis in encysted embryos of *Artemia salina*. Comp Biochem Physiol 20:801–809

Clegg JS (1986) The physical properties and metabolic status of *Artemia* cysts at low water contents: the water replacement hypothesis. In: Leopold AC (ed) Membranes, metabolism and dry organisms. Cornell University Press, New York, pp 169–187

Clegg JS (1997) Embryos of *Artemia franciscana* survive four years of continuous anoxia: the case for complete metabolic rate depression. J Exp Biol 200:467–475

Clegg JS (2007) Protein stability in *Artemia* embryos during prolonged anoxia. Biol Bull 212:74–81

Clegg JS, Conte F (1980) A review of the cellular and developmental biology of *Artemia*. In: Persoone GP, Sorgeloos P, Roels O, Jaspers E (eds) The brine shrimp, *Artemia*. Universa Press, Wetteren, pp 11–54

Clegg JS, Drost-Hansen W (1990) On the biochemistry and cell physiology of water. In: Hochachka PW, Mommsen TP (eds) Biochemistry and molecular biology of fishes. Elsevier, Amsterdam, pp 1–23

Clegg JS, Jackson SA, Liang P, MacRae TH (1995) Nuclear-cytoplasmic translocations of protein p26 during aerobic-anoxic transitions in embryos of *Artemia franciscana*. Exp Cell Res 219:1–7

Clegg JS, Jackson SA, Warner AH (1994) Extensive intracellular translocations of a major protein accompany anoxia in embryos of *Artemia franciscana*. Exp Cell Res 212:77–83

Clegg JS, Willsie JK, Jackson SA (1999) Adaptive significance of a small heat shock/alpha-crystallin protein (p26) in encysted embryos of the brine shrimp, *Artemia franciscana*. Am Zool 39:836–847

Crowe JH (1971) Anhydrobiosis: an unsolved problem. Am Nat 105:563–573

Crowe JH (1975) The physiology of cryptobiosis in tardigrades. Memorie dell'Istituto Italiano di Idrobiologica 32(Suppl):37–59

Crowe LM (2002) Lessons from nature: the role of sugars in anhydrobiosis. Comp Biochem Physiol A Mol Integr Physiol 131:505–513

Crowe JH, Clegg JS (1973) Anhydrobiosis. Dowden, Hutchinson and Ross, Stroudsburg, p 477

Crowe JH, Clegg JS (1978) Dry biological systems. Academic Press, New York

Crowe JH, Madin KAC (1975) Anhydrobiosis in nematodes: evaporative water loss and survival. J Exp Zool 193:323–333

Crowe JH, Crowe LM, Carpenter JF, Wistrom CA (1987) Stabilization of dry phospholipid bilayers and proteins by sugars. Biochem J 242:1–10

Crowe JH, Crowe LM, Carpenter JF, Rudolph AS, Wistrom CA, Spargo BJ, Anchordoguy TJ (1988) Interactions of sugars with membranes. Biochim Biophys Acta 947:367–384

Crowe JH, Hoekstra FA, Crowe LM (1992) Anhydrobiosis. Annu Rev Physiol 54:579–599

Crowe JH, Leslie SB, Crowe LM (1994) Is vitrification sufficient to preserve liposomes during freeze-drying? Cryobiology 31:355–366

Crowe JH, Crowe LM, Petrelski S, Hoekstra FA, Araujo PD, Panek AD (1997) Anhydrobiosis: cellular adaptation to extreme dehydration. In: Dantzler WH (ed) Handbook of physiology. Oxford University Press, New York

Crowe JH, Carpenter JF, Crowe LM (1998) The role of vitrification in anhydrobiosis. Annu Rev Physiol 60:73–103

De Graaf J, Amons R, Möller W (1990) The primary structure of artemin from artemia cysts. Eur J Biochem 193:737–750

Doyère PLN (1842) Memoires sur les Tardigrades. Sur le facilité que possedent les Tardigrades, les rotiferes, les anguillules des toits et quelques autres animalcules, de revenir à la vie après été completement desséchées. Ann Sci Nat Zool Biol Anim 2e:5–35

Drennan P, Smith M, Goldsworthly D, van Staden J (1993) The occurence of trehalose in the leaves of the desiccation-tolerant angiosperm *Myrothamnus flabellifolius*. J Plant Physiol 142:493–496

Ellis RJ (2004) From chloroplasts to chaperones: how one thing led to another. Photosynth Res 80:333–343

Feder ME, Hofmann GE (1999) Heat-shock proteins, molecular chaperones, and the stress response: evolutionary and ecological physiology. Annu Rev Physiol 61:243–282

Fielding MJ (1951) Observations on the length of dormancy in certain plant infecting nematodes. Proc Helminthol Soc Wash 18:110–112

Förster F, Beisser D, Frohme M, Schill RO, Dandekar T (2011a) Tardigrade bioinformatics: molecular adaptations, DNA j-family and dynamical modeling. J Zool Syst Evol Res 49:120–126

Förster F, Beisser D, Grohme M, Liang C, Mali B, Siegl AM, Engelmann JC, Shkumatov AV, Schokraie E, Müller T, Schnölzer M, Schill RO, Frohme M, Dandekar T (2011b) Transcriptome analysis in tardigrade species reveals specific molecular pathways for stress adaptations. Bioinf Biol Insights 6:69–96

Förster F, Liang C, Shkumatov A, Beisser D, Engelmann JC, Schnölzer M, Frohme M, Müller T, Schill RO, Dandekar T (2009) Tardigrade workbench: comparing stress-related proteins, sequence-similar and functional protein clusters as well as RNA elements in tardigrades. BMC Genomics 10:469–479

Franceschi T (1948) Anabiosi nei tardigradi. Boll Mus Ist Biot Univ Genova:47–49

Freeman BC, Morimoto RI (1996) The human cytosolic molecular chaperones hsp90, hsp70 (hsc70) and hdj-1 have distinct roles in recognition of a non-native protein and protein refolding. EMBO J 15:2969–2979

Galau GA, Hughes DW, Dure L III (1986) Abscisic-acid induction of cloned cotton gossypium-hirsutum late embryogenesis-abundant lea messenger rna species. Plant Mol Biol 7:155–170

Gething M-J, Sambrook J (1992) Protein folding in the cell. Nature 355:33–45

Goodey T (1923) Quiescence and reviviscence in nematodes, with special reference to *Tylenchus tritici* and *Tylenchus dipsaci*. J Helminthol 1:47–52

Goyal K, Pinelli C, Maslen SL, Rastogi RK, Stephens E, Tunnacliffe A (2005) Dehydration-regulated processing of late embryogenesis abundant protein in a desiccation-tolerant nematode. FEBS Lett 579:4093–4098

Goyal K, Tisi L, Basran A, Browne J, Burnell A, Zurdo J, Tunnacliffe A (2003) Transition from natively unfolded to folded state induced by desiccation in an anhydrobiotic nematode protein. J Biol Chem 278:12977–12984

Gros L, Saparbaev MK, Laval J (2002) Enzymology of the repair of free radicals-induced DNA damage. Oncogene 21:8905–8925

Grzelezak ZF, Sattalo MH, Hanley-Bowdoin LK, Kennedy TD, Lane BG (1982) Synthesis and turnover of proteins and mRNA in germinating wheat embryos. Can J Biochem Physiol:389–397

Guidetti R, Jönsson KI (2002) Long-term anhydrobiotic survival in semi-terrestrial micrometazonas. J Zool (Lond) 257:181–187

Guzhova I, Krallish I, Khroustalyova G, Margulis B, Rapoport A (2008) Dehydration of yeast: changes in the intracellular content of Hsp70 family proteins. Process Biochem 43:1138–1141

Hanafusa H (1969) Rapid transformation of cells by rous sarcoma virus. Proc Natl Acad Sci 63:318–325

Harrigan PR, Madden TD, Cullis PR (1990) Protection of liposomes during dehydration or freezing. Chem Phys Lipids 52:139–149

Hengherr S, Heyer AG, Brümmer F, Schill RO (2008) Trehalose as protecting agent in aquatic invertebrates during dormancy induced by desiccation. Comp Biochem Physiol A Comp Physiol 151:S34

Hengherr S, Heyer AG, Brümmer F, Schill RO (2011) Trehalose and vitreous states: desiccation tolerance in dormant stages of the crustaceans *Triops* and *Daphnia*. Physiol Biochem Zool 84:147–153

Hengherr S, Schill RO (2011) Dormant stages in freshwater bryozoans – an adaptation to transcend environmental constraints. J Insect Physiol 57:595–601

Hengherr S, Worland MR, Reuner A, Brümmer F, Schill RO (2009) High-temperature tolerance in anhydrobiotic tardigrades is limited by glass transition. Physiol Biochem Zool 82:749–755

Hoekstra FA (1986) Water content in relation to stress in pollen. In: Leopold CA (ed) Membranes, metabolism and dry organisms. Cornell University Press, New York

Ingram J, Bartels D (1996) The molecular basis of dehydration tolerance in plants. Annu Rev Plant Physiol Plant Mol Biol 47:377–403
Jönsson KI, Bertolani R (2001) Facts and fiction about long-term survival in tardigrades. J Zool (Lond) 255:121–123
Jørgensen A, Møbjerg N, Kristensen RM (2007) A molecular study of the tardigrade *Echiniscus testudo* (Echiniscidae) reveals low DNA sequence diversity over a large geographical area. J Limnol 66:77–83
Kalichevsky MT, Jaroszkiewicz EM, Ablett S, Blanshard JMV, Lillford PJ (1992) The glass transition of amylopectin measured by DSC, DMTA and NMR. Carbohydr Polym 18:77–88
Keilin D (1959) The leeuwenhoek lecture. The problem of anabiosis or latent life: history and current concept. Proc R Soc Lond B Biol Sci 150:149–191
Kikawada T, Nakahara Y, Kanamori Y, Iwata K-i, Watanabe M, McGee B, Tunnacliffe A, Okuda T (2006) Dehydration-induced expression of LEA proteins in an anhydrobiotic chironomid. Biochem Biophys Res Commun 348:56–61
Kondo K, Kubo T, Kunieda T (2015) Suggested involvement of PP1/PP2A activity and *de novo* gene expression in anhydrobiotic survival in a tardigrade, *Hypsibius dujardini*, by chemical genetic approach. PLoS One 10:e0144803
Koster KL, Leopold AC (1988) Sugars and desiccation tolerance in seeds. Plant Physiol (Rockv) 88:829–832
Lapinski J, Tunnacliffe A (2003) Anhydrobiosis without trehalose in bdelloid rotifers. FEBS Lett 553:387–390
Lee DL (1961) Two new species of cryptobiotic (anabiotic) freshwater nematodes, *Actinolaimus hintoni* sp. nov. and *Dorylaimus keilini* sp. nov. (Dorylaimidae). Parasitology 51:237–240
Liang P, Amons R, Clegg JS, MacRae TH (1997a) Molecular characterization of a small heat shock/alpha-crystallin protein in encysted *Artemia* embryos. J Biol Chem 272:19051–19058
Liang P, Amons R, MacRae TH, Clegg JS (1997b) Purification, structure and *in vitro* molecular-chaperone activity of *Artemia* p26, a small heatshock a-crystallin protein. Eur J Biochem 243:225–232
Liang P, MacRae TH (1999) The synthesis of a small heat shock/alpha-crystallin protein in *Artemia* and its relationship to stress tolerance during development. Dev Biol 207:445–456
Lindahl T (1993) Instability and decay of the primary structure of DNA. Nature 362:709–715
Madin KAC, Crowe JH (1975) Anhydrobiosis in nematodes: carbohydrate and lipid metabolism during dehydration. J Exp Zool 193:335–342
Mali B, Grohme M, Wełnicz W, Dandekar T, Schnölzer M, Reuter D, Schill RO, Frohme M (2008) Genomic analyses of cryptobiotic tardigrades. In: Wissenschaftliche Beiträge 2008. Wildau
Martinez-Guitarte JL, Planello R, Morcillo G (2007) Characterization and expression during development and under environmental stress of the genes encoding ribosomal proteins L11 and L13 in *Chironomus riparius*. Comp Biochem Physiol B Biochem Mol Biol 147:590–596
Mattimore V, Battista JR (1996) Radioresistance of *Deinococcus radiodurans*: functions necessary to survive ionizing radiation are also necessary to survive prolonged desiccation. J Bacteriol 178:633–637
McGee B, Schill RO, Tunnacliffe A (2004) Hydrophilic proteins in invertebrate anhydrobiosis. Integr Comp Biol 44:679–679
Morimoto RI (1993) Cells in stress: transcriptional activation of heat shock genes. Science 259:1409–1410
Nathan DF, Vos MH, Lindquist S (1997) *In vivo* functions of the *Saccharomyces cerevisiae* Hsp90 chaperone. Proc Natl Acad Sci U S A 94:12949–12956
Neumann S, Reuner A, Brümmer F, Schill RO (2009) DNA damage in storage cells of anhydrobiotic tardigrades. Comp Biochem Physiol A Mol Integr Physiol 153:425–429
Örstan A (1995) Desiccation survival of the eggs of the rotifer *Adineta vaga* (Davis, 1873). Hydrobiologia 313/314:373–375
Örstan A (1998) Factors affecting long-term survival of dry bdelloid rotifers: a preliminary study. Hydrobiologia 387/388:327–331

Parsell DA, Lindquist S (1993) The function of heat-shock proteins in stress tolerance: degradation and reactivation of damaged proteins. Annu Rev Genet 27:437–496
Picard D, Khursheed B, Garabedian MJ, Fortin MG, Lindquist S, Yamamoto KR (1990) Reduced levels of hsp90 compromise steroid receptor action *in vivo*. Nature 348:166–168
Potts M (2001) Desiccation tolerance: a simple process? Trends Microbiol 9:553–559
Prestrelski SJ, Tedeschi N, Arakawa T, Carpenter JF (1993) Dehydration-induced conformational transitions in proteins and their inhibition by stabilizers. Biophys J 65:661–671
Priestley DA (1986) Seed aging. Cornell University Press, New York
Rahm PG (1923) Biologische und physiologische Beiträge zur Kenntnis der Moosfauna. Zeitschrift für allgemeine Physiologie 20:1–34
Ramazzotti G, Maucci W (1983) Il phylum Tardigrada. Memorie dell'Istituto Italiano di ldrobiologia. Istituto Italiano di idrobiologia, Verbana Pallanza, Italien, 1014 pp
Ramløv H, Westh P (2001) Cryptobiosis in the eutardigrade *Adorybiotus* (*Richtersius*) *coronifer*: tolerance to alcohols, temperature and de novo protein synthesis. Zool Anz 240:517–523
Rebecchi L, Guidetti R, Borsari S, Altiero T, Bertolani R (2006) Dynamics of long-term anhydrobiotic survival of lichen-dwelling tardigrades. Hydrobiologia 558:23–30
Reuner A, Hengherr S, Mali B, Förster F, Arndt D, Reinhardt R, Dandekar T, Frohme M, Brümmer F, Schill RO (2009) Stress-response in tardigrades: differential gene expression of molecular chaperones. Cell Stress Chaperones 15:423–430
Ricci C (1998) Anhydrobiotic capabilities of bdelloid rotifers. Hydrobiologia 387/388:321–326
Ricci C (2001) Dormancy patterns in rotifers. Hydrobiologia 446/447:1–11
Ricci C, Pagani M (1997) Desiccation of *Panagrolaimus rigidus* (nematoda): survival, reproduction and the influence on the internal clock. Hydrobiologia 347:1–13
Ricci C, Caprioli M (1998) Stress during dormancy: effect on recovery rates and life-history traits of anhydrobiotic animals. Aquat Ecol 32:353–359
Ricci C, Caprioli M (2005) Anhydrobiosis in bdelloid species, populations and individuals. Integr Comp Biol 45:759–763
Ricci C, Covino C (2005) Anhydrobiosis of *Adineta ricciae*: costs and benefits. Hydrobiologia 546:307–314
Ricci C, Vaghi L, Manzini ML (1987) Desiccation of rotifers (*Macrotrachela quadricornifera*): survival and reproduction. Ecology 68:1488–1494
Sakurai M, Furuki T, Akao K-i, Tanaka D, Nakahara Y, Kikawada T, Watanabe M, Okuda T (2008) Vitrification is essential for anhydrobiosis in an african chironomid, *Polypedilum vanderplanki*. Proc Natl Acad Sci 105:5093–5098
Schill RO, Fritz GB (2008) Desiccation tolerance in embryonic stages of the tardigrade. J Zool (Lond) 276:103–107
Schill RO, Steinbrück GHB, Köhler H-R (2004) Stress gene (*hsp70*) sequences and quantitative expression in *Milnesium tardigradum* (Tardigrada) during active and cryptobiotic stages. J Exp Biol 207:1607–1613
Schokraie E, Hotz-Wagenblatt A, Warnken U, Mali B, Frohme M, Forster F, Dandekar T, Hengherr S, Schill RO, Schnölzer M (2010) Proteomic analysis of tardigrades: towards a better understanding of molecular mechanisms by anhydrobiotic organisms. PLoS One 5: e9502.9501–e9502.9537
Schokraie E, Warnken U, Hotz-Wagenblatt A, Grohme M, Hengherr S, Förster F, Schill RO, Frohme M, Dandekar T, Schnölzer M (2012) Comparative proteome analysis of *Milnesium tardigradum* in early embryonic state versus adults in active and anhydrobiotic state. PLoS One 7:e45682
Schöneich C (1999) Reactive oxygen species and biological aging: amechanistical approach. Exp Gerontol 34:19–34
Schumacher RJ, Hansen WJ, Freeman BC, Alnemri E, Litwack G, Toft DO (1996) Cooperative action of Hsp70, Hsp90, and DnaJ proteins in protein renaturation. Biochemistry 35:14889–14898
Scott P (2000) Resurrection plants and the secrets of eternal leaf. Ann Bot (Lond) 85:159–166

Slade L, Levine H (1991) Beyond water activity: recent advances based on an alternative approach to the assessment of food quality and safety. Crit Rev Food Sci Nutr 30:115–360
Steiner G, Albin FE (1946) Resuscitation of the nematode *Tylenchus polyhypnus* n. sp. after almost 39 years' dormancy. J Wash Acad Sci 36:97–99
Sun WQ, Leopold AC (1997) Cytoplasmic vitrification and survival of anhydrobiotic organisms. Comp Biochem Physiol A Physiol 117:327–333
Tanaka S, Tanaka J, Miwa Y, Horikawa DD, Katayama T, Arakawa K, Toyoda A, Kubo T, Kunieda T (2015) Novel mitochondria-targeted heat-soluble proteins identified in the anhydrobiotic Tardigrade improve osmotic tolerance of human cells. PLoS One 10:e0118272
Tissières A, Mitchell HK, Tracy UM (1974) Protein synthesis in the salivary glands of *Drosophila melanogaster*. J Mol Biol 84:389–398
Tomos D (1992) Life without water. Curr Biol 2:594–596
Tourancheau AB, Morin L, Yang T, Perasso R (1999) Messenger RNA in dormant cells of sterkiella histriomuscorum (oxytrichidae): identification of putative regulatory gene transcripts. Protist 150:137–147
Tunnacliffe A, de Castro AG, Manzanera M (2001) Anhydrobiotic engineering of bacterial and mammalian cells: is intracellular trehalose sufficient? Cryobiology 43:124–132
Tunnacliffe A, Lapinski J (2003) Resurrecting van Leeuwenhoek's rotifers: a reappraisal of the role of disaccharides in anhydrobiosis. Philos Trans R Soc Lond Ser B Biol Sci 358:1755–1771
Tunnacliffe A, Lapinski J, McGee B (2005) A putative LEA protein, but no trehalose, is present in anhydrobiotic bdelloid rotifers. Hydrobiologia 546:315–321
Tunnacliffe A, Wise MJ (2007) The continuing conundrum of the LEA proteins. Naturwissenschaften 94:791–812
van Leeuwenhoek A (1702) On certain animalcules found in the sediments in gutters of the roofs of houses. Letter 144. In: The selected works of Anton van Leeuwenhoek. London, pp 207–213
Vertucci CW, Farrant JM (1995) Acquisition and loss of desiccation. Marcel Dekker, New York
Warner AH, Brunet RT, MacRae TH, Clegg JS (2004) Artemin is an RNA-binding protein with high thermal stability and potential RNA chaperone activity. Arch Biochem Biophys 424:189–200
Watanabe M, Kikawada T, Minagawa N, Yukuhiro F, Okuda T (2002) Mechanism allowing an insect to survive complete dehydration and extreme temperatures. J Exp Biol 205:2799–2802
Webb SJ, Dumasia MD, Bhorjee JS (1965) Bound water, inositol, and the biosynthesis of temperate and virulent bacteriophages by air-dried *Escherichia coli*. Can J Microbiol 11:141–150
Westh P, Ramlov H (1988) Cryptobiosis in arctic tardigrades with special attention to the appearance of trehalose. In: Petersen GH (eds) Bericht über die Grönland-Exkursion des Instituts für Polarökologie. Universität für Polarökologie, Kiel, pp 227–245
Westh P, Ramlov H (1991) Trehalose accumulation in the tardigrade *Adorybiotus coronifer* during anhydrobiosis. J Exp Zool 258:303–311
Wharton DA (2003) The environmental physiology of antarctic terrestrial nematodes: a review. J Comp Physiol B Biochem Syst Environ Physiol 173:621–628
Wharton DA, Aalders O (1999) Desiccation stress and recovery in the anhydrobiotic nematode *Ditylenchus dipsaci* (Nematoda : Anguinidae). Eur J Entomol 96:199–203
Willsie JK, Clegg JS (2002) Small heat shock protein p26 associates with nuclear lamins and HSP70 in nuclei and nuclear matrix fractions from stressed cells. J Cell Biochem 84:601–614
Wolkers WF, McCready S, Brandt WF, Lindsey GG, Hoekstra FA (2001) Isolation and characterization of a D-7 LEA protein from pollen that stabilizes glasses *in vitro*. Biochim Biophys Acta Protein Struct Mol Enzymol 1544:196–206
Wolkers WF, Oldenhof H, Alberda M, Hoekstra FA (1998) A Fourier transform infrared microspectroscopy study of sugar glasses: application to anhydrobiotic higher plant cells. Biochim Biophys Acta 1379:83–96
Wolkers WF, Tetteroo FA, Alberda M, Hoekstra FA (1999) Changed properties of the cytoplasmic matrix associated with desiccation tolerance of dried carrot somatic embryos. An in situ fourier transform infrared spectroscopic study. Plant Physiol 120:153–164

Womersley CZ (1987) A re-evaluation of strategies employed by nematode anhydrobiotes in relation to their natural environment. In: Veech JA, Dickson DW (eds) Vistas on nematology. Society of Nematologists, Hyattsville, pp 165–173

Wright JC (1991) The significance of four xeric parameters in the ecology of terrestrial tardigrada. J Zool (Lond) 224:59–77

Wright JC (2001) Cryptobiosis 300 years on from van Leuwenhoek: what have we learned about tardigrades? Zool Anz 240:563–582

Xu Y, Lindquist S (1993) Heat-shock protein hsp90 governs the activity of pp60v-src kinase. Proc Natl Acad Sci U S A 90:7074–7078

Yamaguchi A, Tanaka S, Yamaguchi S, Kuwahara H, Takamura C (2012) Two novel heat-soluble protein families abundantly expressed in an anhydrobiotic tardigrade. PLoS One 7:e44209

Yoshida Y, Koutsovoulos G, Laetsch DR, Stevens L, Kumar S, Horikawa DD, Ishino K, Komine S, Kunieda T, Tomita M, Blaxter M, Arakawa K (2017) Comparative genomics of the tardigrades *Hypsibius dujardini* and *Ramazzottius varieornatus*. PLoS Biol 15:e2002266

Chapter 11
Environmental Adaptations: Cryobiosis

Steffen Hengherr and Ralph O. Schill

Abstract Tardigrades are well known to withstand very low temperatures in the anhydrobiotic state. However, they even tolerate such low temperatures like −196 °C in the fully hydrated state which is then described with the term cryobiosis. Although this extreme subzero temperature tolerance got quite a lot of attention, there is little knowledge regarding their physiological and biochemical adaptations connected to ecological representative subzero temperatures. General studies on cold tolerance have highlighted some strategies including freeze avoidance, rapid cold hardening and freeze tolerance. Although studies on survival rates, cooling rates and ice formation in tardigrades show high interspecific variations in subzero temperature survival, the water bears seem to tolerate ice formation within their bodies and therefore belong to freeze-tolerant organisms. Calorimetric studies also provide evidence for homogenous ice nucleation, indicating that ice formation is not largely affected by ice-nucleating agents. Ability to tolerate low temperatures and freezing even in embryonic developmental stages further increases the adaptive benefit of tardigrades to cope with low-temperature events.

11.1 Introduction to Low-Temperature Survival

11.1.1 Subzero Temperature Exposure and Its Effects on Living Organisms

Low temperatures, especially below the freezing point of water, are important in biological systems. Since subzero temperatures occur periodically over much of the

S. Hengherr (✉)
Jakob-Friedrich-Schöllkopf School, Kirchheim u. T., Germany
e-mail: hen@jfs.de

R. O. Schill
Institute of Biological Materials and Biomolecular Systems, University of Stuttgart, Stuttgart, Germany
e-mail: ralph.schill@bio.uni-stuttgart.de

R. O. Schill (ed.), *Water Bears: The Biology of Tardigrades*, Zoological Monographs 2,
https://doi.org/10.1007/978-3-319-95702-9_11

globe's surface, water-ice transitions have a profound influence in all ecosystems (Sømme and Meier 1995). On a much finer scale, water-ice transition may cause cellular dehydration, disrupt chemical gradients and eventually kill organisms (Franks 1985).

For evaluating the state of activity and geographic distribution of hydrophilous micrometazoans like tardigrades, temperature is among the most determining factors (Doucet et al. 2009). Many limno-terrestrial tardigrade species, which occupy unstable habitats like mosses, lichens and intertidal zones, are tolerating harsh environmental conditions in any developmental stage (Schill and Fritz 2008) by entering a cryptobiotic state. During anhydrobiotic state, tardigrades, but also other cryptobiotes such as nematodes and rotifers, are showing extraordinary tolerance of physical extremes (Horikawa et al. 2006; Jönsson and Schill 2007; Ramløv and Westh 2001; Seki and Toyoshima 1998). Besides the tolerance during dehydrated state, tardigrades have also been reported to survive exposures down to −196 °C hydrated (Ramløv and Westh 1992; Sømme and Meier 1995; Hengherr et al. 2009, 2010).

Colonization of habitats showing frequent low-temperature periods requires certain adaptations in order to reduce the damaging impact of cold and freezing stress. Physiological studies, predominantly on insects, have highlighted some strategies including freeze avoidance, freeze tolerance including cryoprotective dehydration and rapid cold hardening (Clark and Worland 2008).

11.1.2 Freeze Avoidance

Freeze-avoiding organisms usually keep their body fluids in a liquid state at temperatures below their melting point. However, field and laboratory studies have shown that mortality of some freeze-avoiding insects occurs regularly above their freezing temperature (T_c, temperature of spontaneous freezing). Therefore, this mortality cannot be ascribed to freezing damages (Bale 1996). This prefreeze injury (chilling injury) may be related to membrane transitions and protein conformational changes at subzero temperatures (Ramløv 2000). Seasonality of low temperatures imposes on organisms the ability to undergo physiological adaptations making them tolerant for longer periods of time. But at the same time, they also need an ability to respond to quick low-temperature exposures. Slower hardening response may take days, weeks or even months and is often associated with the entry into a state of diapause or quiescence (Lee et al. 1996; Doucet et al. 2009). Faster hardening process, called rapid cold-hardening process, includes a series of metabolic changes. These changes lead to a reduction of susceptibility to damaging effects of what would otherwise be a lethal temperature and increase overall cold tolerance.

To avoid chilling injuries, the freezing of body fluids or intracellular freezing which would be lethal for freeze-susceptible organisms, freeze-avoiding organisms synthesize small organic molecules such as polyhydric alcohols, sugars and antifreeze proteins (AFPs) (Danks et al. 1994; Doucet et al. 2009; Duman 2001).

Studies with insects have shown that cold hardening is correlated with an accumulation of cryoprotectant glycerol, which reduces water loss and can shield membranes from temperature-induced phase transitions (Lee et al. 1996; Yoder et al. 2006). Although glycerol seems to be the most common polyhydric alcohol in cold-tolerant organisms, other polyols such as mannitol, erythritol, myoinositol and sorbitol have also been reported in several insect species (Lalouette et al. 2007; Watanabe and Tanaka 1998). Although not studied well yet, structural relatedness of glycerol and sorbitol suggests they may contribute to subzero temperature survival in the same manner (Salvucci et al. 2000).

Although many animals avoid water loss by seeking shelter, some animals such as nematodes, insect larvae, Collembola and tardigrades (Grewal et al. 2006; Hengherr et al. 2008a, b; Kikawada et al. 2006; Worland et al. 1998) are tolerant to significant water loss, and the threat of freezing is overcome. In challenges faced through cold or desiccation in terms of increase in solute concentrations and osmotic stress, membrane-damaging effects and reduction in supercooling point (Convey 2000) are very similar. Hence desiccation and cold tolerance are considered to be overlapping adaptations (Ring and Danks 1994). Studies in the Arctic springtail *Onychiurus arcticus* (Worland and Block 2003; Worland et al. 1998) and the Antarctic midge *Belgica antarctica* (Elnitsky et al. 2008) have shown a strategy of cryoprotective dehydration as a survival strategy. In this case, desiccation occurs due to a difference in water vapour pressure between the animals' supercooled body fluids and ice in its surroundings.

One common feature of desiccation- and cold-tolerant organisms is an accumulation of two alpha-linked glucose units, trehalose. Sugars, such as the non-reducing disaccharide trehalose, can additionally protect against cold-induced damages (Overgaard et al. 2008). Many cold-tolerant organisms are sharing the ability to synthesize trehalose. This response might be linked to the capacity of trehalose to replace water molecules and compensating the loss of water and the stabilization of membranes by interacting with the phosphate of phospholipids during cold shock (Crowe et al. 1992; Sakurai et al. 2008).

However, it has been shown that the accumulation of cryoprotectants depresses the freezing point not sufficiently, if ice-nucleating agents are present (Zachariassen 1991). To protect themselves by avoiding the initiation of ice growth, freeze-avoiding organisms eliminate materials from their midguts through the synthesis of antifreeze proteins (AFPs). Thus these organisms can avoid freezing by lowering the freezing point of the body fluids in order to supercool. After the first discovery of antifreeze glycoproteins in fish by DeVries (1971), AFPs have been identified in many other cold-tolerant organisms, predominantly insects (Clark and Worland 2008; Doucet et al. 2009). These proteins are able to depress the freezing point of body fluids in the presence of ice or ice nucleators by inhibiting ice growth (Raymond and DeVries 1977; Duman 2001; Wharton 2003). They are also called thermal hysteresis proteins (THPs) because they lower the non-equilibrium freezing point by adsorbing to ice, but are not affecting the melting point. The phenomenon is termed thermal hysteresis (DeVries 1986).

11.1.3 Freeze Tolerance

In contrast to freeze-avoiding organisms, freeze-tolerant organisms tolerate a controlled ice formation in extracellular body water in a controlled manner to maintain the intracellular contents in an unfrozen state. Small volumes of pure water have the capacity to supercool and do not freeze at temperatures close to their melting point. Only the presence of ice nucleators ensures that water crystallizes near 0 °C. The common small size of freeze-tolerant organisms dictates that they are unlikely to spontaneously freeze until their body fluid drops to −25 °C of the equilibrium freezing point (Lee et al. 1996). However, the crystallization in freeze-tolerant organisms is often triggered at high subzero temperatures (−5 °C to −10 °C) by ice-nucleating agents (INA) in the extracellular fluid (Block 1991; Lee and Costanzo 1998; Ramløv 2000; Storey and Storey 1996) leading to a slow process of ice formation by which the growth of ice crystals can be controlled. This results in smaller and potentially less damaging initial ice crystals upon freezing. INAs may consist of proteins, lipoproteins, salts and even bacteria or fungi (Duman 2001).

As it has been shown in studies on the gall fly *Eurosta solidaginis*, many freeze-tolerant organisms also accumulate polyhydroxy alcohols, sugars and amino acids to maintain the cellular structures in an unfrozen state, to protect proteins and membranes against phase transition and to control the ice fractions size and the minimum cell volume, resulting from freeze concentration and osmotic dehydration (Ramløv 2000; Sinclair et al. 2003; Zachariassen 1985). To inhibit recrystallization, freeze-tolerant organisms sometimes use ice-active proteins and even AFPs. The seemingly anomalous use of AFPs by freeze-tolerant organisms might be explained by the fact that these proteins prevent a size of increase of ice crystals, known as recrystallization, which would otherwise cause physical damage to cells (Doucet et al. 2009; Duman 2001; Knight and Duman 1986; Wharton 2003). Although most freeze-tolerant organisms only tolerate extracellular freezing, it has been shown that the Antarctic nematode *Panagrolaimus davidi* even survives freezing of the intracellular compounds (Smith et al. 2008; Wharton et al. 2005, 2003).

11.2 Tolerance to Subzero Temperatures in Tardigrades

11.2.1 Long-Term Survival Under Cryobiotic Conditions

Tardigrades or water bears are well known to survive freezing in the dehydrated state (Wright 2001), but studies on subzero temperatures have also reported a survival after exposure to −196 °C (Ramløv and Westh 1992; Sømme and Meier 1995). Sømme and Meier (1995) reported a survival of three tardigrade species, *Echiniscus jenningsi*, *Macrobiotus furciger* and *Diphascon chilense*, after 8.3 years in anhydrobiosis under frozen conditions. Although the long-term survival in anhydrobiotic micrometazoans has been a well-studied topic of cryptobiosis, there are only few studies on the long-term survival under cryobiotic conditions.

Newsham et al. (2006) stated that various Antarctic micrometazoans were revived from a liverwort sample, stored frozen at −80 °C for a period of 6 years. Living individuals of the Antarctic nematode *Plectus murrayi* were recovered after 25.5 years of storage under frozen conditions (Kagoshima et al. 2012). Similarly, in a study on the Antarctic tardigrade *Acutuncus antarcticus*, living individuals were retrieved from frozen phytobenthos after 5 and even after 30.5 years (Tsujimoto et al. 2016). To withstand such long periods in low-temperature conditions, which are much longer than any documented life span of tardigrades, ageing should be slowed down or even stopped during cryobiosis. Recent studies confirmed this "Sleeping Beauty" hypothesis for cryobiosis (Sieger, pers. commun), which has also already been proven in anhydrobiotic tardigrades (Hengherr et al. 2008a, b).

11.2.2 Subzero Temperature Survival and the Effects of Cooling Rates in Tardigrades

Based on the current information available, the cooling rate affects the subzero temperature survival of adult tardigrades and their embryonic stages. Ramløv and Westh (1992) used in their study concerning freeze tolerance in the tardigrade *Adorybiotus coronifer* cooling rates between 1 °C min^{-1} and more than 1000 °C min^{-1}. Although laboratory studies have used a variety of cooling rates, usually distributed around 1 °C min^{-1} and faster, to investigate cold hardiness in invertebrates, these cooling rates have been criticized as too fast compared with natural cooling rates (Sinclair et al. 2003). More natural cooling rates with temperature decreases between 3 and 6 °C h^{-1} have been recorded in ecological studies on, e.g. *Drosophila melanogaster* (Kelty and Lee 1999). In the most extensive study available, considering freeze tolerance in tardigrades as general feature, the freeze tolerance of nine limno-terrestrial tardigrade species originating from polar, temperature and tropical regions has been investigated (Hengherr et al. 2009). In this comparative investigation, cooling rates of 9, 7, 5, 3 and 1 °C h^{-1} were applied. Table 11.1 presents the survival rates of the different tardigrade species used in the study of Hengherr et al. (2009). All species, except *Macrobiotus tonollii*, showed a decrease in survival with a reduction in cooling rate from 9 to 5 °C h^{-1} and an increase towards 1 °C h^{-1}. In general, the cooling at the higher rate of 9 °C h^{-1} resulted in higher survival compared with the slower cooling rates of 7, 5 and 3 °C h^{-1}. However, the interspecific comparison of the survival rate after cooling at different rates revealed high variations. *Milnesium tardigradum* showed the highest survival rate at 9 °C h^{-1} and the lowest at 3 °C h^{-1}. Most of the other species showed al lower but generally similar pattern of subzero temperature survival. Only the species *Macrobiotus sapiens* and *Macrobiotus tonollii* showed a total loss of the ability to survive after cooling rates of 3, 5 and 7 °C h^{-1}. These results support earlier

Table 11.1 Survival rates (mean ± s.d.) of the tested tardigrade species after cooling at different rates from room temperature (RT) down to −30 °C and a constant thawing rate of 10 °C h^{-1} up to RT again (modified from Hengherr et al. 2009)

Species	Survival rates [%] after different cooling rates				
	1 °C h^{-1}	3 °C h^{-1}	5 °C h^{-1}	7 °C h^{-1}	9 °C h^{-1}
Macrobiotus sapiens	27.50 ± 10.41	0.00 ± 0.00	0.00 ± 0.00	43.75 ± 14.93	47.5 ± 17.08
Macrobiotus richtersi "group 1"	27.50 ± 5.00	16.25 ± 14.93	32.50 ± 11.90	77.50 ± 10.41	90.00 ± 10.80
Macrobiotus richtersi "group 2"	38.75 ± 11.90	15 ± 10.80	5.00 ± 4.80	7.50 ± 5.00	23.75 ± 8.54
Macrobiotus richtersi "group 3"	15.00 ± 7.07	8.75 ± 2.25	1.25 ± 2.50	3.75 ± 4.79	20.00 ± 4.80
Macrobiotus richtersi	72.50 ± 6. 45	61.25 ± 19.31	55.00 ± 14.72	66.25 ± 8.54	92.50 ± 9.57
Macrobiotus tonollii	76.25 ± 12.50	35.00 ± 8.16	27.50 ± 13.23	0.00 ± 0.00	35.00 ± 8.16
Milnesium tardigradum	86.25 ± 7.50	71.25 ± 20.16	71.25 ± 8.54	86.25 ± 8.54	95.00 ± 4.08
Echiniscus granulatus	60.63 ± 4.27	52.50 ± 8.42	22.50 ± 11.37	46.88 ± 7.47	53.13 ± 3.15
Echiniscus testudo	42.50 ± 4.56	17.50 ± 5.40	19.38 ± 5.54	38.75 ± 6.29	46.88 ± 5.54

investigations on the eutardigrade *R. coronifer*, done by Ramløv and Westh (1992), that the cooling rate has a large influence on the survival of tardigrades.

As climate diagrams of the collection region of the species *Milnesium tardigradum* indicate, low-temperature events can occur on a periodic, sometimes even daily basis in the environment of tardigrades (Sträßer 1998). The population of *Milnesium tardigradum* used in the studies of Hengherr et al. (2009, 2010) has a mean longevity of 82 days and an embryonic developmental time of approximately 5 days (Hengherr et al. 2008a, b). Considering the developmental time and the periodic low-temperature events, the ability to tolerate low temperatures and freezing in different developmental stages increases the adaptive benefit of tardigrades. The longevity data of *Milnesium tardigradum* further indicates that being not freeze tolerant in all developmental stages, any period without freezing temperatures that is too short to pass through the development to the juvenile stage would eliminate the new generation from the population. Regarding this aspect, the freeze tolerance of several developmental stages of *Milnesium tardigradum* was also investigated by Hengherr et al. (2010). The developmental stages were classified according to Suzuki (2003). Stage 1 was the blastula stage, 24 h after egg deposition. Stage 2 was 24 h later and showed morphogenetic movements together with a ventrolateral cleft. After 72 h stage 3 was characterized by an increasing transparency. During stage 4 (96 h), occasional rotatory movements were visible, and during stage 5 (120 h), the development was almost completed.

As applied to the fully developed tardigrades, the developmental stages were exposed to the same different cooling rates. The postfreeze survival and hatching after the exposure to low temperatures are presented in Table 11.2 and Fig. 11.1. Similarly, as reported for the adult specimens, the cooling rate had significant effects on survival (hatching rate). Although the reason for the effect of the cooling rate cannot completely explain the reasons for this effect, it might be due to the physical effects of rapid cooling and ice crystal formation as shown in Fig. 11.2.

All cooling conditions in the embryonic study lead to a delayed development; additionally the developmental stage showed a significant effect on the hatching rate. The survivorship of the stages increases with the age of the stages. The oldest developmental stage 5 even showed a no significant difference to the observed

Table 11.2 Percentage hatching success (means ± s.d.) of the tested embryonic stages of *Milnesium tardigradum* after cooling at different rates from room temperature (RT) down to −30 °C with a constant thawing rate of 10 °C h^{-1} up to RT again (modified from Hengherr et al. 2010)

Stage	1 °C h^{-1}	3 °C h^{-1}	5 °C h^{-1}	7 °C h^{-1}	9 °C h^{-1}
1	12.0 ± 4.5*	9.4 ± 4.7*	3.8 ± 4.3*	7.0 ± 2.1*	5.6 ± 1.3*
2	15.5 ± 3.7*	15.0 ± 3.1*	6.3 ± 2.5*	14.0 ± 2.9*	15.0 ± 5.0*
3	87.0 ± 2.7*	79.0 ± 7.2*	83.1 ± 4.3*	81.0 ± 3.8*	86.5 ± 2.9
4	90.5 ± 3.7*	79.0 ± 6.8*	86.9 ± 4.2*	89.0 ± 7.8*	91.5 ± 0.4
5	97.5 ± 4.3	87.0 ± 1.6*	88.1 ± 3.1*	90.6 ± 6.3	95.0 ± 4.6
Control	99.4 ± 1.3	98.8 ± 1.4	99.4 ± 1.3	100.0 ± 0.0	99.4 ± 1.3

Significant difference to the control is indicated with asterisks

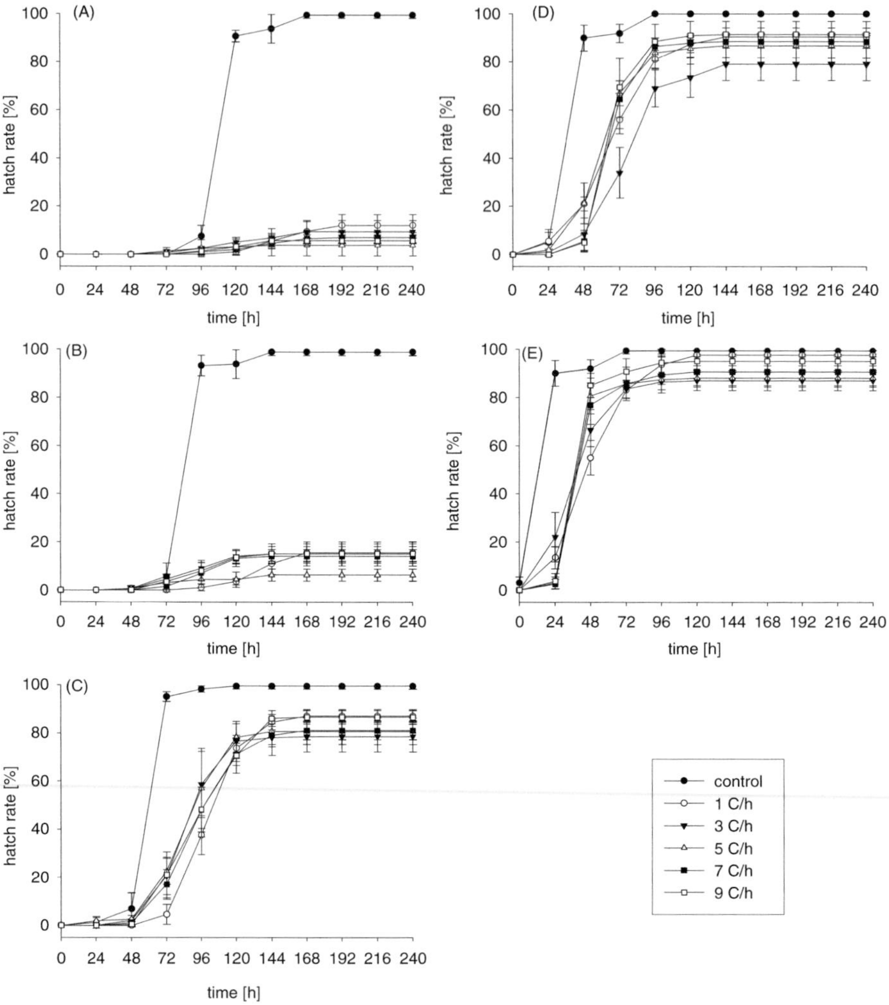

Fig. 11.1 Hatching rates (mean values ± s.d.) of the embryonic stages 1 (**a**), 2 (**b**), 3 (**c**), 4 (**d**) and 5 (**e**) of *Milnesium tardigradum* following a 10-day period after subzero temperature treatment at five different cooling rates (1, 3, 5, 7, 9 °C h^{-1}) (from Hengherr et al. 2010)

control. The increase in freezing tolerance as embryonic development proceeds suggests that enhanced protective mechanisms appear as embryogenesis proceeds. It is not known yet whether adult tardigrades and embryos use similar mechanisms to survive subzero temperatures.

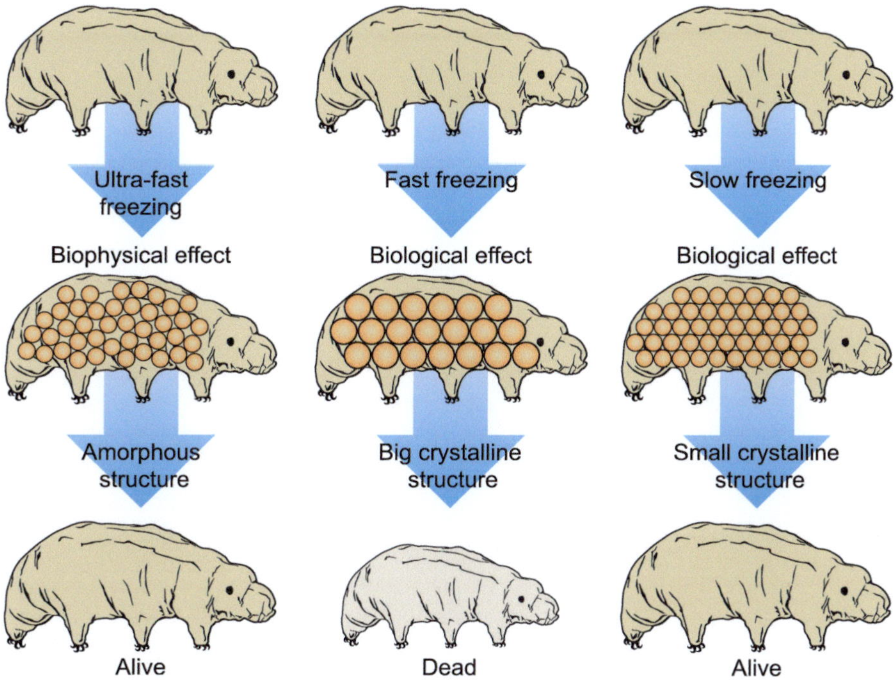

Fig. 11.2 Divergent survival patterns after different cooling rates might be induced due to the biological and physical effects of rapid cooling and ice crystal formation

11.2.3 Ice Nucleation and Supercooling Points in Tardigrades

Considering studies on the survival rates and calorimetric measurements during low-temperature exposure in tardigrades (Halberg et al. 2009; Hengherr et al. 2009, 2010; Westh and Kristensen 1992; Westh et al. 1991), it may be stated that all investigated tardigrade species can tolerate ice formation within their body and therefore belong to the group of freeze-tolerant organisms with a very high potential to supercool and the ability to survive temperatures below the temperature of spontaneous freezing (supercooling point, SCP).

Ice formation in freeze-tolerant organisms usually occurs at moderate-low temperatures (ca. −5 to −10 °C) and is often triggered by ice-nucleating agents (INA) in the extracellular fluid (Block 1991; Duman 2001). Slowing down and controlling the ice formation, INAs are responsible for producing potentially less destructive crystals and therefore reduce freeze injuries. Temperature of crystallization and the behaviour of ice growth are usually studied by using differential scanning calorimetry. Figure 11.3 illustrates a representative thermogram with each peak indicating a freezing event of an individual specimen. The ice formation in tardigrades appears to be a very rapid process with the freezing exotherm lasting usually less than 30 s.

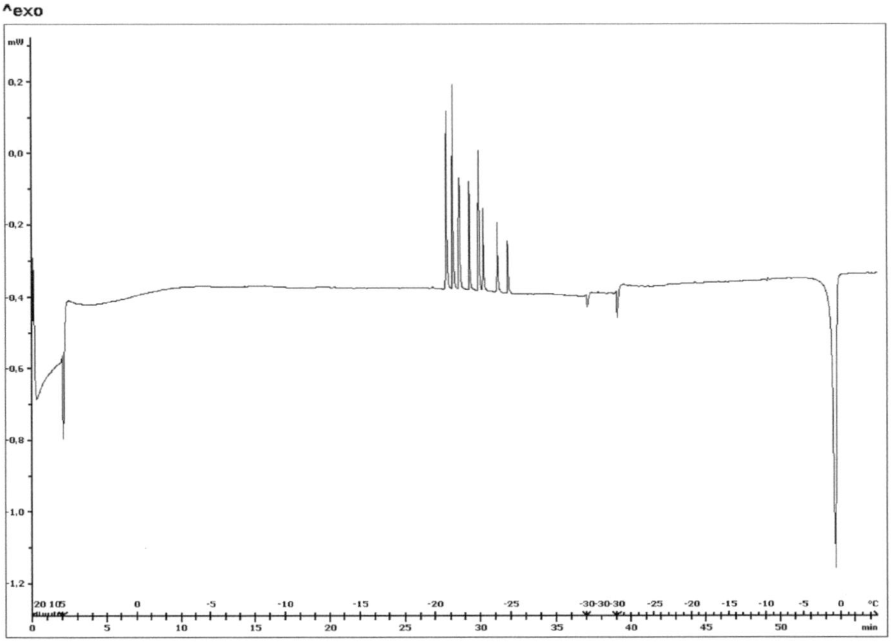

Fig. 11.3 Representative thermogram illustrating the freeze-thaw cycle of starved animals of *Paramacrobiotus richtersi* "group 1". Each peak in the cooling period indicates a freezing exotherm of an individual tardigrade. The peak of the melting endotherm close to 0 °C represents the melting of the frozen body water of all specimens (modified from Hengherr et al. 2009)

Studies on the tardigrade *Richtersius* (*Adorybiotus*) *coronifer*, by Westh and Kristensen (1992), and on the tardigrade *Amphibolus nebulosus* by Westh et al. (1991) showed relatively high crystallization temperatures (−6 to −7 °C) supporting the evidence of INAs in these tardigrade species. A decrease of the SCP to −16 °C after previous heating to 90 °C and DSC analysis of gel filtered body fluids indicates that the ice-nucleating activity is composed of proteins. In contrast, studies carried out on marine and terrestrial eutardigrade species, as well as on terrestrial heterotardigrade species and embryonic stages of *Milnesium tardigradum*, revealed much lower crystallization temperatures (Halberg et al. 2009; Hengherr et al. 2009, 2010). Halberg et al. (2009) described a SCP of −15 °C in the marine eutardigrade *Halobiotus crispae*, and Hengherr et al. (2009, 2010) showed in a comparative study SCPs ranging from −10 °C to −24 °C (Tables 11.3 and 11.4). The data provide evidence that, at least in the species used in these studies, the presence of any physiologically relevant ice nucleator is excluded. Plotting SCPs as a function of water volume together with available data on the nucleation temperature of pure water samples and freeze-avoiding insects (Mackenzie et al. 1977; Wilson et al. 2003; Zachariassen et al. 2004) is pointing out that the SCP of the tardigrades *Macrobiotus, Paramacrobiotus* and *Milnesium tardigradum*, including its developmental stages, fits in the regression line of the earlier studies, indicating a

Table 11.3 Thermal analysis results of tardigrades after a period of starvation, cold acclimatization and feeding (modified from Hengherr et al. 2009)

Species	SCP starved [°C]	SCP cold acclimatized [°C]	SCP fed [°C]	% of body water frozen during freezing exotherm	
				Warm acclimatized	Cold acclimatized
Macrobiotus sapiens	−23.7 ± 3.9[a, b]	−24.2 ± 3.9[a]	n.t.	84.5 ± 6.1	81.1 ± 3.2
Paramacrobiotus richtersi "group 1"	−21.7 ± 2.1[a]	−19.2 ± 2.0[a]	n.t.	84.3 ± 6.1	79.9 ± 4.4
Paramacrobiotus richtersi "group 2"	−20.5 ± 4.9[a]	−21.0 ± 4.9[a]	n.t.	85.5 ± 5.9	82.6 ± 4.2
Paramacrobiotus richtersi "group 3"	−20.4 ± 2.9[a]	−21.3 ± 2.3[a]	−21.5 ± 4.5	86.8 ± 4.0	80.9 ± 4.9
Paramacrobiotus richtersi	−21.6 ± 3.8[a]	−20.6 ± 2.8[a]	−20.9 ± 4.3	85.7 ± 3.7	82.5 ± 6.5
Macrobiotus tonollii	−21.8 ± 3.2[a, b]	−20.9 ± 4.0[a]	n.t.	86.4 ± 3.0	812.0 ± 5.3
Milnesium tardigradum	−22.0 ± 3.5[a, b]	−22.6 ± 2.8[a]	n.t.	85.1 ± 7.8	80.5 ± 4.5
Echiniscus granulatus	−11.5 ± 3.0	−10.3 ± 1.9	n.t.	83.8 ± 8.1	78.0 ± 5.4
Echiniscus testudo	−18.4 ± 2.2	−16.7 ± 3.9	n.t.	84.3 ± 8.2	80.5 ± 4.9

Values are presented as means ± s.d. (n.t., not tested)

[a]Significant difference from SCPs of *Echiniscus granulatus*

[b]Significant difference from SCPs of *Echiniscus testudo*

Table 11.4 Temperature of ice crystallization (T_c), total egg mass and water mass (means ± s.d.) of the tested embryonic stages of *Milnesium tardigradum*

	T_c [°C]	Egg mass [μg]	Water mass [μg]
Stage 1	−21.5 ± 1.9	0.58 ± 0.10	0.46 ± 0.10
Stage 2	−22.3 ± 2.4	0.59 ± 0.08	0.45 ± 0.09
Stage 3	−22.1 ± 2.8	0.58 ± 0.10	0.48 ± 0.10
Stage 4	−21.6 ± 2.5	0.58 ± 0.15	0.50 ± 0.13
Stage 5	−22.0 ± 2.7	0.56 ± 0.09	0.45 ± 0.08

There is no significant difference between the stages (modified from Hengherr et al. 2010)

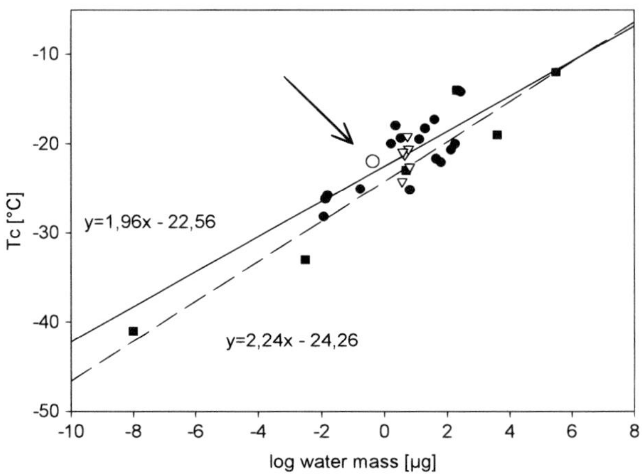

Fig. 11.4 Nucleating temperatures of embryos (mean of all stages) of *Milnesium tardigradum* plotted as a function of the logarithm of the water mass (mean value of all stages). The data point is presented by the white circle (indicated with an arrow). The white triangles present the data points obtained from fully developed freeze-tolerant eutardigrades (Hengherr et al. 2009). The solid regression line represents the linear regression line of the data points (circles) obtained from freeze-avoiding insects. The broken line is the regression line of the data points (squares) of pure water samples observed by Mackenzie et al. (1977) and Wilson et al. (2003) (from Hengherr et al. 2010)

homogenous nucleation and underlining that INAs are not present in these eutardigrades (Fig. 11.4).

It is known that some arthropod species shed their midgut during moulting. This evacuates the complete gut content, which would otherwise initiate ice nucleation at relatively high temperatures, and decreases the SCP (Lee and Costanzo 1998; Worland and Convey 2008).

The available data for eutardigrades of the genus *Paramacrobiotus* (Hengherr et al. 2009) show no difference in SCP distribution and any involvement of the gut content as ice nucleators. However, an effect due to moulting, as observed in Collembola by Worland et al. (2006), cannot be excluded. Considering the fact that limno-terrestrial tardigrades live in habitats with frequently occurring freeze and thaw cycles, it might be energetically essential to retain the gut contents during low-temperature periods. However, Ramazzotti and Maucci (1983) have shown that tardigrades are capable of surviving long periods without food.

Also, an acclimatization of tardigrades to low temperatures does not lead to any changes in the SCP or the melting point (Hengherr et al. 2009), which would suggest the synthesis of low molecular weight cryoprotectants, such as polyols or sugars. However, cold acclimation at 4 °C decreased the amount of frozen water in all tested specimens (Hengherr et al. 2009, 2010; Westh and Kristensen 1992). This reduction may reflect an increase in the amount of "bound" water due to interactions between water and macromolecules, as reported in insect larvae (Storey et al. 1981). Therefore, low concentrations of small carbohydrates may still be involved in the freeze tolerance of tardigrades.

11.2.4 The Future of Cryobiology in Tardigrades

The fast cooling rates tolerated by tardigrades, the fast ice formation and the invariant SCPs may lead to the assumption that these invertebrates do not require a metabolical or biochemical preparation to tolerate subzero temperature exposure. However, we must not exclude it. The increase in survival at slow cooling rates still may indicate that the animals synthesize some ice-active proteins or other cryoprotective compounds to increase their survival ability. Since environmental cooling rates of 0.6 °C h^{-1} in temperate environments are not unusual (Sinclair 2001), even slower (yet to investigate) cooling rates would give tardigrades more time to recruit or synthesize protecting molecules, which might regulate ice growth to conserve cell structure. For this however, further detailed studies in typical field microhabitats are required.

Usually, extracellular ice crystallization will subject cells and tissues to freeze dehydration. Consequently, intracellular solutes will become more concentrated and the cells become osmotically dehydrated. This means that the fast ice growth, observed in tardigrades, presents an enormous shock. Together with cell volume collapse, this shock might be the most likely cause of mortality in unprotected cells. Therefore, the extreme freeze tolerance observed in tardigrades may be related to their tolerance to extreme dehydration and their ability to enter the state of anhydrobiosis, which also requires the ability to deal with a wide variation in cell volumes and osmolality of body fluids. It is likely that a better knowledge of anhydrobiosis in tardigrades may also lead to a better understanding of cryobiosis in tardigrades.

Besides that, rapid ice formation also increases the likelihood of intracellular freezing with its associated problems resulting in cellular damage due to damaging ice crystal growth and rapid changes in cell volume. It is still not known if tardigrades tolerate intracellular freezing. But if it occurs, it could provide a successful strategy for tardigrades to cope with very low temperatures and reduce transmembrane osmotic stress during freezing as it has been demonstrated in the Antarctic nematode *Panagrolaimus davidi* by Wharton et al. (2003). However, studies on tardigrades concerning intracellular freezing, as well as molecular and metabolic investigations at slow cooling rates, have yet to be performed. Together with synergistic investigations in the field of anhydrobiosis, they would provide a deeper insight into the remarkable phenomenon of subzero temperature tolerance in tardigrades.

References

Bale JS (1996) Insect cold hardiness: a matter of life and death. Eur J Entomol 93:369–382

Block W (1991) To freeze or not to freeze - invertebrate survival of subzero temperatures. Funct Ecol 5:284–290

Clark MS, Worland MR (2008) How insects survive the cold: molecular mechanisms - a review. J Comp Physiol B Biochem Syst Environ Physiol 178:917–933

Convey P (2000) How does cold constrain life cycles of terrestrial plants and animals. CryoLetters 21:73–82

Crowe JH, Folkert A, Hoekstra I, Crowe LM (1992) Anhydrobiosis. Annu Rev Physiol 54:579–599

Danks HV, Kukal O, Ring RA (1994) Insect cold-hardiness – insights from the Arctic. Arctic 47:391–404

DeVries AL (1971) Glycoproteins as biological antifreeze agents in Antarctic fishes. Science 172:1152

DeVries AL (1986) Antifreeze glycopeptides and peptides: interactions with ice and water. Methods Enzymol 127:293–303

Doucet D, Walker VK, Qin W (2009) The bugs that came in from the cold: molecular adaptations to low temperatures in insects. Cell Mol Life Sci 66:1404–1418

Duman JG (2001) Antifreeze and ice nucleator proteins in terrestrial arthropods. Annu Rev Physiol 63:327–357

Elnitsky MA, Hayward SAL, Rinehart JP, Denlinger DL, Lee RE (2008) Cryoprotective dehydration and the resistance to inoculative freezing in the Antarctic midge, *Belgica antarctica*. J Exp Biol 211:524–530

Franks F (1985) Biophysics and biochemistry at low temperatures. Cambridge University Press, Cambridge, UK

Grewal PS, Bornstein-Forest S, Burnell AM, Glazer I, Jagdale GB (2006) Physiological, genetic, and molecular mechanisms of chemoreception, thermobiosis, and anhydrobiosis in entomopathogenic nematodes. Biol Control 38:54–65

Halberg KA, Persson D, Ramløv H, Westh P, Møbjerg-Kristensen R, Møbjerg N (2009) Cyclomorphosis in Tardigrada: adaptation to environmental constraints. J Exp Biol 212:2803–2811

Hengherr S, Brümmer F, Schill RO (2008a) Anhydrobiosis in tardigrades and ist effects on longevity traits. J Zool (Lond) 275:216–220

Hengherr S, Heyer AG, Köhler HR, Schill RO (2008b) Trehalose and anhydrobiosis in tardigrades-evidence for divergence in responses to dehydration. FEBS J 275:281–288

Hengherr S, Worland MR, Reuner A, Brümmer F, Schill RO (2009) Freeze tolerance, supercooling points and ice formation: comparative studies on the subzero temperature survival of limno-terrestrial tardigrades. J Exp Biol 212:802–807

Hengherr S, Reuner A, Brümmer F, Schill RO (2010) Ice crystallization and freeze tolerance in embryonic stages of the tardigrade *Milnesium tardigradum*. Comp Biochem Physiol A 156:151–155

Horikawa DD, Sakashita T, Katagiri C, Watanabe M, Kikawada T, Nakahara Y, Hamada N, Wada S, Funayama T, Higashi S (2006) Radiation tolerance in the tardigrade *Milnesium tardigradum*. Int J Radiat Biol 82:843–848

Jönsson KI, Schill RO (2007) Induction of Hsp70 by desiccation, ionising radiation and heat-shock in the eutardigrade *Richtersius coronifer*. Comp Biochem Physiol B Biochem Mol Biol 146:456–460

Kagoshima H, Kito T, Aizu T, Shin-I H, Kanda S, Kobayashi A, Toyoda A, Fujiyama Y, Kohara P, Convey P, Niki H (2012) Multi-decadal survival of an antarctic nematode, *Plectus murrayi*, in a -20°C stored moss sample. CryoLetters 33:280–288

Kelty JD, Lee RE (1999) Induction of rapid cold hardening by cooling at ecologically relevant rates in *Drosophila melanogaster*. J Insect Physiol 45:719–726

Kikawada T, Nakahara Y, Kanamori Y, Iwata KI, Watanabe M, McGee B, Tunnacliffe A, Okuda T (2006) Dehydration-induced expression of LEA proteins in an anhydrobiotic chironomid. Biochem Biophys Res 348:56–61

Knight CA, Duman JG (1986) Inhibition of recrystallization of ice by insect thermal hysteresis proteins: a possible cryoprotective role. Cryobiology 23:256–262

Lalouette L, Kostal V, Colinet H, Gagneul D, Renault D (2007) Cold exposure and associated metabolic changes in adult tropical beetles exposed to fluctuating thermal regimes. FEBS J 274:1759–1767

Lee RE, Costanzo JP (1998) Biological ice nucleation and ice distribution in cold-hardy ectothermic animals. Annu Rev Physiol 60:55–72

Lee YJ, Chung TJ, Park CW, Hahn Y, Chung JH, Lee BL, Han DM, Jung YH, Kim S, Lee Y (1996) Structure and expression of the tenecin 3 gene in *Tenebrio molitor*. Biochem Biophys Res Commun 218:6–11

Mackenzie AP, Derbyshire W, Reid DS (1977) Nonequilibrium freezing behaviour of aqueous systems. Philos Trans R Soc Lond Ser B Biol Sci 278:167–189

Newsham KK, Maslen NR, Mcinnes S (2006) Survival of Antarctic soil metazoans at -80°C for six years. CryoLetters 27:269–280

Overgaard J, Tomcala A, Sørensen JG, Holmstrup M, Krogh PH, Simek P, Kost lV (2008) Effects of acclimation temperature on thermal tolerance and membrane phospholipid composition in the fruit fly *Drosophila melanogaster*. J Insect Physiol 54:619–629

Ramazzotti G, Maucci W (1983) The phylum tardigrada. Mem Ist Ital Idrob 41:1–1012

Ramløv H (2000) Aspects of natural cold tolerance in ectothermic animals. Hum Reprod 15:26–46

Ramløv H, Westh P (1992) Survival of the cyptobiotic Eutardigrade *Adorybiotus coronifer* during cooling to -196°C: effect of cooling rate, trehalose level, and short-term acclimation. Cryobiology 29:125–130

Ramløv H, Westh P (2001) Cryptobiosis in the eutardigrade *Adorybiotus* (*Richtersius*) *coronifer*: tolerance to alcohols, temperature and de novo protein synthesis. Zool Anz 240:517–523

Raymond JA, DeVries AL (1977) Adsorption inhibition as a mechanism of freezing resistance in polar fishes. Proc Natl Acad Sci U S A 74:2589–2593

Ring RA, Danks HV (1994) Desiccation and cryoprotection—overlapping adaptations. CryoLetters 15:181–190

Sakurai M, Furuki T, Akao K, Tanaka D, Nakahara Y, Kikawada T, Watanabe M, Okuda T (2008) Vitrification is essential for anhydrobiosis in an African chironomid, *Polypedilum vanderplanki*. Proc Natl Acad Sci U S A 105:5093–5098

Salvucci ME, Strecher DS, Henneberry TJ (2000) Heat shock proteins in whiteflies, an insect that accumulates sorbitol in response to heat stress. J Thermal Biol 25:363–371

Schill RO, Fritz GB (2008) Desiccation tolerance in embryonic stages of the tardigrade *Milnesium tardigradum*. J Zool (Lond) 276:103–107

Seki K, Toyoshima M (1998) Preserving tardigrades under pressure. Nature 395:853–854

Sinclair BJ (2001) Field ecology of freeze tolerance: interannual variation in cooling rates, freeze-thaw and thermal stress in the microhabitat of the alpine cockroach *Celatoblatta quinquemaculata*. Oikos 93:286–293

Sinclair BJ, Vernon P, Klok CJ, Chown SL (2003) Insects at low temperatures: an ecological perspective. Trends Ecol Evol 18:257–262

Smith T, Wharton DA, Marshall CJ (2008) Cold tolerance of an Antarctic nematode that survives intracellular freezing: comparisons with other nematode species. J Comp Physiol B Biochem Syst Environ Physiol 178:93–100
Sømme L, Meier T (1995) Cold tolerance in Tardigrada from Dronning Maud Land, Antarctica. Polar Biol 15:221–224
Storey KB, Storey JM (1996) Natural freezing survival in animals. Annu Rev Ecol Syst 27:365–386
Storey KB, Baust JG, Buescher P (1981) Determination of water "bound" by soluble subcellular components during low-temperature acclimation in the gall fly larva, *Eurosta solidagensis*. Cryobiology 18:315–321
Sträßer M (1998) Klimadiagramm-Atlas der Erde, Bd. 1. Dortmunder Vertrieb für Bau und Planungsliteratur. Dortmund, Germany
Suzuki AC (2003) Life history of *Milnesium tardigradum* Doyère (Tardigrada) under a rearing environment. Zool Sci (Tokyo) 20:40–57
Tsujimoto M, Satoschi I, Hiroshi K (2016) Recovery and reproduction of an Antarctic tardigrade retrieved from a moss sample frozen for over 30 years. Cryobiology 72:78–81
Watanabe M, Tanaka K (1998) Adult diapause and cold hardiness in *Aulacophora nigripennis* (Coleoptera: Chrysomelidae). J Insect Physiol 44:1103–1110
Westh P, Kristensen RM (1992) Ice formation in the freeze-tolerant eutardigrades *Adorybiotus coronifer* and *Amphibolus nebulosus* studied by differential scanning calorimetry. Polar Biol 12:693–699
Westh P, Kristiansen J, Hvidt A (1991) Ice-nucleating activity in the freeze-tolerant tardigrade *Adorybiotus coronifer*. Comp Biochem Physiol A Mol Physiol 99:401–404
Wharton DA (2003) The environmental physiology of Antarctic terrestrial nematodes: a review. J Comp Physiol B 173:621–628
Wharton DA, Goodall G, Marshall CJ (2003) Freezing survival and cryoprotective dehydration as cold tolerance mechanisms in the antarctic nematode *Panagrolaimus davidi*. J Exp Biol 206:215–221
Wharton DA, Downes MF, Goodall G, Marshall CJ (2005) Freezing and cryoprotective dehydration in an antarctic nematode (*Panagrolaimus davidi*) visualised using a freeze substitution technique. Cryobiology 50:21–28
Wilson PW, Heneghan AF, Haymet ADJ (2003) Ice nucleation in nature: supercooling point (SCP) measurements and the role of heterogeneous nucleation. Cryobiology 46:88–98
Worland MR, Block W (2003) Desiccation stress at sub-zero temperatures in polar terrestrial arthropods. J Insect Physiol 49:193–203
Worland MR, Convey P (2008) The significance of the moult cycle to cold tolerance in the Antarctic collembolan *Cryptopygus antarcticus*. J Insect Physiol 54:1281–1285
Worland MR, Grubor-Lajsic G, Montiel PO (1998) Partial desiccation induced by sub-zero temperatures as a component of the survival strategy of the Arctic collembolan *Onychiurus arcticus* (Tullberg). J Insect Physiol 44:211–219
Worland MR, Leinaas HP, Chown SL (2006) Supercooling point frequency distributions in Collembola are affected by moulting. Funct Ecol 20:323–329
Wright JC (2001) Cryptobiosis 300 years on from van Leuwenhoek: what have we learned about tardigrades? Zool Anz 240:563–582
Yoder JA, Benoit JB, Denlinger DL, Rivers DB (2006) Stress-induced accumulation of glycerol in the flesh fly, *Sarcophaga bullata*: evidence indicating anti-desiccant and cryoprotectant functions of this polyol and a role for the brain in coordinating the response. J Insect Physiol 52:202–214
Zachariassen KE (1985) Physiology of cold tolerance in insects. Physiol Rev 65:799–832
Zachariassen KE (1991) The water relations of overwintering insects. In: Lee RE, Denlinger DL (eds) Insects at low temperature. Chapman and Hall, London, pp 47–63
Zachariassen KE, Kristiansen E, Pedersen SA, Hammel HT (2004) Ice nucleation in solutions and freeze-avoiding insects - homogeneous or heterogeneous? Cryobiology 48:309–321

Chapter 12
Environmental Adaptations: Radiation Tolerance

K. Ingemar Jönsson, Eliana B. Levine, Andrzej Wojcik, Siamak Haghdoost, and Mats Harms-Ringdahl

Abstract Several studies in different species have documented that tardigrades are among the most radiation-tolerant animals on Earth, surviving doses of ionizing radiation on the order of kGy. Both low-LET and high-LET radiation have been used with no apparent differences in the tolerance of the animals. Tolerance to ionizing radiation in tardigrades also seems to be independent of whether the animal has entered a dry anhydrobiotic state or is hydrated with normal activity. However, when exposed to UV radiation, desiccated tardigrades show a higher tolerance than hydrated animals. Recent studies in several species have shown that tardigrade embryos have considerably lower tolerance to ionizing radiation compared to adults, and embryos in the early stage of development are clearly more sensitive to radiation than those in the late developmental stage. The molecular mechanisms behind radiation tolerance in tardigrades are still largely unclear, but available evidence suggests that mechanisms related to both the avoidance of DNA damage and the repair of damage are involved.

K. I. Jönsson (✉)
Department of Environmental Science and Bioscience, Kristianstad University, Kristianstad, Sweden
e-mail: ingemar.jonsson@hkr.se

E. B. Levine
Department of Radiation Oncology, Division of Molecular and Cellular Oncology, David Geffen school of Medicine at UCLA, Los Angeles, CA, USA

Colscience Institute, Bogotá, Colombia
e-mail: eblevine@mednet.ucla.edu

A. Wojcik · M. Harms-Ringdahl
Centre for Radiation Protection Research, Department of Molecular Biosciences, The Wenner-Gren Institute, Stockholm University, Stockholm, Sweden
e-mail: andrzej.wojcik@su.se; mats.harms-ringdahl@su.se

S. Haghdoost
Department of Molecular Biosciences, The Wenner-Gren Institute, Stockholm University, Stockholm, Sweden

LARIA – CIMAP, University of Caen, Caen, France
e-mail: siamak.haghdoost@su.se

R. O. Schill (ed.), *Water Bears: The Biology of Tardigrades*, Zoological Monographs 2,
https://doi.org/10.1007/978-3-319-95702-9_12

12.1 Introduction

12.1.1 Effects of Radiation on Living Organisms

Organisms show different degrees of radiation tolerance that is largely based on differences in the intrinsic radiosensitivity of their cells. Some, but not all, aspects of this intrinsic radiosensitivity are known. An understanding of the mechanisms behind radiosensitivity requires a short introduction to the radiation biology of cells.

Ionizing radiation interacts with matter through ionizations that can cause chemical changes in molecules and may lead to losses or changes of function in cells or organs. Depending on the ionizing particle, ionization may occur in a dense or sparse manner that is reflected in its linear energy transfer (LET). LET describes the amount of energy lost by a particle traversing a certain distance. Photons and electrons are sparsely ionizing (low-LET radiations), while alpha particles, protons, neutrons and heavy ions are densely ionizing (high-LET radiations).

Cells contain various complex structures and molecules that can all be damaged by radiation. However, within the range of doses that organisms are typically exposed to, it is mainly genetic material—DNA—that is the principal target. DNA contains nearly all of the genetic information of an organism and maintains gene stability; it is therefore essential for cell survival (Hall and Giaccia 2012). DNA lesions induced by radiation can be grouped into three major categories: base damage, single strand breaks and double strand breaks (DSB). DSB are the most detrimental lesions because they disrupt DNA integrity, and pieces of broken strands must be excised in the repair process, which can potentially result in the loss of genetic material.

The effects of ionizing radiation on DNA may arise by either direct or indirect action. In the former case, radiation impacts DNA directly, resulting in the ionization of atoms in the DNA molecule. In the latter case, radiation interacts with water in the cell, resulting in radiolysis. In this process, highly reactive free radicals are formed that can diffuse short distances and damage the DNA, contributing to the damage caused by the direct action of radiation on DNA. The magnitude of the damage is strongly dependent on molecular oxygen inside the cell nucleus because, in its presence, the radical-induced DNA damage is fixed and cannot be reversed by the action of antioxidants. Consequently, the irradiation of cells under dry or hypoxic conditions has a sparing effect, which is stronger for low-LET radiation compared to high-LET radiation (Hall and Giaccia 2012).

DNA damage is sensed by the cell via special sensor molecules (e.g., ataxia-telangiectasia mutated (ATM) kinases) that initiate signalling cascades, leading to cell cycle arrest and repair of the damage (Maréchal and Zou 2013). However, DNA repair is not error-free, and the probability of damage misrepair increases with the complexity of the damage, which increases with LET such that high-LET radiation has a higher biological effectiveness than low-LET radiation. Protracted exposure to radiation causes a sparing effect due to the increasing ability of DNA repair enzymes to cope with lesions because their induction rates decrease. Misrepaired DNA damage can lead to mutations and even cell death.

Apart from ionizing radiation, organisms are also exposed to UV radiation (UVR). UVR can react with DNA molecules and cause DNA damage, but the mode of action is strongly dependent on the UVR wavelength. UVC and the short-wave part of UVB are directly absorbed by DNA, leading to photoinduced reactions that can cause DNA crosslinking. In contrast, UVA and the long wave part of UVB induce DNA damage indirectly via photosensitization reactions, which lead to the formation of reactive oxygen species (ROS) that can react with DNA and other cell components.

Even if the damage to DNA is generally considered to be a primary effect of radiation, the oxidative damage of proteins may play an important role in determining the effects of radiation, as they regulate the DNA repair system and cell death (Krisko and Radman 2013a, b; Radman 2016). Free radicals produced by radiation convert protein amino groups to carbonyl (carbonylation) groups in a dose-dependent manner (Winter and Liehr 1991), resulting in irreversible structural and functional changes to proteins (Reisz et al. 2014). The processes of DNA repair and replication are dependent on the functional efficiency of related proteins, which can be highly affected by radiation due to carbonylation, resulting in mutations, genomic instability and cell death.

Cells undergoing reproduction (mitosis or meiosis) are more sensitive to radiation than non-reproducing cells, a phenomenon known as the "Law of Bergonié and Tribondeau" (Bergonié and Tribondeau 1906), which can be attributed to the fact that the correct transmission of DNA is required at the time of cell division to result in viable new cells. Differentiating cells are also believed to be more sensitive to radiation than differentiated cells as are organs with rapidly dividing cells such as the small intestine or bone marrow (Nicolay et al. 2015). Therefore, within an individual, susceptibility to radiation generally varies among different kinds of cells and tissues.

Due to the abundance of ionizing radiation and UVR on Earth, cells have developed mechanisms that protect them from the associated detrimental effects. These include antioxidant molecules and DNA repair proteins as well as pigmentation and shielding for UVR. Different organisms differ greatly in their tolerance to ionizing radiation, and generally, among animals, mammals are the most sensitive (LD_{50} in the range of several Gy), while invertebrates are more radioresistant (Table 12.1). Very large differences exist in the resistance of invertebrates, with LD_{50} values ranging up to several kGy (Table 12.1). The most radiation-resistant animals known are those that are able to desiccate, suggesting a common mechanism that maintains the integrity of DNA damaged by radiation and dehydration. Tardigrades belong to this group of exceptionally desiccation- and radiation-tolerant animals.

It should be noted that the concepts of "tolerance" and "resistance" are often used interchangeably in the radiation literature, and that a variety of measures for responses to radiation have been used in studies of tardigrades, including both short-term and long-term animal survival, the ability of adults to produce eggs,

Table 12.1 Tolerance to gamma radiation of some animal groups

Organism	Tolerated dose or LD_{50} for survival	Author
Homo sapiens	$LD_{50/30d} = 2.5–4.5$ Gy	Bolus (2001)
Mouse	$LD_{50/30d} = 4.5$ Gy	Bolus (2001)
Goldfish	$LD_{50/30d} = 8$ Gy	Bolus (2001)
Cockroach	$LD_{50/30d} = 50$ Gy	Bolus (2001)
Cysts of the *Artemia franciscana* (Crustacea)	LD_{50} = ca. 5000 Gy	Iwasaki (1964)
Caenorhabditis elegans (Nematoda)	Doses greater than 1000 Gy required to reduce mean life span	Johnson and Hartman (1988)
Fruit fly (*Drosophila melanogaster*, Insecta)	$LD_{50} = 1238–1339$ Gy	Parashar et al. (2008)
Chironomus ramosus (midge, Insecta)	$LD_{50} = 2000–2300$ Gy	Datkhile et al. (2009)
African chironomid *Polypedilum vanderplanki* (Insecta)	$LD_{50} = 4400$ Gy in dehydrated larvae. $LD_{50} = 2000$ Gy in hydrated larvae	Watanabe et al. (2006)
Rotifers	No effect on adult survival up to 1120 Gy	Gladyshev and Meselson (2008)
Tardigrades	$LD_{50} = 1270–5000$ Gy	Hashimoto and Kunieda (2017)

and the ability of eggs to hatch and give rise to viable offspring. A majority of the studies on the effects of radiation on tardigrades have been based on the measurement of general phenotypic performance (e.g., body movements) rather than effects at the cellular or molecular level. Therefore, reports of "tolerance/resistance" do not necessarily mean that the animals are unaffected by exposure or escape damage. There is a need for studies comparing general phenotypic performance with effects at the cellular and molecular levels. As will be clear from the following review, the level of tolerance is also dependent on life cycle phases within a species. A summary of the studies on radiation tolerance in tardigrades is given in Table 12.2.

12.2 Tolerance to Ionizing Radiation in Tardigrades

12.2.1 Tolerance to Low-LET Radiation

The first experiments on radiation tolerance in tardigrades were reported by May et al. (1964) in which specimens of *Macrobiotus areolatus* were exposed to X-ray and UV radiation. The results showed a very high and similar tolerance to X-ray radiation in both desiccated and hydrated tardigrades (Fig. 12.1). The acute dose response, measured as the LD_{50} one day post-irradiation, was 5.7 kGy in animals irradiated in a desiccated state and 5.4 kGy in active, hydrated animals. The animals were followed for 20–30 days under semi-cultured conditions (dish with water and

Table 12.2 Studies related to radiation tolerance in tardigrades

Species	Study type	Radiation type	Author
		Low-LET	
Echiniscoides sigismundi	Tolerance, adults	γ-Radiation	Jönsson et al. (2016a)
Milnesium cf. *tardigradum*	Tolerance, adults, juveniles	γ-Radiation	Horikawa et al. (2006)
Milnesium cf. *tardigradum*	Tolerance, embryos	γ-Radiation	Beltrán-Pardo et al. (2013a)
Milnesium cf. *tardigradum*	DNA repair gene, rad51	γ-Radiation	Beltrán-Pardo et al. (2013b)
Hypsibius dujardini	Tolerance, adults and embryos	γ-Radiation	Beltrán-Pardo et al. (2015)
Hypsibius dujardini	Tolerance, adults; Bystander effects	γ-Radiation	Fernandez et al. (2016)
Hypsibius dujardini	DNA repair gene, rad51	γ-Radiation	Beltrán-Pardo et al. (2013b)
Macrobiotus areolatus	Tolerance, adults	X-Radiation	May et al. (1964)
Macrobiotus cf. *harmsworthi*	DNA repair gene, rad51	γ-Radiation	Beltrán-Pardo et al. (2013b)
Richtersius coronifer	Tolerance, adults	γ-Radiation	Jönsson et al. (2005)
Richtersius coronifer	Tolerance, embryos	γ-Radiation	Jönsson et al. (2013)
Richtersius coronifer	Tolerance, adults	X-Radiation	Jönsson and Wojcik (2017)
Richtersius coronifer	DNA repair gene, hsp70	γ-Radiation	Jönsson and Schill (2007)
		High-LET	
Milnesium cf. *tardigradum*	Tolerance, adults	α-Radiation	Horikawa et al. (2006)
Ramazzottius varieornatus	Tolerance, adults	α-Radiation	Horikawa et al. (2008)
Ramazzottius varieornatus	Tolerance, embryos	α-Radiation	Horikawa et al. (2012)
Richtersius coronifer	Tolerance, adults	Protons	Nilsson et al. (2010)
Richtersius coronifer	Tolerance, adults	Iron ions (Fe), helium ions (He)	Jönsson and Wojcik (2017)
		UV-radiation	
Milnesium tardigradum	Tolerance, adults and embryos	Vacuum UV–UVA (116.5–400 nm) in space	Jönsson et al. (2008)
Hypsibius dujardini	Tolerance, adults	UVC (254 nm)	Horikawa et al. (2013)
Ramazzottius oberhaeuseri	Tolerance, adults	UVB (312 nm)	Altiero et al. (2011)
Ramazzottius varieornatus	Tolerance, adults	UVC (254 nm)	Horikawa et al. (2013)

(continued)

Table 12.2 (continued)

Species	Study type	Radiation type	Author
Macrobiotus areolatus	Tolerance, adults	UVC (254 nm)	May et al. (1964)
Paramacrobiotus richtersi	Tolerance, adults	UVB (312 nm)	Altiero et al. (2011)
Richtersius coronifer	Tolerance, adults and embryos	Vacuum UV–UVA (116.5–400 nm) in space	Jönsson et al. (2008)

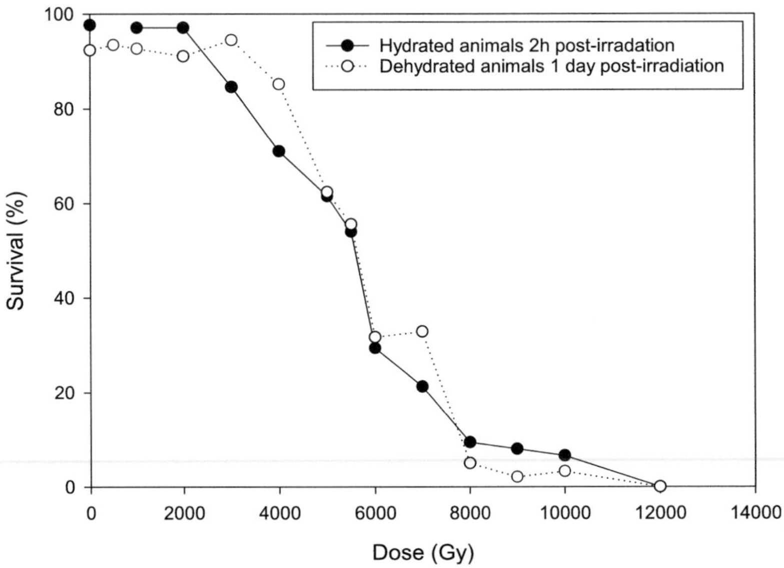

Fig. 12.1 Dose response of *Macrobiotus areolatus* irradiated with X-rays in a hydrated and dehydrated state, based on data from May et al. (1964), which was the first study of radiation tolerance in tardigrades. Specimens from irradiated dehydrated animals were dried on filter paper and then kept under high humidity (ca. 90% RH) for at least 1 week before irradiation. It is not clear from the original paper if the animals were hydrated immediately after irradiation. The hydrated animals were irradiated in 1 mm water

pieces of moss). For many years, references to the May et al. (1964) study (written in French) did not recognize the high tolerance of active tardigrades but usually interpreted the tolerance to radiation as a consequence of the dry anhydrobiotic state. Thus, it was not until a second study on radiation tolerance in tardigrades was reported by Jönsson et al. (2005) that the similar tolerance of tardigrades in the hydrated state was recognized. This study used gamma ray radiation (^{137}Cs) and observed no effects on the survival or fecundity of the eutardigrade *R. coronifer* with a dose of 1 kGy, but their eggs did not develop into juveniles and were therefore probably sterile (however, the occurrence of resting eggs in tardigrades complicates interpretation; Altiero et al. 2010). LD_{50} values one day after irradiation were 3.0 and 4.7 kGy for the anhydrobiotic and hydrated states, respectively. Horikawa et al. (2006) reported similar results in adult ("medium and large-sized") *Milnesium tardigradum* after gamma-irradiation (^{60}Co), with LD_{50} values close to those in the study by May et al. (1964). Few irradiated animals produced eggs, and no eggs hatched. A high tolerance to gamma radiation has also been reported in adult *Hypsibius dujardini*[1] (LD_{50} ca. 4.2 kGy 48 h post-irradiation, Beltrán-Pardo et al. 2015; see also Fernandez et al. 2016), a freshwater species that is desiccation tolerant but more sensitive to rapid dehydration compared to the limno-terrestrial species (e.g., *M. tardigradum* and *R. coronifer*) and requires preconditioning at high relative humidity to survive exposure to low humidity conditions (Wright 1989; Kondo et al. 2015). Only one study on radiation tolerance in a heterotardigrade species has been reported in which hydrated adult specimens of the marine species *Ecíniscoides sigismundi* were exposed to gamma radiation up to 5 kGy (Jönsson et al. 2016a). This species inhabits barnacle shells in the tidal zone where the tardigrades are exposed to recurrent periods of desiccation and shows a high tolerance to desiccation (Hygum et al. 2016). Although some specimens of *E. sigismundi* survived for 7 days after exposure to 4 kGy, with an LD_{50} value 7 days post-irradiation of ca. 1.5 kGy, the overall tolerance was lower compared to previous studies on limno-terrestrial/ limnic species.

12.2.2 Tolerance to High-LET Radiation

Studies on high-LET radiation tolerance in tardigrades have so far been performed using alpha, proton, helium ion, and iron ion irradiations. Horikawa et al. (2006) studied the short-term effects (up to 48 h post-irradiation) of alpha (^{4}He) radiation in *M. tardigradum* and found no effects on survival from irradiation up to 4 kGy, but a relatively sharp decline in survival at higher doses. Additionally, tardigrades irradiated at higher doses in a hydrated state tended to have higher survival rates than those irradiated in a desiccated state. Similar results were reported by Horikawa et al. (2008) for the eutardigrade *Ramazzottius varieornatus* in which alpha irradiation with 4 kGy resulted in 100% survival of the hydrated animals with slightly lower (but non-significant) survival of the desiccated animals. Two studies on the tolerance to high-LET radiation in *R. coronifer* have been reported. The first one reported (Nilsson et al. 2010) a very high tolerance to proton (microbeam) irradiation up to a

[1]Note that the "*Hypsibius dujardini*" strain used in most (possibly all) studies on development and stress tolerance referred to in this chapter has recently been redescribed by Gasiorek et al. (2018) as a new species, *Hypsibius exemplaris*, distinguished from the *Hypsibius dujardini sensu stricto*.

dose of 10 kGy above which survival declined sharply. Up to that level, the dose response was much less marked but tended to increase with time after irradiation (from 4 h to 48 h post-irradiation). The second study showed no effect on the survival of *R. coronifer* from irradiation with iron ions up to 2 kGy or with helium ions up to 1 kGy, and these dose levels represented the highest doses used in the experiment (Jönsson and Wojcik 2017). There was a dose-dependent tendency of reduced egg production for both Fe and He irradiation, but the differences were not statistically confirmed. These studies suggest that there is little difference in the tolerance of tardigrades to low-LET and high-LET radiation, an unexpected result since high-LET radiation is predicted to create more complex damage in tissues and to have a higher relative biological effectiveness (RBE) than low-LET radiation. Also consistent with the low-LET studies, the study on proton irradiation by Nilsson et al. (2010) showed a sharp break-point in the dose response to radiation with rapid declines in survival above a certain dose. This suggests that the ultimate molecular effects (e.g., integrity of DNA) of radiation in tardigrades are not linearly related to dose, but that damage can be avoided or repaired up to a certain level above which mortality increases rapidly.

12.3 Radiation Tolerance in the Embryonic and Juvenile Stages of Tardigrades

Three studies on the tolerance of tardigrade eggs to low-LET radiation have been reported thus far (Jönsson et al. 2013; Beltrán-Pardo et al. 2013a, 2015). As tissues under rapid proliferation are more sensitive to radiation, these studies expectedly documented a lower tolerance to radiation in eggs compared to adults. However, they also showed that tolerance to radiation changes dramatically in the course of embryonic development. In *R. coronifer*, irradiation with a 500 Gy dose of gamma ray radiation resulted in the near complete mortality of eggs that were in the early or middle stages of development, while eggs in the late stage were not affected by the irradiation (Jönsson et al. 2013). Similar results were reported by Beltrán-Pardo et al. (2013a) in *M.* cf. *tardigradum*, but in this species, both the middle and late developmental stages were very tolerant and had no sign of a dose response (Fig. 12.2a). In contrast, eggs in the early developmental stage showed a clear dose response above 50 Gy. The study by Beltrán-Pardo et al. (2013a) also showed that irradiation has a delaying effect on development, but a dose response (i.e., more delay with higher dose) was observed only in eggs that were in the early developmental stage. Additionally, the limnic species *H. dujardini* showed similar patterns of high radiation sensitivity in the early stage of egg development as well as delayed development (Beltrán-Pardo et al. 2015, Fig. 12.2b). These changes in the tolerance to radiation over the course of development may arise in two ways. Higher sensitivity in the early developmental stage may be a direct result of higher cell proliferation and a higher proportion of undifferentiated cells, both of which are factors

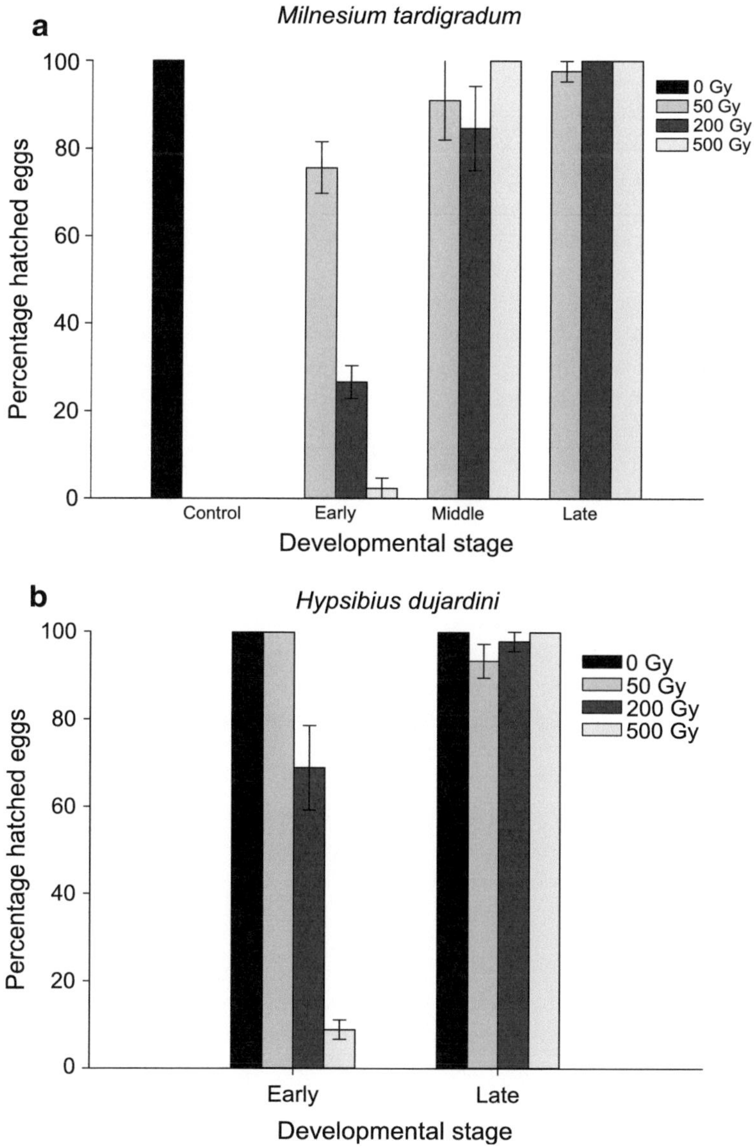

Fig. 12.2 Dose response of eggs in different developmental stages in (**a**) *Milnesium tardigradum* and (**b**) *Hypsibius dujardini* exposed to gamma radiation. Figure (**a**) is a modified version of figure 2 in Beltrán-Pardo et al. (2013a). Figure (**b**) is based on data from Beltrán-Pardo et al. (2015)

known to be connected to high radiation sensitivity (Bergonié and Tribondeau 1906). There is also evidence that the rate of cell division in developing tardigrade embryos is highest in the earliest phase (corresponding to the early stage in the three studies above) and then declines considerably (Gabriel et al. 2007). The second possible explanation, which does not exclude a role for the first, is that specific molecular pathways for tolerance (i.e., related to DNA repair) are activated in the course of embryo development. Although the up-regulation of such mechanisms is expected based on the tolerance observed in adult tardigrades, we are not aware of any studies documenting the composition of proteins in the different stages of egg development in tardigrades. However, Schokraie et al. (2012) reported that 24% of all proteins found in *M. tardigradum* were unique to the egg stage and not found in adults, but only eggs in the early developmental stage (24 h post-deposition, blastula stage) were used in the analysis. Additionally, desiccation tolerance in tardigrade eggs seems to change over the course of development, from low tolerance in the earliest stage to high tolerance towards the end of development (Schill and Fritz 2008), consistent with the suggestion that radiation and desiccation tolerance in tardigrades rely on the same mechanism (Jönsson 2003), as previously proposed for radiation-tolerant prokaryotes (Mattimore and Battista 1996).

The study by Horikawa et al. (2012) is the only study on the high-LET (^{4}He) radiation of tardigrade eggs thus far, and it documented a considerably higher tolerance in anhydrobiotic eggs (LD_{50} = 1690 Gy) compared to eggs irradiated in a hydrated state (LD_{50} = 509 Gy) for *Ramazzottius varieornatus*. Irradiated anhydrobiotic eggs were not affected by radiation up to a dose of 1 kGy, but hatchability then declined to 30% at 2 kGy, whereas the hatchability of hydrated eggs declined to 5% at 1 kGy. The irradiations in this study were performed during the middle stage of egg development.

Only one study thus far has reported tolerance to radiation in juvenile tardigrades. Horikawa et al. (2006) irradiated both desiccated and hydrated immature individuals (second or third instars without developed ovaries) of *M. tardigradum* with gamma radiation (1–4 kGy) and followed their survival and reproduction over 31 days under cultured conditions. The results showed a relatively linear dose-dependent decline in short-term survival (24 h) up to 4 kGy (highest dose), but at both 3 kGy and 4 kGy, the tardigrades irradiated in an active hydrated state had significantly higher survival rates (62 and 36%, respectively) than the tardigrades irradiated in a desiccated state (20 and 11%, respectively). The short-term dose-dependent mortality pattern continued over a longer time period, but no tardigrades irradiated at or above 2 kGy survived the whole 31-day period. Very few eggs were produced by the irradiated animals (as they developed into adults), and no eggs hatched, suggesting that irradiation in the juvenile stage results in sterility.

12.4 Tolerance to UV Radiation in Tardigrades

There are few studies on tolerance to UV radiation in tardigrades, and May et al. (1964) reported the seminal study also in this area. Using UVC (254 nm), they showed that the exposure of *M. areolatus* for up to 6 h did not have any effect on the revival of desiccated animals. In contrast, active hydrated tardigrades showed high sensitivity to UV. Exposure for 1 h reduced their revival rate to 59% compared to

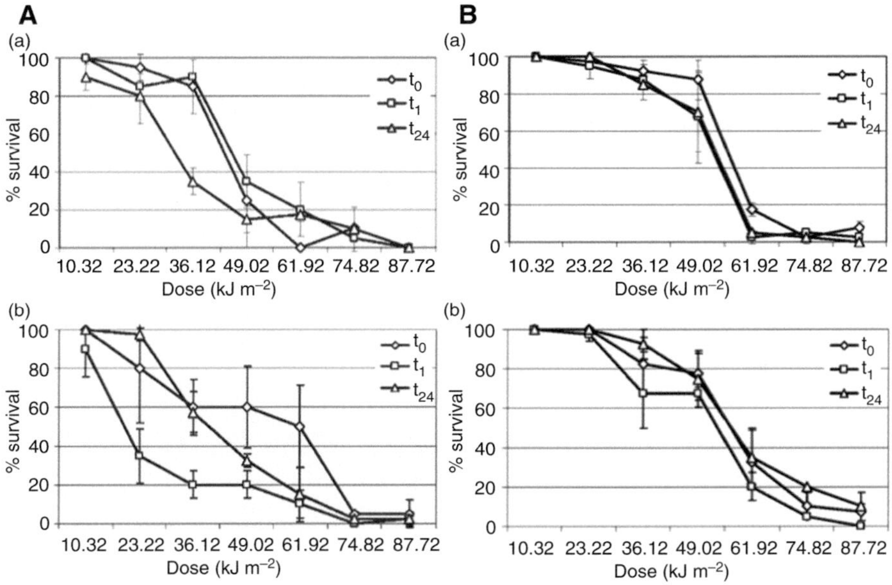

Fig. 12.3 Dose response of survival to UVB radiation (312 nm) in hydrated (**A**) and dehydrated (**B**) tardigrade species *Ramazzottius oberhaeuseri* (**a**) and *Paramacrobiotus richtersi* (**b**). The three dose-response lines in each figure represent survival estimates at different times after irradiation (t_0 = end of irradiation, t_0 = 1 h after irradiation, and t_{24} = 24 h after irradiation). Reprinted from Altiero et al. (2011) with permission

non-irradiated animals, and at 2 h of exposure, revival was reduced to 9%. Longer exposure did not result in any viable animals. Unfortunately, May et al. (1964) did not provide any information on the doses that resulted from the exposure times, and it is therefore difficult to evaluate these results. However, their study clearly indicated that tardigrades are more sensitive to UV irradiation in the hydrated state than the desiccated state, which contrasts to the results obtained for both high-LET and low-LET ionizing radiation. Similar differences in tolerance to UVC (254 nm) radiation between anhydrobiotic and hydrated animals have recently been reported by Horikawa et al. (2013) for *Ramazzottius varieornatus*. Likewise, Altiero et al. (2011) reported that dehydrated animals of the tardigrades *Ramazzottius oberhaeuseri* and *Paramacrobiotus richtersi* were slightly more tolerant to UVB (312 nm) than hydrated animals (Fig. 12.3). The survival of these species after similar exposure times was considerably higher than that reported in the study by May et al. (1964), which can probably be explained by the shorter and more damaging wavelength used in the latter study. Altiero et al. (2011) also found that high temperature (35 °C) had a strong negative impact on UV radiation survival in *Paramacrobiotus richtersi*. Apart from these UV radiation studies performed in the laboratory, the tolerance of tardigrades to UV radiation has also been evaluated under space conditions (see below).

12.5 Radiation Exposure of Tardigrades Under Space Conditions

The extreme tolerance of tardigrades to both desiccation and radiation has led to an interest in using this group as an animal model organism for exposure to space conditions (Bertolani et al. 2001; Jönsson 2007; Horikawa 2008). Tardigrades were first exposed to space conditions at low Earth orbit in connection with the FOTON-M3 mission in 2007 in which several different experiments were included (TARDIS, Jönsson et al. 2008; RoTaRad, Persson et al. 2011; TARSE, Rebecchi et al. 2009, 2011). The radiation environment in space (outside the atmosphere of the Earth) is characterized by a combination of ionizing solar (SCR) and galactic (GCR) cosmic radiation, both of which consist mainly of protons (85–95%) and alpha particles (5–15%) (Horneck et al. 2006). In addition, approximately 1% of GCR consists of heavier nuclei of "High charge Z and high Energy" (HZE). Although it represents a very low percentage of the total radiation in space, the high energy and material penetration of HZE makes it a considerable concern for biological material exposed in space, particularly during long-term flights. However, for biological tissues, the most challenging radiation is solar UV radiation, which directly damages cell components, including DNA (Horneck 2003).

The deleterious effect of UV radiation was very clear in the TARDIS experiment (Jönsson et al. 2008) in which samples of *R. coronifer* and *M. tardigradum* exposed to very high doses of UV (ca. 7.6×10^6 J m^{-2} for the full UV spectrum of 116.5–400 nm) suffered high mortality, while samples sheltered from UV but exposed to space vacuum and cosmic radiation were practically unaffected by the 10 days of exposure. However, some animals of *M. tardigradum* exposed to UVA and UVB managed to survive and reproduce, becoming the first animals to ever survive under "open space" conditions, and their offspring did not show any signs of reduced fitness (Jönsson et al. 2016b). Compared to the doses of ionizing radiation that tardigrades have survived in laboratories, the absorbed dose of cosmic radiation (mainly protons) from short-term exposures in space, such as during the FOTON-M3 mission, is very low, usually measured in the range of a few Gray (but depending on the level of shielding) and is therefore not expected to affect radiation-tolerant tardigrades. In the TARSE experiment, both desiccated and active hydrated tardigrades of the species *Paramacrobiotus richtersi* (previously *Macrobiotus richtersi*) were exposed to low doses of cosmic ionizing radiation (estimated at 1.9 mGy; Rebecchi et al. 2011). Neither effects on survival nor DNA damage were observed in comparison to laboratory controls on Earth in either the hydrated or desiccated tardigrades (Rebecchi et al. 2009, 2011), but the tardigrades exposed to space conditions had elevated levels of heat-shock proteins (Rebecchi et al. 2011).

In the RoTaRad experiment, adults of *R. coronifer* exposed to 4 Gy of cosmic radiation had considerably reduced survival compared to controls as well as animals exposed to lower doses in flight, and embryos of this species exposed to space conditions had lower survival (Persson et al. 2011). These findings contrast with the results from the TARDIS experiment in the same species as well as the much higher doses that this species has tolerated in previous laboratory experiments (e.g., Jönsson et al. 2005), and the explanation for these different results is unclear.

Tardigrades have also been used in a study of simulated Martian conditions, including exposure to UV radiation in the spectrum of 200–400 nm with a total radiation dose over 40 days (12 h per Martian sol) of 3.6×10^5 kJ m^{-2} of which about 9% represented radiation from the UVC spectrum (Johnson et al. 2011). This study included samples of desiccated *R. varieornatus*, and survival 6 days after rehydration was estimated to be 70% in the UV exposed samples and 92% in the samples sheltered from UV. However, under the UV exposure conditions used in the experiment, the tardigrades were embedded in 5 mm regolith, and it is likely that this layer shielded them from most of the UV radiation.

12.6 Molecular Mechanisms of Radiation Tolerance in Tardigrades

The mechanisms behind the documented radiation tolerance in tardigrades are not well understood, and few studies have addressed this question directly. In principle, tolerance to environmental agents may arise either from mechanisms that protect the organism from damage (e.g., by shielding layers or stabilizing cell components) or from mechanisms that repair damage. Both kinds of mechanisms may be part of an organism's evolved tolerance. Extensive DNA damage (Double Strand Breaks) from ionizing radiation in some other radio-resistant organisms, such as *Deinococcus radiodurans* and bdelloid rotifers (Battista et al. 1999; Gladyshev and Meselson 2008), indicates they have no structural protection against direct DNA damage. Desiccation could potentially prevent much of this damage since the deleterious effect of ionizing radiation on tissues is mostly due to interactions with free radicals produced when radiation interacts with water (see Introduction). However, since hydrated tardigrades seem to cope as well as dehydrated tardigrades do with radiation, DNA repair may be an important component of their tolerance mechanism (Jönsson 2003). In addition, studies in tardigrades using comet assays have shown that desiccation itself give rises to single and double strand breaks in DNA (Neumann et al. 2009).

Investigations into the role of DNA repair in tardigrade radiation tolerance should be focused on the following two aspects: (1) direct verification that DNA is indeed damaged by radiation and that a process of repair occurs and (2) indirect verification by documenting the activity of molecular components known to be involved in DNA repair processes. For both aspects, few studies on tardigrades have been reported. Only one report on radiation-induced DNA damage in tardigrades has been published in which active adults of *M. tardigradum* were irradiated with UVB ($\lambda = 312$ nm) at a dose of 3.735 J m^{-2} (Neumann et al. 2009), and damage to DNA was estimated by comet assays. The irradiation did not affect survival or reproduction of the tardigrades but resulted in significant fragmentation of their

DNA (ca. 17% DNA in comet tail) compared to controls (0.44% DNA in comet tail). Repair of the damage induced by irradiation was not studied, but changes in the level of DNA fragmentation after rehydration of desiccated tardigrades indicated the presence of repair processes.

Some evidence related to molecular processes of DNA repair induced by radiation in tardigrades has also been reported. Jönsson and Schill (2007) showed that the heat-shock protein HSP70, a highly conserved protein found in a variety of organisms, from bacteria to humans (Thacker 1999), was strongly induced by gamma irradiation (500 Gy) in hydrated *R. coronifer* (and to a lower extent in dehydrated animals). HSP70 is known to maintain genomic stability under stress conditions by avoiding telomere instability, apoptosis induction and the frequency of chromosomal aberrations (Pandita et al. 2004; Hunt et al. 2004). However, HSP70 may also function as a heat-shock chaperon, refolding and protecting damaged proteins (Mayer and Bukau 2005). Another protein known to be associated with DNA repair, RAD51, was shown to be induced by gamma radiation in *M.* cf. *tardigradum*, implying a specific response related to the activation of the homologous DNA repair pathway (Beltrán-Pardo et al. 2013b). Horikawa et al. (2013) also showed evidence of DNA repair in the tardigrade *R. varieornatus*, which was irradiated in a hydrated state with 2.5 kJ m^{-2} of UVC, by observing the post-irradiation reduction of thymine dimers in DNA; the induction of the *phrA* gene was also observed 18 h after UVC irradiation. The photolyase protein encoded by the *phrA* gene is involved in the nucleotide repair of cyclobutane pyrimidine dimers (CPDs) when activated by light (Eker et al. 2009).

To the extent that radiation tolerance and desiccation tolerance are based on the same underlying mechanisms (e.g., Mattimore and Battista 1996; Jönsson 2003), considerably more molecular studies of potential relevance for understanding radiation tolerance are available from studies on tardigrades exposed to desiccation. In particular, proteomic and transcriptomic analyses within the German research project Funcrypta (http://www.funcrypta.de/www/en/) have provided much new information about the molecular pathways involved in tolerance to desiccation, although an understanding of the mechanism is still far from complete. These studies have revealed several metabolic pathways associated with the down-regulation of DNA replication (inhibition of cell division) and the up-regulation of DNA repair molecules (e.g., ubiquitin and the DnaJ Family), some of which are unique to tardigrades and others that are shared with some other animals, including *C. elegans*, *D. melanogaster*, and humans (Wang et al. 2014; Förster et al. 2012). Comparisons between the more desiccation tolerant species *M. tardigradum* and the less tolerant *H. dujardini* also indicated that *M. tardigradum* has broader metabolic adaptations for stress tolerance. Differences in molecular expression between tardigrade species with high preconditioning requirements for desiccation (*H. dujardini* and *P. richtersi*) and species with relatively little preconditioning needs (*M. tardigradum*) was also reported by Boothby et al. (2017). They showed that *H. dujardini* and *P. richtersi* had high expression levels of CASH and SAHS (TDPs or intrinsically disordered proteins), while in *M. tardigradum*, there was no expression of TDPs, but the protein CASH was constitutively expressed. Thus, different molecular strategies seem to have evolved in species adapted to rapid versus slow desiccation. However, the extent to which tolerance to desiccation is related to tolerance to radiation, and relies on the same molecular mechanisms, is still unclear. More studies comparing desiccation and radiation tolerance in tardigrades adapted to different humidity conditions, including analyses of phylogenetic backgrounds and molecular stress responses, will be needed to evaluate this hypothesis. In addition, as discussed in Jönsson et al. (2016a), the concept of desiccation tolerance includes several aspects that should be considered, including the *rate of desiccation*, *final level of desiccation*, and *length of time in the desiccated state*, all of which may be influenced by different molecular mechanisms.

Antioxidant mechanisms that protect against ROS created both at desiccation (from increased ionic concentrations) and from radiation (from indirect interactions between radiation and water molecules) are also important for desiccation and radiation tolerance (Rebecchi 2013). Rizzo et al. (2010) found increased superoxide dismutase, glutathione peroxidase and glutathione activity in desiccated tardigrades in comparison with hydrated tardigrades as well as higher percentages of polyunsaturated fatty acids and thiobarbituric acid reactive substances. In a transcriptomic analysis of *M. tardigradum*, pathway proteins corresponding to peroxidase and superoxide dismutase (SOD) were reported (Förster et al. 2012). The role of protein carbonylation in mediating the effects of radiation was mentioned earlier, and the possibility that the observed radiation tolerance in tardigrades largely relies on molecular mechanisms that protect cells from carbonylation will likely be one of the main research thrusts in future studies on the radiation tolerance of tardigrades.

In a study based on the genome sequence analysis of *R. varieornatus*, a species with high tolerance to both desiccation and radiation, Hashimoto et al. (2016, see also Hashimoto and Kunieda 2017) reported a large number (16) of SOD genes that may be involved in detoxifying reactive oxygen species as well as four different genes within the *MRE11* gene family, which is known to be connected to the repair of DNA double-strand breaks. They also documented a tardigrade-unique protein, Dsup, which showed an association with nuclear DNA, indicating a possible role in repairing or protecting DNA from damage. When Dsup was inserted into human embryonic kidney cells (HEK293), DNA damage from X-rays (10 Gy) was reduced by up to 40%, suggesting that the Dsup protein protected the DNA from strand breaks. Irradiated Dsup-expressing cells also had improved viability compared to non-transfected cells. Comparisons between the genomes of *R. varieornatus* and *H. dujardini* has also revealed a possible Dsup orthologue in *H. dujardini* (Hashimoto and Kunieda 2017). Although details of the mechanism by which Dsup protects DNA remains to be documented, this unique study suggests that tardigrades have evolved specific molecular mechanisms that are able to prevent DNA from being damaged even by ionizing radiation. It also showed, for the first time, evidence that these molecular mechanisms can be transferred to human cells, which is potentially important for medical treatments such as radiation therapy.

The molecular studies on desiccation and radiation tolerance in tardigrades are consistent with the results from other radiation-tolerant invertebrates. For instance, Gusev et al. (2010) reported DNA fragmentation and repair following desiccation and radiation with both low-LET and high-LET, the increased expression of DNA repair enzymes (*Rad23* and *Rad51*) after both low-LET and high-LET radiation, and the up-regulation of antioxidant genes by low-LET but not high-LET radiation in the chironomid *Polypedilum vanderplanki*. Krisko et al. (2012) also reported evidence of antioxidant mechanisms that prevent protein carbonylation in radiation-tolerant rotifers.

12.7 Conclusions

Our understanding of radiation tolerance in tardigrades has improved considerably over the last decade, with consistent results concerning the levels of tolerance in both adult tardigrades and in embryos. From these studies, it is clear that tardigrades belong to the most radiation-tolerant animals on Earth. Although revealing the molecular mechanisms behind this tolerance has just begun, the similar tolerance of desiccated and hydrated tardigrades together with reports on the molecular pathways connected to DNA repair suggest that DNA repair represents an important component of the adaptive system for stress tolerance in tardigrades. However, the report on a DNA-associated protein (Dsup) that seems to partially prevent DNA from being damaged by ionizing radiation suggests that tardigrades have also evolved other mechanisms to maintain DNA integrity, and mechanisms that protect proteins from carbonylation deserve further investigation. Many questions about radiation tolerance in tardigrades remain to be answered. Why do tardigrades tolerate much higher levels of radiation than they would ever be exposed to under natural conditions? To what extent is tolerance to desiccation and to radiation connected, and are they based on the same molecular pathways? What is the pattern of tolerance to radiation among different phylogenetic lineages within the tardigrade phylum? These and other questions will hopefully be answered in future efforts to reveal the mechanisms by which tardigrades demonstrate remarkable tolerance to radiation.

Acknowledgements The current review was financially supported by the Swedish Space Research Board (grant 87/11 to AW).

References

Altiero T, Bertolani R, Rebecchi L (2010) Hatching phenology and resting eggs in tardigrades. J Zool 280:290–296

Altiero T, Guidetti R, Caselli V, Cesari M, Rebecchi L (2011) Ultraviolet radiation tolerance in hydrated and desiccated eutardigrades. J Zool Syst Evol Res 49(Suppl 1):104–110

Battista JR, Earl AM, Park M-J (1999) Why is *Deinococcus radiodurans* so resistant to ionizing radiation? Trends Microbiol 7:362–365

Beltrán-Pardo E, Jönsson KI, Wojcik A, Haghdoost S, Harms-Ringdahl M, Bermúdez-Cruz RM, Bernal Villegas JE (2013a) Effects of ionizing radiation on embryos of the tardigrade *Milnesium* cf. *tardigradum* at different stages of development. PLoS One 8(9):e72098

Beltrán-Pardo EA, Jönsson KI, Wojcik A, Haghdoost S, Bermúdez Cruz RM, Bernal Villegas JE (2013b) Sequence analysis of the DNA-repair gene *rad51* in the tardigrades *Milnesium* cf. *tardigradum*, *Hypsibius dujardini* and *Macrobiotus* cf. *harmsworthi*. J Limnol 72(s1):80–91

Beltrán-Pardo E, Jönsson KI, Harms-Ringdahl M, Haghdoost S, Wojcik A (2015) Differences in tolerance to gamma radiation in the tardigrade *Hypsibius dujardini* from embryo to adult correlate inversely with cellular proliferation. PLoS One 10(7):e0133658

Bergonié J, Tribondeau L (1906) De Quelques Résultats de la Radiotherapie et Essai de Fixation d'une Technique Rationnelle. C R Acad Sci 143:983–985

Bertolani R, Rebecchi L, Jönsson KI, Borsari S, Guidetti R, Altiero T (2001) Tardigrades as a model for experiences of animal survival in the space. Micrograv Space Station Util 2:211–212

Bolus NE (2001) Basic review of radiation biology and terminology. J Nucl Med Technol 29 (2):67–73

Boothby TC, Tapia H, Brozena AH, Piszkiewicz S, Smith AE, Giovannini I, Rebecchi L, Pielak GJ, Koshland D, Goldstein B (2017) Tardigrades use intrinsically disordered proteins to survive desiccation. Mol Cell 65(6):975–984.e5

Datkhile KD, Dongre TK, Mukhopadhyaya R, Nath BB (2009) Gamma radiation tolerance of a tropical species of midge, *Chironomus ramosus* Chaudhuri (Diptera: Chironomidae). Int J Radiat Biol 85:495–503

Eker AP, Quayle C, Chaves I, van der Horst GT (2009) DNA repair in mammalian cells: direct DNA damage reversal: elegant solutions for nasty problems. Cell Mol Life Sci 66(6):968–980

Fernandez C, Vasanthan T, Kissoon N, Karam G, Duquette N, Seymour C, Stone JR (2016) Radiation tolerance and bystander effects in the eutardigrade species *Hypsibius dujardini* (Parachaela: Hypsibiidae). Zool J Linn Soc 178:919–923

Förster F, Beisser D, Grohme MA, Liang C, Mali B, Siegl AM, Engelmann JC, Shkumatov AV, Schokraie E, Müller T, Schnölzer M, Schill RO, Frohme M, Dandekar T (2012) Transcriptome analysis in tardigrade species reveals specific molecular pathways for stress adaptations. Bioinform Biol Insights 6:69–96

Gabriel WN, McNuff R, Patel SK, Gregory TR, Jeck WR, Jones CD, Goldstein B (2007) The tardigrade *Hypsibius dujardini*, a new model for studying the evolution of development. Dev Biol 312:545–559

Gąsiorek P, Stec D, Morek W, Michalczyk Ł (2018) An integrative redescription of *Hypsibius dujardini* (Doyère, 1840), the nominal taxon for Hypsibioidea (Tardigrada: Eutardigrada). Zootaxa 4415(1):45

Gladyshev E, Meselson M (2008) Extreme resistance of bdelloid rotifers to ionizing radiation. Proc Nat Acad Sci USA 105:5139–5144

Gusev O, Nakahara Y, Vanyagina V, Malutina L, Cornette R, Sakashita T, Hamada N, Kikawada T, Kobayashi Y, Okuda T (2010) Anhydrobiosis-associated nuclear DNA damage and repair in the sleeping chironomid: linkage with radioresistance. PLoS One 5(11):e14008

Hall EJ, Giaccia AJ (2012) Radiobiology for the radiologist. Lippincott Williams & Wilkins, Philadelphia, PA, 576 p

Hashimoto T, Kunieda T (2017) DNA protection protein, a novel mechanism of radiation tolerance: lessons from tardigrades. Life 7(2):26

Hashimoto T, Horikawa DD, Saito Y, Kuwahara H, Kozuka-Hata H, Shin-I T, Minakuchi Y, Ohishi K, Motoyama A, Aizu T, Enomoto A, Kondo K, Tanaka S, Hara Y, Koshikawa S, Sagara H, Miura T, Yokobori S-I, Miyagawa K, Suzuki Y, Kubo T, Oyama M, Kohara Y, Fujiyama A, Arakawa K, Katayama T, Toyoda A, Kunieda T (2016) Extremotolerant tardigrade genome and improved radiotolerance of human cultured cells by tardigrade-unique protein. Nat Commun 7:12808

Horikawa DD (2008) The tardigrade *Ramazzottius varieornatus* as a model animal for astrobiological studies. Biol Sci Space 22(3):93–98

Horikawa DD, Sakashita T, Katagiri C, Watanabe M, Kikawada T, Nakahara Y, Hamada N, Wada S, Funayama T, Higashi S, Kobayashi Y, Okuda T, Kuwabara M (2006) Radiation tolerance in the tardigrade *Milnesium tardigradum*. Int J Rad Biol 82:843–848

Horikawa DD, Kunieda T, Abe W, Watanabe M, Nakahara Y, Yukuhiro F, Sakashita T, Hamada N, Wada S, Funayama T, Katagiri C, Kobayashi Y, Higashi S, Okuda T (2008) Establishment of a rearing system of the extremotolerant tardigrade *Ramazzottius varieornatus*: a new model animal for astrobiology. Astrobiology 8(3):549–556

Horikawa DD, Yamaguchi A, Sakashita T, Tanaka D, Hamada N, Yukuhiro F, Kuwahara H, Kunieda T, Watanabe M, Nakahara Y, Wada S, Funayama T, Katagiri C, Higashi S, Yokobori S-I, Kuwabara M, Rothschild LJ, Okuda T, Hashimoto H, Kobayashi Y (2012) Tolerance of anhydrobiotic eggs of the tardigrade *Ramazzottius varieornatus* to extreme environments. Astrobiology 12:283–289

Horikawa DD, Cumbers J, Sakakibara I, Rogoff D, Leuko S, Harnoto R, Arakawa K, Katayama T, Kunieda T, Toyoda A, Fujiyama A, Rothschild LJ (2013) Analysis of DNA repair and protection in the tardigrade *Ramazzottius varieornatus* and *Hypsibius dujardini* after exposure to UVC radiation. PLoS One 8:e64793

Horneck G (2003) Could life travel across interplanetary space? Panspermia revisited. In: Rothschild LJ, Lister AM (eds) Evolution on planet earth: the impact of the physical environment. Academic Press, London, pp 109–127

Horneck G, Baumstark-Khan C, Facius R (2006) Chap 7: Radiation biology. In: Clément G, Slenzka K (eds) Fundamentals of space biology research on cells, animals, and plants in space. Published jointly by Springer and Microcosm Press, El Segundo, CA, 375 p

Hunt CR, Dix DJ, Sharma GG, Pandita RK, Gupta A, Funk M, Pandita TK (2004) Genomic instability and enhanced radiosensitivity in Hsp70.1- and Hsp70.3-deficient mice. Mol Cell Biol 24(2):899–911

Hygum TL, Clausen LKB, Halberg KA, Jørgensen A, Møbjerg N (2016) Tun formation is not a prerequisite for desiccation tolerance in the marine tidal tardigrade *Echiniscoides sigismundi*. Zool J Linn Soc 178(4):907–911

Iwasaki T (1964) Sensitivity of *Artemia* eggs to the γ-irradiation. I. Hatchability of encysted dry eggs. J Radiat Res 5:69–75

Johnson TE, Hartman PS (1988) Radiation effects on life span in *Caenorhabditis elegans*. J Gerontol 43:B137–B141

Johnson AP, Pratt LM, Vishnivetskaya T, Pfiffner S, Bryan RA, Dadachova E, Whyte L, Radtke K, Chan E, Tronick S, Borgonie G, Mancinelli RL, Rothschild LJ, Rogoff DA, Horikawa DD, Onstott TC (2011) Extended survival of several organisms and amino acids under simulated Martian surface conditions. Icarus 211:1162–1178

Jönsson KI (2003) Causes and consequences of excess resistance in cryptobiotic metazoans. Physiol Biochem Zool 76:429–435

Jönsson KI (2007) Tardigrades as a potential model organism in space research. Astrobiology 7:757–766

Jönsson KI, Schill RO (2007) Induction of Hsp70 by desiccation, ionising radiation and heat-shock in the eutardigrade *Richtersius coronifer*. Comp Biochem Physiol B Biochem Mol Biol 146:456–460

Jönsson KI, Wojcik A (2017) STARLIFE IX: tolerance to X-rays and heavy ions (Fe, He) in the tardigrade *Richtersius coronifer* and the bdelloid rotifer *Mniobia russeola*. Astrobiology 17 (2):163–167

Jönsson KI, Harms-Ringdahl M, Torudd J (2005) Radiation tolerance in the tardigrade *Richtersius coronifer*. Int J Rad Biol 81:649–656

Jönsson KI, Rabbow E, Schill RO, Harms-Ringdahl M, Rettberg P (2008) Tardigrades survive exposure to space in low Earth orbit. Curr Biol 18:R729–R731

Jönsson KI, Beltrán-Pardo E, Haghdoost S, Wojcik A, Bermúdez-Cruz RM, Bernal Villegas JE, Harms-Ringdahl M (2013) Tolerance to gamma-irradiation in eggs of the tardigrade *Richtersius coronifer* depends on stage of development. J Limnol 72(s1):73–79

Jönsson KI, Hygum TL, Andersen KN, Clausen LKB, Møbjerg N (2016a) Tolerance to gamma radiation in the marine heterotardigrade, *Echiniscoides sigismundi*. PLoS One 11(12):e0168884

Jönsson KI, Schill RO, Rabbow E, Rettberg P, Harms-Ringdahl M (2016b) The fate of the TARDIS offspring: no intergenerational effects of space exposure. Zool J Linn Soc 178:924–930

Kondo K, Kubo T, Kunieda T (2015) Suggested involvement of PP1/PP2A activity and *de novo* gene expression in anhydrobiotic survival in a tardigrade, *Hypsibius dujardini*, by chemical genetic approach. PLoS One 10(12):e0144803

Krisko A, Radman M (2013a) Biology of extreme radiation resistance: the way of *Deinococcus radiodurans*. Cold Spring Harb Perspect Biol 5:a012765

Krisko A, Radman M (2013b) Phenotypic and genetic consequences of protein damage. PLoS Genet 9(9):e1003810

Krisko A, Leroy M, Radman M, Meselson M (2012) Extreme anti-oxidant protection against ionizing radiation in bdelloid rotifers. Proc Natl Acad Sci USA 109(7):2354–2357

Maréchal A, Zou L (2013) DNA damage sensing by the ATM and ATR kinases. Cold Spring Harb Perspect Biol 5(9):a012716

Mattimore V, Battista JR (1996) Radioresistance of *Deinococcus radiodurans*: functions necessary to survive ionizing radiation are also necessary to survive prolonged desiccation. J Bacteriol 178:633–637

May RM, Maria M, Guimard J (1964) Actions différentielles des rayons x et Ultraviolets sur le tardigrade *Macrobiotus areolatus*, à l'état actif et desséché. Bull Biol France Belgique 98:349–367

Mayer MP, Bukau B (2005) Hsp70 chaperones: cellular functions and molecular mechanism. Cell Mol Life Sci 62(6):670–684

Neumann S, Reuner A, Brümmer F, Schill RO (2009) DNA damage in storage cells of anhydrobiotic tardigrades. Comp Biochem Physiol A Mol Integr Physiol 153:425–429

Nicolay NH, Perez RL, Saffrich R, Huber PE (2015) Radio-resistant mesenchymal stem cells: mechanisms of resistance and potential implications for the clinic. Oncotarget 6 (23):19366–19380

Nilsson EJC, Jönsson KI, Pallon J (2010) Tolerance to proton irradiation in the eutardigrade *Richtersius coronifer* – a nuclear microprobe study. Int J Radiat Biol 86:1–8

Pandita TK, Higashikubo R, Hunt CR (2004) HSP70 and genomic stability. Cell Cycle 3 (5):591–592

Parashar V, Frankel S, Lurie AG, Rogina B (2008) The effects of age on radiation resistance and oxidative stress in adult *Drosophila melanogaster*. Radiat Res 169:707–711

Persson D, Halberg KA, Jørgensen A, Ricci C, Møbjerg N, Kristensen RM (2011) Extreme stress tolerance in tardigrades: surviving space conditions in low earth orbit. J Zool Syst Evol Res 49:90–97

Radman M (2016) Protein damage, radiation sensitivity and aging. DNA Repair 44:186–192

Rebecchi L (2013) Dry up and survive: the role of antioxidant defences in anhydrobiotic organisms. J Limnol 72(s1):62–72

Rebecchi L, Altiero T, Guidetti R, Cesari M, Bertolani R, Negroni M, Rizzo AM (2009) Tardigrade resistance to space effects: first results of experiments on the LIFE-TARSE mission on FOTON-M3 (September 2007). Astrobiology 9(6):581–591

Rebecchi L, Altiero T, Cesari M, Bertolani R, Rizzo AM, Corsetto PA, Guidetti R (2011) Resistance of the anhydrobiotic eutardigrade *Paramacrobiotus richtersi* to space flight (LIFE–TARSE mission on FOTON-M3). J Zool Syst Evol Res 49(Suppl 1):98–103

Reisz JA, Bansal N, Qian J, Zhao W, Furdui CM (2014) Effects of ionizing radiation on biological molecules—mechanisms of damage and emerging methods of detection. Antioxid Redox Signal 21(2):260–291

Rizzo AM, Negroni M, Altiero T, Montorfano G, Corsetto P, Berselli P, Berra B, Guidetti R, Rebecchi L (2010) Antioxidant defences in hydrated and desiccated states of the tardigrade *Paramacrobiotus richtersi*. Comp Biochem Physiol B Biochem Mol Biol 156:115–121

Schill RO, Fritz GB (2008) Desiccation tolerance in embryonic stages of the tardigrade. J Zool 276:103–107

Schokraie E, Warnken U, Hotz-Wagenblatt A, Grohme MA, Hengherr S, Förster F, Schill RO, Frohme M, Dandekar T, Schnölzer M (2012) Comparative proteome analysis of *Milnesium tardigradum* in early embryonic state *versus* adults in active and anhydrobiotic state. PLoS One 7:e45682

Thacker J (1999) A surfeit of RAD51-like genes? Trends Genet 15:166–168

Wang C, Grohme MA, Mali B, Schill RO, Frohme M (2014) Towards decrypting cryptobiosis - analyzing anhydrobiosis in the tardigrade *Milnesium tardigradum* using transcriptome sequencing. PLoS One 9(3):e92663

Watanabe M, Sakashita T, Fujita A, Kikawada T, Horikawa DD, Nakahara Y, Wada S, Funayama T, Hamada N, Kobayashi Y, Okuda T (2006) Biological effects of anhydrobiosis in an African chironomid, *Polypedilum vanderplanki* on radiation tolerance. Int J Radiat Biol 82:587–592

Winter ML, Liehr JG (1991) Free radical-induced carbonyl content in protein of estrogen-treated hamsters assayed by sodium boro[^{3}H]hydrid reduction. J Biol Chem 266:14446–14450

Wright JC (1989) Desiccation tolerance and water-retentive mechanisms in tardigrades. J Exp Biol 142:267–292

Chapter 13
Molecular Biology in Tardigrades

Thomas C. Boothby

Abstract Molecular biology, a term first coined in the 1930s, can be viewed as a set of techniques and approaches, as well as a subdiscipline within biology. Molecular approaches have and continue to be used in nearly every area of biological study today, including genetics, biochemistry, biophysics, cell and developmental biology, physiology, and evolutionary biology. The adoption of molecular techniques to tardigrade research has helped to propel these fascinating animals from obscure biological novelties to important emerging models.

13.1 Introduction

While the unique biology of tardigrades has fascinated scientists for centuries, at the most basic level, the molecular building blocks, nucleic acids, proteins, lipids, and carbohydrates that make up these animals are no different than other organisms. This makes tardigrades amenable to study using classic molecular techniques. The use and adaptation of molecular techniques for understanding the basic biology of tardigrades has risen dramatically within the past two decades, increasing the utility of tardigrades as emerging model organisms. As in other model systems, the use of molecular techniques in tardigrades often crosses over with other classically defined areas of study, such as genetics, cell biology, developmental biology, ecology, physiology, and evolutionary biology. For example, DNA extraction and sequencing in tardigrades was pioneered by researchers interested in tardigrade systematics and phylogeny (see Chap. 3 of this book). Molecular techniques have been adapted for tardigrades that allow researchers to isolate and study the building blocks listed above. This chapter presents an overview of techniques that have been adapted to and employed in the study of tardigrades and the importance of these and new techniques to the field. The details of these studies will not be discussed extensively here, because other chapters in this book highlight many of the pertinent findings.

T. C. Boothby (✉)
Department of Chemistry, University of North Carolina at Chapel Hill, Chapel Hill, NC, USA
e-mail: tboothby@live.unc.edu

R. O. Schill (ed.), *Water Bears: The Biology of Tardigrades*, Zoological Monographs 2,
https://doi.org/10.1007/978-3-319-95702-9_13

The concluding section of this chapter highlights existing molecular techniques and approaches that have yet to be employed, but show promise, for studying tardigrades.

13.2 Isolation of Biomolecules from Tardigrades

Classical molecular techniques such as the polymerase chain reaction (PCR), reverse transcription PCR (RT-PCR), and western blotting require the isolation of nucleic acids or proteins from biological samples. In addition, the extraction of lipids and carbohydrates from tardigrades is of interest to researchers, especially those studying stress tolerance (Westh and Ramløv 1991; Ingemar Jönsson et al. 2005; Hengherr et al. 2008; Jönsson and Persson 2010; Rizzo et al. 2010, 2015; Hashimoto et al. 2016; Hygum et al. 2017). Researchers have adapted a variety of methods for isolating biological macromolecules from tardigrades, including DNA, RNA, and proteins from both bulk multi-animal samples and individual specimens, to provide insight into tardigrade biology.

13.2.1 Bulk DNA and RNA Isolation

A number of studies have used various methods for extracting DNA or RNA from bulk tardigrade samples (Schill 2007; Mali et al. 2010; Tenlen et al. 2013; Wang et al. 2014; Boothby et al. 2015; Smith et al. 2016; Koutsovoulos et al. 2016). As in other model organisms, these techniques serve as the starting point for most molecular studies. The ability to isolate high-quality DNA has facilitated studies in tardigrades such as species barcoding, large genome and transcriptome sequencing projects, reverse genetics, and studies examining evolution, development, and physiology (Blaxter et al. 2005; Mali et al. 2010; Bertolani et al. 2011; Tenlen et al. 2013; Wang et al. 2014; Sarkies et al. 2015; Boothby et al. 2015; Smith et al. 2016; Levin et al. 2016; Koutsovoulos et al. 2016). Figures 13.1 and 13.2 provide simple protocols for bulk extraction and purification of DNA and RNA from tardigrades. Depending on the downstream application, additional purification and quality control steps may be necessary.

13.2.2 Extracting DNA and RNA from Single Specimens

For some applications, it is not absolutely necessary to purify large amounts of DNA or RNA, and simple lysates will suffice. Additionally, in some instances limited access to specimens or experimental constraints or design may require the use of a few or even a single specimen. Unpurified extracts from single or small pools of tardigrades can be used for simple amplification procedures such as PCR and

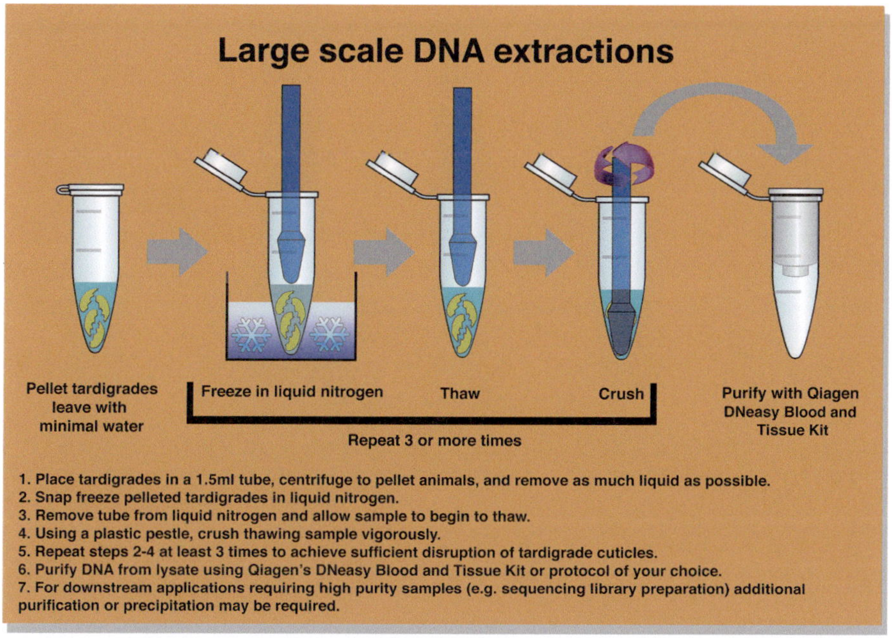

Fig. 13.1 A simple method for bulk DNA extraction from tardigrades. Animals in a 1.5 ml microcentrifuge tube are pelleted by centrifugation and left in a minimum of culture medium. A plastic pestle is placed in the open tube, and the sample is rapidly frozen in liquid nitrogen or a dry ice bath. The frozen sample is thawed and simultaneously homogenized with the pestle. The freeze/thaw/crush steps are repeated until homogenization of the sample is achieved (visual inspection of the sample with a dissection microscope can be used to insure tardigrade cuticles have been completely disrupted). Homogenized lysate is then used as input with Qiagen's DNeasy Blood and Tissue Kit (Cat# 69506). Additional cleanup steps such as ethanol precipitation may be required for downstream applications requiring higher-purity samples Protocol from (Boothby et al. 2015)

RT-PCR. A variety of protocols have been adapted from other invertebrate systems (mostly nematodes) for this purpose (Blaxter et al. 2004; Blaxter et al. 2005; Schill 2007; Bertolani et al. 2011; Tenlen et al. 2013; Levin et al. 2016).

Figure 13.3 demonstrates a quick and inexpensive protocol adapted from (Tenlen et al. 2013; Nayak) for obtaining lysates from tardigrades. The protocol works well for obtaining extracts to be used in PCR or RT-PCR from 1 to 10 tardigrades at various life stages, ranging from adults to embryos.

13.2.3 Protein Extraction

Along with purifying nucleic acids from tardigrades, protein extraction has been employed by researchers to study the basic biology (Rebecchi et al. 2009; Schokraie et al. 2010, 2011, 2012; Rizzo et al. 2010; Obinata et al. 2011; Yamaguchi et al.

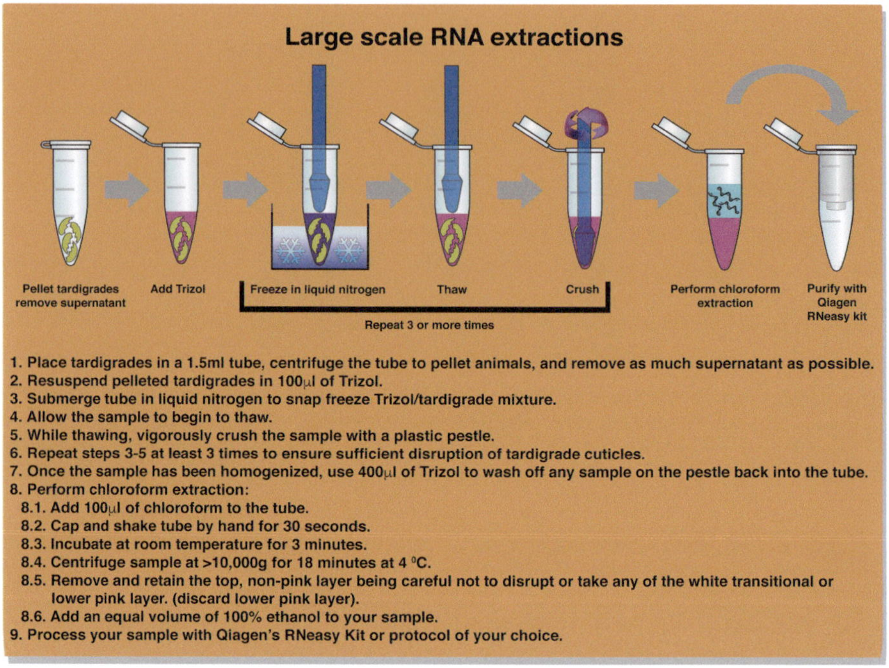

Fig. 13.2 A simple method for total RNA extraction from tardigrades. Tardigrades are added to 1.5 ml microcentrifuge tube, pelleted by centrifugation, and left in a minimum of culture media. 100 μl of Trizol is used to resuspend pelleted animals. The microcentrifuge tube, with a plastic pestle placed in it, is rapidly frozen in liquid nitrogen, thawed, and homogenized with the pestle. Freeze/thaw/homogenization steps are repeated until the sample is adequately homogenized. Excess sample remaining on the pestle is washed back into the microcentrifuge tube using 400 μl of Trizol. To perform chloroform extraction, add 100 μl of chloroform to the sample. Shake the sample for 30 s to insure complete mixing. Let the tube sit at room temperature for 3 min and then centrifuge at 10,000×*g* or more for 18 min at 4 °C. After centrifugation RNA will be present in the top, non-pink layer. This top layer is carefully removed and placed in a new 1.5 ml microcentrifuge tube. An equal volume of 100% ethanol is added to this new tube and the sample processed with Qiagen's RNeast Kit (Cat# 74104). Include optional on-column DNase digestion step for applications sensitive to small amounts of DNA contamination

2012; Tanaka et al. 2015; Hering et al. 2016). As with nucleic acid extraction, extracting proteins allows researchers to ask targeted questions about the abundance and activity of specific tardigrade proteins (Rebecchi et al. 2009; Rizzo et al. 2010; Obinata et al. 2011; Tanaka et al. 2015; Hering et al. 2016) as well as offering avenues for pursuing large-scale proteomic studies (Schokraie et al. 2010, 2011, 2012; Yamaguchi et al. 2012). In general, simple extraction protocols have been employed to procure proteins from tardigrades, and these methods should suffice for most applications. Some downstream applications, such as mass spectrometry will benefit from additional extraction and/or cleanup, such as a combination of methods used by Hurkman and Tanaka (1986) and the "Perfect-Focus" cleanup kit (Genotech Inc. Cat #786–124) (Boothby unpublished data). Figure 13.4 details a basic

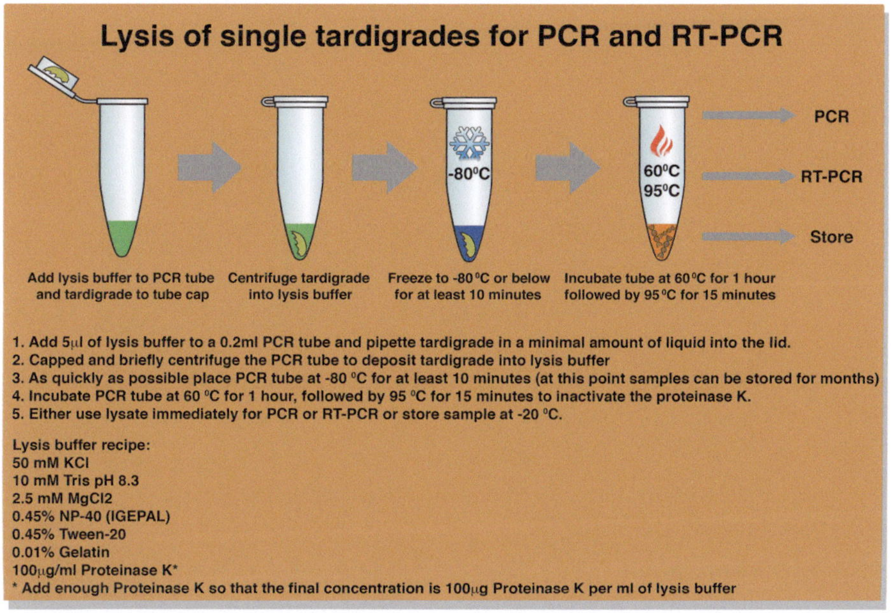

Fig. 13.3 A simple method for generating lysates for single tardigrades or embryos for PCR or RT-PCR. One to ten tardigrades/embryos are placed in a minimum of culture media (<5 μl) to the lid of a 0.2 ml PCR tube. 5 μl of lysis buffer is added to the tube; the tube is capped and briefly centrifuged to move the tardigrade(s) into the lysis buffer. The PCR tube is immediately placed at −80 °C for at least 10 min. Following freezing, samples are incubated at 60 °C for 1 h and then 95 °C for 15 min. Crude lysate can be used immediately for PCR or RT-PCR reactions or frozen and stored for later use [Adapted from Nayak; Tenlen et al. (2013)]

extraction protocol that was employed by Schokraie et al. (2010) for extracting tardigrade proteins for shotgun proteomics.

13.2.4 Lipid and Carbohydrate Extraction

Along with proteins and nucleic acids, lipids and carbohydrates have also been extracted from tardigrades. The ability to obtain high-quality lipid and carbohydrate extracts is of great importance to researchers interested in the cryptobiotic survival abilities, as changes in the content of these molecules have been linked to stress tolerance in other cryptobiotic organisms (Crowe et al. 1984; Hoekstra et al. 1989; Holmstrup et al. 2002; Erkut et al. 2011; Tapia and Koshland 2014).

Two recent studies report the extraction and quantification of fatty acid composition from tardigrades. The first study (Rizzo et al. 2010) examined changes in fatty acid composition in hydrated and desiccated tardigrades. As with protein and DNA extraction, the authors adopted an existing chloroform/methanol technique (Folch et al. 1957). The results indicate that the percent of polyunsaturated fatty acids, and

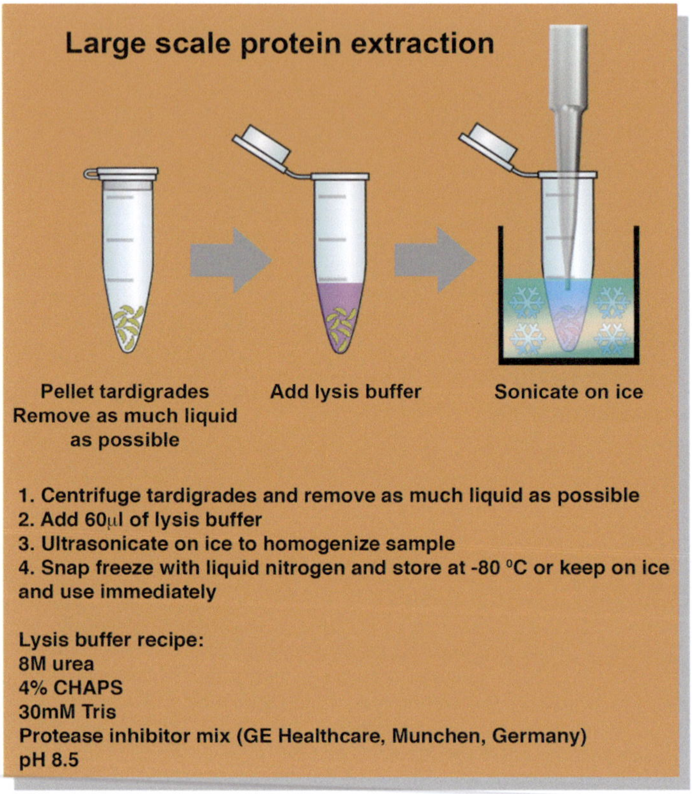

Fig. 13.4 A simple protocol for extracting proteins from tardigrades. Tardigrades are pelleted and as much culture media as possible is removed. The tardigrade pellet is resuspended in 60 μl of lysis buffer. Animals are disrupted via ultrasonication. Lysates are frozen with liquid nitrogen and stored at −80 °C or used immediately for downstream procedures [from Schokraie et al. (2010)]

specifically arachidonic acid, increases in desiccated animals relative to controls (Rizzo et al. 2010). The second study examined fatty acid composition in two different tardigrade species exposed to the stresses of space flight (Rizzo et al. 2015). The results indicate that differences in fatty acid composition exist between tardigrade species and are subject to change under stress conditions (Rizzo et al. 2015).

Carbohydrates have also been extracted from tardigrades (Westh and Ramløv 1991; Hengherr et al. 2008; Jönsson and Persson 2010). The primary focus of this work was on assessing the trehalose content of hydrated and desiccated specimens, because this disaccharide is known to or is implicated in mediating desiccation tolerance in several other anhydrobiotic organisms (Crowe et al. 1984; Erkut et al. 2011; Tapia and Koshland 2014). However, carbohydrates in addition to trehalose are suspected to play roles in stress tolerance, e.g., sucrose in plants (Hoekstra et al.

1989), and the field would benefit from more extensive surveys of carbohydrate content in tardigrades.

Combined, these studies show that tardigrades are amenable to detailed study using existing lipid and carbohydrate extraction protocols.

13.3 Tardigrades in the Omics Era

The ability to extract and purify various macromolecules from tardigrades has led to an explosion of omics studies. Researchers have sequenced the genomes, transcriptomes, and proteomes of several tardigrade species (Mali et al. 2010; Schokraie et al. 2010, 2011, 2012; Yamaguchi et al. 2012; Wang et al. 2014; Sarkies et al. 2015; Boothby et al. 2015, 2017; Smith et al. 2016; Levin et al. 2016; Koutsovoulos et al. 2016; Hashimoto et al. 2016). These studies have mainly, but not exclusively, been used to contribute to our understanding of tardigrade physiology, evolution, and phylogeny, and such omics data will continue to inform nearly every subdiscipline of tardigrade research.

It is important to remember and recognize the work of researchers using sequencing, nucleic acid, and protein techniques to study tardigrades prior to the adoption of next-generation and high-throughput approaches. Expressed sequences tags (EST) and barcoding projects gave the first glimpses of tardigrade genomic and transcript sequences, while targeted protein-based studies were used to identify, localize, and quantify levels of specific proteins (Blaxter et al. 2004, 2005; Rebecchi et al. 2009; Mali et al. 2010; Rizzo et al. 2010; Bertolani et al. 2011; Tanaka et al. 2015; Hering et al. 2016). Only the handful of tardigrade species routinely kept in laboratory culture are amenable to omics approaches requiring large nucleic acid or protein inputs. The advent of ultralow input techniques for next-generation sequencing and proteomics means that these technologies are or soon will be used, for the study of *individual* specimens (Hashimshony et al. 2012; Hughes et al. 2014; Levin et al. 2016).

13.3.1 Next-Generation Sequencing

This approach offers researchers access to vast amounts of data at dramatically reduced cost per base pair compared with older sequencing methods (Wetterstrand 2016). Researchers can now get high-quality data, numbering in the billions of bases, in less than 2 weeks for a few thousand dollars. Next-generation sequencing has and will continue to allow researchers to rapidly sequence complete tardigrade genomes and transcriptomes, biological replicates for comparative studies, and different conditions or developmental stages (Wang et al. 2014; Sarkies et al. 2015; Boothby et al. 2015, 2017; Levin et al. 2016; Koutsovoulos et al. 2016; Hashimoto et al. 2016).

A good example of the completeness of transcriptomes that can be generated with next-generation sequencing versus traditional EST sequencing is highlighted by two studies of the tardigrade *Milnesium tardigradum* (Mali et al. 2010; Wang et al. 2014). The first study (Mali et al. 2010) sequenced ~10,000 ESTs from tardigrades in active and inactive (desiccated) states, which were assembled into 3283 unique transcripts. The second study (Wang et al. 2014) used a combination of Sanger, 454, and Illumina technology to sequence RNA from hydrated, desiccating, desiccated, as well as desiccated and then rehydrated specimens, which resulted in a transcriptome assembly containing 79,064 transcripts. One method for assessing the completeness of a transcriptome or genome, beyond simply assessing the raw number of transcripts/genes assembled, is to assess the representation of core eukaryotic genes (CEGs) in the assembly (Parra et al. 2007). Comparing the percent of CEGs represented in the EST-based transcriptome (21.77%) to the next-generation transcriptome (93.55%), one can easily see the degree to which the completeness differs (Percentages were estimated by TCB for this chapter. Percentages include partial gene sequences).

Besides being able to generate large amounts of sequencing data at relatively low costs, next-generation sequencing (as well as other omics approaches) offers an unbiased means of identifying molecular mediators of different biological processes. With regard to tardigrades, this may be particularly pertinent to the study of stress tolerance. For example, the disaccharide trehalose has long been considered a key molecular player in desiccation tolerance. While this sugar is indeed essential for desiccation tolerance in some systems (Crowe et al. 1984; Erkut et al. 2011; Tapia and Koshland 2014), other anhydrobiotic organisms, such as rotifers, do not appear to make this sugar, even in response to drying (Lapinski and Tunnacliffe 2003).

Because trehalose promotes desiccation tolerance in other organisms, numerous studies have been conducted to examine levels of trehalose in both hydrated and desiccated tardigrades (Westh and Ramløv 1991; Hengherr et al. 2008; Jönsson and Persson 2010). The picture that emerges is complicated, with some studies detecting trehalose in certain species, but not in others. Even studies examining the same species differ on the presence or absence of trehalose. One point of interest is that in all cases where the sugar is detected, it is detected at low levels relative to other animals that rely on trehalose to survive desiccation (Westh and Ramløv 1991; Hengherr et al. 2008; Jönsson and Persson 2010). The question as to the functional significance of trehalose to tardigrade desiccation tolerance remains unanswered. Several omics (both next-generation sequencing and proteomics) studies have examined changes in gene expression and protein abundance in tardigrades exposed to drying. None of these reports noted increases in the expression of genes involved in trehalose synthesis (Mali et al. 2010; Wang et al. 2014; Boothby et al. 2017). Instead they note increases in fatty acid-binding proteins, cytochrome *c* oxidase subunit 1, protease inhibitors, as well as heat shock proteins, antioxidant enzymes, and aquaporins or the abundance of intrinsically disordered proteins (Mali et al. 2010; Schokraie et al. 2011; Yamaguchi et al. 2012; Wang et al. 2014; Boothby et al. 2017). These studies do not rule out the possibility that trehalose plays a functional role in tardigrade desiccation tolerance, but they do indicate that there are likely other important mediators of desiccation tolerance. Thus, while there is clear value in targeted studies examining the presence or role of specific genes or gene products, omics style approaches can be used for the unbiased detection of known or novel candidates for additional in-depth studies.

In addition to transcriptome studies of gene expression during stress, next-generation sequencing has been used to sequence small RNAs and detailed embryonic time course transcriptomes of *H. dujardini* (Sarkies et al. 2015; Levin et al. 2016) as well as multiple *H. dujardini* genome assemblies (Boothby et al. 2015; Koutsovoulos et al. 2016). Additionally, the genome of *Ramazzottius varieornatus* has been published (Hashimoto et al. 2016). Transcriptome and whole-genome assemblies provide researchers with in-depth views of which genes tardigrades have gained and lost over the course of their evolution and how the expression of these genes varies between developmental stages and under different environmental conditions.

13.3.2 Shotgun and Targeted Proteomics

The central dogma of molecular biology states that the flow of genetic information in cells goes from DNA to RNA to proteins. While various nonprotein-based molecules (e.g., ribozymes) also carryout cellular functions, proteins are predominantly viewed as the functional product of most genes. Advances in proteomics and next-generation sequencing have shed light on the fact that there are a number of different posttranscriptional and posttranslational mechanisms that influence the abundance, localization, and function of proteins within cells (Vogel and Marcotte 2012). Because of this, mRNA abundance estimates generated from next-generation RNA sequencing may not always correlate with protein abundance or activity (Taniguchi et al. 2010; Vogel and Marcotte 2012). In this light, proteomics offers an important and additional level of insight into gene expression and regulation.

Proteomic techniques have been applied to tardigrades mainly with the goal of identifying potential mediators of stress tolerance (Schokraie et al. 2010, 2011, 2012; Yamaguchi et al. 2012). These studies have highlighted the putative role of conserved heat shock proteins (Schokraie et al. 2011) as well as novel phylum-specific heat soluble proteins (Yamaguchi et al. 2012; Tanaka et al. 2015) in tardigrade stress tolerance.

13.3.3 Ultralow Input Omics Techniques

In recent years, there has been a decrease in both the cost of next-generation sequencing and other omics approaches as well as in the amount of input material (DNA, RNA, protein, etc.) required for these techniques. While early next-generation protocols called for microgram levels of nucleic acids as input, the advent of ultralow input sequencing has spawned protocols and technologies that require only picogram inputs, allowing researchers to sequence minuscule amounts of starting material, including single cells (for review: Kolodziejczyk et al. 2015; Gawad et al. 2016).

Researchers have already begun to apply such techniques to tardigrades. Recently Levin et al., performed RNAseq on libraries derived from single *H. dujardini* embryos spanning different developmental stages (Levin et al. 2016) using the single-cell sequencing technique, CEL-seq (Hashimshony et al. 2012). In addition to examining *H. dujardini* embryos, the authors looked at temporal transcriptomes from nine additional species from different phyla. They found that early and late phases of embryonic development display conserved gene expression across phyla and that these phases are bridged by a "mid-developmental transition," when species-specific genes are upregulated. The mid-developmental transition overlaps the phylotypic period for several of the species. Thus, the authors propose the exciting possibility that phyla can be defined as groups of species whose gene expression at the mid-developmental transition diverges from other species but is conserved within the group (Levin et al. 2016).

As omics technologies utilizing ultralow DNA, RNA, and protein inputs improve, sequencing and proteomic analysis of individual tardigrades or even individual cells will become more accessible and commonplace. Much as next-generation sequencing has, for the most part, replaced EST sequencing or microarrays for expression studies, ultralow input omics promises to overtake techniques requiring large inputs, which will greatly expand the range of biological questions accessible using tardigrades.

13.4 Disrupting Gene Function

With the expansion of omics techniques in tardigrades has come the identification of vast numbers of conserved and novel genes. Additionally, the use of antibodies and in situ hybridization probes allow localization of gene products at both the protein and transcript level (Gabriel and Goldstein 2007; Smith and Jockusch 2014; Tanaka et al. 2015; Hering et al. 2016; Smith et al. 2016). While these approaches have and will continue to provide researchers with insight into the basic biology of tardigrades, these techniques can only hint at the functional role(s) and biological significance of genes and their products.

To gain insight into gene function, researchers often turn to either forward or reverse genetic approaches. Forward approaches consist of using natural or induced mutations to identify the genetic basis of a phenotype. To date, no forward screens have been conducted in tardigrades. However, the availability of tardigrade sequencing data and the amenability of these animals to (re)sequencing and cryopreservation make forward genetic screens a possible and likely productive approach (Gabriel et al. 2007).

While forward genetic screens are employed in many fields of biology, the development of reverse genetic approaches makes targeted studies of gene function possible. As opposed to forward genetic approaches, which examine a phenotype to uncover the underlying genetic basis of a mutation, reverse genetic approaches target specific known genes to examine the phenotypic effect of perturbing that gene's function. Reverse genetic techniques include directed point mutations, deletions, or insertions in an organism's genome or the disruption of a gene's function through the targeted destruction or silencing of gene products using techniques such as RNA interference (RNAi), morpholinos, or through ectopic expression of a gene.

To date the only reverse genetic technique employed in tardigrades is RNAi (Tenlen et al. 2013). RNAi was first demonstrated by Fire and Mellow who showed that double-stranded RNA injected into the nematode worm, *Caenorhabditis elegans*, resulted in the potent and specific destruction, or "knockdown," of targeted transcripts, thus interfering with the function of a specific gene (Fire et al. 1998). In tardigrades, as in other organisms, RNAi works by introducing double-stranded RNA (dsRNA) into cells, where endogenous RNAi machinery processes the dsRNA and ultimately uses the processed dsRNA to destroy complimentary transcripts within the cell (for review: Hannon 2002). Some organisms, including tardigrades, display a systemic RNAi response, where dsRNA is transported into and between cells, facilitating the spread of the RNAi effect to additional cells or the whole organism (Winston et al. 2002; Shannon et al. 2008; Tenlen et al. 2013).

Using dsRNA injected into the gut of *H. dujardini* specimens, Tenlen et al. (2013) demonstrated that RNAi works in tardigrades and is specific, since targeting different genes produced different phenotypes and resulted in reduced levels of target but not control transcripts (Tenlen et al. 2013). RNAi has also been used to assess the functional significance of specific genes in tardigrade stress tolerance (Boothby et al. 2017).

The first step in performing RNAi is to obtain high-quality dsRNA with the same sequence as a portion of the transcript one wishes to target. The best method is to synthesize the dsRNA in vitro, which involves using RT-PCR to produce a complementary DNA (cDNA) from the target mRNA. cDNA is then cloned into a plasmid for storage and propagation. Amplifying the target out of the plasmid using primers with T7 RNA polymerase promotor sequences on their 5′ ends then allows in vitro transcription and annealing to produce dsRNA (Fig. 13.5).

Once one has high-quality dsRNA, the next step is to introduce this dsRNA into the tardigrade. Currently the only published method for introducing dsRNA in tardigrades is microinjection, but other methods such as soaking, feeding, chemical transfection, or electroporation maybe be more efficient. Using microinjection, Tenlen et al. (2013) found that introduction of dsRNA into the gut of the tardigrade *H. dujardini* leads to the spread of the RNAi effect to embryos, suggesting that tardigrades, like *C. elegans*, have a systemic RNAi response. While microinjection is by no means a high-throughput technique, a proficient injectionist can inject ~10–12 tardigrades in an hour. The physical disruption of injection has little or no effect on the tardigrade survival (Tenlen et al. 2013).

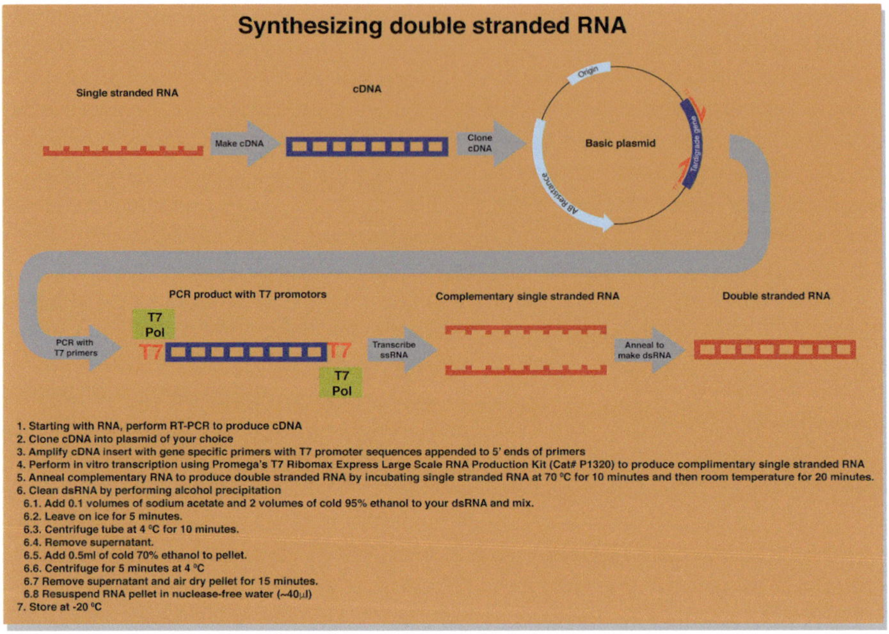

Fig. 13.5 An overview of in vitro double-stranded RNA synthesis for RNA interference. Reverse transcription PCR is performed to generate cDNA of your target. cDNA is then cloned into a bacterial plasmid for propagation and storage. Vectors exist with integrated RNA polymerase promoters (typically T7, T3, or SP6 promoters). If your vector does not already have RNA polymerase promoters, perform PCR with primers specific to your cDNA with T7 promotor sequences appended to their 5′ ends. Check PCR product on a gel to insure a single product of the correct size. If multiple amplicons were generated, perform gel extraction to isolate the correct-sized product. Perform in vitro transcription, Promega's T7 RiboMax Express Large Scale RNA Production Kit works well (Cat# P1320). Single-stranded RNA is annealed by incubating complementary single-stranded RNA at 70 °C for 10 min and then allowing the product to reach room temperature by leaving the mixture at ambient temperature for 20 min. Double-stranded RNA is then cleaned using ethanol precipitation. Double-stranded RNA integrity can be check on a 2% agarose gel. Double-stranded RNA should be stored at −20 °C at as high a concentration as possible, with dilutions for RNAi injections being made and used as needed

Care should be taken to insure that phenotypes resulting from RNAi are specific. Introducing large amounts of dsRNA into cells could have toxic or unanticipated off-target effects that might be confused with specific effects of knocking down a target transcript (Jackson et al. 2003; Grimm et al. 2006). To control non-specific toxicity, a control dsRNA targeting a non-tardigrade or nonessential gene should be used at the same concentration as experimental dsRNA to insure effects are not the result of dsRNA toxicity.

The sequence of the dsRNA should be carefully chosen to reduce the risk of knocking down transcripts with similar sequences. In addition, it is optimal to test at least two non-overlapping dsRNAs targeting the same transcript. Knockdowns using non-overlapping dsRNAs should produce similar phenotypes at similar concentrations. If they do not, one or both of the dsRNAs could be targeting unintended transcripts, increasing or decreasing the severity and variety of phenotypes.

Since RNAi does not knockout or remove a gene, its effects are both concentration and time dependent. Enough time should be given to allow RNAi to take effect before phenotypes are scored. Likewise, the RNAi effect will dissipate with time, so phenotypes should be assessed before the dsRNA is completely degraded and the animals recover their initial levels of endogenous transcripts. Factors influencing the timing and duration of RNAi include the amount of endogenous transcript, the range of endogenous transcript level needed to maintain the normal phenotype, the concentration of dsRNA introduced, as well as the redundancy of the target transcript with transcripts from other genes.

The efficacy of an RNAi experiment depends on the level of knockdown. This level will likely vary by target transcript and will be subject to the concentration of dsRNA used and effectiveness of introduction (injection). The efficacy of a knockdown can be assessed by examining relative levels of target RNA and/or protein in experimental and control animals. Performing RT-PCR to assess relative levels of the target transcript can assess efficacy. Primers for RT-PCR should be chosen such that they do not amplify a region of the transcript that is also contained within the dsRNA, or else amplification of both the transcript and the dsRNA could make detection of a knockdown difficult or impossible. Other methods such as qRT-PCR or northern blotting could also be used to quantify levels of RNA after treatment. Assessment of RNAi efficiency can also be carried-out by examining relative levels of protein encoded by the target RNA using western blotting or similar approaches.

One final note is that RNAi can produce different levels of knockdown within an experimental group of animals resulting in different phenotypes. It is therefore important to establish a good scoring system for reporting both the effects of RNAi and the variability of phenotypes.

To date, RNAi remains the only reverse genetic approach adapted for use in tardigrades. Despite its limitations, RNAi provides researchers with an established method for assessing the function of specific genes in tardigrades. Coupled with the vast amount of new transcriptomic and genomic data, this technique promises to substantially increase the utility of tardigrades as model systems.

13.5 Future Perspectives

Tardigrade research has seen the adaptation of a number of molecular techniques that have facilitated research in many areas of study including evolutionary biology, developmental and cell biology, physiology, genetics, taxonomy, and ecology, positioning tardigrades as a valuable emerging model system.

With advances in sequencing technology has come a plethora of information, including whole genomes from multiple species with thousands of genes homologous to those of other organisms as well as novel, tardigrade-specific genes.

Researchers will be tasked with uncovering the function of these genes in tardigrades. To this end, both forward and reverse genetics hold significant promise.

Given that one can obtain sufficient material to perform next-generation sequencing and resequencing from a single tardigrade, forward genetics could be an efficient method for generating stable lines with known mutations. Forward genetics have been used to great effect in other model systems, such as *D. melanogaster*, *C. elegans*, and *S. cerevisiae*, and the applicability of forward screens for studying the biology of tardigrades has been previously suggested (Gabriel et al. 2007). Given that many tardigrade species can remain viable when frozen, as well as their relatively short generation times, producing and maintaining stocks of mutant lines appear to be a technical possibility (Gabriel et al. 2007).

An alternate to forward genetics for generating mutant lines is to perform targeted genome editing. This involves making a specific genetic deletion, insertion, or point mutation in the genome of germline cells, so that the mutation is passed on to subsequent generations. A variety of methods have been established in other systems toward this end, such as clustered regulatory interspaced short palindromic repeats (CRISPR)-, zinc finger nucleases (ZFNs)-, and transcription-activator like-effector nucleases (TALENs)-based approaches (Miller et al. 2007; Urnov et al. 2010; Mussolino et al. 2011; Sander and Joung 2014). The recent explosion in the use of CRISPR/Cas technology highlights the efficiency, reprogramability, and lower overall cost of this technique. CRISPR/Cas strategies allow the insertion or deletion of essentially any sequence from or into any portion of a genome, a significant advantage over random insertional methods and has been used in a wide variety of organisms, including fungi, plants, and both vertebrate and invertebrate animals (Doudna and Charpentier 2014). Given the relative ease, widespread applicability, and power of CRISPR/Cas genome editing technology, attempts to adapt this technique for use in tardigrades should be made.

A key point in tardigrades regarding the development of CRISPR/Cas genome editing is the need to make heritable changes to the genome of germ cells. In tardigrades, the cuticle surrounding eggs is extremely hard, making microinjection of embryos difficult (Boothby, personal observation). In *C. elegans*, injections for CRISPR genome editing can be performed by injecting into the germline of worms (Dickinson et al. 2013), which is syncytial. It is not clear whether the germline of tardigrades is also syncytial, but if it is, microinjection or electroporation may be an effective means of getting the necessary reagents into the germline and ultimately oocytes, before eggshell deposition.

Through the hard work and dedication of researchers worldwide, the use and adaptation of molecular approaches to tardigrades have increased dramatically within the past few decades. These techniques have and will continue to promote the use of tardigrades as important model systems and provide novel avenues for studying the basic biology of these fascinating animals.

References

Bertolani R, Rebecchi L, Giovannini I, Cesari M (2011) DNA barcoding and integrative taxonomy of *Macrobiotus hufelandi* CAS Schultze 1834, the first tardigrade species to be described, and some related species. Zootaxa 2997:e36

Blaxter M, Elsworth B, Daub J (2004) DNA taxonomy of a neglected animal phylum: an unexpected diversity of tardigrades. Proc R Soc B Biol Sci 271:S189–S192. https://doi.org/10.1098/rsbl.2003.0130

Blaxter M, Mann J, Chapman T et al (2005) Defining operational taxonomic units using DNA barcode data. Philos Trans R Soc B Biol Sci 360:1935–1943. https://doi.org/10.1098/rstb.2005.1725

Boothby TC, Tenlen JR, Smith FW et al (2015) Evidence for extensive horizontal gene transfer from the draft genome of a tardigrade. Proc Natl Acad Sci 112:15976–15981. https://doi.org/10.1073/pnas.1510461112

Boothby TC, Tapia H, Brozena AH et al (2017) Tardigrades use intrinsically disordered proteins to survive desiccation. Mol Cell 65:975–984. https://doi.org/10.1016/j.molcel.2017.02.018

Crowe JH, Crowe LM, Chapman D (1984) Preservation of membranes in anhydrobiotic organisms: the role of trehalose. Science 223:701–703

Dickinson DJ, Ward JD, Reiner DJ, Goldstein B (2013) Engineering the *Caenorhabditis elegans* genome using Cas9-triggered homologous recombination. Nat Methods 10:1028–1034. https://doi.org/10.1038/nmeth.2641

Doudna JA, Charpentier E (2014) The new frontier of genome engineering with CRISPR-Cas9. Science 346:1258096. https://doi.org/10.1126/science.1258096

Erkut C, Penkov S, Khesbak H et al (2011) Trehalose renders the dauer larva of *Caenorhabditis elegans* resistant to extreme desiccation. Curr Biol 21:1331–1336. https://doi.org/10.1016/j.cub.2011.06.064

Fire A, SiQun X, Montgomery MK et al (1998) Potent and specific genetic interference by double-stranded RNA in *Caenorhabditis elegans*. Nature 391:806–811

Folch J, Lees M, Stanley GHS (1957) A simple method for the isolation and purification of total lipides from animal tissues. J Biol Chem 226:497–509

Gabriel WN, Goldstein B (2007) Segmental expression of Pax3/7 and engrailed homologs in tardigrade development. Dev Genes Evol 217:421–433. https://doi.org/10.1007/s00427-007-0152-5

Gabriel WN, McNuff R, Patel SK et al (2007) The tardigrade *Hypsibius dujardini*, a new model for studying the evolution of development. Dev Biol 312:545–559. https://doi.org/10.1016/j.ydbio.2007.09.055

Gawad C, Koh W, Quake SR (2016) Single-cell genome sequencing: current state of the science. Nat Rev Genet 17:175–188. https://doi.org/10.1038/nrg.2015.16

Grimm D, Streetz KL, Jopling CL et al (2006) Fatality in mice due to oversaturation of cellular microRNA/short hairpin RNA pathways. Nature 441:537–541. https://doi.org/10.1038/nature04791

Hannon GJ (2002) RNA interference. Nature 418:244–251

Hashimoto T, Horikawa DD, Saito Y et al (2016) Extremotolerant tardigrade genome and improved radiotolerance of human cultured cells by tardigrade-unique protein. Nat Commun 7:12808. https://doi.org/10.1038/ncomms12808

Hashimshony T, Wagner F, Sher N, Yanai I (2012) CEL-Seq: single-cell RNA-Seq by multiplexed linear amplification. Cell Rep 2:666–673. https://doi.org/10.1016/j.celrep.2012.08.003

Hengherr S, Heyer AG, Köhler H-R, Schill RO (2008) Trehalose and anhydrobiosis in tardigrades - evidence for divergence in responses to dehydration: trehalose and anhydrobiosis in tardigrades. FEBS J 275:281–288. https://doi.org/10.1111/j.1742-4658.2007.06198.x

Hering L, Bouameur J-E, Reichelt J et al (2016) Novel origin of Lamin-derived cytoplasmic intermediate filaments in tardigrades. eLife 5:e11117

Hoekstra FA, Crowe LM, Crowe JH (1989) Differential desiccation sensitivity of corn and *Pennisetum* pollen linked to their sucrose contents. Plant Cell Environ 12:83–91

Holmstrup M, Hedlund K, Boriss H (2002) Drought acclimation and lipid composition in *Folsomia candida*: implications for cold shock, heat shock and acute desiccation stress. J Insect Physiol 48:961–970

Hughes CS, Foehr S, Garfield DA et al (2014) Ultrasensitive proteome analysis using paramagnetic bead technology. Mol Syst Biol 10:757–757. https://doi.org/10.15252/msb.20145625

Hurkman WJ, Tanaka CK (1986) Solubilization of plant membrane proteins for analysis by two-dimensional gel electrophoresis. Plant Physiol 81:802–806

Hygum TL, Fobian D, Kamilari M et al (2017) Comparative investigation of copper tolerance and identification of putative tolerance related genes in tardigrades. Front Physiol 8:95. https://doi.org/10.3389/fphys.2017.00095

Ingemar Jönsson K, Harms-Ringdahl M, Torudd J (2005) Radiation tolerance in the eutardigrade *Richtersius coronifer*. Int J Radiat Biol 81:649–656. https://doi.org/10.1080/09553000500368453

Jackson AL, Bartz SR, Schelter J et al (2003) Expression profiling reveals off-target gene regulation by RNAi. Nat Biotechnol 21:635–637

Jönsson KI, Persson O (2010) Trehalose in three species of desiccation tolerant tardigrades. Open Zool J 3:1–5

Kolodziejczyk AA, Kim JK, Svensson V et al (2015) The technology and biology of single-cell RNA sequencing. Mol Cell 58:610–620. https://doi.org/10.1016/j.molcel.2015.04.005

Koutsovoulos G, Kumar S, Laetsch DR et al (2016) No evidence for extensive horizontal gene transfer in the genome of the tardigrade *Hypsibius dujardini*. Proc Natl Acad Sci:201600338. https://doi.org/10.1073/pnas.1600338113

Lapinski J, Tunnacliffe A (2003) Anhydrobiosis without trehalose in *bdelloid* rotifers. FEBS Lett 553:387–390. https://doi.org/10.1016/S0014-5793(03)01062-7

Levin M, Anavy L, Cole AG et al (2016) The mid-developmental transition and the evolution of animal body plans. Nature 531:637–641. https://doi.org/10.1038/nature16994

Mali B, Grohme MA, Förster F et al (2010) Transcriptome survey of the anhydrobiotic tardigrade *Milnesium tardigradum* in comparison with *Hypsibius dujardini* and *Richtersius coronifer*. BMC Genomics 11:168

Miller JC, Holmes MC, Wang J et al (2007) An improved zinc-finger nuclease architecture for highly specific genome editing. Nat Biotechnol 25:778–785. https://doi.org/10.1038/nbt1319

Mussolino C, Morbitzer R, Lutge F et al (2011) A novel TALE nuclease scaffold enables high genome editing activity in combination with low toxicity. Nucleic Acids Res 39:9283–9293. https://doi.org/10.1093/nar/gkr597

Nayak S Single worm PCR. http://genetics.wustl.edu/tslab/protocols/genomic-stuff/single-worm-pcr/. Accessed 12 May 2016

Obinata T, Ono K, Ono S (2011) Detection of a troponin I-like protein in non-striated muscle of the tardigrades (water bears). BioArchitecture 1:96–102. https://doi.org/10.4161/bioa.1.2.16251

Parra G, Bradnam K, Korf I (2007) CEGMA: a pipeline to accurately annotate core genes in eukaryotic genomes. Bioinformatics 23:1061–1067. https://doi.org/10.1093/bioinformatics/btm071

Rebecchi L, Altiero T, Guidetti R et al (2009) Tardigrade resistance to space effects: first results of experiments on the LIFE-TARSE mission on FOTON-M3 (September 2007). Astrobiology 9:581–591

Rizzo AM, Negroni M, Altiero T et al (2010) Antioxidant defences in hydrated and desiccated states of the tardigrade *Paramacrobiotus richtersi*. Comp Biochem Physiol B Biochem Mol Biol 156:115–121. https://doi.org/10.1016/j.cbpb.2010.02.009

Rizzo AM, Altiero T, Corsetto PA et al (2015) Space flight effects on antioxidant molecules in dry tardigrades: the TARDIKISS experiment. Biomed Res Int 2015:1–7. https://doi.org/10.1155/2015/167642

Sander JD, Joung JK (2014) CRISPR-Cas systems for editing, regulating and targeting genomes. Nat Biotechnol 32:347–355. https://doi.org/10.1038/nbt.2842

Sarkies P, Selkirk ME, Jones JT et al (2015) Ancient and novel small RNA pathways compensate for the loss of piRNAs in multiple independent nematode lineages. PLoS Biol 13:e1002061. https://doi.org/10.1371/journal.pbio.1002061

Schill RO (2007) Comparison of different protocols for DNA preparation and PCR amplification of mitochondrial genes of tardigrades. J Limnol 66:164–170

Schokraie E, Hotz-Wagenblatt A, Warnken U et al (2010) Proteomic analysis of tardigrades: towards a better understanding of molecular mechanisms by anhydrobiotic organisms. PLoS One 5:e9502. https://doi.org/10.1371/journal.pone.0009502

Schokraie E, Hotz-Wagenblatt A, Warnken U et al (2011) Investigating heat shock proteins of tardigrades in active versus anhydrobiotic state using shotgun proteomics: investigating heat shock proteins of tardigrades. J Zool Syst Evol Res 49:111–119. https://doi.org/10.1111/j.1439-0469.2010.00608.x

Schokraie E, Warnken U, Hotz-Wagenblatt A et al (2012) Comparative proteome analysis of *Milnesium tardigradum* in early embryonic state versus adults in active and anhydrobiotic state. PLoS One 7:e45682. https://doi.org/10.1371/journal.pone.0045682

Shannon AJ, Tyson T, Dix I et al (2008) Systemic RNAi mediated gene silencing in the anhydrobiotic nematode *Panagrolaimus superbus*. BMC Mol Biol 9:58. https://doi.org/10.1186/1471-2199-9-58

Smith FW, Jockusch EL (2014) The metameric pattern of *Hypsibius dujardini* (Eutardigrada) and its relationship to that of other panarthropods. Front Zool 11(1):66

Smith FW, Boothby TC, Giovannini I et al (2016) The compact body plan of tardigrades evolved by the loss of a large body region. Curr Biol 26:224–229. https://doi.org/10.1016/j.cub.2015.11.059

Tanaka S, Tanaka J, Miwa Y et al (2015) Novel mitochondria-targeted heat-soluble proteins identified in the anhydrobiotic tardigrade improve osmotic tolerance of human cells. PLoS One 10:e0118272. https://doi.org/10.1371/journal.pone.0118272

Taniguchi Y, Choi PJ, Li G-W et al (2010) Quantifying *E. coli* proteome and transcriptome with single-molecule sensitivity in single cells. Science 329:533–538

Tapia H, Koshland DE (2014) Trehalose is a versatile and long-lived chaperone for desiccation tolerance. Curr Biol 24:2758–2766. https://doi.org/10.1016/j.cub.2014.10.005

Tenlen JR, McCaskill S, Goldstein B (2013) RNA interference can be used to disrupt gene function in tardigrades. Dev Genes Evol 223:171–181. https://doi.org/10.1007/s00427-012-0432-6

Urnov FD, Rebar EJ, Holmes MC et al (2010) Genome editing with engineered zinc finger nucleases. Nat Rev Genet 11:636–646. https://doi.org/10.1038/nrg2842

Vogel C, Marcotte EM (2012) Insights into the regulation of protein abundance from proteomic and transcriptomic analyses. Nat Rev Genet 13:227–232

Wang C, Grohme MA, Mali B et al (2014) Towards decrypting cryptobiosis—analyzing anhydrobiosis in the tardigrade *Milnesium tardigradum* using transcriptome sequencing. PLoS One 9:e92663. https://doi.org/10.1371/journal.pone.0092663

Westh P, Ramløv H (1991) Trehalose accumulation in the tardigrade *Adorybiotus coronifer* during anhydrobiosis. J Exp Zool 258:303–311

Wetterstrand KA (2016) DNA sequencing costs: data from the NHGRI Genome Sequencing Program (GSP). www.genome.gov/sequencingcosts. Accessed 23 May 2016

Winston WM, Molodowitch C, Hunter CP (2002) Systemic RNAi in *C. elegans* requires the putative transmembrane protein SID-1. Science 295:2456–2459

Yamaguchi A, Tanaka S, Yamaguchi S et al (2012) Two novel heat-soluble protein families abundantly expressed in an anhydrobiotic tardigrade. PLoS One 7:e44209. https://doi.org/10.1371/journal.pone.0044209

Chapter 14
Field and Laboratory Methods

Peter Degma

Abstract Tardigrades live in diverse habitats. Fundamental research on tardigrades is based on habitat sampling, sample processing, tardigrade culture and preparation for descriptive microscopy, identification and molecular studies. These methods are described and referenced in this chapter which I hope it will be a starting point especially for new tardigradologists.

14.1 Introduction

Tardigrades live in diverse terrestrial, freshwater and marine habitats, from the poles to the tropics and from alpine through coastal to abyssal depths (McInnes 1994; Kaczmarek et al. 2014, 2015a, b, 2016; Meyer 2013). The phylum Tardigrada consists currently of more than 1250 known species (Degma et al. 2018).

Basic research on tardigrade diversity, taxonomy, phylogenetics and distribution is based on habitat sampling, sample extraction, culturing tardigrades if necessary, processing specimens for microscopy (light and electron microscopy) and molecular studies, specimens observation, identification and occasional description. The methods of these operations are scattered in number of papers or more or less briefly described in a few book chapters (e.g. Nelson et al. 2015). The chapter focuses on the sequence of these operations more thoroughly (if any not described exhaustively references at relevant papers are added) and is aimed mainly at new tardigradologists.

14.2 Collection of Tardigrades

Collection comprises of sampling, extraction, fixation and staining.

P. Degma (✉)
Department of Zoology, Comenius University in Bratislava, Bratislava, Slovakia
e-mail: degma1@uniba.sk

R. O. Schill (ed.), *Water Bears: The Biology of Tardigrades*, Zoological Monographs 2,
https://doi.org/10.1007/978-3-319-95702-9_14

14.2.1 Sampling of Habitats

Methods vary with habitat and objective. For example, ecological studies require equally sized quantitative samples for subsequent statistical evaluation. Sample number is usually estimated via analysis of preliminary sampling (Eckblad 1991).

14.2.1.1 Sampling of Terrestrial Habitats

Mosses growing on various terrestrial substrata are optimal habitats for many terrestrial tardigrades. These mosses are collected by hand scraping from a substrate, i.e. tree bark, dead wood, rock, soil, roof, wall, etc. (for information on collecting from the tree canopy, see Chappell et al. 2015). Tardigrades are not uniformly distributed in moss cushions, which can be an issue when attempting to subsample larger moss cushions (Miller et al. 1994; Degma et al. 2011). Furthermore, 'pooling' more than one smaller sample could spuriously associate free laid eggs with the wrong adults of a similar species (e.g. finding more than one *Macrobiotus* species from the *hufelandi* group).

We put each moss sample into a small well-labelled paper bag immediately after collection to allow slow natural air-drying so resident tardigrades have enough time to enter an anhydrobiotic stage. We write basic sample data (sample code, date, locality and/or GPS coordinates, altitude, collector and substratum type) on the empty paper bag prior taking a moss sample. It should also be noted whether the sample constitutes one unbroken moss (solid sample) or combines several smaller sample units (e.g. Steiner 1994; de Peluffo et al. 2006). Additional ecological data can be entered, against the samples code number, in a separate note book.

We can also collect tardigrades from other terrestrial habitats which provide damp or wet conditions at least from time to time, for example, liverworts, lichens, club mosses, terrestrial algae, soil, leaf litter, touchwood (unpublished) and cushion-forming vascular plants (e.g. moss campion *Silene acaulis* (L.) Jacq., mossy cyphel *Minuartia sedoides* (L.) Hiern.). Most terrestrial habitats are easy to collect by hand or using basic pocket knife or similar appropriate tools.

For more precise ecological studies, we remove solid quantitative samples using a corer or delineate an area with a prescribed quadrat so each sample has the same area or volume; samples can be subdivided into upper and deeper subsamples (e.g. Dastych 1980; Guidetti and Bertolani 2001; Harada and Ito 2006; Nelson and Bartels 2013).

Samples taken from terrestrial habitats can be stored dry in paper bags in room temperature, though preservation of dry samples at −20 °C to −80 °C is preferred (e.g. McInnes and Convey 2005; Bartels and Nelson 2006; Faurby et al. 2011).

For methods of collection from wind-transported sediments, see Nkem et al. (2006).

14.2.1.2 Sampling of Freshwater Habitats

True limnic tardigrades co-occur with some limnoterrestrial species on submerged vegetation, in sediments, periphyton or among algal encrustations in lentic or weakly running lotic waters. We collect these using the same methods as for collecting freshwater meiofauna. Sediment samples can be collected with a 'turkey baster', modified Patalas sampler, piston or gravity corers, while periphyton samples can be scraped from the substrate with any sharp tool. We can sample benthic sediments via a dip net or rinse aquatic plants in a 1000 μm metal sieve into a 60 μm screen or dip net. Submerged and waterfall-/spring-inundated mosses and freshwater sponges can be sampled by hand, while zooplankton samples can be collected using a fine mesh net (e.g. Van Rompu and De Smet 1991; Janiec 1996; Bartels and Nelson 2006; Gibson et al. 2007; Pilato et al. 2010; Kaczmarek et al. 2012a; Nelson et al. 2015; Tałanda and Zawierucha 2017).

Samples have to be kept wet until tardigrades can be extracted, usually being transported in plastic bags or jars. Samples may be preserved in situ via the addition of formalin, to a final concentration of ca. 4%. Fresh/non-preserved material could be refrigerated and processed as quickly as possible or frozen.

Cryoconite holes in polar and mountain glaciers provide a freshwater habitat for tardigrades, and the cryoconite sediment is best collected via a pipette and concentrated by decanting off excess water (Dastych et al. 2003; Zawierucha et al. 2016b).

Transient and semi-permanent pools of water collected by plants (e.g. teasel—*Dipsacus* spp. and hollow tree trunks) can also be sampled by pipette.

14.2.1.3 Sampling of Marine Habitats

Marine tardigrades are present from coastal intertidal to abyssal depths on surfaces or within sediments. Collections can be made via coring using corers of defined surface and length, dredges, epibenthic sled, grabs and scuba diving with appropriate tools. Sandy samples can be taken also by digging a hole to beach groundwater level and taking the sediment saturated with seawater in a plastic bucket (e.g. Kristensen and Higgins 1984a, b; Pollock 1989; Gallo D'Addabbo et al. 2001; Hansen et al. 2001; Jørgensen and Kristensen 2001; Albuquerque et al. 2007; Accogli et al. 2011; Romano et al. 2011). The type of equipment used dictates the quality of sampling, i.e. coring can be subdivided into layers, while dredging disturbs them.

Tardigrades may also be sampled as meiofauna on mussels (*Mytilus* spp.) (e.g. Eibye-Jacobsen 1996), barnacles, algae, lichens and sediments (e.g. Faurby et al. 2012).

Marine samples should be examined in seawater or preserved with buffered formalin.

14.2.2 *Sample Processing*

14.2.2.1 Extraction of Tardigrades from a Sample

The extraction process is basically the same for terrestrial, limnic and marine samples, comprising two steps: (1) concentration of a sample into a small volume aqueous suspension and (2) collecting tardigrades and their eggs from the suspension.

Concentration of a Sample Content

The described concentration method for terrestrial samples is very efficient (based on Degma 2010):

1. Submerge dry terrestrial sample in tap water or double-distilled/filtered water which could prevent possible sample contamination, refrigerate for ca. 24 h (even if sample is many years old) to allow time for the tardigrades to revive from an anhydrobiotic state.
2. Put the sample into the top section of a series of two or three sieves (grading form coarse to fine: 1200, 250 and 40 μm mesh).
3. Tear the sample into small fragments (so that water can wash all specimens, eggs and exuvia from each fragment), and wash each fragment in water by rapid shaking as briskly as possible using solid tweezers. After shaking and squeezing remaining water into the beaker, put each fragment aside for subsequent drying and storing them in the original paper bag for potential repeated analysis and/or future identification, or fix the analysed material before throwing it away to prevent introduction non-native animals into our nature. Finally rinse the beaker with more distilled/filtered water.
4. Wash the remaining fine material through sieve series with copious distilled/filtered water to concentrate Tardigrada on the finest sieve.
5. Rinse sediment from a finest sieve into one or several 7 cm Petri dishes using a wash bottle of distilled water, leaving only a thin layer of sediment so tardigrades can be seen under a stereomicroscope.

In the absence of an appropriate series of sieves, the material should be poured through a 1 mm mesh sieve to separate larger detritus particles. Residual sediment and tardigrades remain in the beaker/graduated cylinder bottom after 30 min sedimentation, and excess water can be carefully decanted leaving a sample that can be washed into Petri dishes for observation (Dastych 1985).

Steps 3 and 4 of can be substituted by using a floatation technique. Samples are prepared by producing a watery suspension by short bursts with a food blender. The suspension is then washed through a coarse sieve to remove larger debris and resulting suspension floated above a concentration of sucrose solution or colloidal silica (Ludox AM™ to extracting tardigrades from the terrestrial/freshwater material and Ludox TM™/Ludox HS-40 for marine tardigrades). The samples are centrifuged (1000 rpm for 1 min), and the supernatant passed through a fine mesh sieve (ca 30–60 μm). The sieve is flushed with distilled/filtered water to remove sugar or silica colloid and the sediment washed into a Petri dish for examination (compare with McInnes and Ellis-Evans 1987; Bartels and Nelson 2006; Fujimoto 2015; Nelson et al. 2015). A slightly modified floatation procedure using OptiPrep™ is described in Sands et al. (2008a).

Baermann funnels can be used to concentrate tardigrades from soil or leaf litter by placing the wet sample on Baermann funnel to allow active tardigrades to migrate from desiccating substrate downwards into a water-filled funnel and hose (e.g. Guidetti and Bertolani 2001; Boeckner and Proctor 2005; Hohberg et al. 2011).

Sandy sediment samples are stirred in distilled/filtered water, the sand allowed to settle and the water decanted through a 40 μm mesh and rinsed into a Petri dish (Nelson et al. 2015).

Marine tardigrades can be dislodged from their habitat by anesthetizing with (ca. 7%) $MgCl_2$ solution or using a freshwater shock technique—short agitation of small amounts of material in fresh water and quick decantation of water through a 30–60 μm fine sieve or plankton net after settling out the heavier particles. The procedure should be repeated several times. Contents of the sieve can be reimmersed in seawater where the meiofauna regains osmotic equilibrium and either sorted under a stereomicroscope or fixed (e.g. Kristensen and Higgins 1984a, b, 1989; Boesgaard and Kristensen 2001; D'Addabbo et al. 2007).

Collecting Tardigrades from the Concentrate

Tardigrades and their eggs can be collected from concentrated sample as follows (based on Degma 2010):

1. Spread content of the fine mesh sieve in a Petri dish and examine under a stereomicroscope using reflected or dark-field illumination at 25–-80 × magnification. Marking the base of the Petri dish with tramlines or squares (e.g. Stec et al. 2015) can aid observation and reduce repetition. Transfer specimens to a watch glass/Stender dish/embryo dish with a small amount of water via a micropipette, Irwin loop (Schram and Davison 2012), small glass Pasteur pipette or Wheaton 0.4 μL pipette (Nichols et al. 2001) after cleaning surroundings with entomological pin (Figs 14.1 and 14.2).
2. Add a single small drop of glacial acetic acid to fix and straighten specimens in a few minutes (Fig. 14.3).
3. Fix all remaining material to prevent future cross-contamination from sample waste.

In some special cases, we can collect tardigrades directly from their habitat, for example, marine tardigrades on barnacles, using a stereomicroscope (Kristensen and Hallas 1980; Perry and Miller 2015).

Fig. 14.1 Living tardigrades *Macrobiotus polonicus* in detritus on a Petri dish after their concentration on the 40 μm sieve

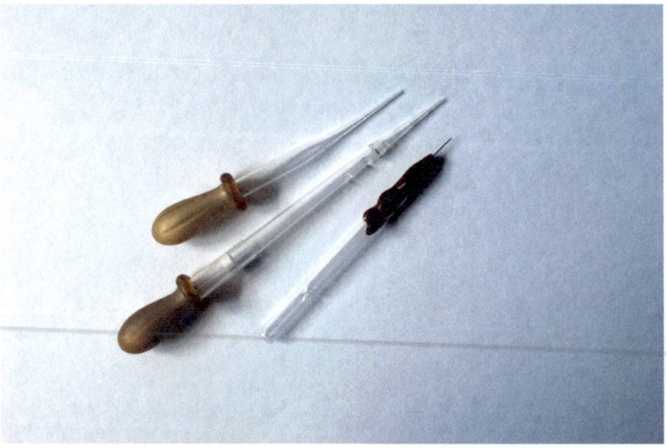

Fig. 14.2 Some handmade preparatory aids: micropipettes made from eye dropper and commercial micropipette tips and adjusted entomological pin

Fig. 14.3 Collected and fixed specimens (adults, juveniles and an egg) of *Macrobiotus polonicus* on the watch glass

14.2.2.2 Fixation and Staining of Tardigrades

Specimens can be fixed individually after collecting from a Petri dish, as described above, or fixed as concentrated whole samples.

Specimens can be fixed or postfixed in acetic acid, Carnoy's fluid, boiling absolute alcohol, warm or boiling water, dimethyl sulfoxide (DMSO), trialdehyde, buffered glutaraldehyde or buffered 4–7% formalin (seawater solution in case of marine samples) (see Horning et al. 1978; Dastych 1984; Boesgaard and Kristensen 2001; Hansen et al. 2001; Tumanov 2004; Gallo D'Addabbo et al. 2005; Bartels and Nelson 2006; Guil et al. 2007; Grothman 2011; Gross et al. 2014). Morek et al. (2016b) recommended thermal relaxation of live specimens before mounting them instead using any fixative.

Staining tardigrades in a sample, e.g. with rose bengal, can make them more obvious and aid their collection from a Petri dish under stereomicroscope (Boesgaard and Kristensen 2001; Hansen et al. 2001; Fujimoto et al. 2013; Kharkevych and Sergeeva 2013). For identification of gender, tardigrades are first fixed in Carnoy's fluid (methanol/acetic acid, 3:1), and then a drop of acetic lactic orcein or acetic carmine is added (e.g. Rebecchi et al. 2000; Guidetti et al. 2016).

14.3 Culturing of Tardigrades

Rearing tardigrades in laboratory cultures provides material for experiments as well as for reproduction, life history, taxonomic and phylogenetic studies. Some aquatic and terrestrial species are commercially available (Nelson et al. 2015).

Several freshwater and limnoterrestrial eutardigrades have been successfully cultured in laboratories. Adults or eggs are placed in small open or closed Petri dishes or plastic culture dishes lined with a thin (4–5 mm) layer of 1.2–3.0% agar or with a bottom scratched with fine sandpaper to allow tardigrade locomotion. Tardigrades can be maintained in a thin film/layer of water (usually commercial bottled spring water) plus a food source, which is frequently changed. Vessels can be kept at constant temperatures from 3 °C to 25 °C or simply at room temperature, in complete darkness or a regular 12 h/12 h (light/dark) photoperiod. New laid eggs are transferred to a separate plastic dish but without any food until they hatch (see culture details of different Tardigrada in papers cited in the next paragraph).

Proven tardigrade culturing conditions include those fed on algae: *Diphascon* and *Isohypsibius* (Altiero and Rebecchi 2001; Poprawa 2011; Bingemer et al. 2016; Kosztyła et al. 2016 or green algae and rotifers—Ito et al. 2016), *Acutuncus* (Kagoshima et al. 2013; Altiero et al. 2015; Tsujimoto et al. 2015), *Ramazzottius* (Horikawa et al. 2008), *Thulinius* (Hejnol and Schnabel 2005; Sobczyk et al. 2015; Kosztyła et al. 2016) and *Dactylobiotus* (Poprawa 2005a). Those fed on rotifers, nematodes and/or algae are the following: *Milnesium* (adults, rotifers/nematodes; juveniles, algae—Suzuki 2003; Förster et al. 2012; Beltrán-Pardo et al. 2013;

Kosztyła et al. 2016), *Hypsibius* (rotifers/algae—Gabriel et al. 2007; Melo et al. 2015; Hyra et al. 2016; Kosztyła et al. 2016), *Macrobiotus* (nematodes/algae—Altiero and Rebecchi 2001; Hengherr et al. 2009; rotifers and algae—Stec et al. 2015), *Mesobiotus* (rotifers and algae—Mapalo et al. 2016) and *Paramacrobiotus* (nematodes/rotifers/algae—Altiero and Rebecchi 2001; Hengherr et al. 2009; Schill 2013; Melo et al. 2015; Kosztyła et al. 2016).

14.4 Processing of Tardigrades for Microscopy and Molecular Taxonomy

14.4.1 Preparation of Whole Specimens for Light Microscopy

Fixed material can be cleaned of adherent debris in an ultrasonic bath prior to mounting on slides (Kristensen and Higgins 1989). The following mounting method is based on Degma (2010):

1. Transfer a fixed specimen with minimal fixative to the centre of clean microscope slide, and note its colour.
2. Quickly remove redundant fixative with a thin micropipette.
3. Before evaporation of rest of liquid, use entomological pin for orientation of the specimen ideally with dorsal or lateral surface uppermost to display buccal cavity armature (Macrobiotidae), transverse rows of gibbosities (*Isohypsibius*) and apophyses for the insertion of the stylet muscles (Hypsibioidea and Isohypsibioidea). When just dry and affixed to the glass (Fig. 14.4), immediately add a drop of Hoyer's or another water-soluble mounting medium and cover with glass cover slip. Use a bit diluted medium with distilled water—it diffuses under

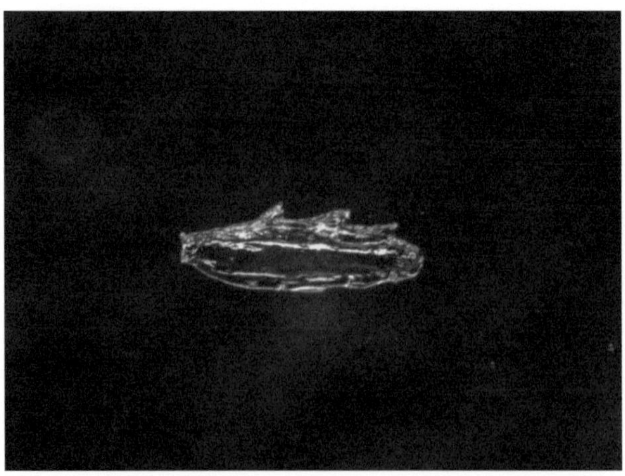

Fig. 14.4 Just-dried and affixed tardigrade *Macrobiotus polonicus* on a slide

cover slip more quickly and ordinarily does not manage to unstick specimen from the slide.

4. Carefully mark the position of the specimen(s) with a ring of permanent marker pen dots on cover slip.
5. Immediately note presence and colour or absence of eyes via bright field/phase contrast as eyes frequently dissolve in mounting media. Identify all slides from same sample with the same batch code and a unique slide identifier. It is suitable to note information concerning each slide including preliminary identification and info about eyes and body colour into a separate notebook.
6. Store slides flat in dark, dry and dustless environment for 15–20 days in room temperature to prevent medium becoming crystalline (c.f. Horning et al. 1978—dry at 50 °C for a month; Meyer 2009—dry at 65 °C for several hours).
7. Mark the position of the specimen(s) on the lower surface of the slide with permanent marker pen. Seal the cover slip to avoid medium crystallization with epoxy paint, clear nail varnish or asphalt varnish (some media—e.g. Hoyer's—enable slide remounting after their rehydration).
8. Identify the slide with two labels noting (1) code and collection data (locality, date of collection of the sample, name of the collector, habitat, e.g. moss, lichen, etc.) and (2) specimen identification, name of preparator, year of preparation, developmental state (egg, juvenile, etc.), mounting medium, presence of eyes and taxonomical status (e.g. paratype(s)).

Hoyer's medium (e.g. version in Nelson et al. 2015 or Morek et al. 2016b), Faure's medium (see in Ramazzotti and Maucci 1983) (chloral hydrate, one of main components of these media, is restricted substance in some countries) or polyvinyl lactophenol are the most frequently used tardigrade mounting media; water-soluble alternatives include polyvinyl alcohol and (for marine Tardigrada) glycerol (Villora-Moreno 1996; Boesgaard and Kristensen 2001; Rubal et al. 2013; Bartels et al. 2015).

It is recommended to mount each specimen (adult, juvenile or egg) individually; however, described preparation method allows us easy mounting more specimens on a slide if necessary.

14.4.2 Preparation of Whole Specimens for Scanning Electron Microscopy (SEM)

Specimens and eggs prepared for SEM are first fixed in osmium tetroxide, glutaraldehyde, warm water, Bouin's medium or ethanol and then dehydrated in an ethanol/acetone series (some authors omitted acetone series or substituted it with its single concentration) prior to critical point drying in liquid CO_2. Dried specimens are mounted on stubs; coated with palladium, gold or platinum and observed via SEM (Dastych 1992; Rho et al. 1997; Chang and Rho 1998; Guidetti et al. 2000; Poprawa 2005b; Suzuki 2008; Vicente et al. 2008; Trygvadóttir and Kristensen 2011; Gross et al. 2014; Stec et al. 2015; Bąkowski et al. 2016; Hansen et al. 2017; Hygum et al. 2016). Repeated application of absolute ethanol instead of using ethanol-graded series and liquid carbon dioxide is an alternative of tardigrade dehydration (Bertolani et al. 2014).

Alternative SEM-appropriate techniques include cryo-preparation (Robinson et al. 1996) and the preparation of the bucco-pharyngeal apparatus (Eibye-Jacobsen 2001; Guidetti et al. 2013a; Gąsiorek et al. 2016).

14.4.3 Preparation of Tardigrade Sections for Light Microscopy and Transmission Electron Microscopy (TEM)

Specimens are fixed/postfixed in glutaraldehyde and osmium tetroxide solutions, buffered with sodium cacodylate or sodium phosphate and then dehydrated in an ethanol/acetone series prior to embedding in resin for cutting semi-thin or ultra-thin sections with an ultramicrotome. Semi-thin sections are usually stained with methylene/toluidine blue for light microscopy and ultra-thin sections with uranyl acetate and lead citrate for TEM-examination (Eibye-Jacobsen 1997; Guidetti et al. 2000; Rebecchi et al. 2000; Jørgensen and Kristensen 2001; Greven et al. 2005; Avdonina et al. 2007; Greven 2007; Rocha et al. 2007; Rost-Roszkowska et al. 2013; Hyra et al. 2016).

14.4.4 Preparation of Specimens for Molecular Taxonomy

Four DNA fragments are currently used to study the relationships between tardigrades. The small (18S) and the large (28S) ribosome subunit rRNA units are usually used for resolving phylogenetic relationships between genera, families and orders. Cytochrome oxidase subunit I (COI), a protein coding mitochondrial DNA fragment, is widely used as a standard barcode gene for interspecific comparisons, while the internal transcribed spacer (ITS-2) non-coding nuclear DNA fragment is suitable for intraspecific comparisons and separating sibling and cryptic species (Nelson et al. 2015; Stec et al. 2015). Sequences are prepared from a single entire animal, egg or embryo, optimally from live or anhydrobiotic material (Schill 2007).

First step of DNA isolation from single tardigrades is followed by polymerase chain reaction (PCR), amplification of a region of the mitochondrial DNA (mtDNA)—cytochrome c oxidase subunit I (COI) gene—details of which can be found in Guidetti et al. (2005, 2009, 2013b), Jørgensen et al. (2007, 2011), Schill (2007), Faurby et al. (2008, 2011), Cesari et al. (2009, 2011, 2016), Guil and Giribet (2009, 2011), Rebecchi et al. (2009), Bertolani et al. (2011a, b), Michalczyk et al. (2012), Kagoshima et al. (2013), Vicente et al. (2013), Stec et al. (2015, 2016b),

Gąsiorek et al. (2016), Kosztyła et al. (2016), Mapalo et al. (2016) and Zawierucha et al. (2016a). The most frequently used primers are those of Folmer et al. (1994).

For protocols for amplification of the 18S rRNA gene subunits and primers, see Nichols et al. (2006), Kiehl et al. (2007), Schill and Steinbrück (2007), Sands et al. (2008b), Guidetti et al. (2009, 2013b, 2014), Jørgensen et al. (2010, 2011), Calloway et al. (2011), Guil and Giribet (2011), Guil et al. (2013), Kagoshima et al. (2013), Vicente et al. (2013), Bertolani et al. (2014), Dabert et al. (2014, 2015), Stec et al. (2015), Cesari et al. (2016), Fujimoto et al. 2016, Gąsiorek et al. (2016), Mapalo et al. (2016) and Zawierucha et al. (2016a).

For PCR amplification of the 28S rRNA subunits protocols and primers, see Jørgensen et al. (2010, 2011), Guil and Giribet (2011), Guil et al. (2013), Kagoshima et al. (2013), Bertolani et al. (2014), Dabert et al. (2014, 2015), Guidetti et al. (2014), Fujimoto et al. (2016), Stec et al. (2015, 2016b), Gąsiorek et al. (2016), Mapalo et al. (2016) and Zawierucha et al. (2016a).

For PCR amplification of the internal transcriber spacer ITS2 protocols, see Jørgensen et al. (2007), Schill et al. (2010), Wełnicz et al. (2011), Michalczyk et al. (2012) and Stec et al. (2015).

It is important to prepare slide mounts of paragenophores (i.e. specimens collected from the same sample as those sequenced) and if possible slides with hologenophores (i.e. same specimens which were used for molecular analysis) (Pleijel et al. 2008). Tardigrade hologenophores can be egg shells from which embryos were used for sequencing or egg shells of newly hatched juveniles used for molecular analysis. Hologenophores may also be obtained via photographs of live animals, which will be used for molecular analyses, though it is difficult to attribute specimens to species via in vivo observations (Cesari et al. 2009, 2011, 2013; Bertolani et al. 2011b). Dabert et al. (2008) described and later Dabert et al. (2014) firstly used for Tardigrada a nondestructive method for extraction a total genomic DNA, while an exoskeleton of a specimen (thus hologenophore) can be mounted on a slide for further study.

14.5 Observation, Identification and Description of Tardigrada

Identification of tardigrades is based on observation and may require measuring taxonomically important characters of prepared or in vivo adults, juveniles and eggs using a light microscope (LM) and appropriate identification key(s). LM observation of tardigrades can be via Köhler and dark-field illumination but more usually with phase contrast (often essential) or a differential interference contrast. Live specimens rarely yield sufficient clarity for diagnosis, though useful for observing eyes and stylets that are often dissolved by mounting media. Thus preliminary observation of living specimens is a valuable practice (Jørgensen and Kristensen 2001).

Epifluorescence microscopy is useful for imaging the autofluorescence from tardigrade cuticle and interpreting structures which are not easily seen with other microscopic techniques (Perry et al. 2015). Confocal laser scanning microscope, with its three-dimensionally imaging is a useful addition to the optical examination of tardigrades (Russell et al. 2001).

Important structures can be measured via a calibrated ocular micrometer or attached digital camera. Optimal results are obtained via differential interference contrast as phase contrast measurements may be hampered by halo effect.

The first step of identification is using an identification key to higher (generic) taxa (e.g. Ramazzotti and Maucci 1983). For freshwater and limnoterrestrial tardigrades, use Pilato and Binda (2010), Degma (2010) and Trygvadóttir and Kristensen (2011) and for marine genera Fontoura et al. (2017).

Species-level keys include Ramazzotti and Maucci (1983), but more recent sources are available for *Bryodelphax* (Fontoura et al. 2008; Kaczmarek et al. 2012b), *Milnesium* (Morek et al. 2016a), *Bertolanius* (Hansen et al. 2017), *Doryphoribius* (Michalczyk and Kaczmarek 2010), *Pseudobiotus* (Chang et al. 2007), *Thulinius* (Kaczmarek et al. 2010), *Minibiotus* (Claxton 1998), *Paramacrobiotus* (Kaczmarek et al. 2017), *Dactylobiotus* (Kaczmarek et al. 2012a), *Diphascon pingue* species group (Fontoura and Pilato 2007), *Hypsibius dujardini* species group (Gąsiorek et al. 2018), *Mesobiotus harmsworthi* species group (Kaczmarek et al. 2011), *Macrobiotus hufelandi* species group (Kaczmarek and Michalczyk 2017) and the marine genera *Archechiniscus* and *Anisonyches* (Bartels et al. 2018), *Batillipes* (Santos et al. 2018), *Megastygarctides* (Ürkmez et al. 2018), *Styraconyx* (Kristensen and Higgins 1984a).

It is essential to check whether the taxonomic key used includes all currently described taxa and to complete the specimen identification by comparing with the descriptions and/or type specimens of those species missing from the key. The checklist of tardigrade taxa (Guidetti and Bertolani 2005) was updated (Degma and Guidetti 2007) and is available online as a continuously updated checklist (Degma et al. 2018).

Gender determination in Eutardigrada requires staining of the specimens (see above)—in Heterotardigrada the shape of gonoporus is sufficient, see e.g. Miller et al. (1999). Limno terrestrial species show limited sexual dimorphism, so this is rarely needed for taxonomy but provides additional data on the reproductive strategy of the taxa.

Tardigrade taxonomy is based on external and internal cuticular structures and percent ratios of specific structures (*pt*, *ptd* and *pbf* indices are basic indices—e.g. Guidetti and Bertolani 2005).

Descriptions of new taxa may require hand-drawn figures to highlight important structures which are poorly resolved in photomicrographs even after their adjustment via appropriate software. This can be achieved by using a *camera lucida* or drawing tube attached to a light microscope.

Traditional descriptions of new morphospecies have focused on qualitative characters, morphometry of type specimens and documentation (photomicrographs, electron micrographs and drawings). Needful number of measured type specimens depends on the accuracy of mean value of each character we want to achieve (Eckblad 1991; for estimating of optimal sample size, see also Stec et al. 2016a). Morphometric data can be recorded in own tables or using templates available from the Tardigrada Register (http://www.tardigrada.net/register/submit.htm; Michalczyk and Kaczmarek 2013). Bartels et al. (2011) demonstrated that morphometric traits in tardigrades are often allometric rather than isometric, and they recommended using a normalization technique to eliminate body size effects (for practical use of this technique, see also Meyer 2016). More recently, studies are using integrated taxonomy, which combines traditional alpha taxonomy with genetic analysis (i.e. sequences of four standard genes mentioned in previous section) (Guidetti et al. 2014; Stec et al. 2015; Gąsiorek et al. 2016).

Acknowledgments I thank Dr. Ralph Schill (University of Stuttgart) for inviting me to join the project and two anonymous reviewers for the valuable suggestions, which significantly helped me to improve the manuscript.

References

Accogli G, Gallo M, D'Addabbo R, Hansen JG (2011) Diversity and ecology of the marine tardigrades along the Apulian Coast. J Zool Syst Evol Res 49(Suppl. 1):53–57

Albuquerque EF, Brandão Pinto AP, d'Alcântara de Queiroz Perez A, Gomes Veloso V (2007) Spatial and temporal changes in interstitial meiofauna on a sandy ocean beach of South America. Braz J Oceanogr 55:121–131

Altiero T, Rebecchi L (2001) Rearing tardigrades: results and problems. Zool Anz 240:217–221

Altiero T, Giovannini I, Guidetti R, Rebecchi L (2015) Life history traits and reproductive mode of the tardigrade *Acutuncus antarcticus* under laboratory conditions: strategies to colonize the Antarctic environment. Hydrobiologia 761:277–291

Avdonina AM, Biserova NM, Bertolani R, Rebecchi L (2007) Ultrastructure of the digestive system of *Ramazzottius tribulosus* and *Macrobiotus richtersi* (Eutardigrada) in relationship with diet. J Limnol 66(Suppl. 1):5–11

Bąkowski M, Roszkowska M, Gawlak M, Kaczmarek Ł (2016) *Macrobiotus naskreckii* sp. nov., a new tardigrade (Eutardigrada: Macrobiotidae) of the *hufelandi* group from Mozambique. Ann Zool 66(2):155–164

Bartels PJ, Nelson DR (2006) A large-scale, multihabitat inventory of the phylum Tardigrada in the Great Smoky Mountains National Park, USA: a preliminary report. Hydrobiologia 558:111–118

Bartels PJ, Nelson DR, Exline RP (2011) Allometry and the removal of body size effects in the morphometric analysis of tardigrades. J Zool Syst Evol Res 49(Suppl. 1):17–25

Bartels PJ, Fontoura P, Nelson DR (2015) New records of marine tardigrades from Moorea, French Polynesia, with the description of *Styraconyx turbinarium* sp. nov. (Arthrotardigrada, Halechiniscidae). Zootaxa 3955(3):389–402

Bartels PJ, Fontoura P, Nelson DR (2018) Marine tardigrades of the Bahamas with the description of two new species and updated keys to the species of *Anisonyches* and *Archechiniscus*. Zootaxa 4420:43–70

Beltrán-Pardo E, Jönsson KI, Wojcik A, Haghdoost S, Harms-Ringdahl M, Bermúdez-Cruz RM, Bernal Villegas JE (2013) Effects of ionizing radiation on embryos of the tardigrade *Milnesium* cf. *tardigradum* at different stages of development. PLoS One 8(9):e72098. https://doi.org/10.1371/journal.pone.0072098

Bertolani R, Biserov V, Rebecchi L, Cesari M (2011a) Taxonomy and biogeography of tardigrades using an integrated approach: new results on species of the *Macrobiotus hufelandi* group. Invertebr Zool 8:23–36

Bertolani R, Rebecchi L, Giovannini I, Cesari M (2011b) DNA barcoding and integrative taxonomy of *Macrobiotus hufelandi* C.A.S. Schultze, 1834, the first tardigrade species to be described, and some related species. Zootaxa 2997:19–36

Bertolani R, Guidetti R, Marchioro T, Altiero T, Rebecchi L, Cesari M (2014) Phylogeny of Eutardigrada: New molecular data and their morphological support lead to the identification of new evolutionary lineages. Mol Phylogenet Evol 76:110–126

Bingemer J, Hohberg K, Schill RO (2016) First detailed observations on tardigrade mating behaviour and some aspects of the life history of *Isohypsibius dastychi* Pilato, Bertolani & Binda 1982 (Tardigrada, Isohypsibiidae). Zool J Linn Soc Lond 178:856–862

Boeckner MJ, Proctor HC (2005) Water-bears from Rocky mountains: a first look at Alberta's tardigrade fauna. Can Field Nat 119:586–588

Boesgaard TM, Kristensen RM (2001) Tardigrades from Australian marine caves. With a redescription of *Actinarctus neretinus* (Arthrotardigrada). Zool Anz 240:253–264

Calloway S, Miller WR, Johansson C, Whiting J (2011) Tardigrades of North America: *Oreella chugachii*, a new species (Heterotardigrada, Echiniscoide, Oreellidae) from Alaska. Proc Biol Soc Wash 124:28–39

Cesari M, Bertolani R, Rebecchi L, Guidetti R (2009) DNA barcoding in Tardigrada: the first case study on *Macrobiotus macrocalix* Bertolani & Rebecchi 1993 (Eutardigrada, Macrobiotidae). Mol Ecol Resour 9:699–706

Cesari M, Giovannini I, Bertolani R, Rebecchi L (2011) An example of problems associated with DNA barcoding in tardigrades: a novel method for obtaining voucher specimens. Zootaxa 3104:42–51

Cesari M, Guidetti R, Rebecchi L, Giovannini I, Bertolani R (2013) A DNA barcoding approach in the study of tardigrades. J Limnol 72(Suppl. 1):182–198

Cesari M, McInnes SJ, Bertolani R, Rebecchi L, Guidetti R (2016) Genetic diversity and biogeography of the south polar water bear *Acutuncus antarcticus* (Eutardigrada: Hypsibiidae) - evidence that it is a truly pan-Antarctic species. Invertebr Syst 30:635–649

Chang CY, Rho HS (1998) Three new tardigrade species associated with barnacles from the Thai coast of Andaman Sea. Korean J Biol Sci 2(3):323–331

Chang CY, Kaczmarek Ł, Lee JM, Michalczyk Ł (2007) *Pseudobiotus spinifer*, a new tardigrade species (Eutardigrada: Hypsibiidae) from Nakdong River, South Korea, with a redescription of *P. vladimiri* Biserov, Dudichev & Biserova. Zool Sci 24:623–629

Chappell BM, Parry D, Miller WR, Lowman MD (2015) Tardigrades of the canopy: *Doryphoribius gibber* Beasley and Pilato, 1987 (Eutardigrada: Parachela: Hypsibiidae) new records from eastern Kansas, U.S.A. Trans Kans Acad Sci 118:48–52

Claxton SK (1998) A revision of the genus *Minibiotus* (Tardigrada: Macrobiotidae) with descriptions of eleven new species from Australia. Rec Aust Mus 50:125–160

D'Addabbo R, Gallo M, De Leonardis C, Sandulli R, De Zio Grimaldi S (2007) Further studies on the marine tardigrade fauna from Sardinia (Italy). J Limnol 66(Suppl. 1):56–59

Dabert J, Ehrnsberger R, Dabert M (2008) *Glaucalges tytonis* sp. n. (Analgoidea, Xolalgidae) from the barn owl *Tyto alba* (Strigiformes, Tytonidae): compiling morphology with DNA barcode data for taxon descriptions in mites (Acari). Zootaxa 1719:41–52

Dabert M, Dastych H, Hohberg K, Dabert J (2014) Phylogenetic position of the enigmatic clawless eutardigrade genus *Apodibius* Dastych, 1983 (Tardigrada), based on 18S and 28S rRNA sequence data from its type species *A. confusus*. Mol Phylogenet Evol 70:70–75

Dabert M, Dastych H, Dabert J (2015) Molecular data support the dispersal ability of the glacier tardigrade *Hypsibius klebelsbergi* Mihelčič, 1959 across the environmental barrier (Tardigrada). Entomol Mitt Zool Mus Hambg 17(194):233–240

Dastych H (1980) Niesporczaki (Tardigrada) Tatrzańskiego parku narodowego. Monogr Fauny Pol 9:1–233

Dastych H (1984) The Tardigrada from Antarctic with descriptions of several new species. Acta Zool Cracov 27(19):377–436

Dastych H (1985) West Spitsbergen Tardigrada. Acta Zool Cracov 28(3):169–214

Dastych H (1992) *Paradiphascon manningi* gen. n. sp. n., a new water-bear from South Africa, with the erecting of a new subfamily Diphasconinae (Tardigrada). Mitt Hambg Zool Mus Inst 89:125–139

Dastych H, Kraus H, Thaler K (2003) Redescription and notes on the biology of the glacier tardigrade *Hypsibius klebelsbergi* Mihelčič, 1959 (Tardigrada), based on material from the Ötztal Alps, Austria. Mitt Hambg Zool Mus Inst 100:73–100

de Peluffo MCM, Peluffo JR, Rocha AM, Doma L (2006) Tardigrade distribution in a medium-sized city in Central Argentina. Hydrobiologia 558:141–150

Degma P (2010) Moss dwelling Tardigrada - from sampling to their identification. In: European Distributed Institute of Taxonomy. EDIT Summer School 2010, Madeira. http://edit.africamuseum.be/wp8_summer_school_elearning/mod/resource/view.php?id=55

Degma P, Guidetti R (2007) Notes to the current checklist of Tardigrada. Zootaxa 1579:41–53

Degma P, Katina S, Sabatovičová L (2011) Horizontal distribution of moisture and Tardigrada in a single moss cushion. J Zool Syst Evol Res 49(Suppl. 1):71–77

Degma P, Bertolani R, Guidetti R (2018) Actual checklist of Tardigrada species, p 48. http://www.tardigrada.modena.unimo.it/miscellanea/Actual%20checklist%20of%20Tardigrada.pdf. Accessed 20 Aug 2018

Eckblad JV (1991) How many samples should be taken? Bioscience 41:346–348

Eibye-Jacobsen J (1996) On the nature of pharyngeal muscle-cells in the Tardigrada. Zool J Linn Soc Lond 116:123–138

Eibye-Jacobsen J (1997) Development, ultrastructure and function of the pharynx of *Halobiotus crispae* Kristensen, 1982 (Eutardigrada). Acta Zool (Stockh) 78(4):329–347

Eibye-Jacobsen J (2001) A new method for making SEM preparations on the tardigrade buccopharyngeal apparatus. Zool Anz 240:309–319

Faurby S, Jönsson KI, Rebecchi L, Funch P (2008) Variation in anhydrobiotic survival of two eutardigrade morphospecies: a story of cryptic species and their dispersal. J Zool 275:139–145

Faurby S, Jørgensen A, Kristensen RM, Funch P (2011) Phylogeography of North Atlantic intertidal tardigrades: refugia, cryptic speciation and the history of the Mid-Atlantic Islands. J Biogeogr 38:1613–1624

Faurby S, Jørgensen A, Kristensen RM, Funch P (2012) Distribution and speciation in marine intertidal tardigrades: testing the roles of climatic and geographical isolation. J Biogeogr 39 (9):1596–1607

Folmer O, Black M, Hoeh W, Lutz R, Vrijenhoek R (1994) DNA primers for amplification of mitochondrial cytochrome *c* oxidase subunit I from diverse metazoan invertebrates. Mol Mar Biol Biotechnol 3(5):294–299

Fontoura P, Pilato G (2007) *Diphascon* (*Diphascon*) *faialense* sp. nov. a new species of Tardigrada (Eutardigrada, Hypsibiidae) from the Azores and a key to the species of the *D. pingue* group. Zootaxa 1589:47–55

Fontoura P, Pilato G, Lisi O (2008) Echiniscidae (Tardigrada, Heterotardigrada) from Faial and Pico Islands, the Azores, with the description of two new species. Zootaxa 1693:49–61

Fontoura P, Bartels PJ, Jørgensen A, Kristensen RM, Hansen JG (2017) A dichotomous key to the genera of the Marine Heterotardigrades (Tardigrada). Zootaxa 4294:1–45

Förster F, Beisser D, Grohme MA, Liang C, Mali B, Siegl AM, Engelmann JC, Shkumatov AV, Schokraie E, Müller T, Schnölzer M, Schill RO, Frohme M, Dandekar T (2012) Transcriptome analysis in tardigrade species reveals specific molecular pathways for stress adaptations. Bioinform Biol Insights 6:69–96

Fujimoto S (2015) *Quisarctus yasumurai* gen. et sp. nov. (Arthrotardigrada: Halechiniscidae) from a submarine cave, off Iejima, Ryukyu Islands, Japan. Zootaxa 3948(1):145–150

Fujimoto S, Miyazaki K, Suzuki AC (2013) A new marine tardigrade, *Tanarctus diplocerus* (Arthrotardigrada: Halechiniscidae) from Japan. J Mar Biol Assoc U K 93(4):955–961

Fujimoto S, Jørgensen A, Hansen JG (2016) A molecular approach to arthrotardigrade phylogeny (Heterotardigrada, Tardigrada). Zool Scr 46:496–505

Gabriel WN, McNuff R, Patel SK, Ryan Gregory T, Jeck WR, Jones CD, Goldstein B (2007) The tardigrade *Hypsibius dujardini*, a new model for studying the evolution of development. Dev Biol 312:545–559

Gallo D'Addabbo M, De Zio Grimaldi S, Sandulli R (2001) Heterotardigrada of two submarine caves in S. Domino Island (Tremiti Islands) in the Mediterranean Sea with the description of two new species of Stygarctidae. Zool Anz 240:361–369

Gallo D'Addabbo M, Sandulli R, de Zio Grimaldi S (2005) A new Batillipedidae (Tardigrada, Heterotardigrada) from the Orosei Gulf, Sardinia, Tyrrhenian Sea. Zool Anz 243:219–225

Gąsiorek P, Stec D, Morek W, Zawierucha K, Kaczmarek Ł, Lachowska-Cierlik D, Michalczyk Ł (2016) An integrative revision of *Mesocrista* Pilato, 1987 (Tardigrada: Eutardigrada: Hypsibiidae). J Nat Hist 50(45–46):2803–2828

Gąsiorek P, Stec D, Morek W, Michalczyk Ł (2018) An integrative redescription of *Hypsibius dujardini* (Doyère, 1840), the nominal taxon for Hypsibioidea (Tardigrada: Eutardigrada). Zootaxa 4415:45–75

Gibson JAE, Cromer L, Agius JT, McInnes SJ, Marley NJ (2007) Tardigrade eggs and exuviae in Antarctic lake sediments: insights into Holocene dynamics and origins of the fauna. J Limnol 66 (Suppl. 1):65–71

Greven H (2007) Comments on the eyes of tardigrades. Arthropod Struct Dev 36:401–407

Greven H, Dastych H, Kraus H (2005) Notes on the integument of the glacier-dwelling tardigrade *Hypsibius klebelsbergi* Mihelcic, 1959 (Tardigrada). Mitt Hambg Zool Mus Inst 102:11–20

Gross V, Miller WR, Hochberg R (2014) A new tardigrade, *Mutaparadoxipus duodigifinis* gen. nov., sp. nov. (Heterotardigrada: Arthrotardigrada), from the Southeastern United States. Zootaxa 3835(2):263–272

Grothman GT (2011) Tardigrades of Fish Creek Provincial Park, Alberta, Canada: a preliminary survey. Can Field Nat 125:22–26

Guidetti R, Bertolani R (2001) The tardigrades of Emilia (Italy). III. Piane di Mocogno (Northern Apennines). Zool Anz 240:377–383

Guidetti R, Bertolani R (2005) Tardigrade taxonomy: an updated check list of the taxa and a list of characters for their identification. Zootaxa 845:1–46

Guidetti R, Rebecchi L, Bertolani R (2000) Cuticle structure and systematics of the Macrobiotidae (Tardigrada, Eutardigrada). Acta Zool (Stockh) 81:27–36

Guidetti R, Gandolfi A, Rossi V, Bertolani R (2005) Phylogenetic analysis of Macrobiotidae (Eutardigrada, Parachela): a combined morphological and molecular approach. Zool Scr 34:235–244

Guidetti R, Schill RO, Bertolani R, Dandekar T, Wolf M (2009) New molecular data for tardigrade phylogeny, with the erection of *Paramacrobiotus* gen. nov. J Zool Syst Evol Res 47:315–321

Guidetti R, Bertolani R, Rebecchi L (2013a) Comparative analysis of the tardigrade feeding apparatus: adaptive convergence and evolutionary pattern of the piercing stylet system. J Limnol 72(Suppl. 1):24–35

Guidetti R, Peluffo JR, Rocha AA, Cesari M, Moly de Peluffo MC (2013b) The morphological and molecular analyses of a new South American urban tardigrade offer new insights on the biological meaning of the *Macrobiotus hufelandi* group of species (Tardigrada: Macrobiotidae). J Nat Hist 47(37–38):2409–2426

Guidetti R, Rebecchi L, Cesari M, McInnes SJ (2014) *Mopsechiniscus franciscae*, a new species of a rare genus of Tardigrada from continental Antarctica. Polar Biol 37:1221–1233

Guidetti R, Rebecchi L, Bertolani R, Jönsson KI, Kristensen RM, Cesari M (2016) Morphological and molecular analyses on *Richtersius* (Eutardigrada) diversity reveal its new systematic position and lead to the establishment of a new genus and a new family within Macrobiotoidea. Zool J Linn Soc Lond 178:834–845

Guil N, Giribet G (2009) Fine scale population structure in the *Echiniscus blumi-canadensis* series (Heterotardigrada, Tardigrada) in an Iberian mountain range—when morphology fails to explain genetic structure. Mol Phylogenet Evol 53:606–613

Guil N, Giribet G (2011) A comprehensive molecular phylogeny of tardigrades – adding genes and taxa to a poorly resolved phylum-level phylogeny. Cladistics 27:1–29

Guil N, Guidetti R, Machordom A (2007) Observations on the "*tenuis* group" (Eutardigrada, Macrobiotidae) and description of a new *Macrobiotus* species. J Nat Hist 41(41–44):2741–2755

Guil N, Jørgensen A, Giribet G, Kristensen RM (2013) Congruence between molecular phylogeny and cuticular design in Echiniscoidea (Tardigrada, Heterotardigrada). Zool J Linn Soc Lond 169:713–736

Hansen JG, Jørgensen A, Kristensen RM (2001) Preliminary studies of the tardigrade fauna of the Faroe Bank. Zool Anz 240:385–393

Hansen JG, Kristensen RM, Bertolani R, Guidetti R (2017) Comparative analyses of *Bertolanius* species (Eohypsibiidae; Eutardigrada) with the description of *Bertolanius birnae* sp. nov. from northern polar regions. Polar Biol 40:123–140

Harada H, Ito MT (2006) Soil-inhabiting tardigrade communities in forests of Central Japan. Hydrobiologia 558:119–127

Hejnol A, Schnabel R (2005) The eutardigrade *Thulinia stephaniae* has an indeterminate development and the potential to regulate early blastomere ablations. Development 132:1349–1361

Hengherr S, Worland MR, Reuner A, Brümmer F, Schill RO (2009) Freeze tolerance, supercooling points and ice formation: comparative studies on the subzero temperature survival of limno-terrestrial tardigrades. J Exp Biol 212:802–807

Hohberg K, Russell DJ, Elmer M (2011) Mass occurrence of algal-feeding tardigrade *Apodibius confusus*, in the young soils of a post-mining site. J Zool Syst Evol Res 49(Suppl. 1):62–65

Horikawa DD, Kunieda T, Abe W, Watanabe M, Nakahara Y, Yukuhiro F, Sakashita T, Hamada N, Wada S, Funayama T, Katagiri C, Kobayashi Y, Higashi S, Okuda T (2008) Establishment of a rearing system of the extremotolerant tardigrade *Ramazzottius varieornatus*: a new model animal for astrobiology. Astrobiology 8:549–556

Horning DS, Schuster RO, Grigarick AA (1978) Tardigrada of New Zealand. N Z J Zool 5:185–280

Hygum TL, Clausen LKB, Halberg KA, Jørgensen A, Møbjerg N (2016) Tun formation is not a prerequisite for desiccation tolerance in the marine tidal tardigrade *Echiniscoides sigismundi*. Zool J Linn Soc Lond 178:907–911

Hyra M, Poprawa I, Włodarczyk A, Student S, Sonakowska L, Kszuk-Jendrysik M, Rost-Roszkowska MM (2016) Ultrastructural changes in the midgut epithelium of *Hypsibius dujardini* (Doyère, 1840) (Tardigrada, Eutardigrada, Hypsibiidae) in relation to oogenesis. Zool J Linn Soc Lond 178:897–906

Ito M, Saigo T, Abe W, Kubo T, Kunieda T (2016) Establishment of an isogenic strain of the desiccation-sensitive tardigrade *Isohypsibius myrops* (Parachela, Eutardigrada) and its life history traits. Zool J Linn Soc Lond 178:863–870

Janiec K (1996) The comparison of freshwater invertebrates of Spitsbergen (Arctic) and King George Island (Antarctic). Pol Polar Res 17:173–202

Jørgensen A, Kristensen RM (2001) A new tanarctid arthrotardigrade with buoyant bodies. Zool Anz 240:425–439

Jørgensen A, Møbjerg N, Kristensen RM (2007) A molecular study of the tardigrade *Echiniscus testudo* (Echiniscidae) reveals low DNA sequence diversity over a large geographical area. J Limnol 66(Suppl. 1):77–83

Jørgensen A, Faurby S, Hansen JG, Møbjerg N, Kristensen RM (2010) Molecular phylogeny of Arthrotardigrada (Tardigrada). Mol Phylogenet Evol 54:1006–1015

Jørgensen A, Møbjerg N, Kristensen RM (2011) Phylogeny and evolution of the Echiniscidae (Echiniscoidea, Tardigrada) – an investigation of the congruence between molecules and morphology. J Zool Syst Evol Res 49(Suppl. 1):6–16

Kaczmarek Ł, Michalczyk Ł (2017) The *Macrobiotus hufelandi* group (Tardigrada) revisited. Zootaxa 4363:101–123

Kaczmarek Ł, Bertolani R, Nędzyńska-Stygar M (2010) *Thulinius saltursus* comb. nov.: a new systematic position for *Isohypsibius saltursus* Schuster, Toftner & Grigarick, 1978 (Eutardigrada: Hypsibiidae) and a key for the genus *Thulinius*. Zootaxa 2483:58–64

Kaczmarek Ł, Gołdyn B, Prokop ZM, Michalczyk Ł (2011) New records of Tardigrada from Bulgaria with the description of *Macrobiotus binieki* sp. nov. (Eutardigrada: Macrobiotidae) and a key to the species of the *harmsworthi* group. Zootaxa 2781:29–39

Kaczmarek Ł, Schabetsberger R, Litwin M, Michalczyk Ł (2012a) A new freshwater eutardigrade from Fiji and Vanuatu (Oceania), with remarks on the genus *Dactylobiotus*. N Z J Zool 39:311–318

Kaczmarek Ł, Zawierucha K, Smykla J, Michalczyk Ł (2012b) Tardigrada of the Revdalen (Spitsbergen) with the descriptions of two new species: *Bryodelphax parvuspolaris* (Heterotardigrada) and *Isohypsibius coulsoni* (Eutardigrada). Polar Biol 35:1013–1026

Kaczmarek Ł, Michalczyk Ł, McInnes SJ (2014) Annotated zoogeography of non-marine Tardigrada. Part I: Central America. Zootaxa 3763(1):1–62

Kaczmarek Ł, Bartels PJ, Roszkowska M, Nelson DR (2015a) The zoogeography of marine Tardigrada. Zootaxa 4037:1–189

Kaczmarek Ł, Michalczyk Ł, McInnes SJ (2015b) Annotated zoogeography of non-marine Tardigrada. Part II: South America. Zootaxa 3923:1–107

Kaczmarek Ł, Michalczyk Ł, McInnes SJ (2016) Annotated zoogeography of non-marine Tardigrada. Part III: North America and Greenland. Zootaxa 4203:1–249

Kaczmarek Ł, Gawlak M, Bartels PJ, Nelson DR, Roszkowska M (2017) Revision of the genus *Paramacrobiotus* Guidetti et al., 2009 with the description of a new species, re-descriptions and a key. Ann Zool 67:627–656

Kagoshima H, Imura S, Suzuki AC (2013) Molecular and morphological analysis of an Antarctic tardigrade, *Acutuncus antarcticus*. J Limnol 72(Suppl. 1):15–23

Kharkevych KO, Sergeeva NG (2013) Deep-water Tardigrada of the Istanbul strait's (Bosporus) outlet area of the Black Sea. Vestn Zool 47(3):17–27

Kiehl E, Dastych H, D'Haese J, Greven H (2007) The 18S rDNA sequences support polyphyly of the Hypsibiidae (Eutardigrada). J Limnol 66(Suppl. 1):21–25

Kosztyła P, Stec D, Morek W, Gąsiorek P, Zawierucha K, Michno K, Ufir K, Małek D, Hlebowicz K, Laska A, Dudziak M, Frohme M, Prokop ZM, Kaczmarek Ł, Michalczyk Ł (2016) Experimental taxonomy confirms the environmental stability of morphometric traits in a taxonomically challenging group of microinvertebrates. Zool J Linn Soc Lond 178:765–775

Kristensen RM, Hallas TE (1980) The tidal genus *Echiniscoides* and its variability, with erection of Echiniscoididae fam. n. (Tardigrada). Zool Scr 9:113–127

Kristensen RM, Higgins RP (1984a) Revision of *Styraconyx* (Tardigrada: Halechiniscidae), with descriptions of two new species from Disko Bay, West Greenland. Smithson Contrib Zool 391:1–40

Kristensen RM, Higgins RP (1984b) A new family of Arthrotardigrada (Tardigrada: Heterotardigrada) from the Atlantic coast of Florida, U.S.A. Trans Am Microsc Soc 103:295–311

Kristensen RM, Higgins RP (1989) Marine Tardigrada from the Southeastern United States coastal waters I. *Paradoxipus orzeliscoides* n. gen., n. sp. (Arthrotardigrada: Halechiniscidae). Trans Am Microsc Soc 108(3):262–282

Mapalo MA, Stec D, Mirano-Bascos D, Michalczyk Ł (2016) *Mesobiotus philippinicus* sp. nov., the first limnoterrestrial tardigrade from the Philippines. Zootaxa 4126(3):411–426

McInnes SJ (1994) Zoogeographic distribution of terrestrial/freshwater tardigrades from current literature. J Nat Hist 28(2):257–352

McInnes SJ, Convey P (2005) Tardigrade fauna of the South Sandwich Islands, maritime Antarctic. Zootaxa 1058:43–49

McInnes SJ, Ellis-Evans JC (1987) Tardigrades from maritime Antarctic freshwater lakes. In: Bertolani R (ed) Biology of tardigrades, Selected Symposia and Monographs U.Z.I. 1. Mucchi, Modena, pp 111–123

Melo JC, Beltrán-Pardo E, Bernal JE, Kaczmarek Ł (2015) New records of tardigrades from Colombia (Guatavita, Cundinamarca Department). Turk J Zool 39:412–420
Meyer HA (2009) Tardigrades of the Florida Keys. Fla Sci 72(2):142–146
Meyer HA (2013) Terrestrial and freshwater Tardigrada of the Americas. Zootaxa 3747(1):1–71
Meyer HA (2016) Re-description of *Echiniscus cavagnaroi* Schuster & Grigarick, 1966 (Tardigrada: Heterotardigrada: Echiniscoidea: Echiniscidae) from type material, with new records from Hawaii and Bermuda. Zootaxa 4121(5):575–582
Michalczyk Ł, Kaczmarek Ł (2010) Description of *Doryphoribius dawkinsi*, a new species of Tardigrada (Eutardigrada: Hypsibiidae) from the Costa Rican highlands, with the key to the genus *Doryphoribius*. Zootaxa 2393:46–58
Michalczyk Ł, Kaczmarek Ł (2013) The Tardigrada Register: a comprehensive online data repository for tardigrade taxonomy. J Limnol 72(Suppl. 1):175–181
Michalczyk Ł, Wełnicz W, Frohme M, Kaczmarek Ł (2012) Redescriptions of three *Milnesium* Doyère, 1840 taxa (Tardigrada: Eutardigrada: Milnesiidae), including the nominal species for the genus. Zootaxa 3154:1–20
Miller WR, Miller JD, Heatwole HF (1994) Tardigrades of the Australian Antarctic territories: assessing diversity within a sample. Mem Queensl Mus 36:137–145
Miller WR, Claxton SK, Heatwole HF (1999) Tardigrades of the Australian Antarctic Territories: males in the genus *Echiniscus* (Tardigrada: Heterotardigrada). Zool Anz 238:303–309
Morek W, Gąsiorek P, Stec D, Blagden B, Michalczyk Ł (2016a) Experimental taxonomy exposes ontogenetic variability and elucidates the taxonomic value of claw configuration in *Milnesium* Doyère, 1840 (Tardigrada: Eutardigrada: Apochela). Contrib Zool 85:173–200
Morek W, Stec D, Gąsiorek P, Schill R, Kaczmarek Ł, Michalczyk Ł (2016b) An experimental test of eutardigrade preparation methods for light microscopy. Zool J Linn Soc Lond 178:785–793
Nelson DR, Bartels PJ (2013) Species richness of soil and leaf litter tardigrades in the Great Smoky Mountains National Park (North Carolina/Tennessee, USA). J Limnol 72(Suppl. 1):144–151
Nelson DR, Guidetti R, Rebecchi L (2015) Phylum Tardigrada. In: Thorp J, Rogers DC (eds) Ecology and general biology: Thorp and Covich's freshwater invertebrates. Academic Press, pp 347–380
Nichols PB, Romano FA, Nelson DR (2001) Seasonal and altitudinal variation in the distribution and abundance of Tardigrada on Dugger Mountain, Alabama. Zool Anz 240:501–504
Nichols PB, Nelson DR, Garey JR (2006) A family level analysis of tardigrade phylogeny. Hydrobiologia 558:53–60
Nkem JN, Wall DH, Virginia RA, Barrett JE, Broos EJ, Porazinska DL, Adams BJ (2006) Wind dispersal of soil invertebrates in the McMurdo Dry Valleys, Antarctica. Polar Biol 29:346–352
Perry ES, Miller WR (2015) *Echiniscoides wyethi*, a new marine tardigrade from Maine, U.S.A. (Heterotardigrada: Echiniscoidea: Echiniscoididae). Proc Biol Soc Wash 128:103–110
Perry ES, Miller WR, Lindsay S (2015) Looking at tardigrades in a new light: using epifluorescence to interpret structure. J Microsc 257:117–122
Pilato G, Binda MG (2010) Definition of families, subfamilies, genera and subgenera of the Eutardigrada, and keys to their identification. Zootaxa 2404:1–54
Pilato G, Lisi O, Binda MG (2010) Tardigrades of Israel with description of four new species. Zootaxa 2665:1–28
Pleijel F, Jondelius U, Norlinder E, Nygren A, Oxelman B, Schander C, Sundberg P, Thollesson M (2008) Phylogenies without roots? A plea for the use of vouchers in molecular phylogenetic studies. Mol Phylogenet Evol 48:369–371
Pollock LW (1989) Marine interstitial Heterotardigrada from the Pacific coast of the United States, including a description of *Batillipes tridentatus* n. sp. Trans Am Microsc Soc 108:169–189
Poprawa I (2005a) The ovary structure, previtellogenic and vitellogenic stages in parthenogenetic species *Dactylobiotus dispar* (Murray, 1907) (Tardigrada: Eutardigrada). Tissue Cell 37:385–392

Poprawa I (2005b) The structure and the formation of egg shells in the parthenogenetic species *Dactylobiotus dispar* Murray, 1907 (Tardigrada: Eutardigrada). Folia Biol (Cracow) 53 (3–4):173–177

Poprawa I (2011) Ultrastructural studies of the formation of the egg capsule in the hermaphroditic species, *Isohypsibius granulifer granulifer* Thulin, 1928 (Eutardigrada: Hypsibiidae). Zool Sci 28:37–40

Ramazzotti G, Maucci W (1983) Il phylum Tardigrada (III edizione riveduta e aggiornata). Mem Ist Ital Idrobiol 41:1–1016

Rebecchi L, Guidi A, Bertolani R (2000) Maturative pattern of the ovotestis in two hermaphrodite species of eutardigrades. Invertebr Reprod Dev 37:25–34

Rebecchi L, Boschini D, Cesari M, Lencioni V, Bertolani R, Guidetti R (2009) Stress response of a boreo-alpine species of tardigrade, *Borealibius zetlandicus* (Eutardigrada, Hypsibiidae). J Limnol 68:64–70

Rho HS, Moon SY, Chang CY, Kim W (1997) Four freshwater eutardigrades from Korea with description of *Isohypsibius brevitubulatus* n. sp. Korean J Biol Sci 1:271–279

Robinson K, Pugh PJA, McInnes SJ, Llewellyn PJ, Shackley SE (1996) Cryo-preparation of small or lightly attached biological specimens. Microsc Anal November:13

Rocha AM, Izaguirre MF, Moly de Peluffo MC, Peluffo JR, Casco VH (2007) Ultrastructure of the cuticle of *Echiniscus* rufoviridis [Du Bois - Raymond Marcus, (1944) Heterotardigrada]. Acta Microsc 16:16–21

Romano F, Gallo M, D'Addabbo R, Accogli G, Baguley J, Montagna P (2011) Deep-sea tardigrades in the northern Gulf of Mexico with a description of a new species of Coronarctidae (Tardigrada: Arthrotardigrada), *Coronarctus mexicus*. J Zool Syst Evol Res 49(Suppl. 1):48–52

Rost-Roszkowska MM, Poprawa I, Hyra M, Marek-Swędzioł M, Kaczmarek Ł (2013) The fine structure of the midgut epithelium in *Xerobiotus pseudohufelandi* (Iharos, 1966) (Tardigrada, Eutardigrada, Macrobiotidae). J Limnol 72(Suppl. 1):54–61

Rubal M, Veiga P, Fontoura P, Sousa-Pinto I (2013) A new intertidal arthrotardigrade, *Prostygarctus aculeatus* gen. nov., sp. nov. (Tardigrada: Heterotardigrada) from the North of Portugal (Atlantic Ocean). J Limnol 72(Suppl. 1):8–14

Russell PM, Marley NJ, Hockings MF (2001) Do confocal microscopy and tardigrades have a future together? Zool Anz 240:543–548

Sands CJ, Convey P, Linse K, McInnes SJ (2008a) Assessing meiofaunal variation among individuals utilising morphological and molecular approaches: an example using the Tardigrada. BMC Ecol 8:7 http://www.biomedcentral.com/1472-6785/8/7

Sands CJ, McInnes SJ, Marley NJ, Goodall-Copestake WP, Convey P, Linse K (2008b) Phylum Tardigrada: an "individual" approach. Cladistics 24:861–871

Santos E, Rubal M, Veiga P, da Rocha CMC, Fontoura P (2018) *Batillipes* (Tardigrada, Arthrotardigrada) from the Portuguese coast with the description of two new species and a new dichotomous key for all species. Eur J Taxon 425:1–32

Schill RO (2007) Comparison of different protocols for DNA preparation and PCR amplification of mitochondrial genes of tardigrades. J Limnol 66(Suppl. 1):164–170

Schill RO (2013) Life-history traits in the tardigrade species *Paramacrobiotus kenianus* and *Paramacrobiotus palaui*. J Limnol 72(Suppl. 1):160–165

Schill RO, Steinbrück G (2007) Identification and differentiation of Heterotardigrada and Eutardigrada species by riboprinting. J Zool Syst Evol Res 45:184–190

Schill RO, Förster F, Dandekar T, Wolf M (2010) Using compensatory base change analysis of internal transcribed spacer 2 secondary structures to identify three new species in *Paramacrobiotus* (Tardigrada). Org Divers Evol 10:287–296

Schram MD, Davison PG (2012) Irwin loops - a history and method of constructing homemade loops. Trans Kans Acad Sci 115:35–40

Sobczyk M, Michno K, Kosztyła P, Stec D, Michalczyk Ł (2015) Tolerance to ammonia of *Thulinius ruffoi* (Bertolani, 1981), a tardigrade isolated from a sewage treatment plant. Bull Environ Contam Toxicol 95:721–727

Stec D, Smolak R, Kaczmarek Ł, Michalczyk Ł (2015) An integrative description of *Macrobiotus paulinae* sp. nov. (Tardigrada: Eutardigrada: Macrobiotidae: *hufelandi* group) from Kenya. Zootaxa 4052(2):501–526

Stec D, Gąsiorek P, Morek W, Kosztyła P, Zawierucha K, Michno K, Kaczmarek Ł, Prokop ZM, Michalczyk Ł (2016a) Estimating optimal sample size for tardigrade morphometry. Zool J Linn Soc Lond 178:776–784

Stec D, Morek W, Gąsiorek P, Kaczmarek Ł, Michalczyk Ł (2016b) Determinants and taxonomic consequences of extreme egg shell variability in *Ramazzottius subanomalus* (Biserov, 1985) (Tardigrada). Zootaxa 4208(2):176–188

Steiner WA (1994) The influence of air pollution on moss-dwelling animals: 2. Aquatic fauna with emphasis on Nematoda and Tardigrada. Rev Suisse Zool 101:699–724

Suzuki AC (2003) Life history of *Milnesium tardigradum* Doyère (Tardigrada) under a rearing environment. Zool Sci 20:49–57

Suzuki AC (2008) Appearance of males in a thelytokous strain of *Milnesium* cf. *tardigradum* (Tardigrada). Zool Sci 25(8):849–853

Tałanda J, Zawierucha K (2017) First record of water bears (Tardigrada) from sponges (Porifera). Turk J Zool 41:161–163

Trygvadóttir BV, Kristensen RM (2011) Eohypsibiidae (Eutardigrada, Tardigrada) from the Faroe Islands with the description of a new genus containing three new species. Zootaxa 2886:39–62

Tsujimoto M, Suzuki AC, Imura S (2015) Life history of the Antarctic tardigrade, *Acutuncus antarcticus*, under a constant laboratory environment. Polar Biol 38:1575–1581

Tumanov DV (2004) *Macrobiotus kovalevi*, a new species of Tardigrada from New Zealand (Eutardigrada, Macrobiotidae). Zootaxa 406:1–8

Ürkmez D, Ostrowska M, Roszkowska M, Gawlak M, Zawierucha K, Kristensen RM, Kaczmarek Ł (2018) Description of *Megastygarctides sezginii* sp. nov. (Tardigrada: Arthrotardigrada: Stygarctidae) from the Turkish Black Sea coast and a key to the genus *Megastygarctides*. Mar Biol Res 14:1–16

Van Rompu EA, De Smet WH (1991) Contribution to the fresh water Tardigrada from Barentsøya, Svalbard (78°30'N). Fauna Nor Ser A 12:29–39

Vicente F, Michalczyk Ł, Kaczmarek Ł, Boavida M-J (2008) Observations on *Pyxidium tardigradum* (Ciliophora), a protozoan living on Eutardigrada: infestation, morphology and feeding behaviour. Parasitol Res 103:1323–1331

Vicente F, Fontoura P, Cesari M, Rebecchi L, Guidetti R, Serrano A, Bertolani R (2013) Integrative taxonomy allows the identification of synonymous species and the erection of a new genus of Echiniscidae (Tardigrada, Heterotardigrada). Zootaxa 3613(6):557–572

Villora-Moreno S (1996) A new genus and species of the deep-sea family Coronarctidae (Tardigrada) from a submarine cave with a deep-sea like condition. Sarsia 81(4):275–283

Wełnicz W, Grohme MA, Kaczmarek Ł, Schill RO, Frohme M (2011) ITS-2 and 18S rDNA data from *Macrobiotus polonicus* and *Milnesium tardigradum* (Eutardigrada, Tardigrada). J Zool Syst Evol Res 49(Suppl. 1):34–39

Zawierucha K, Kolicka M, Kaczmarek Ł (2016a) Re-description of the Arctic tardigrade *Tenuibiotus voronkovi* (Tumanov, 2007) (Eutardigrada; Macrobiotidea), with the first molecular data for the genus. Zootaxa 4196(4):498–510

Zawierucha K, Ostrowska M, Vonnahme TR, Devetter M, Nawrot AP, Lehmann S, Kolicka M (2016b) Diversity and distribution of Tardigrada in Arctic cryoconite holes. J Limnol 75 (3):545–559

Chapter 15
Tardigrade Taxa

Peter Degma and Roberto Guidetti

Abstract The list of tardigrade taxa up to a sub-generic level, with the type species of each genus and the numbers of lower taxa in each taxon (known in the end of June 2018) is here presented together with the main environments in which taxa were found.

Each listed taxon is characterized by characters common to all lower taxa using predominantly most recent taxa definitions. Taxa definitions were eventually adjusted using original descriptions.

This review of tardigrade taxa underlines as taxonomy and systematics of the phylum are continuously updated by researchers and new species are constantly discovered. Currently, there are 1265 species (2 of them fossil) of Tardigrada belonging to 129 genera and 28 families including one fossil.

15.1 Introduction

More than one thousand Tardigrada species were included in the Tardigrada checklist (Guidetti and Bertolani 2005) with its additions and corrections (Degma and Guidetti 2007). Both sources are joined and continuously updated in the Actual checklist of Tardigrada (Degma et al. 2018) which is freely available for all users on Internet. 1265 species (2 of them fossil) of Tardigrada (129 genera and 28 families including one fossil) were known in the end of June 2018 (Degma et al. 2018).

The list of tardigrade taxa up to a sub-generic level, with the type species of each genus and the numbers of lower taxa in each taxon, arranged according to Degma et al. (2018) is here presented. Each taxon is characterized by characters common to all lower taxa using predominantly most recent taxa definitions. Taxa definitions

P. Degma (✉)
Department of Zoology, Comenius University in Bratislava, Bratislava, Slovakia
e-mail: degma1@uniba.sk

R. Guidetti
Department of Life Sciences, University of Modena and Reggio Emilia, Modena, Italy
e-mail: roberto.guidetti@unimore.it

R. O. Schill (ed.), *Water Bears: The Biology of Tardigrades*, Zoological Monographs 2,
https://doi.org/10.1007/978-3-319-95702-9_15

were eventually adjusted using original descriptions of species. Sources of characterization of each taxon are cited (species descriptions are cited only if they improved the taxon characterization). The names of claw types and types of bucco-pharyngeal apparatuses in the class Eutardigrada are taken from Pilato and Binda (2010), but in adjusted form (some characters are written out of the type description). The main environments in which taxa were found are reported.

15.2 List of Taxa of the Phylum Tardigrada Doyère, 1840 Above Species Level

Within braces {taxon definition}; within parentheses (references); within brackets [number of lower taxa within the considered taxon; for each genus, the type species is also reported]

15.2.1 Class Heterotardigrada Marcus, 1927

marine, limnic and terrestrial species
{Cephalic appendages, lateral cirri *A* included, present. Gonopore separated from anus. Malpighian tubules lacking. Placoids in the pharynx consisting of three $CaCO_3$ elements or three delicate, bar-shaped cuticular structures.} (Kristensen 1987)

15.2.1.1 Order Arthrotardigrada Marcus, 1927

marine species, one limnic species
{Median cirrus usually present, rarely absent. Extremities of the leg digitate, or not digitate, but in such cases the claws fixed directly onto leg extremities and not on papillae.} (Ramazzotti and Maucci 1983)

Archechiniscidae Binda, 1978 [1 genus]
{Without armor but with thin dorsal metameric plates present; cuticle finely pointed. Median cirrus absent or vestigial. Four claws present: internal claws inserted on digits, and bearing hook-like dorsal points; external claws directly inserted on the tarsus, and with a powerful internal accessory point (secondary point).} (Binda 1978; Fontoura et al. 2017)

***Archechiniscus* Schulz, 1953** [*A. marci* Schulz, 1953; 5 species]
{as the family} (for a more detailed diagnosis see Fontoura et al. 2017)

Batillipedidae Ramazzotti, 1962 [1 genus]
{Trapezoid head divided into three parts: two lateroventral triangular lobes bearing all the cephalic cirri (simple with lance-like or frayed tips) and the primary (tubular) and secondary (papillar) clavae, and a central lobate portion with the medial cirrus. Stylet supports present. Legs with four/six digits, according to the age, bearing terminal adhesive discs. Claws absent. Sense organs present on all the legs. Cirrus *E* dorsal to the hind legs. A longitudinal groove generally present between the rosette-like female gonopore and the anus. Single internal seminal receptacle sometimes present, opposite the oviduct. Circular male gonopore covered with a crescent fold. Epicuticular pillars present, dorsally higher than ventrally.} (Gallo D'Addabbo et al. 2005)

Batillipes **Richters, 1909** [*B. mirus* Richters, 1909; 36 species]
{Four stalked digits bearing adhesive ends in the first juvenile instar and six digits in subsequent instars. Cephalic cirri simple (not comprised of distinct sections), but well-developed. Secondary clavae, if present, as dome-shaped papillae. Gonopores dimorphic.} (Pollock 1989)

Coronarctidae Renaud-Mornant, 1974 [2 genera]
{Cylindrical worm-like body. Trunk cuticle smooth, with folds, bearing three somatic cirri (*B*, *C* and *E*) or only one (*E*). Eleven cephalic sense organs present, including large flattened secondary clavae (occasionally occurring as indistinct areas). Three or four claws connected to foot by membranes.} (Villora-Moreno 1996)

Coronarctus **Renaud-Mornant, 1974** [*C. tenellus* Renaud-Mornant, 1974; 7 species]
{Three pairs of somatic cirri with accordion-like scapus. Conical head separated into two parts; anterior triangular part with median cirrus and flattened secondary clavae, posterior part narrower, bearing lateral cirri and primary clavae. Four claws on each leg displaying more or less important heterometry. Stylet supports and placoids absent. Digestive tract with poorly developed diverticula. Female with two seminal receptacles.} (Hansen 2007)

Trogloarctus **Villora-Moreno, 1996** [*T. trionyches* Villora-Moreno, 1996; 1 species]
{Elongated body with cuticular folds. Only one pair of somatic cirri (cirrus *E*) with accordion-like scapus. Eleven cephalic sense organs. Primary clava small and globular, inserted separately from cirrus *A*. Secondary clava flattened and not pronounced. Small cephalic cirri consisting of two parts: scapus and flagellum. External cirri equally distant from the internal cirri and the primary clavae. Spines on all legs. Three claws on each foot, very long on hind legs. Pharyngeal bar shaped placoids in the pharynx. Stylet supports absent.} (Villora-Moreno 1996)

Halechiniscidae Thulin, 1928 [6 subfamilies; 17 genera]
{Sclerotized cuticular plates absent. Cephalic appendages usually consist of unpaired median cirrus, three pairs of cirri and two pairs of clavae, buccal clavae may be indistinguishable or consist of fused secondary and tertiary clavae. Unmodified papillate sensory organs on leg IV. Adult legs terminate in either four digits with

claws or four suction pads with or without claws. Peduncles present only in external digits, or as tarsal cuticular bars or absent. Single-pointed, crescent-shaped claws may bear dorsal spurs and/or minute calcars. A pair of seminal receptacles present in adult females.} (Fujimoto et al. 2016)

Dipodarctinae Pollock, 1995 [1 genus]
{Primary clavae much longer than the lateral cirri and in a more ventral position respect to them; indistinct secondary clavae present as kidney- or sausage-shaped elevations between internal and external cephalic cirri. Feet on legs I–III with at least three (digits I–III) of the four digits short and of equal length; digit IV is more dorsal and can be very long. Two short horizontal strengthening bars (peduncles) inside the tarsus. Foot on leg IV differs strongly from feet of leg I–III. It is of *Tanarctus*-type: with two long, thin, flexible medial digits (digits II and III) with small crescent claws with accessory spines, and two shorter lateral digits (digits I and IV) without accessory spines. Calcar can be present in all claws. Leg IV sensory organ differ from the spines of legs I–III. The leg IV sense organ is an ovoid to club-shaped papilla with a terminal spine.} (Jørgensen et al. 2014)

***Dipodarctus* Pollock, 1995** [*D. borrori* Pollock, 1995; 5 species]
{as the subfamily} (for a more detailed diagnosis see Jørgensen et al. 2014)

Euclavarctinae Renaud-Mornant, 1983 [6 genera]
{Unplated body without projections. Conical head bearing two sets of unequally shaped clavae. Cirri *A* and primary clavae inserted separately. Claws simple or with dorsal spur in the internal digits.} (Renaud-Mornant 1983)

***Clavarctus* Renaud-Mornant, 1983** [*C. falculus* Renaud-Mornant, 1983; 1 species]
{Small cephalic cirri, consisting of a short and swollen scapus and a short and beveled flagellum. Cirrophore absent. Median claws with a large accessory spine.} (Renaud-Mornant 1983)

***Euclavarctus* Renaud-Mornant, 1975** [*E. thieli* Renaud-Mornant, 1975; 2 species]
{Cephalic cirri composed of a cirriform in a funnel, a basal sheath, a large principal part, and a slender distal part. Two pairs of clavae present: primary clavae club-shaped, bent forward; secondary clavae spherical. Claws without spurs.} (Renaud-Mornant 1983)

***Exoclavarctus* Renaud-Mornant, 1983** [*E. dineti* Renaud-Mornant, 1983; 1 species]
{Cephalic cirri with a clear dimorphism: median cirrus and cirrus *A* with a long flagellum, external and internal cirri with a very short flagellum. Cirrophore at the base of the scapus merged into a swollen bulb. The end of scapus gives a bifid appearance to the cephalic appendages. Posterior clavae bent forward. Simple claws.} (Renaud-Mornant 1983)

***Moebjergarctus* Bussau, 1992** [*M. manganis* Bussau, 1992; 1 species]
{Dorsal segmental plates absent. Eleven cephalic sense organs present. Primary clavae club-shaped, bent forward. Secondary clavae spherical. Cephalic cirri separated into three parts: short cirrophore, long segmented scapus, and short flagellum. Spines on the first three pairs of legs. Small clavoid papillae on hind legs. Long cirri

E. Legs separated into four parts: coxa, femur, tibia, and tarsus. Tarsus with four digits. Digits corkscrew-shaped with numerous sharp-edged convolutions. External digits shorter than internal digits. Simple horseshoe-shaped claws retractable into membranous sheath.} (Bussau 1992)

***Parmursa* Renaud-Mornant, 1984** [*P. fimbriata* Renaud-Mornant, 1984; 2 species] {Cuticular aliform expansions supported externally by dorsal cuticular ribs. Trapezoid head, completely proportionate to the arrangement of the cephalic sense organs. Cephalic cirri with a funnel-shaped scapus. Secondary clavae erected. Stylet supports absent.} (Hansen 2007)

***Proclavarctus* Renaud-Mornant, 1983** [*P. fragilis* Renaud-Mornant, 1983; 1 species]
{Cephalic cirri composed by a cirrophore forming a narrow sheath, an elongated scapus, ringed at the base, a bayonet flagellum. Median cirrus and cirrus *A* larger than the internal and external cirri. Posterior clavae erected, anterior clavae spherical. Simple claws.} (Renaud-Mornant 1983)

Florarctinae Renaud-Mornant, 1982 [3 genera]
{Cuticular aliform expansions present. Complete set of cephalic sense organs with secondary clavae transformed as dome-shaped papillae or H-shaped flat sacs. Four digits with claws present in adult. External digits with a hook-shaped pedunculus. Uncus with a calcar externum. Females with two seminal receptacles, each consisting of a spheroid vesicle and an S-shaped genital duct.} (Kristensen 1984)

***Florarctus* Delamare Deboutteville & Renaud-Mornant, 1965** [*F. heimi* Delamare Deboutteville & Renaud-Mornant, 1965; 14 species]
{Three or five lateral and caudal wide aliform lobate expansions which cover the legs more or less completely. Sometimes a short frontal ala present. Alae present in all the developmental stages with the exception of the first-stage larva which is of the *Halechiniscus* type. Caudal ala which can be supported by spiky expansions. Procuticular processes (caestus) frequently present inside the alae. External digits shorter than the internal ones, with claws ending with an articulate portion (avicularia) and hook-shaped pedunculi inside the bases. Internal digits ending with simple claws with short dorsal spurs.} (de Zio Grimaldi et al. 1999)

***Ligiarctus* Renaud-Mornant, 1982** [*L. eastwardi* Renaud-Mornant, 1982; 2 species]
{Narrow head. Wide implantation of the primary clavae that are bent backwards, occupying the reduced lateral edges of the head. With lateral and caudal aliform lobate expansions, sometimes restricted to the caudal aliform expansion. A short frontal ala is frequently present. Procuticular processes (caesti) either absent or present inside the alae. Claws of all four digits or on claws external digits only with a distal notch on the internal side. Only a wing-shaped caudal expansion present.} (Gomes-Júnior et al. 2018)

***Wingstrandarctus* Kristensen, 1984** [*W. corallinus* Kristensen, 1984; 5 species]
{Alae present in all stages except for *Halechiniscus*-type two-clawed larvae, without procuticular hooks or expansions. Unci simple or with a calcar externum. Uncus

externus formed from a single piece. Uncus internus with accessory spine. Hook-shaped peduncle present on digitus externus. With three or no cephalic vesicles with symbiotic bacteria. Seminal receptacles and sperms of the florarctid type. Segmental glands and unicellular epidermal glands present in all segments.} (Kristensen 1984)

Halechiniscinae Thulin, 1928 [2 genera]
{Usually with complete set of cephalic appendages. Cylindrical tibia followed by a small tarsus; claws either simple or with a dorsal spur; peduncles absent. Sense organ of the fourth pair of legs short and non-branched. Vestigial secondary clavae.} (Grimaldi de Zio et al. 1990)

***Chrysoarctus* Renaud-Mornant, 1984** [*C. briandi* Renaud-Mornant, 1984; 2 species]
{Straight tibia, tarsus open as a fan and the four digits without stalk inserted side by side. Simple claws with small external calcar, papilla on the hind legs small.} (Renaud-Mornant 1984)

***Halechiniscus* Richters, 1908** [*H. guiteli* Richters, 1908; 11 species]
{Flat head extending into lateral lobes. Claws either simple or with a distal spur; calcar always absent; peduncles absent. Medial cirrus present; primary clava may be longer than the lateral cirrus.} (Grimaldi de Zio et al. 1990)

Orzeliscinae Schulz, 1963 [4 genera]
{Triangular tarsus bearing four digits. Elongated digits with proximal paddle- or spatula-shaped adhesive organs. Complete set of cephalic appendages present. Secondary clavae occurring as flattened papillae or bent, sausage-shaped structures. Stylet supports present.} (Gross et al. 2014)

***Mutaparadoxipus* Gross, Miller & Hochberg, 2014** [*M. duodigifinis* Gross *et al.*, 2014; 1 species]
{Adhesive pads of two different lengths; legs I–III: digits I–III with both proximal adhesive pad and distal claw, digit IV with adhesive pad only; legs IV: digits II & III with proximal adhesive pad and distal claw, digits I & IV with adhesive pads only; claws II & III of each foot with single accessory point. Complete set of cephalic cirri present; primary clavae and lateral cirri from common base, on a distinct lateral extension; secondary clavae in form of bent sausage-shaped structure lateral to ventral mouth cone. Stylets, stylet supports and placoids present. Paired seminal receptacles with coiled ducts opening lateral to large gonopore. Pointed lateral and caudal alae present.} (Gross et al. 2014)

***Opydorscus* Renaud-Mornant, 1989** [*O. fonsecae* Renaud-Mornant, 1989; 1 species]
{Procuticle containing pillars, 2-8 μm long dorsally, more than 10 μm laterally. Ventral procuticle finely punctuated by small pillars. Head with small expansions surmounted by the insertion of cirri *A*. Complete set of cephalic cirri. Secondary clavae lightly visible around the mouth cone. Stylets and buccal tube calcified in their distal portion. Coxal sensory setae present, those of hind legs with a papillary distal tip. Legs with four paddle-shaped digits inserted on a single tarsus expanded cone bearing a crescent and dorsal "triquetrum". Digits without claws, only the first

leg of females with claws on all digits, and internal claws with bifid tips. Spiral seminal receptacles positioned quite lateral and opened on an expansion of hind legs. Female and male gonopores as other Halechiniscidae.} (Renaud-Mornant 1989)

***Orzeliscus* du Bois-Reymond Marcus, 1952** [*O. belopus* du Bois-Reymond Marcus, 1952; 2 species]
{Four digits without claws but with thin sheet-like adhesive extensions. Cephalic cirri in three parts. Median cirrus and one pair of clavae present.} (du Bois-Reymond Marcus 1952)

***Paradoxipus* Kristensen & Higgins, 1989** [*P. orzeliscoides* Kristensen & Higgins, 1989; 1 species]
{Four similar digits on each foot. Each digit with proximal suction disc and distal claw. Complete set of cephalic cirri present. Primary clavae and lateral cirri arising from distinct lateral expansion. Buccal clavae, in form of bent sausage-shaped structures laterally adjacent to mouth cone. Stylet supports, stylets, and pharyngeal bars present. Paired seminal receptacles with coiled ducts opening lateral to a large female gonopore. Transparent, pointed lateral and caudal alae present.} (Kristensen and Higgins 1989)

Quisarctinae Fujimoto, 2015 [1 genus]
{Body cylindrical. Primary clava and lateral cirrus arise from common cirrophore; primary clava longer than lateral cirrus; secondary clava undeveloped. Leg IV sense organ as papilla. Legs terminate in digits without folds, peduncles, proximal pads, pretarsi or wrinkles; internal digits longer than external digits; each digit terminates in sheathed, small, crescent-shaped claw with minute calcar. Pair of ventrally opening seminal receptacles each with slender, sinuous duct terminating in spherical vesicle.} (Fujimoto 2015)

***Quisarctus* Fujimoto, 2015** [*Q. yasumurai* Fujimoto, 2015; 1 species]
{as the subfamily} (for a more detailed diagnosis see Fujimoto 2015)

Neoarctidae (de Zio Grimaldi, D'Addabbo Gallo & Morone De Lucia, 1992) [1 genus]
{A complete set of cephalic cirri and three pairs of clavae present. Body covered with five thin plates: a cephalic plate, a scapular plate (segmental plate I) which bears two long cirrophores with the primary clavae and lateral cirri *A*, two other plates (segmental plates II and III), and the caudal plate (segmental plate IV). No ventral plates. Median cirrus, internal and external cirri, with the secondary and tertiary clavae, located on the cephalic plate. Stylet supports absent. Legs non digitate, lacking true telescopy. Three claws on all legs attached to the tarsus with a dorsal basal membrane. Claws of different length: two longer, on the middle of the tarsus, and a short lateral one. A very short and thin dorsal bristle on the two medial claws. Sense organs on every leg. Cirrus *E* with a coarse accordion-shaped articulation.} (Bello and de Zio Grimaldi 1998)

***Neoarctus* de Zio Grimaldi, D'Addabbo Gallo & Morone De Lucia, 1992** [*N. primigenius* de Zio Grimaldi *et al.*, 1992; 1 species]
{as the family} (Bello and de Zio Grimaldi 1998)

Neostygarctidae de Zio Grimaldi, D'Addabbo Gallo & De Lucia Morone, 1987 [1 genus]
{Five dorsal plates, each with dorsal spines (dorsal cirri) and always three pairs of lateral trunk processes with cuticular spines. Cuticular spines are also present on the head, the caudal segment, and on the fourth leg pair. Intersegmental plates may also carry middorsal spines. Up to five very thin ventral plates present, sometime seen as folds only. Cephalic region subdivided into six lobes; anterior pair with two ventral internal cirri, and a large dorsal median cirrus present near the base of the anterior lobe; the lateral pair of lobes with ventral external cirri and secondary clavae; the lateral lobes consist of very long common cirrophores for lateral cirri and primary clavae. All six lobes have cuticular spines ordered in a specific pattern; furthermore, two swollen lobes with spines are present dorsal and posterior on the cephalic plate, and similar lateral swollen lobes with spines are present on the caudal plate. Cirrus *E* very large with accordion-like articulation. Sense organs on fourth leg pair, similar to primary clava. Legs with four digits, each with a dorsal cuticular support ending in a triangular, solid claw. All four claws with or without dorsal bristle. Stylet supports always absent. Oval male gonopore situated very close to lobed anus. Female gonopore formed as a six-lobed, midventral rosette on the caudal segment; two seminal receptacles present, located caudal to the female gonopore, and with sinuous duct openings as two cuticular tubes lateral to the gonopore.} (Kristensen et al. 2015)

***Neostygarctus* Grimaldi de Zio, D'Addabbo Gallo & Morone De Lucia, 1982** [*N. acanthophorus* Grimaldi de Zio *et al*., 1982; 3 species]
{as the family} (Kristensen et al. 2015)

Renaudarctidae Kristensen & Higgins, 1984 [2 genera]
{Dorsal and ventral segmental plates present. Complete set of cephalic sensory organs: unpaired median cirrus, paired internal cirri, paired external cirri, paired lateral cirri, paired primary clavae, paired secondary clavae and paired sensory plates. Lateral cirri and primary clavae share common cirrophores; three non-branching placoids inside pharyngeal bulb present; legs with wrinkled digits terminating in hook-shaped, single-pointed claws with flexible accessory filaments; leg IV with two or four digits.} (Fujimoto and Yamasaki 2017)

***Nodarctus* Fujimoto & Yamasaki, 2017** [*N. hallucis* Fujimoto & Yamasaki, 2017; 1 species]
{Punctate dorsal cuticle. Dorsal plates consist of: cephalic plate, intersegmental plate I, subdivided dorsal segmental plate I, two subdivided intersegmental plates II, subdivided dorsal segmental plate II, two subdivided intersegmental plates III, subdivided dorsal segmental plate III, two subdivided intersegmental plates IV and caudal plate; dorsal segmental plates I–III lack lateral processes. Ventral plates consist of: ventral segmental plate I, paired ventral segmental plates II and paired ventral segmental plates III. Acute caudal spike present. Paired, short, acute processes present at level of gonopore. Cephalic cirri each consisting of scapus and tapering flagellum; external cirri each inserted on small lobe; club-shaped primary clavae present; sensory plates (including fused secondary clavae) cover most of area

of ventral cephalic region, each with indentation in anterior portion. Legs I–III of adults terminate in four wrinkled digits, each with hook-shaped, singlepointed claw and flexible accessory filament; papillae situated interior to digits on legs I–III; short spikes on exterior side of legs I–III. Legs IV terminate in two wrinkled digits, each with hook-shaped claw and flexible accessory filament; basal spike and two dorsal swellings on leg IV; papillate leg IV sensory organ present. Cirri *E* present. Non-rosette female gonopore present. Seminal receptacle ducts open ventrally near gonopore, run exteriorly and terminate in spherical vesicles.} (Fujimoto and Yamasaki 2017)

***Renaudarctus* Kristensen & Higgins, 1984** [*R. psammocryptus* Kristensen & Higgins, 1984; 2 species]
{Segmental dorsal plates, two dorsal intersegmental plates, four dorsal intersegmental ridges, and thin paired ventral plates. Each body plate with a pair of lateral spines. Dorsal plate on each leg. Caudal plate with two pairs of lateral spikes, one pair of caudal spikes, and lateral margins continuous with furca-like caudal extensions. Complete set of cephalic sense organs. Leg IV with small ovoid papilla; all other legs lack sensory structures. Cirrus *E* with cirrophore. Female and male with ovoid gonopore papillae adjacent to terminal anal system. Female with latero-ventral projections containing seminal receptacles.} (Hansen et al. 2012)

Stygarctidae Schulz, 1951 [2 subfamilies; 7 genera]
{Dorsal cuticle with five plates (sometime absent): cephalic/head plate, three segmental/body plates (I–III), caudal plate (segmental plate IV). Intersegmental dorsal plates and ventral plates may be present. Head subdivided into lobes, with complete set of cirri and two pairs of clavae. Stylet support absent. Non-digitated legs with four/three/two claws attached with dorsal basal membrane. Leg sense organ (bulb-shaped papillae) present only on hind legs. Seminal receptacles formed as pouches with short straight ducts or with long sinuous ducts close to gonopore.} (Hansen et al. 2012)

Megastygarctidinae Bello & de Zio Grimaldi, 1998 [1 genus]
{Four claws on the first three pairs of legs and two claws on the fourth pair; all claws with small dorsal spur; each claw originating from a short pedestal. Typical dorsal plate pattern of the family (head plate, three body plates, caudal plate), modified by intercalated accessory plates, or dorsal plates may be completely absent. Ventral plates may be present. Posterior part of the head with two lateral lobes bearing the primary clavae. Anterior part of the head sometimes divided into two lobes. Primary and secondary clavae of different shapes. Primary clavae club-shaped. Secondary clavae occurring as ovoid structures or as dome-shaped papillae. Cirrus *E* with a strongly annulated base. Leg sense organs present either on the second, third and fourth pair of legs or only on the fourth pair of legs. Sense organs on the fourth pair of legs always formed as bulb-shaped papillae. The two vesicular seminal receptacles with genital ducts extended outside the body as genital spines or ovoid papillae overlapping the female genital rosette-system.} (Hansen and Kristensen 2006)

***Megastygarctides* McKirdy, Schmidt & McGinty-Bayly, 1976** [*M. orbiculatus* McKirdy *et al.*, 1976; 6 species]
{as the subfamily} (Hansen and Kristensen 2006)

Stygarctinae Schulz, 1951 [6 genera]
{Head divided into five lobes; intermetamerical dorsal plates; ventral plates sometimes present. In adults, all legs with the same number of claws: four, three or two. Sense organs of the first three pair of legs present or absent. Bulb-shaped papilla on fourth leg.} (Bello and de Zio Grimaldi 1998)

***Faroestygarctus* Hansen, Kristensen & Jørgensen, 2012** [*F. dezioae* Hansen *et al.*, 2012; 1 species]
{Semicircular head plate with three main lobes: an anterior lobe bearing the median cirrus, the secondary clavae (semi-globular) and two pairs of cirri (internal and external), and two posterolateral lobes bearing the primary clavae (small and oval), and two lateral cirri. Body plates with lateral processes tapering to flexible spines. Intersegmental dorsal plate between second and third body plates. Caudal plate with a pair of triangular processes. Cirrus *E* with indistinct basal accordion articulation. Each leg with two claws with dorsal spurs. Seminal receptacles as small ventrolateral spheroid vesicles with long sinuous ducts.} (Hansen et al. 2012)

***Mesostygarctus* Renaud-Mornant, 1979** [*M. intermedius* Renaud-Mornant, 1979; 2 species]
{Head plate with three lobes: posterior part of the head with two lateral lobes bearing the primary clavae (elongated and club-shaped), anterior part of the head divided into three lobes by a pair of deep indentations with dorsal cuticular membranes. Body plates with lateral membranous margins and posterior processes. Two intersegmental dorsal plates. Cuticle without epicuticular pillars. Large ventral cephalic plate. Ventral body plates sometime present. Ventral cuticular folds present in each trunk segment. Long tapering cirrus *E*, telescopically inserted on a lateral process. Each leg with four claws with dorsal spurs. Seminal receptacles as large ventral spheroid vesicles or arranged in a spiral internally with looped ducts close to gonopore. Seminal receptacle duct openings with internal cuticular bars and external cuticular pockets.} (Hansen et al. 2012)

***Parastygarctus* Renaud-Debyser, 1965** [*P. higginsi* Renaud-Debyser, 1965; 7 species]
{Head plate with five protruded lobes. Body plates with one/two lateral tapering long processes. Primary and secondary clavae of similar shape, generally elongated. Secondary clavae situated posterior to the external cirrus. Cirrus *E* with accordion-plated articulation. Two intersegmental dorsal plates. Ventral plates sometime present. Each leg with two/four claws, the central pair with long accessory filaments. The vesicles of seminal receptacle located dorso-laterally.} (Hansen et al. 2012)

***Prostygarctus* Rubal, Veiga, Fontoura & Sousa-Pinto, 2013** [*P. aculeatus* Rubal *et al.*, 2013; 1 species]
{Dorsal cuticle strongly sculptured. Intersegmental dorsal plates and ventral plates present. Head plate divided into five lobes; the frontal lobe subdivided into four

smaller lobes. Each body plate with two pairs of lateral processes; the posterior-most lateral processes tapering as flexible spines. Terminal edge of segmental and intersegmental plates with paired dorsal spines directed backwards. A ventral spine present at the level of the third pair of legs. Caudal plate with a very long medial spine and two lateral furca-like processes. Cirrus *E* with ball and double-socket articulation. Complete set of cephalic cirri and two pairs of pedunculated elongated clavae: erect primary clavae and laterally bent secondary clavae juxtaposed to the frontal lobe of the head. Digitate legs with simple claws. Sense organs on hind legs as bulb-shaped papillae.} (Rubal et al. 2013)

***Pseudostygarctus* McKirdy, Schmidt & McGinty-Bayly, 1976** [*P. triungulatus* McKirdy *et al.*, 1976; 5 species]
{Head plate usually with five lobes: posterior part of the head with two lateral lobes bearing primary clavae (elongate club-shaped), anterior part usually divided into five lobes by two pairs of deep indentations with cuticular membranes. Body plate margins with thin cuticular membranes and lateral processes. Two intersegmental dorsal plates. Ventral plates sometime present. Cirrus *E* with "ball and socket" articulation. Each leg with two/three/four claws (simple or with dorsal spurs). Seminal receptacles formed as large ventral spheroid vesicles with looped ducts close to the female gonopore. Seminal receptacle duct openings with internal cuticular bars and external cuticular pockets.} (Hansen et al. 2012)

***Stygarctus* Schulz, 1951** [*S. bradypus* Schulz, 1951; 7 species]
{Body plates extending laterally into funnel-shaped processes. Primary and secondary clavae frequently of the same shape, generally elongate. Intersegmental plates among body plates. Ventral plates present. Cirrus *E* with thin accordion-plated articulation, inserted on the lateral processes of the caudal plate. Each leg with four claws, the central pair of which with a long accessory filament. No sense organs on the first three pairs of legs. Bulb shaped papilla on the fourth leg.} (Bello and de Zio Grimaldi 1998)

Styraconyxidae Kristensen & Renaud-Mornant, 1983 [10 genera]
{Sclerotized cuticular plates absent. Cephalic appendages usually consist of unpaired median cirrus, three pairs of cirri and two pairs of clavae. Buccal clavae may be indistinguishable or third pair of clavae may be present (*Angursa*). Unmodified papillate sensory organs on leg IV. Adult legs terminate in four digits, usually with multipointed claws, rarely single-pointed (*Lepoarctus*, *Pleocola*). Peduncles present in all digits, or in external digits only with heart-shaped pads on internal digits, or absent. A pair of seminal receptacles present in adult females.} (Fujimoto et al. 2016)

***Angursa* Pollock, 1979** [*A. bicuspis* Pollock, 1979; 6 species]
{External digits supported by peduncles, internal digits with proximal pads. Claws with 2 divergent points (hooks), primary hook stronger than secondary hook. Claws retractable into membranous sheath. Dorsal cuticular plates absent. Complete set of cephalic cirri present, sometimes the median cirrus is reduced or absent. Configuration of primary clavae and lateral cirri variable, arising from distinct lateral extension or directly from head. Secondary clavae occasionally present, occurring as indistinct

dome-shaped areas. Sometimes tertiary clavae with appearances similar to those of secondary clavae present, situated ventrally to the secondary clavae. Spines present on legs I, II, III or only on legs IV. Usually clavoid papillae, mostly with spine, on legs IV. Body smooth and slender.} (Bussau 1992)

***Bathyechiniscus* Steiner, 1926** [*B. tetronyx* Steiner, 1926; 1 species]
{Cylindrical body. Each digit with claws with three or more vertically arrayed exposed points, on medial digits at least. Well-developed cephalic cirri and median cirrus present. Single pair of clavae immediately adjacent to lateral cirri with one or both of these structures arising from a distinct lateral expansion of the head, i.e. from a pedestal base. Spines present on the first three legs and papilla or papilla/spine combination on hind legs. Lateral folds of cuticle or alae absent.} (Pollock 1983)

***Lepoarctus* Kristensen & Renaud-Mornant, 1983** [*Pleocola conifera* Renaud-Mornant, 1975; 1 species]
{Epicuticle highly developed forming an "envelope" around the trunk (enlarged bell-shaped epicuticle), but not on the cephalic part. Set of cephalic appendages complete with cylindrical scapus, indistinct secondary clavae. Primary clavae retractable in a membranous epicuticular sheath, and cirrus *E* located in an epicuticular funnel. Spines on legs I; reduced on legs II and vestigial on legs III. Papillae on legs IV with pointed tip. Long buccal tube, stylets and stylet supports very fine, thin pharyngeal bars. Simple claws retractable into membranous sheath.} (Kristensen and Renaud-Mornant 1983)

***Paratanarctus* D'Addabbo Gallo, Grimaldi de Zio, Morone De Lucia & Troccoli, 1992** [*P. kristenseni* D'Addabbo Gallo *et al*., 1992; 1 species]
{Complete or incomplete set of cephalic appendages. Secondary clavae sometimes club-shaped or dome-shaped or indistinguishable. Each adult leg with four digits bearing claws either simple or with accessory hooks. Peduncles sometimes present in the digits. Seminal receptacles generally present. Cirrus *E* often with a basal accordion-shaped articulation.} (D'Addabbo Gallo et al. 1992)

***Pleocola* Cantacuzène, 1951** [*P. limnoriae* Cantacuzène, 1951; 1 species]
{Body smooth and slender. Strongly convex conical head; cirri *A* and clavae implanted in a posterior plan relative to median cirrus. Simple claws without spur. Complete set of cephalic sense organs; primary clavae short and ovoid; secondary clavae indistinct. Spines on legs I, II, III; papillae on legs IV. Cirrus *E* present.} (Cantacuzène 1951; Fontoura et al. 2017)

***Raiarctus* Renaud-Mornant, 1981** [*R. colurus* Renaud-Mornant, 1981; 5 species]
{Dorso-lateral epicuticular coat supported by long thin pillars 5–6 μm regularly arranged in close parallel lines. With 3-pointed claws, with or without heart-shaped pads on the internal digits. With or without peduncles on every tarsus. Minor primary clava with a cone or tube-like shape; the short, secondary clavae either reduced or minor. Body dorso-ventrally flattened. With lateral tube-like openings to the seminal receptacles. Halechiniscid male gonopore (oval papilla with a crescent-shaped opening).} (Jørgensen et al. 2014)

***Rhomboarctus* Renaud-Mornant, 1984** [*R. thomassini* Renaud-Mornant, 1984; 3 species]
{Convex frontal edge; indistinct or lenticular secondary clavae; dorsal cuticle punctuated; cuticular fins supported by medium size pillars, extending on both sides of the body, between primary clavae and the hind legs. More or less evident caudal appendage with cuticular fan with pillars. Peduncles present only on the external digits, proximal pads on the internal ones; all the claws with very thin or vestigial secondary hooks.} (Hansen et al. 2003)

***Styraconyx* Thulin, 1942** [*S. haploceros* Thulin, 1942; 13 species, one of them limnic species—in warm springs with high levels of double fluoride salts]
{Legs with four digits. External digits always supported by peduncles; internal digits with proximal pads, peduncles present or absent. Claws, with three exposed points (hooks), may be retracted within claw sheath. Dorsal cuticular plates absent although grid-like pattern of 18–23 folds sometimes present on epicuticle. Complete set of cephalic cirri always present; configuration of primary clavae and lateral cirri variable, arising from distinct lateral extension or directly from head; secondary clavae occurring as indistinct flat sacs or dome-shaped papillae. Stylet supports, stylets, and placoids present. Spines present on legs I–III, clavoid papillae, each with spine, on fourth legs. Paired seminal receptacles with ducts opening posterior to female gonopore present between legs III and IV.} (Kristensen and Higgins 1984)

***Tetrakentron* Cuénot, 1892** [*T. synaptae* Cuénot, 1892; 1 species]
{Body very flattened in dorsoventral sense. Cephalic appendices short, including the median cirrus, in shape of a short spine; only the cephalic papillae are relatively large. Eyes lacking. Mouth opening on buccal cone and anus subterminal. Legs short, with short tibia and digits. Claws compound (with primary and secondary hooks) and with pedunclulus; those of internal digits with large accessory spine on secondary hook; no spurs. Two types of males: mobile males and nearly sessile dwarf males. Ectoparasitic on borrowing holothurians.} (Van der Land 1975; Kristensen and Hansen 2005)

***Tholoarctus* Kristensen & Renaud-Mornant, 1983** [*T. natans* Kristensen & Renaud-Mornant, 1983; 2 species]
{Barrel-shaped outer epicuticle surrounding the trunk, the body itself is thin and elongated. Complete set of cephalic appendages. Cone-shaped primary clavae and secondary clavae enlarged surrounding both the internal and external cirri. Cirri *A* and *E* partially surrounded by funnel-shaped epicuticle folds. Leg sense organs always present on the first and fourth legs, though much smaller sense organs may be present on the second and third legs. Flagellum on the sense organ of the first leg always very long. Sense organ of the fourth leg partially enclosed in funnel-shaped epicuticular folds. Females have claws with a main hook, a basal accessory hook and a terminal accessory spur. Males and larvae might have simple claws in the external digits. Long buccal tube with two lateral projections at ca. the mid-point. Long and thin stylet supports articulate with the furcae. Placoids consist of a non-calcified cuticular lining, but three small calcified apodemes occur in the anterior part of the pharyngeal bulb. Mouth cone is terminal and protruding. Globular coelomocytes

seem to be attached to the midgut. Sexual dimorphism in body size and claws.} (Jørgensen et al. 2014)

Tanarctidae Renaud-Mornant, 1980 [3 genera]
{Sclerotized cuticular plates absent. Cephalic appendages usually consist of unpaired median cirrus, three pairs of cirri, pair of long primary clavae and pair of secondary clavae, which may be indistinguishable. Sensory organs on leg IV long and often bifurcated. Legs long and telescopic, each with lance-like tibia and conical tarsus. Adults with four clawed digits. Internal digits inserted on pretarsi and longer than external digits. Claws with strong calcar. Placoids with subterminal bulbous branch. A pair of spindleshaped seminal receptacles present in adult females.} (Fujimoto et al. 2016)

***Actinarctus* Schulz, 1935** [*A. doryphorus* Schulz, 1935; 4 species]
{Epicuticular alae (wings) internally supported by long pillars increasing in length from body center to periphery. Body with an external thin, gelatinous membrane, in which dorsal and lateral tubular processes (modified pillars) are immersed. Digits with simple external claws without spurs, internal claws with spurs. Very thin peduncles may be present on all digits. Internal buccal cirri long and connected at base by membrane from epicuticle; median cirrus always present. Primary clava long, buccal clava club-shaped or flat H-shaped structure surrounding mouth cone.} (Boesgaard and Kristensen 2001)

***Tanarctus* Renaud-Debyser, 1959** [*T. tauricus* Renaud-Debyser, 1959; 13 species]
{Without dorso-lateral epicuticle long rod-shaped pillars. All four digits with simple external claws without dorsal spurs, internal claws with microscopic spurs or without spurs. The internal digits may be fused proximal and form a protarsus. The external digits have a basal cuticular fold. The primary clava, when present, is very long and similar to the leg IV appendages, or the leg IV appendages are strongly modified with long spines, leaf-like appendages or balloon-like ampullae. The buccal clavae may consist of fused secondary and tertiary clavae and then appear H-shaped around the mouth, or they can be located rostrally as club-shaped clavae (secondary clavae?). The buccal clavae may be absent or indistinct as a dome- or lens-shaped structure.} (Jørgensen and Kristensen 2001)

***Zioella* Renaud-Mornant, 1987** [*Z. pavonina* Renaud-Mornant, 1987; 1 species]
{All cirri aligned on the front edge of the head. Secondary clavae reduced to fine contours in front of the mouth. Epicuticle smooth, no pillars observed by light microscopy. Papillae on hind legs form after two successive dichotomies eight large membranous and fusiform ampullae.} (Renaud-Mornant 1987)

15.2.1.2 Order Echiniscoidea Richters, 1926

marine, limnic and terrestrial species
{Claws inserted on minuscule papillae positioned at the end of the legs, which are not digitate. Median cirrus generally absent.} (Richters 1926)

Echiniscoididae Kristensen & Hallas, 1980 [2 subfamilies; 3 genera]
marine species
{Unplated marine species. Adults with supranumerous claws (6–13) on each leg. Papilla cephalica (secondary clava) dome shaped or almost flat (indistinct). Other cephalic appendages and leg appendages small (reduced). Cirri *A* and *E* similar in both structure and length. The fourth leg sensory structure (p4) is papillar as the primary clava. Seminal receptacles lacking. External fertilization of eggs.} (Kristensen and Hallas 1980; Møbjerg et al. 2016)

Echiniscoidinae Kristensen & Hallas, 1980 [2 genera]
{Characteristics of the subfamily not defined by Møbjerg et al. (2016) who erected the Isoechiniscoidinae.}

***Anisonyches* Pollock, 1975** [*A. diakidius* Pollock, 1975; 4 species]
{Four claws on each of the first pairs of legs and three claws on the hind legs in adults. Claws with paired basal spurs. Median cirrus present but reduced.} (Bartels et al. 2018)

***Echiniscoides* Plate, 1888** [*Echiniscus sigismundi* M. Schultze, 1865; 8 species]
{More than six claws on each leg in adult specimens. Epicuticle without pillars. Claws are without spurs. Tidal or halophilous forms.} (Kristensen and Hallas 1980, Møbjerg et al. 2016)

Isoechiniscoidinae Møbjerg, Kristensen & Jørgensen, 2016 [1 genus]
{Isonychous claws (identical numbers and type of claws on all legs). Epicuticle with pillars. More than four claws (usually six) on each leg in adults. First instar (larva) with four claws. Secondary clavae (cephalic papillae) larger and more dome-shaped in males than in females. Lateral cirri *A* and cirri *E* very long. Sensory setae on all four leg pairs. Setae of the first leg pair are very small papillar structures with a tiny spine.} (Møbjerg et al. 2016)

***Isoechiniscoides* Møbjerg, Kristensen & Jørgensen, 2016** [*I. sifae* Møbjerg *et al.*, 2016; 2 species]
{as the subfamily} (Møbjerg et al. 2016)

Carphaniidae Binda & Kristensen, 1986 [1 genus]
freshwater species
{Unplated body, eutardigrade-like in shape. Primary and secondary clavae absent. External and internal buccal cirri without cirrophorus. Gonopore and anus perhaps not separated. Two claws on the first three leg pairs and one claw on the hind legs. All claws flexible with two basal spurs.} (Binda and Kristensen 1986)

***Carphania* Binda, 1978** [*C. fluviatilis* Binda, 1978; 1 species]
{as the family} (Binda and Kristensen 1986)

Oreellidae Ramazzotti, 1962 (*sensu* Puglia unpublished thesis 1959) [1 genus]
terrestrial species
{Unplated semi-terrestrial Echiniscoidea with a short caudal median projection. Telescopic retractable leg with four claws in adults and two claws in larvae. Each internal claw with a downwardly directed spur. Sense organs present on legs I and

IV. All head projections, except median cirrus, present. Papillae cephalica (= secondary clavae) located under the cuticle surface. Clavae (= primary clavae) papilla-like in females and juveniles, club-like in males. Buccal apparatus with stylet supports. Females with a pair of sack-like structures (seminal receptacles) caudo-laterally, opening directly to the exterior, but without a duct. Eggs freely laid, ornamented.} (Dastych et al. 1998)

***Oreella* Murray, 1910** [*O. mollis* Murray, 1910; 3 species]
{as the family} (Dastych et al. 1998)

Echiniscidae Thulin, 1928 [15 genera]
terrestrial species (found also in freshwater)
{Seminal receptacles absent. Dorsal plates present. Adult with 4 claws on each leg.} (Kristensen 1987)

***Acanthechiniscus* Vecchi, Cesari, Bertolani, Jönsson, Rebecchi & Guidetti, 2016** [*Echiniscus victor* Ehrenberg, 1853; 5 species]
{Black eyes. Rigid buccal tube. Filaments and/or spines (may be small) present, generally in all lateral positions (*B*, *C*, *D*, *E*); dorsal spines present. Pseudosegmental plate IV' present. Notched collar (spine fringe) on hind legs present.} (Vecchi et al. 2016)

***Antechiniscus* Kristensen, 1987** [*Pseudechiniscus lateromamillatus* Ramazzotti, 1964; 6 species]
{Colour grey to light red. Black eyes, unflexible buccal tube. Unpaired pseudosegmental plates (II' and III') present posterior to paired segmental plates II and III. A small bar-shaped pseudosegmental plate (IV'), slightly overlapped by a large undivided median plate (m3), anterior to the caudal plate (IV). Male small with enlarged primary and secondary clavae.} (Kristensen 1987)

***Bryochoerus* Marcus, 1936** [*Echiniscus intermedius* Murray, 1910; 2 species]
{Unpigmented to rose, red eyes. Short inflexible buccal tube, with $CaCO_3$ encrusted stylet supports. All median plates 1, 2, 3 divided, without pseudosegmental plates, small lateral intersegmental plate present.} (Kristensen 1987)

***Bryodelphax* Thulin, 1928** [*B. parvulus* Thulin, 1928; 23 species]
{Sparsely pigmented. Red granulae eyes. Non-flexible buccal tube with $CaCO_3$ encrusted stylet supports. No lateral or dorsal appendages present except cirrus *A*. Median plates all divided, but the caudal portion of the third median plate may be hidden by the terminal plate. Without pseudosegmental plates. Ventral plates may be present.} (Kristensen 1987; Lisi et al. 2017; due to new diagnosis of the genus *Bryodelphax* which is practically identical with that in *Bryochoerus* both genera will have to be reappraised in future)

***Cornechiniscus* Maucci & Ramazzotti, 1981** [*Echiniscus cornutus* Richters, 1907; 9 species]
{Colour, yellow-red. Black eyes. Flexible buccal tube and paired pseudosegmental plates IV'. Cirri *A* horn-shaped with a common base for the primary clavae, secondary clavae ovoid, domed; internal and external cirry onion-shaped.} (Kristensen 1987)

***Diploechiniscus* Vicente, Fontoura, Cesari, Rebecchi, Guidetti, Serrano & Bertolani, 2013** [*Echiniscus oihonnae* Richters, 1903; 1 species]
{Orange body, dark-brown eyes. Dorsal plates I, II, III, IV (II and III paired), transversally subdivided median plates m1 and m2 and undivided plate m3 present; double sculpture in the dorsal plates, represented (under phase contrast) by dark polygonal and white circular grains; ventral plates present, especially evident in the anterior, head region and around the gonopore; supernumerary dorsal-lateral spines present. Buccal tube long and narrow, with stylet supports.} (Vicente et al. 2013)

***Echiniscus* C.A.S. Schultze, 1840** [*Emydium testudo* Doyère, 1840; 158 species]
{Red-granulae eyes, unflexible buccal tube, without pseudosegmental plates IV'. Median plates, sometimes transversally subdivided, caudal plate notched; ventral plates absent with the exception of ventral subcephalic and/or genital plates which may be present. Primary and secondary clavae papillate. Stylet supports absent or not visible by light microscopy.} (Kristensen 1987, integrated by Gąsiorek et al. 2017)

***Hypechiniscus* Thulin, 1928** [*Echiniscus gladiator* Murray, 1905; 4 species]
{Uncoloured. Ovoid black eyes; rigid buccal tube with a cuticular bulb anterior to a small pharyngeal bulb, stylet supports and ventral plates absent. Head plate large with a suture; scapular plates may be paired; paired segmental plates II and III, caudal plate divided in three pieces by sutures; all three median plates undivided. Both primary and secondary clavae papillate; unpaired middorsal trunk appendage can be present.} (Kristensen 1987)

***Mopsechiniscus* du Bois-Reymond Marcus, 1944** [*Echiniscus imberbis* Richters, 1907; 6 species]
{Red. Black eyes, flexible buccal tube, stylet supports without $CaCO_3$; ventral plates absent. Strongly sculptured dorsal plates. Paired pseudosegmental plates IV'; all median plates undivided. A small rectangular plate with a "thorn" present on each side of the scapular plate (I). Cirrus *A* long stiff spine without cirrophore, primary clavae bent backwards. Internal and external cirri absent, secondary clavae small ovoid dome-shaped in female, large round dome-shaped in male.} (Kristensen 1987)

***Multipseudechiniscus* Schulte & Miller, 2011** [*Pseudechiniscus raneyi* Grigarick, Mihelčič & Schuster, 1964; 1 species]
{Yellow-orange with black eyes. Anterior portion of buccal tube rigid, posterior part of buccal tube flexible. Ventral plates absent. Dorsal segmental plates, each divided into two or more pieces, one or more piece paired; posterior piece of trunk segment plates I, II, and III paired, forming paired pseudosegmental plates; three median plates; anterior piece of cephalic and scapular plates paired; terminal plate divided into three pieces; median plates 1 and 2 divided, median plate 3 undivided; lateral intersegmental plates present. Long, flexible stylet sheaths. Bulb shaped buccal apophysis for muscle attachment present. Thin stylet supports attach about midpoint and anterior to buccal tube thickening.} (Miller et al. 2012)

***Novechiniscus* Kristensen, 1987** [*Parechiniscus armadilloides* R.O. Schuster, 1975; 1 species]
{Eyes red. Rigid buccal tube with flexible part inside the pharyngeal bulb, pharyngeal bulb with a drop-shaped structure located posteriorly on the placoids. All four segmental plates I–IV unpaired with two bar-shaped median plates 1 and a distinct bar-shaped median plate 2; primary and secondary clavae with ring shaped annulus at the base; internal cirri with a large swollen base. Claws on leg IV with secondary spurs on both internal and external claws.} (Kristensen 1987; Rebecchi et al. 2008)

***Parechiniscus* Cuénot, 1926** [*P. chitonides* Cuénot, 1926; 2 species]
{Uncoloured to grey. Small black eyes. Rigid buccal tube, stylet supports absent, 8–9 dorsal plates, unpaired and more or less indistinct; secondary clavae dome shaped.} (Kristensen 1987)

***Proechiniscus* Kristensen, 1987** [*Pseudechiniscus hanneae* Petersen, 1951; 1 species]
{Colour reddish brown. Black eyes, flexible buccal canal, presence of calcified stylet supports. Segmental plates II and III paired with unpaired posterior element; median plates 1, 2 and 3 and the unpaired pseudosegmental plate IV' bar-shaped; with dome-shaped secondary clavae.} (Kristensen 1987)

***Pseudechiniscus* Thulin, 1911** [*Echiniscus suillus* Ehrenberg, 1853; 35 species]
{Colour yellow or red. Black eyes; rigid buccal tube, stylet supports may be present, but very tiny and located close to the margin of pharyngeal bulb. All median plates present, plates 1 and 2 divided. Lateral and dorsal trunk filaments or spines generally absent or reduced in number and/or size. Pseudosegmental plate IV' present. Male with enlarged secondary clavae. Notched collar (spine fringe) on hind legs absent.} (Kristensen 1987; Vecchi et al. 2016)

***Testechiniscus* Kristensen, 1987** [*Echiniscus spitsbergensis* Scourfield, 1897; 4 species]
{Colour deep red to orange. Black/brownish or "lipoid" granular eyes; rigid buccal tube, large cuticular stylet supports, seven rows of sculptured ventral plates. Pseudosegmental plates absent. Paired segmental plates II and III with thinner anterior part. Caudal plate IV with a terminal indentation. Median plates 1 and 2 large, plate 3 small and reduced.} (Kristensen 1987)

15.2.2 Class Mesotardigrada Rahm, 1937

freshwater species
{Six to ten claws not differentiated into principal branch and secondary branch in each leg; claws similar to each other. Peribuccal papillae and cirri *A* present, other head appendages (such as cephalic cirri and clavae) absent. Pharynx with pharyngeal apophyses and placoids.} (Ramazzotti and Maucci 1983; *nomen dubium* according to Grothman et al. 2017)

15.2.2.1 Order Thermozodia Ramazzotti & Maucci, 1983

{as the class} (Ramazzotti and Maucci 1983)

Thermozodiidae Rahm, 1937 [1 genus]
{as the order} (Ramazzotti and Maucci 1983)

***Thermozodium* Rahm, 1937** [*T. esakii* Rahm, 1937; 1 species]
{as the family} (Ramazzotti and Maucci 1983)

15.2.3 Class Eutardigrada Richters, 1926

limno-terrestrial, and marine species; two fossil species
{Without cephalic appendages such as cephalic cirri, lateral cirri *A* or clavae, only peribuccal and lateral papillae can be present; gonopore not separated from anus; two double claws composed of a primary and a secondary branch fused or positioned one behind the other (secondary branch can be reduced or absent, or claws can be strongly reduced or even absent in some genera).} (Bertolani et al. 2014)

15.2.3.1 Order Apochela Schuster, Nelson, Grigarick & Christenberry, 1980

limno-terrestrial species, one fossil species
{Double claws of Milnesiidae type (primary and secondary branches are completely separated). Buccal tube completely rigid, or subdivided into rigid and flexible portions, or completely flexible; ventral lamina absent; six peribuccal papilae and six or only four peribuccal lamellae present; two lateral papillae on the head (cephalic papillae) present; apophyses for the insertion of the stylet muscles, if known, in shape of "very short and very flat ridges" symmetrical with respect to the frontal plane and without caudal processes; elongated pharyngeal bulb without pharyngeal apophyses and placoids. Eggs, if known, are smooth and laid in the exuvium.} (Pilato and Binda 2010; Dastych 2011; Bertolani et al. 2014)

Milnesiidae Ramazzotti, 1962 [4 genera]
{as the order} (Bertolani et al. 2014)

***Bergtrollus* Dastych, 2011** [*B. dzimbowski* Dastych, 2011; 1 species]
{Buccal tube long, flexible but without spiral-like strengthening; telescopically protrusible tubular proboscis ('snout') present; six peribuccal lamellae present; stylet furcae broad, flat spade-like. Eggs unknown.} (Dastych 2011)

***Limmenius* Horning, Schuster & Grigarick, 1978** [*L. porcellus* Horning *et al.*, 1978; 1 species]
{Tube subdivided into an anterior, short, rigid portion (buccal tube) and a long posterior flexible portion (pharyngeal tube) with a spiral thickening (*Limmenius*

model); peribuccal papillae and six peribuccal lamellae present; long and thin stylet supports inserted on the pharyngeal tube; stylet furcae very small (shape undescribed). Eggs not found.} (Claxton 1999; Pilato and Binda 2010)

***Milnesioides* Claxton, 1999** [*M. exsertum* Claxton, 1999; 1 species]
{Buccal tube completely rigid (*Milnesium* model) and very long; six triangular peribuccal lamellae present; mouth at anterior of a long protrusible snout; stylet furcae very small, triangular in shape.} (Claxton 1999, Pilato and Binda 2010)

***Milnesium* Doyère, 1840** [*M. tardigradum* Doyère, 1840; 37 species + 1 fossil species]
{Broad buccal tube completely rigid (*Milnesium* model); six or four large, almost triangular, peribuccal lamellae present; stylet furcae triangular in shape with short branches.} (Pilato and Binda 2010)

15.2.3.2 Order Parachela Schuster, Nelson, Grigarick & Christenberry, 1980

limno-terrestrial and marine species, one fossil species
{Complete double claws on the legs with secondary claw branch joined to the primary branch (secondary branch normally developed, reduced, or absent), or double claws completely lost. Buccal tube completely rigid, or subdivided into rigid and annulated flexible portions (pharyngeal tube); ventral lamina present or absent; peribuccal and cephalic papillae absent; pharyngeal bulb with or without placoids. Stylet furcae typically shaped (proximal portion with two branches ending with thickened, swollen, rounded condyles, generally laterally compressed, see Guidetti et al. 2012), with some exceptions. Smooth eggs laid in the exuvium or ornamented eggs laid freely.} (Marley et al. 2011; Bertolani et al. 2014)

Superfamily Eohypsibioidea Bertolani & Kristensen, 1987

{Double claws similar in shape and arranged asymmetrically with respect to the median plane of the leg (conventionally described as sequence 2121); claws may be of similar size; double claws of Eohypsibiidae type (= *Bertolanius* type): subdivided into three distinct sections, basal section, secondary branch and primary branch (rigidly joined to the secondary branch), one on top of the other and separated by septa; the angles between the main and secondary branch different in external and internal claws—an acute angle (about 45°) is formed by the external claw, and an almost right angle (about 80°) by the internal claw; internal claws can rotate on their base up to 180°; lunules present. Buccal tube completely rigid or caudally annulated; ventral lamina absent; 14 peribuccal lamellae present; pharyngeal apophyses and placoids present. Eggs laid freely (or in the old exuvium) and surrounded by an ornamented shell.} (Pilato and Binda 2010; Trygvadóttir and Kristensen 2011; Bertolani et al. 2014)

Eohypsibiidae Bertolani & Kristensen, 1987 [3 genera]
{as the superfamily} (Bertolani et al. 2014)

Austeruseus **Trygvadóttir & Kristensen, 2011** [*A. faeroensis* Trygvadóttir & Kristensen, 2011; 3 species]
{Rigid straight buccal tube; 2 to 6 hook-shaped appendages for insertion of the stylet muscles present on the sides of the buccal tube; the mouth tubular; ventral cuticular enforcement of the buccal tube as a thin line (instead of ventral lamina). Eggs laid freely.} (Trygvadóttir and Kristensen 2011)

Bertolanius **Özdikmen, 2008** [*Isohypsibius smreczynskii* Węglarska, 1970; 8 species]
{Bucco-pharyngeal apparatus of the *Bertolanius* model: tube rigid; apophyses for the insertion of the stylet muscles in the shape of "ridges" symmetrical with respect to the frontal plane; caudal processes of both these apophyses pointing backwards and sideways; trumpet-shaped mouth; stylet furcae typically shaped. Eggs laid freely.} (Pilato and Binda 2010; Trygvadóttir and Kristensen 2011)

Eohypsibius **Kristensen, 1982** [*E. nadjae* Kristensen, 1982; 2 species]
{An extremely slender body with reduced legs and claws. Bucco-pharyngeal apparatus of the *Eohypsibius* model: buccal tube followed by an annulated pharyngeal tube; apophyses for the insertion of the muscles of the stylets form "wingshaped ridges" symmetrical with respect to the frontal plane; caudal processes of both these apophyses pointing backwards and sideways; very large, trumpet-shaped mouth; stylet furcae of the *Eohypsibius* type (arched branches, without particularly swollen apices). Eggs laid freely or laid in the old exuvium and always strongly sculptured.} (Pilato and Binda 2010; Trygvadóttir and Kristensen 2011)

Superfamily Hypsibioidea Pilato, 1969

{Double claws asymmetrical with respect to the median plane of the leg (sequence 2121); double claws with different shapes: double claws of the same leg can be different in the size and shape—the *Hypsibius* type (the external, or posterior, with the secondary branch forming a continuous curve with its basal tract and the primary branch connected to the basal tract with a flexible part while the internal claw has a rigid structure—a solid unit of basal section, secondary branch and primary branch), the *Ramazzottius* type (external claws with basal section longer than the secondary branch; primary branch very long and slender and connected to basal tract with an evident, thin, flexible tract; internal claws short and stout, unified into a single rigid element); double claws of the same leg can be similar in size and shape—the *Calohypsibius* type (small, rigid, in frontal view with a base as large as the sum of the primary and secondary branch widths, without a suture between the two branches), the *Microhypsibius* type (small, rigid, with an evident thin basal tract continuous with the primary branch; the secondary branch rigidly joined to the primary branch at a distance from the base of the claw). The genera *Acutuncus*, *Mixibius* and *Sarascon* with internal and/or external claws of the *Isohypsibius* type (see the superfamily Isohypsibioidea). Buccal tube completely rigid, or caudally annulated and always without ventral lamina; peribuccal lamellae absent in all genera; stylet supports present with one exception (*Astatumen*); stylet furcae mostly typically shaped, or different. Smooth eggs laid in exuvium, or laid freely and surrounded with an ornamented shell.} (Pilato and Binda 2010; Marley et al. 2011; Bertolani et al. 2014)

Calohypsibiidae Pilato, 1969 [1 genus]
{Double claws of the *Calohypsibius* type; lunules absent. Bucco-pharyngeal apparatus of the *Calohypsibius* type: buccal tube completely rigid; apophyses for the insertion of the stylet muscles on the buccal tube asymmetrical with respect to the frontal plane; the dorsal apophysis subdivided into two portions: the anterior portion as stumpy hook with a blunt caudal apex, the caudal portion as short longitudinal thickening; ventral apophysis as a very slightly prominent ridge; both of these apophyses with two processes, but those of the dorsal apophysis point backwards and sideways while those of the ventral apophysis point laterally; pharyngeal apophyses and placoids present; stylet furcae typically shaped. Smooth eggs laid in the exuvium.} (Pilato and Binda 2010; Bertolani et al. 2014)

***Calohypsibius* Thulin, 1928** [*Macrobiotus ornatus* Richters, 1900; 5 species]
{as the family} (Pilato and Binda 2010)

Hypsibiidae Pilato, 1969 [5 subfamilies; 14 genera]
{Double claws of the *Hypsibius* type (except genera *Acutuncus*, *Mixibius* and *Sarascon*); external and posterior claws always clearly larger than internal and anterior claws; lunules present or absent. Buccal tube rigid or followed by a flexible pharyngeal tube; dorsal and ventral apophyses for the insertion of the stylet muscles on the buccal tube are symmetrical or asymmetrical with respect to the frontal plane; pharyngeal apophyses and placoids present or absent. Eggs smooth (but rarely weakly ornamented) laid within the exuvium, in some cases eggs ornamented laid free.} (Pilato and Binda 2010; Bertolani et al. 2014; Gąsiorek et al. 2018)

Diphasconinae Dastych, 1992 [1 genus]
{Lunules absent. Bucco-pharyngeal apparatus of the *Diphascon* type: buccal tube followed by an annulated pharyngeal tube, with a cuticular thickening between them (often drop-shaped, sometimes barely evident); apophyses for the insertion of the stylet muscles in the shape of "semilunar hooks" shaped symmetrical with respect to the frontal plane, caudal processes of both these apophyses pointing backwards and sideways; stylet furcae typically shaped; pharyngeal apophyses present; pharyngeal bulb with always 3 macroplacoids in a line (and sometimes with a microplacoid and/or septulum). Smooth eggs laid in the exuvium.} (Pilato and Binda 2010; Bertolani et al. 2014)

***Diphascon* Plate, 1888** [*D. chilenense* Plate, 1888; 40 species]
{as the subfamily} (Bertolani et al. 2014)

Hypsibiinae Pilato, 1969 [2 genera]
{Lunules present or absent. Buccal tube completely rigid; pharyngeal apophyses and placoids present.} (Pilato and Binda 2010; Bertolani et al. 2014)

***Borealibius* Pilato, Guidetti, Rebecchi, Lisi, Hansen & Bertolani, 2006** [*Macrobiotus zetlandicus* Murray, 1907; 1 species]
{Lunules absent. Bucco-pharyngeal apparatus of the *Borealibius* type: buccal tube rigid; dorsal and ventral walls of the buccal cavity very thickened, their caudal margins form a prominent transverse crest; apophyses for the insertion of the stylet muscles stumpy, almost cylindrical in shape with a large rounded (or bilobed) extremity diverging from the buccal tube wall; both apophyses with two caudal processes pointing backwards and sideways; branches of the stylets' furcae of the *Borealibius* type (furcae branches very robust, with thickened, swollen, and rounded apices). Smooth eggs laid in the exuvium.} (Pilato and Binda 2010)

***Hypsibius* Ehrenberg, 1848** [*Macrobiotus dujardini* Doyère, 1840; 43 species]
{Lunules absent or present but difficult to see. Bucco-pharyngeal apparatus of the *Hypsibius* type: buccal tube rigid; apophyses for the insertion of the stylet muscles in the form of "semilunar hooks" symmetrical with respect to the frontal plane; both apophyses with two caudal processes pointing backwards and sideways; stylet furcae typically shaped. Eggs laid in the exuvium, generaly smooth but with fine ornamentation in some species.} (Ramazzotti and Maucci 1983; Pilato and Binda 2010; Pilato et al. 2012)

Itaquasconinae Bartoš, 1964 [8 genera]
{Lunules present or absent. Rigid buccal tube followed by an annulated flexible pharyngeal tube, without cuticular thickening between them or entire tube rigid (*Parascon* and *Sarascon*), placoids very long and in line or absent. Eggs, if known, smooth and laid in the exuvium.} (Pilato and Binda 2010; Bertolani et al. 2014)

***Adropion* Pilato, 1987** [*Diphascon scoticum* Murray, 1905; 20 species]
{Lunules absent. Apophyses for the insertion of the stylet muscles in the shape of "semilunar hooks" shaped symmetrical with respect to the frontal plane, caudal processes of both these apophyses pointing backwards and sideways; stylet furcae typically shaped; pharyngeal apophyses present or rarely absent; placoids present.} (Pilato and Binda 2010; Bertolani et al. 2014)

***Astatumen* Pilato, 1997** [*Itaquascon trinacriae* Arcidiacono, 1962; 4 species]
{Lunules absent. Bucco-pharyngeal apparatus of the *Astatumen* type: very short rigid buccal tube; stylet supports absent; the apophyses for the insertion of the stylet muscles as long as a buccal tube, and in the shape of "wide and flat ridges" symmetrical with respect to the frontal plane; caudal processes of both these apophyses very slender and pointing sideways; very small stylet furcae of the *Astatumen* type (furcae branches very short and taper at their apices); pharyngeal apophyses absent; no placoids, or only one undivided placoid present.} (Pilato and Binda 2010)

***Bindius* Pilato, 2009** [*B. triquetrus* Pilato, 2009; 1 species]
{Lunules absent. Apophyses for the insertion of the stylet muscles asymmetrical with respect to the frontal plane; the ventral apophysis in the shape of a "semilunar hook", the dorsal apophysis very prominent, triangular with rectilinear dorsal margin; caudal processes of both these apophyses very slender and pointing sideways;

stylet furcae typically shaped; pharyngeal apophyses and placoids present. Eggs unknown.} (Pilato and Binda 2010)

***Itaquascon* de Barros, 1939** [*I. umbellinae* de Barros, 1939; 12 species]
{Lunules absent. Bucco-pharyngeal apparatus of the *Itaquascon* type: the apophyses for the insertion of the stylet muscles clearly shorter as a buccal tube; these apophyses in the shape of "wide and flat ridges" symmetrical with respect to the frontal plane, caudal processes of both these apophyses very slender, almost invisible, and pointing sideways; very small stylet furcae of the *Itaquascon* type (furcae branches very short and without swollen apices); pharyngeal apophyses absent; no placoids, or only one undivided placoid present). Eggs smooth and laid in the exuvium.} (Pilato and Binda 2010)

***Mesocrista* Pilato, 1987** [*Diphascon spitzbergense* Richters, 1903; 2 species]
{Lunules absent. Bucco-pharyngeal apparatus of the *Mesocrista* type: apophyses for the insertion of the stylet muscles in the shape of "wide and flat ridges" symmetrical with respect to the frontal plane, caudal processes of both these apophyses pointing backwards and sideways); stylet furcae of the *Mesocrista* type (the basal portion of the furcae not very enlarged and with branches with thickened, swollen, and rounded apices); pharyngeal apophyses absent; placoids present.} (Pilato and Binda 2010)

***Parascon* Pilato & Binda, 1987** [*P. schusteri* Pilato & Binda, 1987; 2 species]
{Lunules absent. Bucco-pharyngeal apparatus of the *Parascon* type: apophyses for the insertion of the stylet muscles in the shape of "wide and flat ridges" symmetrical with respect to the frontal plane, caudal processes of both these apophyses very slender, almost invisible, and pointing sideways); stylet furcae very small, of the *Itaquascon* type (see *Itaquascon*); pharyngeal apophyses absent; only one undivided placoid present. Eggs never found.} (Pilato and Binda 2010)

***Platicrista* Pilato, 1987** [*Diphascon angustatum* Murray, 1905; 6 species]
{Lunules generally absent (reported only in hind legs of one species). Bucco-pharyngeal apparatus of the *Platicrista* type: apophyses for the insertion of the stylet muscles in the shape of "wide and flat ridges" symmetrical with respect to the frontal plane, caudal processes of both these apophyses slender and pointing sideways); stylet furcae of the *Platicrista* type (furcae branches pointing backwards instead of backwards and sideways, in lateral view with a spoon-like shape and tapered at their apices); pharyngeal apophyses absent; placoids present.} (Pilato and Binda 2010)

***Sarascon* Guil, Rodrigo & Machordom, 2014** [*S. hortensiae* Guil *et al.*, 2014; 1 species]
{Claws of each leg very different from each other; external claw of the *Hypsibius* type with extremely long and slender primary branch; internal claw of the *Isohypsibius* type (claws similar to those in *Ramajendas*, Ramazzottiidae); lunules present. Bucco-pharyngeal apparatus of the *Parascon* type: apophyses for the insertion of the stylet muscles in the shape of "not wide flat ridges" symmetrical with respect to the frontal plane, caudal processes of both these apophyses slender and pointing sideways; stylet furcae very small, of the *Itaquascon* type (see

Itaquascon); pharyngeal apophyses and placoids absent. Eggs unknown.} (Guil et al. 2015)

Pilatobiinae Bertolani, Guidetti, Marchioro, Altiero, Rebecchi & Cesari, 2014 [1 genus]
{Lunules absent, rarely present. Bucco-pharyngeal apparatus of the *Pilatobius* type: buccal tube followed by an annulated pharyngeal tube, with a cuticular thickening between them (often drop-shaped); apophyses for the insertion of the stylet muscles in the shape of "semilunar hooks" shaped symmetrical with respect to the frontal plane, caudal processes of both these apophyses pointing backwards and sideways; stylet furcae typically shaped; pharyngeal apophyses present; pharyngeal bulb roundish or slightly oval, always containing two macroplacoids similar in length and in rows that look as parentheses, and a septulum. Smooth eggs laid in the exuvium.} (Pilato and Binda 2010; Bertolani et al. 2014)

Pilatobius **Bertolani, Guidetti, Marchioro, Altiero, Rebecchi & Cesari, 2014** [*Diphascon bullatum* Murray, 1905; 25 species]
{as the subfamily} (Bertolani et al. 2014)

Incerta subfamilia {according to Bertolani et al. 2014} [2 genera]
{External claws of the *Isohypsibius* type, internal ones different. Buccal tube rigid; two caudal processes of both apophyses for the insertion of the stylets muscles pointing backwards and sideways; stylet furcae typically shaped; pharyngeal apophyses and placoids present.}

Acutuncus **Pilato & Binda, 1997** [*Macrobiotus antarcticus* Richters, 1904; 1 species]
{External claws of the *Isohypsibius* type, internal claws of the *Hypsibius* type; lunules absent. Bucco-pharyngeal apparatus of the *Hypsibius* type: apophyses for the insertion of the stylets muscles in form of "acute hooks" symmetrical with respect to the frontal plane; a median cuticular thickening caudal to both these apophyses present, the dorsal more evident. Eggs laid freely; egg shell provided with processes.} (Pilato and Binda 2010)

Mixibius **Pilato, 1992** [*Isohypsibius saracenus* Pilato, 1973; 10 species]
{External claws of the *Isohypsibius* type, internal claws of modified *Isohypsibius* type with the secondary branch and the basal portion forming an angle slightly larger than 90°; lunules present or absent. Bucco-pharyngeal apparatus of the *Mixibius* type: apophyses for the insertion of the stylet muscles in the shape of "hooks" slightly asymmetrical with respect to the frontal plane, the ventral apophysis similar, but not identical, to the "semilunar hook" of *Hypsibius,* dorsal apophysis more stumpy with a blunt and swollen caudal apex, a short median cuticular thickening caudal to both these apophyses present (the ventral one slightly visible). Eggs unknown.} (Pilato and Binda 2010; Pilato et al. 2010; Lisi et al. 2014)

Microhypsibiidae Pilato, 1998 [2 genera]
{Double claws of the *Microhypsibius* type; lunules absent. Buccal tube completely rigid; dorsal and the ventral apophyses for the insertion of the stylet muscles on the

buccal tube asymmetrical with respect to the frontal plane, and with two very thin caudal processes pointing backwards and sideways; pharyngeal apophyses and placoids present; stylet furcae typically shaped. Smooth eggs laid in the exuvium.} (Pilato and Binda 2010; Bertolani et al. 2014)

***Fractonotus* Pilato, 1998** [*Hypsibius ornatus* f. *caelatus* Marcus, 1928; 1 species]
{A paired elliptical organ on the head present. Dorsal apophysis for the insertion of the stylet muscles in the shape of a stumpy hook with a blunt caudal apex followed by a median, longitudinal, cuticular thickening; the ventral apophysis as a very slightly prominent ridge with no hook; a slightly evident thickening on the lateral walls of the buccal tube, behind the stylet supports, present.} (Pilato and Binda 2010)

***Microhypsibius* Thulin, 1928** [*M. truncatus* Thulin, 1928; 4 species]
{Body terminated bluntly. A paired elliptical organ on the head absent. Dorsal apophysis for the insertion of the stylet muscles in the shape of a "semilunar hooks" followed by a short cuticular thickening gradually decreasing in height; the ventral apophysis as a "ridge with an evident blunt hook". Three macroplacoids present.} (Kristensen 1982a; Pilato and Binda 2010)

Ramazzottiidae Sands, McInnes, Marley, Goodall-Copestake, Convey & Linse, 2008 [3 genera]
{Double claws of *Ramazzottius* type or *Hypsibius* type (or combined with claws of *Isohypsibius* type); lunules present or absent. Buccal tube completely rigid or divided into anterior rigid buccal tube and posterior annulated pharyngeal tube without drop-shaped cuticular thickening between them; apophyses for the insertion of the stylet muscles on the buccal tube, if known, with caudal processes pointing backwards and sideways; pharyngeal apophyses and placoids present; stylet furcae, if described, typically shaped. Smooth eggs laid in the exuvium or ornamented ones laid freely.} (Pilato and Binda 2010; Bertolani et al. 2014)

***Cryoconicus* Zawierucha, Stec, Lachowska-Cierlik, Takeuchi, Z. Li & Michalczyk, 2018** [*C. kaczmareki* Zawierucha *et al.*, 2018; 2 species]
{Juveniles light-brown, adults intensely dark-brown. Claws of the *Ramazzottius* type, but without accessory points; the posterior primary branch almost uniform in diameter from the base to the curving; wide, semi-transparent cuticular bars under claws I–III. Buccal apparatus with asymmetrical apophyses for the insertion of the stylet muscles and two macroplacoids; microplacoid and septulum absent.} (Zawierucha et al. 2018)

***Hebesuncus* Pilato, 1987** [*Hypsibius conjungens* Thulin, 1911; 4 species]
{Double claws of the *Ramazzottius* type; lunules absent. The tube consists of rigid anterior buccal tube and flexible posterior pharyngeal tube; apophyses for the insertion of the stylet muscles in the shape of "blunt hooks" asymmetrical with respect to the frontal plane; dorsal apophysis short, thick, and with caudal apex stumpy and prominent; ventral apophysis long, slender, whose height decreases rapidly but not abruptly. Eggs laid freely; egg shell with processes.} (Pilato and Binda 2010; Bertolani et al. 2014)

***Ramazzottius* Binda & Pilato, 1986** [*Macrobiotus oberhaeuseri* Doyère, 1840; 29 species]
{A paired elliptical organ present on the head. Double claws of the *Ramazzottius* type; very small, reduced, lunules present or absent. Bucco-pharyngeal apparatus of the *Hypsibius* type: buccal tube rigid; apophyses for the insertion of the stylet muscles in the shape of "blunt hooks" asymmetrical with respect to the frontal plane, the dorsal apophysis shorter and stumpy, with caudal apex clearly prominent, ventral apophysis longer and more slender, with a less developed caudal apex; a median cuticular thickening caudal to both these apophyses present. Eggs laid freely; egg shell with processes.} (Pilato and Binda 2010; Bartels et al. 2011)

Superfamily Isohypsibioidea Sands, McInnes, Marley, Goodall-Copestake, Convey & Linse, 2008

{Double claws with similar shape and size on each leg, and asymmetrical with respect to the median plane of the leg (sequence 2121) or completely absent (*Apodibius*, family Isohypsibiidae). Secondary branch can be reduced to a spur (*Haplomacrobiotus*, family Hexapodibiidae) or absent (simple claws—*Haplohexapodibius*, *Haplomacrobiotus*, family Hexapodibiidae); when normally developed double claw of *Isohypsibius* type (secondary branch of the claw inserted perpendicularly on the claw basal tract), or *Hexapodibius* type (very short, without common basal tract, with a base as large as the sum of the primary and secondary branch widths, and with an evident suture between primary and secondary branch). Buccal tube completely rigid, or caudally annulated with or without ventral lamina; peribuccal lamellae present or absent, but never 10 or 14 in number, as in Macrobiotoidea or in Eohypsibioidea; stylet furcae typically shaped; pharyngeal apophyses and placoids present. Eggs with smooth shell laid within the exuvium.} (Pilato and Binda 2010; Pilato 2013; Bertolani et al. 2014)

Hexapodibiidae Cesari, Vecchi, Palmer, Bertolani, Pilato, Rebecchi & Guidetti, 2016 [4 genera]
{Double claws of *Hexapodibius* type or with only the main branch, with the secondary branch being reduced or sometimes absent on some legs. Lunules and other cuticular thickenings absent on the legs in the known species. Bucco-pharyngeal apparatus of the *Hexapodibius* type: buccal tube rigid with ventral lamina, very narrow anterior portion of ventral lamina continues abruptly in a larger triangular portion with two processes pointing backwards and sideways; peribuccal lamellae absent, peribuccal papulae present.} (Cesari et al. 2016)

***Haplohexapodibius* Pilato & Beasley, 1987** [*H. seductor* Pilato & Beasley, 1987; 1 species]
{The claws on the first three pairs of legs without secondary branches (simple claws), hind legs reduced and without claws. Eggs never found.} (Pilato and Binda 2010)

***Haplomacrobiotus* May, 1948** [*H. hermosillensis* May, 1948; 2 species]
{Claws with the secondary branches reduced to a thin and short spur or secondary branches absent (simple claws). Eggs unknown.} (Pilato and Binda 2010)

***Hexapodibius* Pilato, 1969** [*H. micronyx* Pilato, 1969; 6 species]
{Claws of the first three pairs of legs of the *Hexapodibius* type. Hind legs reduced and without claws. Peribuccal lamellae absent.} (Pilato and Binda 2010)

***Parhexapodibius* Pilato, 1969** [*Hypsibius lagrecai* Binda & Pilato, 1969; 5 species]
{Claws of the *Hexapodibius* type; those of the hind legs may be shorter than the others, sometimes with the secondary branch reduced; in some species only one double claw is present on the hind legs.} (Pilato and Binda 2010)

Isohypsibiidae Sands, McInnes, Marley, Goodall-Copestake, Convey & Linse, 2008 [11 genera]
{Characteristics of the family not defined by Cesari et al. (2016) who erected the Hexapodibiidae.}

***Apodibius* Dastych, 1983** [*A. confusus* Dastych, 1983; 3 species]
{Short bilobate legs. Claws and lunules absent. Bucco-pharyngeal apparatus of the *Apodibius* type: buccal tube rigid with ventral lamina; very narrow anterior portion of a ventral lamina continues abruptly into a larger triangular portion with two processes pointing backwards and sideways; peribuccal lamellae absent. Eggs never found.} (Pilato and Binda 2010)

***Dastychius* Pilato, 2013** [*Isohypsibius improvisus* Dastych, 1984; 1 species]
{Claws of the *Isohypsibius* type; lunules present. Bucco-pharyngeal apparatus of the *Isohypsibius* type: buccal tube rigid without ventral lamina; dorsal and ventral apophyses for the insertion of the stylet muscles in shape of long, continuous ridges tailing off caudally and almost reaching the level of the stylet supports, at about a quarter the length of the ridged apophyses an incision and septum present; peribuccal lamellae absent.} (Pilato 2013)

***Doryphoribius* Pilato, 1969** [*Hypsibius doryphorus* Binda & Pilato, 1969; 40 species]
{Claws of the *Isohypsibius* type; lunules present or absent. Bucco-pharyngeal apparatus of the *Doryphoribius* type: buccal tube rigid with ventral lamina, which anterior portion with triangular with the vertex turned backwards, leading to a broad posterior triangular portion with two processes pointing backwards and sideways, peribuccal lamellae absent).} (Pilato and Binda 2010; Lisi 2011)

***Eremobiotus* Biserov, 1992** [*E. ovezovae* Biserov, 1992; 3 species]
{Claws of the *Isohypsibius* type on the first three pairs of legs, hind leg claws modified with two branches of each claw joined to one another for a long portion of their length constituting a large common tract by which the two distal tracts diverge forming an angle of almost 180°; lunules present. Bucco-pharyngeal apparatus of the *Isohypsibius* type (see *Isohypsibius*).} (Biserov 1992; Pilato and Binda 2010)

***Halobiotus* Kristensen, 1982** [*H. crispae* Kristensen, 1982; 4 marine species] {Cyclomorphosis is known in this genus, therefore the taxonomic details refer to the active stage. Claws of the *Isohypsibius* type; lunules present. Two pairs of dome-shaped organs present on the head (anterior papillae cephalicae and posterior lateral temporalia). Bucco-pharyngeal apparatus of the *Hypsibius* type: rigid buccal tube without ventral lamina; apophyses for the insertion of the stylet muscles in form of "semilunar hooks" symmetrical with respect to the frontal plane; a long median cuticular thickening caudally to these apophyses, caudal processes of both apophyses pointing backwards and sideways; six peribuccal sensory structures (papulae) present in the active stage.} (Kristensen 1982b; Pilato and Binda 2010)

***Isohypsibius* Thulin, 1928** [*I. prosostomus* Thulin, 1928; 128 species] {Claws of the *Isohypsibius* type; lunules present or absent. Bucco-pharyngeal apparatus of the *Isohypsibius* type: rigid buccal tube without ventral lamina; dorsal and ventral apophyses for the insertion of the stylet muscles as a "ridge" shaped symmetrical with respect to the frontal plane, caudal processes of both apophyses pointing backwards and sideways; peribuccal lamellae absent. Eggs of many species unknown.} (Pilato and Binda 2010)

***Paradiphascon* Dastych, 1992** [*P. manningi* Dastych, 1992; 1 species] {Claws of the *Isohypsibius* type; lunules present. Head provided with one median and two lateral lobes with flat and rounded apices. Bucco-pharyngeal apparatus of the *Paradiphascon* type: buccal tube rigid followed by an annulated, flexible pharyngeal tube, a large cuticular thickening present between the buccal tube and the pharyngeal tube; ventral lamina absent; apophyses for the insertion of the stylet muscles "triangular ridges" shaped symmetrical with respect to the frontal plane, caudal processes of both these apophyses pointing backwards and sideways; peribuccal lamellae absent.} (Dastych 1992; Pilato and Binda 2010)

***Pseudobiotus* Nelson, 1980** [*P. kathmanae* Nelson, Marley & Bertolani, 1999; 7 species] {Claws of the *Isohypsibius* type with very flexible primary branch; lunules absent. Bucco-pharyngeal apparatus of the *Isohypsibius* type (see *Isohypsibius*); about 30 peribuccal lamellae present; apophyses for the insertion of the stylet muscles as dorsal and ventral double crests in form of two connected expansions (the anterior longer than the posterior).} (Nelson et al. 1999; Pilato and Binda 2010)

Ramajendas Pilato & Binda, 1990 [*Hypsibius renaudi* Ramazzotti, 1972; 4 species, one of them marine] {Claws of each leg very different from each other; external claw of the *Hypsibius* type with extremely long and slender primary branch; internal claw of the *Isohypsibius* type; lunules present. Bucco-pharyngeal apparatus of the *Isohypsibius* type: buccal tube rigid; dorsal and a ventral apophysis for the insertion of the stylet muscles as a very long "ridge" and symmetrical with respect to the frontal plane. Smooth eggs laid in the exuvium.} (Pilato & Binda 2010)

***Thalerius* Dastych, 2009** [*T. konradi* Dastych, 2009; 1 species]
{Claws of each leg very different from each other; double claws of the *Hypsibius* type, external ones with extremely long and slender primary branch; the bases of external claws expanded posteriorly, each with roundish, deep incision; the sides of the incision with a well formed tooth; lunules present in internal claws only. Rigid buccal tube; apophyses for the insertion of the stylet muscles unknown; stylet furcae undescribed. Eggs unknown.} (Dastych 2009)

***Thulinius* Bertolani, 2003** [*Isohypsibius stephaniae* Pilato, 1974; 6 species, one of them marine]
{Claws of the *Isohypsibius* type; lunules present or absent. Bucco-pharyngeal apparatus of the *Isohypsibius* type (see *Isohypsibius*); 12 peribuccal lamellae present.} (Pilato and Binda 2010)

Superfamily Macrobiotoidea Thulin, 1928

{Double claws symmetrical with respect to the median plane of the leg (sequence 2112); double claw of each leg similar in shape and size; each double claw usually characterized by the presence of a peculiar stalk (peduncle) with cylindrical or laminar shape (the only exception is *Xerobiotus* type of claws in *Biserovus* and *Xerobiotus*, Macrobiotidae). Ten peribuccal lamellae or papulae or velum in a buccal opening; buccal tube completely rigid, or caudally annulated, strengthened by a ventral lamina; stylet furcae typically shaped with some exceptions (*Insuetifurca*, Macrobiotidae and *Adorybiotus* and *Richtersius*, Richtersiidae); pharyngeal apophyses and placoids present. Eggs laid freely and always surrounded by an ornamented shell.} (Pilato and Binda 2010; Pilato and Lisi 2011; Bertolani et al. 2014)

Macrobiotidae Thulin, 1928 [13 genera]
{Double claws Y-shaped (both branches form an evident common basal tract of variable length). Double claws of *Macrobiotus* (=*hufelandi*) type (the primary and the secondary branches rigidly joined to each other, forming an acute angle, for a variable but long common basal tract provided with a cylindrical stalk), of *Mesobiotus* type (the primary and the secondary branches rigidly joined to each other, forming an acute angle, for a variable but long common basal tract provided with an internal septum defining a distal part terminating with a cylindrical stalk), or *Xerobiotus* type (the whole double claw equally sclerified; the basal section not subdivided into a basal stem and a distinct distal portion, no separating septa present; the primary and secondary branches rigidly joined to each other forming an acute angle), or *Calcarobiotus* type (the basal section of each double claw, with or without basal spurs, subdivided into a thin flexible stem and a wide distal section in the shape of an upside-down triangle distally delimited by a septum; primary and secondary branches similar in shape and size), or *tenuis* type (with a stalk, a long common tract formed by the joined primary and secondary branches, and distal sections of both the secondary branch—shorter than the primary—and primary branch forming almost a right angle). Buccal tube completely rigid, or caudally annulated. The anterior portion of the ventral lamina with two caudal processes pointing backwards and

sideways. Epicuticular layer compact, without pillar-like structures.} (Pilato and Binda 2010; Pilato and Lisi 2011; Bertolani et al. 2014)

***Biserovus* Guidetti & Pilato, 2003** [*Pseudodiphascon bindae* Christenberry & Higgins, 1979; 1 species]
{Claws of the *Xerobiotus* type; lunules absent. Cuticle without pores. Bucco-pharyngeal apparatus of the *Biserovus* type: buccal tube rigid followed by an annulated, flexible pharyngeal tube; 10 peribuccal lamellae present; buccal armature with one system of dorsal and ventral transverse ridges (the mid-dorsal ridge with the shape of a small triangle with a vertex pointing backwards) in the posterior portion of buccal cavity. Eggs never found.} (Guidetti and Pilato 2003; Pilato and Binda 2010)

***Calcarobiotus* Dastych, 1993** [*C. filmeri* Dastych, 1993; 2 sub-genera]
{Claws of the *Calcarobiotus* type at least in the first three pairs of legs; lunules present. Cuticle without pores. Bucco-pharyngeal apparatus of the *Macrobiotus* type (see *Macrobiotus*).} (Pilato and Binda 2010)

***Calcarobiotus* (*Calcarobiotus*) Dastych, 1993** [8 species]
{Claws in the hind legs of the *Calcarobiotus* type.} (Guidetti and Bertolani 2001)

***Calcarobiotus* (*Discrepunguis*) Guidetti & Bertolani, 2001** [2 species]
{Claws in the hind legs of the *hufelandi* type.} (Guidetti and Bertolani 2001)

***Famelobiotus* Pilato, Binda & Lisi, 2004** [*F. scalicii* Pilato *et al.*, 2004; 1 species]
{Claws of the *hufelandi* type; lunules present. Cuticle without pores. Bucco-pharyngeal apparatus of the *Macrobiotus* type: buccal tube completely rigid without the posterior bend; 10 peribuccal lamellae present; buccal armature with two systems of dorsal and ventral transverse ridges, instead of only one, in the posterior portion of the buccal cavity. Eggs never found.} (Pilato and Binda 2010)

***Insuetifurca* Guidetti & Pilato, 2003** [*Pseudodiphascon arrowsmithi* Kathman & Nelson, 1989; 4 species]
{Claws of modified *hufelandi* type—the basal section of claws subdivided into a very short stem and a large distal portion; lunules present. Cuticle without pores. Bucco-pharyngeal apparatus of the *Biserovus* type: buccal tube rigid followed by an annulated, flexible pharyngeal tube; 10 peribuccal lamellae present; buccal armature with one system of dorsal and ventral transverse ridges in the posterior portion of buccal cavity; stylet furcae of the *Insuetifurca* type (the branches arched and converging backwards). Eggs never found.} (Guidetti and Pilato 2003; Pilato and Binda 2010)

***Macrobiotus* C.A.S. Schultze, 1834** [*M. hufelandi* C.A.S. Schultze, 1834; 107 species]
{Claws of the *hufelandi* type; lunules present. Cuticle with or without pores. Bucco-pharyngeal apparatus of the *Macrobiotus* type: buccal tube completely rigid without the posterior bend; 10 peribuccal lamellae present; buccal armature with one system of dorsal and ventral transverse ridges (sometimes reduced or subdivided into distinct teeth) in the posterior portion of buccal cavity. Egg shell typically with processes (few exceptions known).} (Pilato and Binda 2010)

***Mesobiotus* Vecchi, Cesari, Bertolani, Jönsson, Rebecchi & Guidetti, 2016** [*Macrobiotus harmsworthi* Murray, 1907; 60 species]
{Claws of the *Mesobiotus* type; lunules present. Cuticle without pores (one exception known). Bucco-pharyngeal apparatus of the *Macrobiotus* type (see *Macrobiotus*); three roundish macroplacoids arranged along a curved line; microplacoid clearly close (less than its length) to the third macroplacoid. Eggs with conical or hemispherical processes, generally with pointed tips.} (Vecchi et al. 2016)

***Minibiotus* R.O. Schuster, 1980** [*Macrobiotus intermedius* Plate, 1888; 48 species]
{Claws of the *hufelandi* type; lunules present. Cuticle with or without pores. Bucco-pharyngeal apparatus of the *Macrobiotus* type: buccal tube completely rigid, often with a posterior bend; peribuccal lamellae absent, 10 peribuccal papulae present (sometimes very difficult to see); buccal armature very simple; stylet supports inserted at anterior position (73% or less of the buccal tube length).} (Pilato and Binda 2010)

***Minilentus* Guidetti & Pilato, 2003** [*Pseudodiphascon dubius* Schuster & Toftner, 1982; 1 species]
{Claws of the *hufelandi* type; lunules present. Cuticle with pores. Bucco-pharyngeal apparatus of the *Biserovus* type: buccal tube rigid followed by an annulated, flexible pharyngeal tube; buccal tube narrow; peribuccal lamellae absent, 10 peribuccal papulae present (sometimes very difficult to see); buccal armature very simple. Eggs never found.} (Pilato and Binda 2010)

***Paramacrobiotus* Guidetti, Schill, Bertolani, Dandekar & Wolf, 2009** [*Macrobiotus richtersi* Murray, 1911; 2 sub-genera]
{Claws of the *hufelandi* type; lunules present. Cuticle without pores (one exception known). Bucco-pharyngeal apparatus of the *Macrobiotus* type: buccal tube completely rigid without the posterior bend; buccal armature with one system of dorsal and ventral transverse ridges (sometimes reduced or subdivided into distinct teeth) in the posterior portion of buccal cavity; three macroplacoids present in pharyngeal bulb; microplacoid, if present, distant from the last macroplacoid more than its length. Eggs, laid freely, with conical processes and areolated egg-shell.} (Pilato and Binda 2010)

***Paramacrobiotus (Amicrobiotus)* Marley, Kaczmarek, Gawlak, Bartels, Nelson, Roszkowska, Stec & Degma, 2018** [12 species]
{Microplacoid absent.} (Marley et al. 2018)

***Paramacrobiotus (Paramacrobiotus)* Guidetti, Schill, Bertolani, Dandekar & Wolf, 2009** [26 species]
{Microplacoid present, with two lateral "wings".} (Kaczmarek et al. 2017)

***Pseudohexapodibius* Bertolani & Biserov, 1996** [*Hexapodibius degenerans* Biserov, 1990; 1 species]
{Claws on the first three pairs of legs of the *Xerobiotus* type; claws on the fourth pair of legs reduced to granular thickenings or absent; lunules absent. Cuticle without

pores. Bucco-pharyngeal apparatus of the *Macrobiotus* type (see *Macrobiotus*). Eggs never found.} (Pilato and Binda 2010)

***Schusterius* Kaczmarek & Michalczyk, 2006** [*Macrobiotus tridigitus* R.O. Schuster, 1983; 1 species]
{Claws of the *Calcarobiotus* type without basal spurs in their basal section and with extremely developed, long and thin accessory points connected to their main branches by a flexible portion; lunules present. Cuticle without pores. Bucco-pharyngeal apparatus of the *Macrobiotus* type (see *Macrobiotus*). Eggs unknown.} (Kaczmarek and Michalczyk 2006; Pilato and Binda 2010)

***Tenuibiotus* Pilato & Lisi, 2011** [*Macrobiotus tenuis* Binda & Pilato, 1972; 13 species]
{Claws of the *tenuis* type; lunules present. Cuticle with or without pores. Bucco-pharyngeal apparatus of the *Macrobiotus* type: buccal tube narrow and completely rigid without the posterior bend; 10 peribuccal lamellae present; small buccal cavity often with poorly developed buccal armature.} (Pilato and Binda 2010; Pilato and Lisi 2011)

***Xerobiotus* Bertolani & Biserov, 1996** [*Macrobiotus pseudohufelandi* Iharos, 1966; 3 species]
{Claws of the *Xerobiotus* type; lunules present only on the hind legs. Cuticle without pores. Bucco-pharyngeal apparatus of the *Macrobiotus* type: buccal tube completely rigid without the posterior bend; 10 peribuccal lamellae present; buccal armature with one system of dorsal and ventral transverse ridges with the tendency of joining of dorsal ridges.} (Pilato and Binda 2010; Pilato et al. 2011)

Murrayidae Guidetti, Rebecchi & Bertolani, 2000 [3 genera]
{Double claws V-shaped (*Murrayon* type) or L-shaped (*Dactylobiotus* type, *Macroversum* type), with the two branches diverging immediately after a short common basal section. Cuticle without pores. Bucco-pharyngeal apparatus of the *Macrobiotus* type: buccal tube rigid; anterior portion of the ventral lamina with a ventral hook (in *Macroversum* should be confirmed); 10 peribuccal lamellae present. Epicuticular layer with pillar-like structures, sometime visible only at ultrastructural level.} (Pilato and Binda 2010; Bertolani et al. 2014)

***Dactylobiotus* R.O. Schuster, 1980** [*Macrobiotus grandipes* Schuster, Toftner & Grigarick, 1978; 17 species]
{Claws of the *Dactylobiotus* type (= *macronyx* type) (the basal section of the double claw as a trapezoidal lamina; the secondary branch clearly shorter than the primary branch and inserted near the base of the latter; the two branches form an almost right angle); each pair of double claws connected by a cuticular bar; lunules absent.} (Pilato and Binda 2010)

***Macroversum* Pilato & Catanzaro, 1988** [*M. mirum* Pilato & Catanzaro, 1988; 1 species]
{Claws of the *Macroversum* type (the basal section of the double claw subdivided into a thin, flexible, stem and a distal portion not very sclerotised; the primary and the

secondary branches connected to each other for a short portion and form an almost right angle); lunules present—in each leg the two lunules connected to each other by a cuticular bar. Eggs never found.} (Pilato and Binda 2010)

***Murrayon* Bertolani & Pilato, 1988** [*Macrobiotus pullari* Murray, 1907; 8 species] {Claws of the *Murrayon* type (= *pullari* type) (the basal section of the double claw as a trapezoidal lamina; the primary and the secondary branches joined to each other for a very short portion and form an acute angle); lunules present.} (Pilato and Binda 2010)

Richtersiidae Guidetti, Rebecchi, Bertolani, Jönsson, Kristensen & Cesari, 2016 [3 genera]
{Double claws Y-shaped (*hufelandi* type) with the two branches forming an evident common tract of variable length. Large lunules with dense and large teeth on all legs. Cuticle with pores (at least in a phase of the life cycle). Buccal tube completely rigid with a cuticular thick on the anterior, dorsal wall (which can form a large apophysis). Transverse crests in the buccal armature absent. Two macroplacoids in the pharynx.} (Guidetti et al. 2016)

***Adorybiotus* Maucci & Ramazzotti, 1981** [*Macrobiotus granulatus* Richters, 1903; 1 species]
{Cuticle with pores. Bucco-pharyngeal apparatus of the *Adorybiotus* type: buccal tube without a posterior bend; both the anterior portion of dorsal wall of the buccal tube and the anterior portion of the ventral lamina with a longitudinal ridge-shaped thickening; a continuous peribuccal velum present instead of distinct lamellae or papulae; robust stylet furcae of the *Richtersius* type (see in *Richtersius*).} (Pilato and Binda 2010)

***Diaforobiotus* Guidetti, Rebecchi, Bertolani, Jönsson, Kristensen & Cesari, 2016** [*Macrobiotus islandicus* Richters, 1904; 1 species]
{Cuticle with pores. Bucco-pharyngeal apparatus of the *Macrobiotus* type: buccal tube without the posterior bend; 10 peribuccal lamellae present; a dorsal thickening present in the anterior portion of the buccal tube, in conjunction with a large tooth on the internal surface of the tube; some strong, scattered round teeth present posterior to the second band of teeth of the buccal armature; stylet furcae typically shaped.} (Guidetti et al. 2016)

***Richtersius* Pilato & Binda, 1989** [*Macrobiotus coronifer* Richters, 1903; 1 species]
{Cuticle with pores only in the newborns. Bucco-pharyngeal apparatus of the *Richtersius* type: buccal tube without the posterior bend; both the dorsal wall of the buccal tube and the anterior portion of the ventral lamina with a longitudinal ridge with a large, stout, hook; a continuous peribuccal velum present instead of distinct lamellae or papulae; buccal armature as a wide band of dense minute teeth in the anterior portion of the buccal cavity; robust stylet furcae of the *Richtersius* type (the branches, very robust, have thickened but neither swollen nor rounded apices).} (Pilato 1972; Pilato and Binda 2010; Guidetti et al. 2016)

Incerta Superfamilia {according to Marley et al. 2011 and Bertolani et al. 2014}

Beornidae Cooper, 1964 [fossil family; 1 genus]
{Legs telescopable, each leg with four claws (or 2 two-branched claws?), and with a short, anterior, flattened, apical cuticular extension or spine, but without lateral or basal papillae. Mouth without discernible palps or other appendages, dorsally and laterally enclosed by a well-defined cuticular frontal element which caps and delineates the "head".} (Cooper 1964)

***Beorn* Cooper, 1964** [*B. leggi* Cooper, 1964; 1 fossil species]
{as the family} (Cooper 1964)

Necopinatidae Ramazzotti & Maucci, 1983 [1 genus]
{Claws absent in each leg, those of the first three pairs substituted by a small cuticular forceps without lunules. Bucco-pharyngeal apparatus of the *Necopinatum* type: buccal tube rigid without ventral lamina; apophyses for the insertion of the stylet muscles in the shape of "ridges" symmetrical with respect to the frontal plane; very narrow anterior portion of this lamina continues abruptly in a larger triangular portion with two very robust processes pointing backwards and sideways; the caudal extremity of the buccal tube with an evident, dorsal, thickening; peribuccal lamellae absent, other peribuccal structures not visible but their absence needs to be confirmed); stylet furcae typically shaped; pharyngeal apophyses and placoids present. Egg smooth laid within the exuvium.} (Pilato and Binda 2010, Bertolani et al. 2014)

***Necopinatum* Pilato, 1971** [*N. mirabile* Pilato, 1971; 1 species]
{as the family} (Pilato and Binda 2010)

Acknowledgement We thank anonymous reviewer for the valuable suggestions which significantly helped us to improve the chapter.

References

Bartels PJ, Fontoura P, Nelson DR (2018) Marine tardigrades of the Bahamas with the description of two new species and updated keys to the species of *Anisonyches* and *Archechiniscus*. Zootaxa 4420:43–70

Bartels PJ, Nelson DR, Kaczmarek Ł, Michalczyk Ł (2011) *Ramazzottius belubellus*, a new species of Tardigrada (Eutardigrada: Parachela: Hypsibiidae) from the Great Smoky Mountains National Park (North Carolina, USA). Proc Biol Soc Wash 124:23–27

Bello G, de Zio Grimaldi S (1998) Phylogeny of the genera of the Stygarctidae and related families (Tardigrada: Heterotardigrada). Zool Anz 237:171–183

Bertolani R, Guidetti R, Marchioro T, Altiero T, Rebecchi L, Cesari M (2014) Phylogeny of Eutardigrada: New molecular data and their morphological support lead to the identification of new evolutionary lineages. Mol Phylogenet Evol 76:110–126

Binda MG (1978) Risistemazione di alcuni tardigradi con l'istituzione di un nuovo genere di Oreellidae e della nuova famiglia Archechiniscidae. Animalia 5:307–314

Binda MG, Kristensen RM (1986) Notes on the genus *Oreella* (Oreellidae) and the systematic position of *Carphania fluviatilis* Binda, 1978 (Carphanidae fam. nov., Heterotardigrada). Animalia 13:9–20

Biserov VI (1992) A new genus and three new species of tardigrades (Tardigrada: Eutardigrada) from the USSR. Boll Zool 59:95–103

Boesgaard TM, Kristensen RM (2001) Tardigrades from Australian marine caves. With a redescription of *Actinarctus neretinus* (Arthrotardigrada). Zool Anz 240:253–264

Bussau C (1992) New deep sea Tardigrada (Arthrotardigrada, Halechiniscidae) from a manganese nodule area of the eastern South Pacific. Zool Scr 21:79–92

Cantacuzène MA (1951) Tardigrade marin noveau, commensal de *Limnoria lignorum* (Rathke). C R Hebd Seances Acad Sci 232:1699–1700

Cesari M, Vecchi M, Palmer A, Bertolani R, Pilato G, Rebecchi L, Guidetti R (2016) What if the claws are reduced? Morphological and molecular phylogenetic relationships of the genus *Haplomacrobiotus* May, 1948 (Eutardigrada, Parachela). Zool J Linn Soc Lond 178:819–827

Claxton SK (1999) *Milnesioides exsertum* gen. n. sp. n., a new tardigrade from Australia (Tardigrada: Milnesiidae). Zool Anz 238:183–190

Cooper KW (1964) The first fossil tardigrade: *Beorn leggi* Cooper, from cretaceous amber. Psyche 71:41–48

Dastych H (1992) *Paradiphascon manningi* gen. n. sp. n., a new water-bear from South Africa, with the erecting of a new subfamily Diphasconinae (Tardigrada). Mitt Hamb Zool Mus Inst 89:125–139

Dastych H (2009) *Thalerius konradi* gen. nov., sp. nov., a new tardigrade from the periglacial area of the Ötztal Alps, Austria (Tardigrada). Contrib Nat Hist 12:391–402

Dastych H (2011) *Bergtrollus dzimbowski* gen. n., sp. n., a remarkable new tardigrade genus and species from the nival zone of the Lyngen Alps, Norway (Tardigrada: Milnesiidae). Entomol Mitt Zool Mus Hamburg 15(186):335–359

Dastych H, McInnes SJ, Claxton SK (1998) *Oreella mollis* Murray, 1910 (Tardigrada): a redescription and revision of *Oreella*. Mitt Hamb Zool Mus Inst 95:89–113

D'Addabbo Gallo M, Grimaldi de Zio S, Morone de Lucia MR, Troccoli A (1992) Halechiniscidae and Echiniscoididae from the Western Mediterranean Sea (Tardigrada: Heterotardigrada). Cah Biol Mar 33:299–318

Degma P, Guidetti R (2007) Notes to the current checklist of Tardigrada. Zootaxa 1579:41–53

Degma P, Bertolani R, Guidetti R (2018) Actual checklist of Tardigrada species, p 48. http://www.tardigrada.modena.unimo.it/miscellanea/Actual%20checklist%20of%20Tardigrada.pdf. Accessed 20 Aug 2018

de Zio Grimaldi S, Lamarca A, D'addabbo Gallo M, Pietanza R (1999) Florarctinae of Asdhu Island, Maldives, Indian Ocean (Tardigrada, Heterotardigrada). Ital J Zool 66:383–391

du Bois-Reymond Marcus E (1952) On South American Malacopoda. Bol Fac Fil Ciên Letr Univ S Paulo Zool 17:189–209

Fontoura P, Bartels PJ, Jørgensen A, Kristensen RM, Hansen JG (2017) A dichotomous key to the genera of the marine Heterotardigrades (Tardigrada). Zootaxa 4294(1):1–45

Fujimoto S (2015) *Quisarctus yasumurai* gen. et sp. nov. (Arthrotardigrada: Halechiniscidae) from a submarine cave, off Iejima, Ryukyu Islands, Japan. Zootaxa 3948:145–150

Fujimoto S, Yamasaki H (2017) A new genus and species of Renaudarctidae (Heterotardigrada: Arthrotardigrada) from Ryukyu Archipelago, Japan. Mar Biol Res 13(3):288–299

Fujimoto S, Jørgensen A, Hansen JG (2016) A molecular approach to arthrotardigrade phylogeny (Heterotardigrada, Tardigrada). Zool Scr 46:496–505

Gallo D'Addabbo M, Sandulli R, de Zio Grimaldi S (2005) A new Batillipedidae (Tardigrada, Heterotardigrada) from the Orosei Gulf, Sardinia, Tyrrhenian Sea. Zool Anz 243:219–225

Gąsiorek P, Stec D, Morek W, Michalczyk Ł (2017) An integrative redescription of *Echiniscus testudo* (Doyère, 1840), the nominal taxon for the class Heterotardigrada (Ecdysozoa: Panarthropoda: Tardigrada). Zool Anz 270:107–122

Gąsiorek P, Stec D, Morek W, Michalczyk Ł (2018) An integrative redescription of *Hypsibius dujardini* (Doyère, 1840), the nominal taxon for Hypsibioidea (Tardigrada: Eutardigrada). Zootaxa 4415:45–75

Gomes-Júnior E, Santos É, da Rocha CMC, Santos PJP, Fontoura P (2018) A new species of *Ligiarctus* (Tardigrada, Arthrotardigrada) from the Brazilian continental shelf, Southwestern Atlantic Ocean. Mar Biodivers 48:5–12

Grimaldi de Zio S, D'Addabbo Gallo M, Morone De Lucia MR (1990) Revision of the genus *Halechiniscus* (Halechiniscidae, Arthrotardigrada). Cah Biol Mar 31:271–279

Gross V, Miller WR, Hochberg R (2014) A new tardigrade, *Mutaparadoxipus duodigifinis* gen. nov., sp. nov. (Heterotardigrada: Arthrotardigrada), from the Southeastern United States. Zootaxa 3835:263–272

Grothman GT, Johansson C, Chilton G, Kagoshima H, Tsujimoto M, Suzuki AC (2017) Gilbert Rahm and the status of Mesotardigrada Rahm, 1937. Zool Sci 34:5–10

Guidetti R, Bertolani R (2001) An evolutionary line of the Macrobiotinae (Tardigrada, Macrobiotidae): *Calcarobiotus* and related species. Ital J Zool 68(3):229–233

Guidetti R, Bertolani R (2005) Tardigrade taxonomy: an updated check list of the taxa and a list of characters for their identification. Zootaxa 845:1–46

Guidetti R, Pilato G (2003) Revision of the genus *Pseudodiphascon* (Tardigrada, Macrobiotidae), with the erection of three new genera. J Nat Hist 37:1679–1690

Guidetti R, Altiero T, Marchioro T, Amadè LS, Avdonina AM, Bertolani R, Rebecchi L (2012) Form and function of the feeding apparatus in Eutardigrada (Tardigrada). Zoomorphology 131 (2):127–148

Guidetti R, Rebecchi L, Bertolani R, Jönsson KI, Kristensen RM, Cesari M (2016) Morphological and molecular analyses on *Richtersius* (Eutardigrada) diversity reveal its new systematic position and lead to the establishment of a new genus and a new family within Macrobiotoidea. Zool J Linn Soc Lond 178:834–845

Guil N, Rodrigo E, Machordom A (2015) Soil tardigrade biodiversity with the description of a new eutardigrade genus and its phylogenetic position. Syst Biodivers 13:234–256

Hansen JG (2007) The deep sea elements of the Faroe Bank tardigrade fauna with a description of two new species. J Limnol 66:12–20

Hansen JG, Kristensen RM (2006) The 'hyena female'of tardigrades and descriptions of two new species of *Megastygarctides* (Arthrotardigrada: Stygarctidae) from Saudi Arabia. Hydrobiologia 558(1):81–101

Hansen JG, D'Addabbo Gallo M, de Zio Grimaldi S (2003) A comparison of morphological characters whitin the genus *Rhomboarctus* (Tardigrada: Heterotardigrada) with the description of two new species. Zool Anz 242:83–96

Hansen JG, Kristensen RM, Jørgensen A (2012) The armoured marine tardigrades (Arthrotardigrada, Tardigrada). Scientia Danica, Series B Biologica, vol 2. The Royal Danish Academy of Sciences and Letters, p 91

Jørgensen A, Kristensen RM (2001) A new tanarctid arthrotardigrade with buoyant bodies. Zool Anz 240:425–439

Jørgensen A, Boesgaard TM, Møbjerg N, Kristensen RM (2014) The tardigrade fauna of Australian marine caves: With descriptions of nine new species of Arthrotardigrada. Zootaxa 3802:401–443

Kaczmarek Ł, Michalczyk Ł (2006) Redescription of *Macrobiotus tridigitus* Schuster, 1983 and the erection of a new genus of Tardigrada (Eutardigrada, Macrobiotidae). J Nat Hist 40:1223–1229

Kaczmarek Ł, Gawlak M, Bartels PJ, Nelson DR, Roszkowska M (2017) Revision of the genus *Paramacrobiotus* Guidetti *et al.*, 2009 with the description of a new species, re-descriptions and a key. Ann Zool 67:627–656

Kristensen RM (1982a) New aberrant eutardigrades from homothermic springs on Disko Island, West Greenland. In: Proceedings of the 3rd international symposium on the Tardigrada, Johnson City, TN, USA, pp 203–220

Kristensen RM (1982b) The first record of cyclomorphosis in Tardigrada based on a new genus and species from Arctic meiobenthos. J Zool Syst Evol Res 20:249–270

Kristensen RM (1984) On the biology of *Wingstrandarctus corallinus* nov. gen. et spec., with notes on the symbiontic bacteria in the subfamily Florarctinae (Arthrotardigrada). Vidensk Meddr Dansk Naturh Foren 145:201–218

Kristensen RM (1987) Generic revision of the Echiniscidae (Heterotardigrada), with a discussion of the origin of the family. In: Bertolani R (ed) Biology of Tardigrades. Selected Symposia and Monographs U.Z.I. 1. Mucchi, Modena, pp 261–335

Kristensen RM, Hallas TE (1980) The tidal genus *Echiniscoides* and its variability, with erection of Echiniscoididae fam. n. (Tardigrada). Zool Scr 9:113–127

Kristensen RM, Hansen JG (2005) Tardigrada (water bears). Chapter 5 Minor groups and fossils. In: Rohde K (ed) Marine parasitology. CSIRO, Melbourne, pp 230–235

Kristensen RM, Higgins RP (1984) Revision of *Styraconyx* (Tardigrada: Halechiniscidae), with descriptions of two new species from Disko Bay, West Greenland. Smithson Contr Zool 391:1–40

Kristensen RM, Higgins RP (1989) Marine Tardigrada from the Southeastern United States coastal waters. I. *Paradoxipus orzeliscoides* n. gen., n. sp. (Arthrotardigrada: Halechiniscidae). Trans Am Microsc Soc 108:262–282

Kristensen RM, Renaud-Mornant J (1983) Existence d'Arthrotardigrades semi-benthiques de genres nouveaux de la sous-famille des Styraconyxinae, subfam. nov. Cah Biol Mar 24:337–353

Kristensen RM, Sørensen MV, Hansen JG, Zeppilli D (2015) A new species of *Neostygarctus* (Arthrotardigrada) from the Condor Seamount in the Azores, Northeast Atlantic. Mar Biodivers 45:453–467

Lisi O (2011) Remarks on *Doryphoribius flavus* (Iharos, 1966), and description of three new species (Tardigrada, Hypsibiidae). Zootaxa 2834:17–32

Lisi O, Sabella G, Pilato G (2014) *Mixibius parvus* sp. nov. and *Diphascon* (*Diphascon*) *ziliense* sp. nov., two new species of Eutardigrada from Sicily. Zootaxa 3802:459–468

Lisi O, Daza A, Londoño R, Quiroga S (2017) Echiniscidae from the Sierra Nevada de Santa Marta, Colombia, new records and a new species of *Bryodelphax* Thulin, 1928 (Tardigrada). ZooKeys 703:1–14

Marley NJ, McInnes SJ, Sands CJ (2011) Phylum Tardigrada: A re-evaluation of the Parachela. Zootaxa 2819:51–64

Marley NJ, Kaczmarek Ł, Gawlak M, Bartels PJ, Nelson DR, Roszkowska M, Stec D, Degma P (2018) A clarification for the subgenera of *Paramacrobiotus* Guidetti, Schill, Bertolani, Dandekar and Wolf, 2009, with respect to the International Code of Zoological Nomenclature. Zootaxa 4407:130–134

Miller WR, Schulte R, Johansson C (2012) Tardigrades of North America: further description of the genus *Multipseudechiniscus* Schulte and Miller, 2011 (Heterotardigrada: Echiniscoidea: Echiniscidae) from California. Proc Biol Soc Wash 125:153–164

Møbjerg N, Kristensen RM, Jørgensen A (2016) Data from new taxa infer *Isoechiniscoides* gen. nov. and increase the phylogenetic and evolutionary understanding of echiniscoidid tardigrades (Echiniscoidea: Tardigrada). Zool J Linn Soc Lond 178:804–818

Nelson DR, Marley NJ, Bertolani R (1999) Re-description of the genus *Pseudobiotus* (Eutardigrada, Hypsibiidae) and of the new type species *Pseudobiotus kathmanae* sp. n. Zool Anz 238:311–317

Pilato G (1972) Structure, intraspecific variability and systematic value of the buccal armature of eutardigrades. J Zool Syst Evol 10:65–78

Pilato G (2013) The taxonomic value of the structures for the insertion of the stylet muscles in the Eutardigrada, and description of a new genus. Zootaxa 3721:365–378

Pilato G, Binda MG (2010) Definition of families, subfamilies, genera and subgenera of the Eutardigrada, and keys to their identification. Zootaxa 2404:1–54

Pilato G, Lisi O (2011) *Tenuibiotus*, a new genus of Macrobiotidae (Tardigrada). Zootaxa 2761:34–40

Pilato G, Lisi O, Binda MG (2010) Tardigrades of Israel with description of four new species. Zootaxa 2665:1–28

Pilato G, Kiosya Y, Lisi O, Inshina V, Biserov V (2011) Annotated list of Tardigrada records from Ukraine with the description of three new species. Zootaxa 3123:1–31

Pilato G, Kiosya Y, Lisi O, Sabella G (2012) New records of Eutardigrada from Belarus with the description of three new species. Zootaxa 3179:39–60

Pollock LW (1983) A closer look at some marine Heterotardigrada II. The morphology and taxonomy of *Bathyechiniscus*, with a description of *B. craticulus* n. sp. from the Caribbean. Bull Mar Sci 33:109–117

Pollock LW (1989) Marine interstitial Heterotadrigrada from the Pacific coast of the United States, including a description of *Butillipes tridentatus* n. sp. Trans Am Microsc Soc 108:169–189

Ramazzotti G, Maucci W (1983) Il phylum Tardigrada (III edizione riveduta e aggiornata). Mem Ist Ital Idrobiol 41:1–1016

Rebecchi L, Altiero T, Eibye-Jacobsen J, Bertolani R, Kristensen RM (2008) A new discovery of *Novechiniscus armadilloides* (Schuster, 1975) (Tardigrada, Echiniscidae) from Utah, USA with considerations on non-marine Heterotardigrada phylogeny and biogeography. Org Divers Evol 8:58–65

Renaud-Mornant J (1983) Tardigrades abyssaux nouveaux de la sous-famille des Euclavarctinae n. subfam. (Arthrotardigrada, Halechiniscidae). Bull Mus Natn Hist Nat Paris 5:201–219

Renaud-Mornant J (1984) Halechiniscidae (Heterotardigrada) de la campagne Benthedi, canal du Mozambique. Bull Mus Natn Hist Nat Paris 6:67–88

Renaud-Mornant J (1987) Halechiniscidae nouveaux de sables coralliens tropicaux (Tardigrada, Arthrotardigrada). Bull Mus Natn Hist Nat Paris, 4e Serie Miscellanea 9:353–373

Renaud-Mornant J (1989) *Opydorscus*, un nouveau genre d'Orzeliscinae et sa signification phylogénique (Tardigrada, Arthrotardigrada). Bull Mus Natn Hist Nat Paris, 4e Serie Miscellanea 11:763–771

Richters F (1926) Tardigrada. Handbuch der Zoologie, III, pp 1–68

Rubal M, Veiga P, Fontoura P, Sousa-Pinto I (2013) A new intertidal arthrotardigrade, *Prostygarctus aculeatus* gen. nov., sp. nov. (Tardigrada: Heterotardigrada) from the North of Portugal (Atlantic ocean). Proceedings of the 12th international symposium on Tardigrada. J Limnol 72(Suppl. 1):8–14

Trygvadóttir BV, Kristensen RM (2011) Eohypsibiidae (Eutardigrada, Tardigrada) from the Faroe Islands with the description of a new genus containing three new species. Zootaxa 2886:39–62

Van der Land J (1975) The parasitic marine tardigrade *Tetrakentron synaptae*. Mem Ist Ital Idrobiol 32 Suppl.:413–423

Vecchi M, Cesari M, Bertolani R, Jönsson KI, Rebecchi L, Guidetti R (2016) Integrative systematic studies on tardigrades from Antarctica identify new genera and new species within Macrobiotoidea and Echiniscoidea. Invertebr Syst 30:303–322

Vicente F, Fontoura P, Cesari M, Rebecchi L, Guidetti R, Serrano A, Bertolani R (2013) Integrative taxonomy allows the identification of synonymous species and the erection of a new genus of Echiniscidae (Tardigrada, Heterotardigrada). Zootaxa 3613:557–572

Villora-Moreno S (1996) A new genus and species of the deep-sea family Coronarctidae (Tardigrada) from a submarine cave with a deep-sea like condition. Sarsia 81:275–283

Zawierucha K, Stec D, Lachowska-Cierlik D, Takeuchi N, Li Z, Michalczyk Ł (2018) High mitochondrial diversity in a new water bear species (Tardigrada: Eutardigrada) from mountain glaciers in central Asia, with the erection of a new genus *Cryoconicus*. Ann Zool 68:179–201

Index

R. O. Schill (ed.), *Water Bears: The Biology of Tardigrades*, Zoological Monographs 2,
https://doi.org/10.1007/978-3-319-95702-9